Symposia Board

C. Fred Fox, Ph.D., Director
Professor of Microbiology, University of California, Los Angeles

Members

The members of the board advise the director in matters of policy and in identification of topics for future symposia.

Charles Arntzen, Ph.D.
Director, Plant Science and Microbiology
Dupont

Ronald Cape, Ph.D., M.B.A.
Chairman
Cetus Corporation

Ralph Christoffersen, Ph.D.
Executive Director of Biotechnology
Upjohn Company

John Cole, Ph.D.
Vice-President of Research
and Development
Triton Biosciences

Pedro Cuatrecasas, M.D.
Vice President of Research
Glaxo, Inc.

J. Eugene Fox, Ph.D.
Director
ARCO Plant Cell Research Institute

L. Patrick Gage, Ph.D.
Director of Exploratory Research
Hoffman-La Roche, Inc.

Luis Glaser, Ph.D.
Executive Vice President
University of Miami

Gideon Goldstein, M.D., Ph.D.
Vice President, Immunology
Ortho Pharmaceutical Corp.

Ernest Jaworski, Ph.D.
Director of Biological Sciences
Monsanto Corp.

Irving S. Johnson, Ph.D.
Vice President of Research
Lilly Research Laboratories

Paul Marks, M.D.
President
Sloan-Kettering Memorial Institute

David W. Martin, Jr., M.D.
Vice-President of Research
Genentech, Inc.

Hugh O. McDevitt, M.D.
Professor of Medical Microbiology
Stanford University School of Medicine

Dale L. Oxender, Ph.D.
Professor of Biological Chemistry
University of Michigan

Mark L. Pearson, Ph.D.
Director of Molecular Biology
E.I. du Pont de Nemours and Company

George Poste, Ph.D.
Vice President and Director of Research
and Development
Smith, Kline and French Laboratories

William Rutter, Ph.D.
Professor of Biochemistry
University of California, San Francisco

Donald Steiner, M.D.
Professor of Biochemistry
University of Chicago

Sidney Udenfriend, Ph.D.
Member
Roche Institute of Molecular Biology

Norman Weiner, M.D.
Vice President for Pharmaceutical
Discovery
Abbott Laboratories

UCLA Symposia on Molecular and Cellular Biology, New Series

Series Editor, C. Fred Fox

RECENT TITLES

Volume 22
  Cellular and Molecular Biology of Plant Stress, Joe L. Key and Tsune Kosuge, *Editors*
Volume 23
  Membrane Receptors and Cellular Regulation, Michael P. Czech and C. Ronald Kahn, *Editors*
Volume 24
  Neurobiology: Molecular Biological Approaches to Understanding Neuronal Function and Development, Paul O'Lague, *Editor*
Volume 25
  Extracellular Matrix: Structure and Function, A. Hari Reddi, *Editor*
Volume 26
  Nuclear Envelope Structure and RNA Maturation, Edward A. Smuckler and Gary A. Clawson, *Editors*
Volume 27
  Monoclonal Antibodies and Cancer Therapy, Ralph A. Reisfeld and Stewart Sell, *Editors*
Volume 28
  Leukemia: Recent Advances in Biology and Treatment, David W. Golde and Robert Peter Gale, *Editors*
Volume 29
  Molecular Biology of Muscle Development, Charles Emerson, Donald Fischman, Bernardo Nadal-Ginard, and M.A.Q. Siddiqui, *Editors*
Volume 30
  Sequence Specificity in Transcription and Translation, Richard Calendar and Larry Gold, *Editors*

Volume 31
  Molecular Determinants of Animal Form, Gerald M. Edelman, *Editor*
Volume 32
  Papillomaviruses: Molecular and Clinical Aspects, Peter M. Howley and Thomas R. Broker, *Editors*
Volume 33
  Yeast Cell Biology, James Hicks, *Editor*
Volume 34
  Molecular Genetics of Filamentous Fungi, William Timberlake, *Editor*
Volume 35
  Plant Genetics, Michael Freeling, *Editor*
Volume 36
  Options for the Control of Influenza, Alan P. Kendal and Peter A. Patriarca, *Editors*
Volume 37
  Perspectives in Inflammation, Neoplasia, and Vascular Cell Biology, Thomas Edgington, Russell Ross, and Samuel Silverstein, *Editors*
Volume 38
  Membrane Skeletons and Cytoskeletal–Membrane Associations, Vann Bennett, Carl M. Cohen, Samuel E. Lux, and Jiri Palek, *Editors*
Volume 39
  Protein Structure, Folding, and Design, Dale L. Oxender, *Editor*
Volume 40
  Biochemical and Molecular Epidemiology of Cancer, Curtis C. Harris, *Editor*

Yeast Cell Biology

Yeast Cell Biology

Proceedings of a Cetus-UCLA Symposium on
Yeast Cell Biology, Held in Keystone,
Colorado, April 9-15, 1985

Editor

James Hicks
Scripps Research Clinic and Foundation
La Jolla, California

Alan R. Liss, Inc. • New York

Address all Inquiries to the Publisher
Alan R. Liss, Inc., 41 East 11th Street, New York, NY 10003

Copyright © 1986 Alan R. Liss, Inc.

Printed in the United States of America

Under the conditions stated below the owner of copyright for this book hereby grants permission to users to make photocopy reproductions of any part or all of its contents for personal or internal organizational use, or for personal or internal use of specific clients. This consent is given on the condition that the copier pay the stated per-copy fee through the Copyright Clearance Center, Incorporated, 27 Congress Street, Salem, MA 01970, as listed in the most current issue of "Permissions to Photocopy" (Publisher's Fee List, distributed by CCC, Inc.), for copying beyond that permitted by sections 107 or 108 of the US Copyright Law. This consent does not extend to other kinds of copying, such as copying for general distribution, for advertising or promotional purposes, for creating new collective works, or for resale.

Library of Congress Cataloging-in-Publication Data

Cetus-UCLA Symposium on Yeast Cell Biology (1985 : Keystone, Colo.)
  Yeast cell biology.

  (UCLA symposia on molecular and cellular biology ; new ser. v. 33)
  Includes index.
  1. Yeast fungi—Cytology—Congresses. I. Hicks, James. II. Title. III. Series. [DNLM: 1. Yeasts—cytology—congresses. W3 U17N new ser. v. 33 / QW 180.5.Y3 C423 1985y]
  QK617.5.C47 1985   589.2'33   86-10658
  ISBN 0-8451-2632-6

Contents

Contributors xi
Preface
    James Hicks xix

I. THE CYTOSKELETON

Overview: Why the Yeast Cytoskeleton?
    Lorraine Pillus and Frank Solomon 3
Genetics of the Yeast Cytoskeleton
    James H. Thomas, Peter Novick, and David Botstein 13

II. CONTROL OF THE CELL CYCLE

Cellular Morphogenesis in the Yeast Cell Cycle
    J.R. Pringle, S.H. Lillie, A.E.M. Adams, C.W. Jacobs, B.K. Haarer, K.G. Coleman, J.S. Robinson, L. Bloom, and R.A. Preston ... 47
Adenylate Cyclase and GTP-Binding Proteins in Yeast and Vertebrate Cells
    Henry R. Bourne, Gerald F. Casperson, Naomi Walker, Kathleen Sullivan, and Dan C. Medynski 81
Role of Cyclic AMP in Cell Division
    Kunihiro Matsumoto, Isao Uno, and Tatsuo Ishikawa 101
Functions of the *CYR1* and *RAS* Genes in Yeast
    Isao Uno, Kunihiro Matsumoto, and Tatsuo Ishikawa 113
Biochemistry of Yeast *RAS1* and *RAS2* Proteins
    Asao Fujiyama, Nasrollah Samiy, Madan Rao, and Fuyuhiko Tamanoi 125
Genetics of Spindle Pole Body Regulation
    Peter Baum, Loretta Goetsch, and Breck Byers 151
Gene Expression During Sporulation
    Stephen Kurtz, Janice Rossi, and Susan Lindquist 159

III. DNA REPLICATION AND CHROMOSOME STRUCTURE

Yeast Chromosomal DNA Replication
    J.L. Campbell, L.M. Johnson, A.Y.S. Jong, M. Budd, K. Sweder, and F. Srienc 173
Regulation of DNA Replication Initiation in Yeast
    Pratima Sinha, Clarence Chan, Gregory Maine, Steve Passmore, DaMing Ren, and Bik-Kwoon Tye 193

Structure and Organization of Yeast Chromosome III
Carol S. Newlon, R.P. Green, K.J. Hardeman, K.E. Kim, L.R.
Lipchitz, T.G. Palzkill, S. Synn, and S.T. Woody 211

Structure and Function of Centromeres
Ray Ng, Susan Cumberledge, and John Carbon 225

Genes at Telomeres: Fermentation Gene Families
Marian Carlson 241

Structural and Genetic Analyses of the Yeast Genome
A.J. Lustig, M.G. Goebl, and T.D. Petes 251

Electrophoretic Karyotyping of *Saccharomyces cerevisiae*
Georges F. Carle 271

Control Mechanisms of Chromosome Movement in Mitosis of Fission Yeast
Mitsuhiro Yanagida, Yasushi Hiraoka, Tadashi Uemura, Sanae
Miyake, and Tatsuya Hirano 279

IV. NUCLEAR ORGANIZATION

Genetic Analyses of snRNAs and RNA Processing in Yeast
Christine Guthrie, Nora Riedel, Roy Parker, Harold Swerdlow, and
Bruce Patterson 301

The *REP1* Protein of 2 Micron Circle Is Associated With the Nuclear Matrix
Ling-Chuan Chen Wu, Paul Fisher, and James R. Broach 323

Yeast as a Model System to Dissect the Relationship Between Chromatin Structure and Gene Expression
David S. Gross, Christopher Szent-Gyorgyi, and William T. Garrard . . . 345

Are Specific DNA Sequences Associated With Residual Nuclei?
Judith A. Potashkin and Joel A. Huberman 367

V. MACROMOLECULAR TRAFFIC I: PROTEIN LOCALIZED IN THE NUCLEUS

Identification of a Nuclear Localization Signal of Yeast Ribosomal Protein L3
Robert B. Moreland, Hong Gil Nam, Lynna Hereford, and Howard
M. Fried 379

Entry of a Procaryotic Endonuclease Into the Nucleus of *Saccharomyces cerevisiae*
Jasper Rine and Georjana Barnes 395

Nuclear Protein Localization in Saccharomyces cerevisiae
Pamela Silver 415

Nuclear Protein Localization Signals in Yeast
Michael N. Hall 421

VI. MACROMOLECULAR TRAFFIC II: SECRETION, ENDOCYTOSIS, AND THE CELL SURFACE

The Role of Clathrin in Yeast Cell Growth and Protein Transport
Gregory S. Payne and Randy Schekman 429

Endocytosis in Yeast: Relationship to Other Cellular Pathways
Howard Riezman, Yolande Chvatchko, and Isabelle Howald 443

Endocytosis in *Saccharomyces cerevisiae* Internalization of Soluble and Particulate Markers Into Cells and Spheroplasts
Marja Makarow 451

Post-Translational Processing Events in the Maturation of Yeast Pheromone Precursors
Robert Fuller, Anthony Brake, Rachel Sterne, Riyo Kunisawa, Debra Barnes, Monica Flessel, and Jeremy Thorner 461

Calmodulin and Other Calcium-Binding Proteins in Yeast
Trisha N. Davis and Jeremy Thorner 477

Genetics of Vacuolar Proteases
E.W. Jones, C. Moehle, M. Kolodny, M. Aynardi, F. Park, L. Daniels, and S. Garlow 505

Translocation, Sorting and Transport of Yeast Vacuolar Glycoproteins
Tom H. Stevens, Elizabeth G. Blachly, Craig P. Hunter, Joel H. Rothman, and Luis A. Valls 519

Maturation and Secretion of the M_1-dsRNA Encoded Killer Toxin in *S. cerevisiae*
S.L. Sturley, S.D. Hanes, V. Burn, and K.A. Bostian 537

Coordinate Regulation of Phospholipid Synthesis in Yeast
Brenda Loewy, Jeanne Hirsch, Margaret Johnson, and Susan Henry 551

VII. MACROMOLECULAR TRAFFIC III: IMPORT OF PROTEINS INTO MITOCHONDRIA

Sorting of Cytoplasmic Proteins for Assembly in Mitochondria
Michael G. Douglas 569

Hybrid Protein Detection and Subcellular Localization for Targeting Analysis During Mitochondrial Biogenesis
Linda Marshall-Carlson, Jerry Lynn Allen, and Michael G. Douglas . . . 581

In Vivo Targeting and Assembly of F_1 ATPase β-Subunit Deletions
Alessio Vassarotti, Cynthia Smagula, and Michael G. Douglas 593

Import of a *PUT2-LACZ* Hybrid Protein Into the Mitochondrial Matrix of *Saccharomyces cerevisiae*
Marjorie C. Brandriss and Karen A. Krzywicki 601

Do Mitochondria and Nuclei Share Transfer RNA Modification Enzymes?
Diana Najarian, Steven Ellis, Melitta Dihanich, Michael Morales, Anita Hopper, and Nancy Martin 613

VIII. MODELS FOR DEVELOPMENT

Specialized Cell Types in Yeast: Their Use in Addressing Problems in Cell Biology
Ira Herskowitz 625

Index 657

Contributors

A.E.M. Adams, Division of Biological Sciences, The University of Michigan, Ann Arbor, MI 48109 [47]

Jerry Lynn Allen, Department of Biochemistry, University of Texas Health Science Center, San Antonio, TX 78284, and Dallas, TX 75235 [581]

M. Aynardi, Department of Biological Sciences, Carnegie-Mellon University, Pittsburgh, PA 15213 [505]

Debra Barnes, Department of Biochemistry and Molecular Parasitology Group, School of Public Health, University of California, Berkeley, CA 94720 [461]

Georjana Barnes, Department of Biochemistry, University of California, Berkeley, CA 94720 [395]

Peter Baum, Department of Genetics, University of Washington, Seattle, WA 98195 [151]

Elizabeth G. Blachly, Institute of Molecular Biology, University of Oregon, Eugene, OR 97403 [519]

L. Bloom, Division of Biological Sciences, The University of Michigan, Ann Arbor, MI 48109 [47]

K.A. Bostian, Division of Biology and Medicine, Section of Biochemistry, Brown University, Providence, RI 02912 [537]

David Botstein, Department of Biology, Massachusetts Institute of Technology, Cambridge, MA 02139 [13]

Henry R. Bourne, Departments of Pharmacology and Medicine, University of California, San Francisco, CA 94143 [81]

Anthony Brake, Chiron Research Laboratories, Emeryville, CA 94608 [461]

Marjorie C. Brandriss, Department of Microbiology, UMDNJ-New Jersey Medical School, Newark, NJ 07103 [601]

James R. Broach, Department of Molecular Biology, Princeton University, Princeton, NJ 08544 [323]

M. Budd, Divisions of Biology and Chemistry, California Institute of Technology, Pasadena, CA 91125 [173]

V. Burn, Division of Biology and Medicine, Section of Biochemistry, Brown University, Providence, RI 02912 [537]

The number in brackets is the opening page number of the contributor's article.

Contributors

Breck Byers, Department of Genetics, University of Washington, Seattle, WA 98195 **[151]**

J.L Campbell, Divisions of Biology and Chemistry, California Institute of Technology, Pasadena, CA 91125 **[173]**

John Carbon, Department of Biological Sciences, University of California, Santa Barbara, CA 93106 **[225]**

Georges F. Carle, Department of Genetics, Washington University School of Medicine, Saint Louis, MO 63110 **[271]**

Marian Carlson, Department of Human Genetics and Development and Institute for Cancer Research, Columbia University, College of Physicians and Surgeons, New York, NY 10032 **[241]**

Gerald F. Casperson, Departments of Pharmacology and Medicine, University of California, San Francisco, CA 94143 **[81]**

Clarence Chan, Section of Biochemistry, Molecular and Cell Biology, Cornell University, Ithaca, NY 14853 **[193]**

Yolande Chvatchko, Membrane Unit, Swiss Institute for Experimental Cancer Research, CH-1066 Epalinges, Switzerland **[443]**

K.G. Coleman, Division of Biological Sciences, The University of Michigan, Ann Arbor, MI 48109 **[47]**

Susan Cumberledge, Department of Biological Sciences, University of California, Santa Barbara, CA 93106 **[225]**

L. Daniels, Department of Biological Sciences, Carnegie-Mellon University, Pittsburgh, PA 15213 **[505]**

Trisha N. Davis, Department of Biochemistry, University of California, Berkeley, CA 94720 **[477]**

Melitta Dihanich, Department of Biological Chemistry, Hershey Medical School, Penn State University, Hershey, PA **[613]**

Michael G. Douglas, Department of Biochemistry, University of Texas Health Science Center, San Antonio, TX 78284, and Dallas, TX 75235 **[569, 581, 593]**

Steven Ellis, Department of Biochemistry, Southwestern Graduate School, University of Texas Health Science Center at Dallas, Dallas, TX 75235 **[613]**

Paul Fisher, Department of Pharmacology, State University of New York, Stony Brook, NY 11794 **[323]**

Monica Flessel, Department of Biochemistry, University of California, Berkeley, CA 94720 **[461]**

Howard M. Fried, Department of Biochemistry, University of North Carolina, Chapel Hill, NC 27514 **[379]**

Asao Fujiyama, Department of Biochemistry and Molecular Biology, The University of Chicago, Chicago, IL 60637; Cold Spring Harbor Laboratory, Cold Spring Harbor, NY 11724 **[125]**

Robert Fuller, Department of Biochemistry, University of California, Berkeley, CA 94720 **[461]**

S. Garlow, Department of Biological Sciences, Carnegie-Mellon University, Pittsburgh, PA 15213 **[505]**

William T. Garrard, Department of Biochemistry, The University of Texas Health Science Center, Dallas, TX 75235 **[345]**

Contributors

M.G. Goebl, Department of Molecular Genetics and Cell Biology, The University of Chicago, Chicago, IL 60637 **[251]**

Loretta Goetsch, Department of Genetics, University of Washington, Seattle, WA 98195 **[151]**

R.P. Green, Department of Biology, University of Iowa, Iowa City, IA 52242 **[211]**

David S. Gross, Department of Biochemistry, The University of Texas Health Science Center, Dallas, TX 75235 **[345]**

Christine Guthrie, Department of Biochemistry and Biophysics, University of California, San Francisco, San Francisco, CA 94143 **[301]**

B.K. Haarer, Division of Biological Sciences, The University of Michigan, Ann Arbor, MI 48109 **[47]**

Michael N. Hall, Department of Biochemistry and Biophysics, University of California, San Francisco, CA 94143 **[421]**

S.D. Hanes, Division of Biology and Medicine, Section of Biochemistry, Brown University, Providence, RI 02912 **[537]**

K.J. Hardeman, Department of Biology, University of Iowa, Iowa City, IA 52242; Department of Microbiology, UMDNJ-New Jersey Medical School, Newark, NJ 07103 **[211]**

Susan Henry, Departments of Genetics and Molecular Biology, Albert Einstein College of Medicine, Bronx, NY 10461 **[551]**

Lynna Hereford, Dana Farber Cancer Institute, Boston, MA 02115 **[379]**

Ira Herskowitz, Department of Biochemistry and Biophysics, University of California, San Francisco, San Francisco, CA 94143 **[625]**

James Hicks, Scripps Research Clinic and Foundation, La Jolla, CA **[xix]**

Tatsuya Hirano, Department of Biophysics, Faculty of Science, Kyoto University, Sakyo-ku, Kyoto 606, Japan **[279]**

Yasushi Hiraoka, Department of Biophysics and Biophysics, University of California, San Francisco, CA 94143; Department of Biophysics, Faculty of Science, Kyoto University, Sakyo-ku, Kyoto 606, Japan **[279]**

Jeanne Hirsch, Departments of Genetics and Molecular Biology, Albert Einstein College of Medicine, Bronx, NY 10461 **[551]**

Anita Hopper, Department of Biological Chemistry, Hershey Medical School, Penn State University, Hershey, PA **[613]**

Isabelle Howald, Membrane Unit, Swiss Institute for Experimental Cancer Research, CH-1066 Epalinges, Switzerland **[443]**

Joel A. Huberman, Department of Cell and Tumor Biology, Roswell Park Memorial Institute, Buffalo, NY 14263 **[367]**

Craig P. Hunter, Institute of Molecular Biology, University of Oregon, Eugene OR 97403 **[519]**

Tatsuo Ishikawa, Institute of Applied Microbiology, University of Tokyo, Bunkyo-ku, Tokyo 113, Japan **[101, 113]**

C.W. Jacobs, Division of Biological Sciences, The University of Michigan, Ann Arbor, MI 48109 **[47]**

L.M. Johnson, Divisions of Biology and Chemistry, California Institute of Technology, Pasadena, CA 91125 **[173]**

Margaret Johnson, Departments of Genetics and Molecular Biology, Albert Einstein College of Medicine, Bronx, NY 10461 **[551]**

E.W. Jones, Department of Biological Sciences, Carnegie-Mellon University, Pittsburgh, PA 15213 **[505]**

A.Y.S. Jong, Divisions of Biology and Chemistry, California Institute of Technology, Pasadena, CA 91125 **[173]**

K.E. Kim, Department of Biology, University of Iowa, Iowa City, IA 52242 **[211]**

M. Kolodny, Department of Biological Sciences, Carnegie-Mellon University, Pittsburgh, PA 15213 **[505]**

Karen A. Krzywicki, Department of Microbiology, UMDNJ-New Jersey Medical School, Newark, NJ 07103 **[601]**

Riyo Kunisawa, Department of Biochemistry, University of California, Berkeley, CA 94720 **[461]**

Stephen Kurtz, Department of Molecular Genetics and Cell Biology, The University of Chicago, Chicago, IL 60637 **[159]**

S.H. Lillie, Division of Biological Sciences, The University of Michigan, Ann Arbor, MI 48109 **[47]**

Susan Lindquist, Department of Genetics and Cell Biology, The University of Chicago, Chicago, IL 60637 **[159]**

L.R. Lipchitz, Department of Biology, University of Iowa, Iowa City, IA 52242 **[211]**

Brenda Loewy, Departments of Genetics and Molecular Biology, Albert Einstein College of Medicine, Bronx, NY 10461 **[551]**

A.J. Lustig, Department of Molecular Genetics and Cell Biology, The University of Chicago, Chicago, IL 60637 **[251]**

Gregory Maine, BioTechnica International, Inc., Cambridge, MA 02140 **[193]**

Marja Makarow, Recombinant DNA Laboratory, University of Helsinki, SF-00380 Helsinki, Finland **[451]**

Linda Marshall-Carlson, Department of Biochemistry, University of Texas Health Science Center, San Antonio, TX 78284 **[581]**

Nancy Martin, Department of Biochemistry, Southwestern Graduate School, University of Texas Health Science Center at Dallas, Dallas, TX **[613]**

Kunihiro Matsumoto, Department of Industrial Chemistry, Tottori University, Tottori-shi, Tottori 680, Japan; DNAX Research Institute of Molecular and Cellular Biology, Palo Alto, CA 94304 **[101, 113]**

Dan C. Medynski, Departments of Pharmacology and Medicine, University of California, San Francisco, CA 94143 **[81]**

Sanae Miyake, Department of Biophysics, Faculty of Science, Kyoto University, Sakyo-ku, Kyoto 606, Japan **[279]**

C. Moehle, Department of Biological Sciences, Carnegie-Mellon University, Pittsburgh, PA 15213 **[505]**

Contributors xv

Michael Morales, Department of Biochemistry, Southwestern Graduate School, University of Texas Health Science Center at Dallas, Dallas, TX 75235 **[613]**

Robert B. Moreland, Dana Farber Cancer Institute, Boston, MA 02115 **[379]**

Diana Najarian, Department of Biochemistry, Southwestern Graduate School, University of Texas Health Science Center at Dallas, Dallas TX 75235 **[613]**

Hong Gil Nam, Department of Chemistry, University of North Carolina, Chapel Hill, NC 27514 **[379]**

Carol S. Newlon, Department of Biology, University of Iowa, Iowa City, IA 52242; Department of Microbiology, UMDNJ-New Jersey Medical School, Newark, NJ 07103 **[211]**

Ray Ng, Department of Biological Sciences, University of California, Santa Barbara, CA 93106 **[225]**

Peter Novick, Department of Biology, Massachusetts Institute of Technology, Cambridge, MA 02139 **[13]**

T.G. Palzkill, Department of Biology, University of Iowa, Iowa City, IA 52242; Department of Microbiology, UMDNJ-New Jersey Medical School, Newark, NJ 07103 **[211]**

F. Park, Department of Biological Sciences, Carnegie-Mellon University, Pittsburgh, PA 15213 **[505]**

Roy Parker, Department of Biochemistry and Biophysics, University of California, San Francisco, San Francisco, CA 94143 **[301]**

Steve Passmore, Section of Biochemistry, Molecular and Cell Biology, Cornell University, Ithaca, NY 14853 **[193]**

Bruce Patterson, Department of Biochemistry and Biophysics, University of California, San Francisco, San Francisco, CA 94143 **[301]**

Gregory S. Payne, Department of Biochemistry, University of California, Berkeley, CA 94720 **[429]**

T.D. Petes, Department of Molecular Genetics and Cell Biology, The University of Chicago, Chicago, IL 60637 **[251]**

Lorraine Pillus, Department of Biology and Center for Cancer Research, Massachusetts Institute of Technology, Cambridge, MA 02139 **[3]**

Judith A. Potashkin, Cold Spring Harbor Laboratory, Cold Spring Harbor, NY 11724 **[367]**

R.A. Preston, Division of Biological Sciences, The University of Michigan, Ann Arbor, MI 48109 **[47]**

J.R. Pringle, Division of Biological Sciences, The University of Michigan, Ann Arbor, MI 48109 **[47]**

Madan Rao, Genetics Program, State University of New York at Stony Brook, Stony Brook, NY 11794; Cold Spring Harbor Laboratory, Cold Spring Harbor, NY 11724 **[125]**

DaMing Ren, Fudan University, Shanghai, People's Republic of China **[193]**

Nora Riedel, Department of Biochemistry and Biophysics, University of California, San Francisco, San Francisco, CA 94143 **[301]**

Howard Riezman, Membrane Unit, Swiss Institute for Experimental Cancer Research, CH-1066 Epalinges, Switzerland, **[443]**

Contributors

Jasper Rine, Department of Biochemistry, University of California, Berkeley, CA 94720 **[395]**

J.S. Robinson, Division of Biological Sciences, The University of Michigan, Ann Arbor, MI 48190 **[47]**

Janice Rossi, Department of Molecular Genetics and Cell Biology, The University of Chicago, Chicago, IL 60637 **[159]**

Joel H. Rothman, Institute of Molecular Biology, University of Oregon, Eugene, OR 97403 **[519]**

Nasrollah Samiy, Cold Spring Harbor Laboratory, Cold Spring Harbor, NY 11724 **[125]**

Randy Schekman, Department of Biochemistry, University of California, Berkeley, CA 94720 **[429]**

Pamela Silver, Department of Biochemistry and Molecular Biology, Harvard University, Cambridge, MA 02138 **[415]**

Pratima Sinha, IMTECH, 1383 Sector 33-C, Chandigarh 160031, India **[193]**

Cynthia Smagula, Department of Biochemistry, University of Texas Health Science Center, San Antonio, TX 78284, and Dallas, TX 75235 **[593]**

Frank Solomon, Department of Biology and Center for Cancer Research, Massachusetts Institute of Technology, Cambridge, MA 02139 **[3]**

F. Srienc, Divisions of Biology and Chemistry, California Institute of Technology, Pasadena, CA 91125 **[173]**

Rachel Sterne, Department of Biochemistry, University of California, Berkeley, CA 94720 **[461]**

Tom H. Stevens, Institute of Molecular Biology, University of Oregon, Eugene, OR 97403 **[519]**

S.L. Sturley, Division of Biology and Medicine, Section of Biochemistry, Brown University, Providence, RI 02912 **[537]**

Kathleen Sullivan, Departments of Pharmacology and Medicine, University of California, San Francisco, CA 94143 **[81]**

K. Sweder, Divisions of Biology and Chemistry, California Institute of Technology, Pasadena, CA 91125 **[173]**

Harold Swerdlow, Department of Biochemistry and Biophysics, University of California, San Francisco, San Francisco, CA 94143 **[301]**

S. Synn, Department of Biology, University of Iowa, Iowa City, IA 52242 **[211]**

Christopher Szent-Gyorgyi, Department of Biochemistry, The University of Texas Health Science Center, Dallas, TX 75235 **[345]**

Fuyuhiko Tamanoi, Department of Biochemistry and Molecular Biology, University of Chicago, Chicago, IL 60637; Cold Spring Harbor Laboratory, Cold Spring Harbor, NY 11724 **[125]**

James H. Thomas, Department of Biology, Massachusetts Institute of Technology, Cambridge, MA 02139 **[13]**

Jeremy Thorner, Department of Biochemistry, University of California, Berkeley, CA 94720 **[461, 477]**

Bik-Kwoon Tye, Section of Biochemistry, Molecular and Cell Biology, Cornell University, Ithaca, NY 14853 **[193]**

Tadashi Uemura, Department of Biophysics, Faculty of Science, Kyoto University, Sakyo-ku, Kyoto 606, Japan **[279]**

Isao Uno, Institute of Applied Microbiology, University of Tokyo, Bunkyo-ku, Tokyo 113, Japan **[101, 113]**

Luis A. Valls, Institute of Molecular Biology, University of Oregon, Eugene, OR 97403 **[519]**

Alessio Vassarotti, Department of Biochemistry, University of Texas Health Science Center, San Antonio, TX 78284; Laboratoire de l'Hérédité Cytoplasmique des Plantes Cultivées, Université de Louvain, Belgium **[593]**

Naomi Walker, Departments of Pharmacology and Medicine, University of California, San Francisco, CA 94143 **[81]**

S.T. Woody, Department of Biology, University of Iowa, Iowa City, IA 52242 **[211]**

Ling-Chuan Chen Wu, Department of Pharmacology, State University of New York, Stony Brook, NY 11794 **[323]**

Mitsuhiro Yanagida, Department of Biophysics, Faculty of Science, Kyoto University, Sakyo-ku, Kyoto 606, Japan **[279]**

Preface

The papers collected in this volume are drawn mainly from addresses and posters presented at the Cetus-UCLA Symposium on Yeast Cell Biology at Keystone, Colorado in April 1985. The purpose of that symposium was to showcase the ways in which the power of yeast genetics is being applied to the study of cell structure and whole-cell processes. The recurring theme of the sessions was that with a few obvious exceptions yeast, which used to be referred to as a "lower" eukaryote and is now more often called a "small" eukaryote, exhibits all of the cellular structures and functions of "large" eukaryotes such as mammals and, moreover, that yeast proteins often exhibit primary sequence homology with their well-studied counterparts isolated from multicellular organisms.

The genetic approach available in yeast complements traditional methods in cell biology and comes with its own special powers and limitations. We hope that this volume will provide an introduction to yeast for cell biologists as well as a useful reference for those already working on yeast.

In the interest of providing such an introduction, the authors have been encouraged, wherever appropriate, to provide reviews of the important problems in each research area and to outline the various ways that molecular and classical genetic techniques can be used to study those problems in yeast. This approach is exemplified in the first three articles on the yeast cytoskeleton and its role in the cell division cycle. In Chapter 1 of the first Section, Pillus and Solomon provide an evaluation of yeast from the standpoint of the cell biologist interested in understanding the role of each component of the cytoskeleton. Next, Novick, Thomas and Botstein present the theory and practice of two distinct plans for such studies; these being in one case the identification of all genes involved in a complex process (such as mitosis) and, alternatively, the specific alteration of a known gene (such as tubulin) to identify the gene products that directly interact with it. Finally, both Novick, et al and Pringle, et al demonstrate the integration of genetic studies with direct observation using fluorescent labels and electron microscopy.

The advent of yeast as a subject for cell biology was hastened by the discovery that a number of eukaryotic structural and even regulatory proteins were not only present in yeast but were sufficiently similar in DNA or primary protein sequence that their structural genes could be easily cloned. In particular, the identification of yeast genes homologous to the *ras* oncogenes from Harvey and Kirsten Sarcoma viruses (DeFeo-Jones, et al, Nature *306*:707, 1983; Powers, et al, Cell *36*:607, 1983) has focused a renewed attention on the yeast cell division cycle as a model for understanding the general nature of eucaryotic cell proliferation and has, in fact, led to a deeper understanding of the cell division cycle (*cdc*) mutations uncovered by Hartwell and coworkers in previous years (Hartwell, L.H., Bacteriol. Rev. *38*:164, 1974). The role of the *ras* guanine nucleotide binding proteins in signal transduction through cyclic AMP is reviewed by Bourne, et al, and the genetics of cyclic AMP metabolism in yeast is reviewed by Matsumoto, et al and Uno, et al.

Two other areas of intense research activity concern the structure and replication of yeast chromosomes and the general traffic pattern of macromoles through the nucleus and cytoplasm. Ever since it was finally determined (after some years of controversy) that yeast DNA was packaged into a recognizable chromatin structure with *bona fide* histones (albeit lacking histone Hl) it has been assumed that yeast would be the appropriate model system for studying eucaryotic DNA replication. Yeast is the only eucaryotic organism from which all of the operating parts of the chromosome, including centromeres, teleomeres and replication origins can be isolated and analyzed. The Chapters in Section IV reflect the range of experimental approaches being focused on this major biological problem.

Movement of macromolecules is the major business of the cell. Regulatory proteins made in the cytoplasm must be transported to the nucleus. Some proteins are directed to the cell surface, and beyond, through the secretory pathway while others are directed to form subcellular structures, the most elaborate of course being the mitochondrion. Finally, proteins must eventually be destroyed and recycled. The mechanisms that govern this traffic are also susceptible to the combined biochemical and genetic studies possible in yeast as reflected by the Chapters in Section V–VII. Once again, yeast has been shown to contain the elements expected of complex cell processes including endocytosis (Reizman, et al and Makarow, M.) and clathrin in the secretory pathway (Payne and Schekman).

The closing article highlights the ways that all of these separate studies impinge on the complex series of events that make up the mating and sexual cycles of Saccharomyces yeasts. As described here by Ira Herskowitz, genetic studies of mating have uncovered mutations that affect secretion, hormone receptors, proteins modification and processing, chromatin structure, protein localization and RNA processing. Integration of studies from all of these disciplines will be necessary to understand even a single such pathway in its entirety.

Inevitably, not every presentation at a meeting becomes part of a proceedings volume, and some important areas of cellular research in yeast are not

Preface xxi

covered in detail here. Fortunately, excellent reviews of topics such as yeast cytology (B. Byers), the genetics of cell cycle (J. Pringle and L. Hartwell), meiosis and sporulation (R. Easton-Esposito and S. Klapholz), the cell wall (C. Ballou) and secretion (R. Schekman and P. Novick), among others are available in *The Molecular Biology of the Yeast Saccaromyces*, edited by J.N. Stratern, E.W. Jones and J.R. Broach (Cold Spring Harbor Laboratory Monograph Series, Cold Spring Harbor, New York, 1981).

In addition, several specific observations, reported at the Symposium but not presented in this volume deserve mention here:

Ubiquitin, a protein used to "tab" proteins for turnover, has been found in yeast by Varshavsky and co-workers (Ozkaynah, et al., Nature *312*:663–665, 1984). It is apparently made as a polyprotein from tandem copies in a single gene. Current studies in that laboratory indicate that other proteins, identified by sequence homology, may be naturally made as fusions with ubiquitin-like "tails" already in place (A. Varshavsky, personal communication).

Another development, described at the Symposium by Jenness and Hartwell is the identification of a cell surface receptor for the alpha mating pheromone (Jenness and Hartwell, Cell *35*:521–529, 1983). Since then the primary sequence for both mating factor receptors has been published (Nakayama, et al, EMBO Journal, *4*:2643–2648, 1985; Burkholder and Hartwell, Nuc. Ac. Res., *13*:8463–8475, 1985) and they appear to be membrane proteins, each with seven hydrophobic membrane-spanning domains reminiscent of the structure of mammalian rhodopsin.

Organizing this meeting was an unusually pleasant experience as these things go, due to the excellent administrative work by the UCLA Symposia staff, Robin Yeaton and Betty Handy, and the tireless assistance of Regina Schwarz and Laurie Lowman at Cold Spring Harbor Laboratory. The program was outlined with the help and advice of Jasper Rine, Jeremy Thorner, Ira Herskowitz and Elizabeth Jones, among others. Special thanks go to Betty Handy for finding the band that played at the post-banquet party and again for her patience and attention to the details of editing this volume.

I am grateful to all of the authors for the careful preparation of their manuscripts and to Cold Spring Harbor Laboratory for permission to reprint the articles by Novick, et al and Pringle, et al, originally published in *Molecular Biology of the Cytoskeleton* (Borisy, G.G., Cleveland, D.W. and Murphy, D.G., eds, 1984).

Finally, I would like to thank Cetus Corporation for their generous sponsorship of this meeting, and Rohm and Haas Company for their support of the one day joint session. Additional financial support was received from: Anheuser-Busch companies; E.I. du Pont de Nemours & Company; Phillips Petroleum Company, Research and Development; The Upjohn Company, and Merck Sharp & Dohme Research Laboratories.

James Hicks

I. THE CYTOSKELETON

OVERVIEW: WHY THE YEAST CYTOSKELETON?

Lorraine Pillus and Frank Solomon

Department of Biology and Center for Cancer Research
Massachusetts Institute of Technology
Cambridge, MA 02139

ABSTRACT One of the attractions that the cytoskeleton holds for biologists is its diversity. Related structures occur in a large repertoire of arrangements in different organisms, different cells or even different parts of the same cell, with an equally wide repertoire of putative functions. However, an unwelcome result of this complexity has been a diffusion of effort. Some substantial fraction of the information gathered in this area may turn out to be too particular, too restricted. Focus on a single organism, accessible biochemically and genetically as well as phenomenologically might accelerate and unify the field, just as focus on E. coli has provided so much useful insight into other cells. For cytoskeleton, the organism with the proper balance of advantages and drawbacks could be yeast.

INTRODUCTION

The study of cytoskeleton has been dominated, if not excusively occupied, by descriptive studies. The first observations were anatomic - the specific stains that allowed tracing of the nervous system, we know now are bound to specific elements of neuronal cytoskeleton. The field entered the first of its modern eras with Keith Porter's elegant application of the electron microscope to eukaryotic cells, so that otherwise obscured filamentous elements became apparent, reproducibly so, to enthusiasts and skeptics alike. Refinements of microscopy - immunofluoresence, whole mounts and 3-D - have propagated the appreciation of the beauty of cytoskeletal organelles.

Those descriptive studies have been complemented by a seemingly endless series of drug and mechanical intervention experiments in which probes have been used to interfere with this or that cellular process[1]. The combination of these two sorts of experiment has produced a consensus view of the function of cytoskeletal elements, albeit at rather low resolution. For example, microtubules appear to be important for the maintenance and the assumption of assymetric morphology; microfilaments, for motile events at the membrane; and intermediate filaments - because there is no specific drug for them as yet - responsible for nothing. But how these elements function, how they interact with the components of the cell whose displacement they are supposed to effect, and how the cell switches from one state to another - sessile to mitotic, for example - are mechanistic questions without answers.

To that end, increasing emphasis has been placed upon biochemical analysis of cytoskeleton. Clearly, the major constituents of each of the known cytoskeletal fibers have been identified and characterized, and various techniques have been used to identify and characterize minor players as well. These efforts, which are discussed below in a more appropriate context, have in turn led to closer surmises about the function of cytoskeletal elements.

The great difficulty with these studies, as with previous biochemical studies of biological problems, comes in extrapolating from in vitro activity to true in vivo function. Indeed, as studies of other systems like metabolism and DNA synthesis have shown, it is entirely possible to make wrong guesses about in vivo function, even from the tightest of biochemical experiments.

In the past, this situation has been resolved by using experimental organisms that permit combinations of physiological, biochemical, and genetic experimentation. In this way, the phenomenological observations can be studied at the level of mechanism, and the molecules identified as participants can be assigned precise roles in vivo by analysis of the phenotypes of mutations in the relevant genes. The organism of choice has been the bacterium, because it has all the requisite properties and because it is simpler than most other organisms.

Sadly, bacteria don't have cytoskeletons, and in fact most of the organisms used for studying cytoskeletal function do not have a readily accessible genetics. There is a genetics of mice, certainly, but it is slow to harvest. The genetics of animal cells - selection of

Overview: Why the Yeast Cytoskeleton? 5

somatic mutants of cultured cells - have been developed. These systems are limited by an inherently cumbersome method of complementation analysis. New innovations in this area - specifically the ability to introduce genes along with selectable markers - will make a difference in the near future. It is unlikely, however, that these systems, or others like Dictyostelium, will offer the sort of precise, rapid genetic analysis that enabled scientists to study, for example, E. coli with the confidence of constancy of genome and reproducibility of result that enabled bacteriology to proceed so rapidly, and with so few ambiguities and errors.

This Conference offers yeast as a possible candidate for the E. coli of eukaryotic cell biology. The flexibility and power of its genetics are clear, as is its secure position as a true eukaryote (after some early debate on the matter). There are other candidates as well, of course. The cilium of Chlamydomonas is arguably the most successful case yet of a cytoskeletal organelle which has been studied genetically as well as biochemically and phenomenologically[2]. There are already interesting reports of relevant mutations in other systems with well-characterized genetics, such as Drosophila and Aspergillus. Particularly for studying cytoskeleton, yeast has distinct advantages as well as some disadvantages. We will discuss those, as well as spell out the approaches that seem likely to work well in this organism.

APPROACHES TO MOLECULAR ANALYSIS OF CYTOSKELETON, IN YEAST AND ELSEWHERE

Several sorts of approaches have produced information about the components of cytoskeletal organelles in other eukaryotic cells. Two of them have already been tried on yeast:

Isolation and dissection of the organelle.

The first cytoskeletal structure to be analyzed at the level of molecules was the sarcomere. Huxley and colleagues took advantage of the abundance and stability of this organelle after extraction of muscle. In experiments described in great detail in textbooks, they found that they could selectively extract the sarcomere to release different components of the structure, and correlate that with release of specific molecules. Eventually, they could even reconstitute aspects of both structure and function. This approach has the distinct advantage of analyzing a situation as it exists in the cytoplasm of a cell, so that it should be less sensitive to the vicissitudes of in vitro artifact than many sorts of biochemical experiment (see below).

Huxley's example has been emulated in many situations. In the case of microfilament structures, the most striking example has been the analysis by selective extraction of the microvillus. Here again, use of specific reagents to release specific proteins, and at the same time to cause the disappearance of specific structural elements, has provided a detailed map of this structure[3]. Similar approaches have been taken to the actin filaments of platelets[4].

Isolation of all but the most stable of microtubule structures at first was problematic. Most microtubules appeared to disassemble if cells were opened at all to extracelluar media. Therefore, techniques like those used by Huxley on sarcomeres, and by Steck and others on the red blood cell[5], and Spudich and colleagues on cultured cells[6], for many years could not be extended to microtubules.

This problem can be solved, however, by using non-ionic detergents and extraction buffers which meet the more stringent stability requirements of microtubules. In independent efforts, Osborn and Weber's laboratory[7] and our laboratory[8] found such conditions. Our laboratory has turned those observations into the basis for an assay of microtubule structure. It is described in detail elsewhere[9], but briefly relies upon detergent extraction of two populations of cells: one with microtubules assembled, the other, after pre-treatment with microtubule depolymerizing drugs, with microtubules disassembled. We then look for proteins that remain behind after extraction if and only if there are intact microtubules, and which are selectively and quantitatively released from the cells when the microtubules are depolymerized by a subsequent extraction under destabilizing conditions. This assay has detected both ubiquitous and structure-specific microtubule components, and both assembled and unassembled components, in our laboratory and now others[10-14]. As confirmation, proteins identified by this procedure can be localized to cytoplasmic microtubules by histochemical techniques[15]. In a modification of this approach, spindles have been isolated and used, unrefined, as antigen. At least one of the antibodies elicited by such an experiment appears to stain a spindle component[16].

The advantage of this approach is that it makes rather few assumptions about the nature of the interactions between microtubule components that may be relatively minor and the major components of the microtubule. In recent studies, presented at this meeting, we have successfully worked out the parameters of this analysis for the yeast Saccharomyces cerevisiae. In turn, that has enabled us to identify for the first time microtubule associated proteins in this yeast, as well as to begin an analysis of the assembled and unassembled pools of microtubule components[17]. The future of this work is discussed below.

In vitro reconstitution of cytoskeletal organelles.

The second part of the analysis of the sarcomere - in vitro reconstitution - has been applied successfully to several sorts of microfilament structures. In this way, relatively minor proteins - minor compared to actin - have been identified which interact with either the globular or filamentous forms of that protein, and change their properties[18]. In the study of microtubules, proteins that co-assemble with tubulin have been identified, some of which appear to change the assembly properties of tubulin under defined conditions[19]. Whether or not the properties exhibited by these proteins in vitro reflect accurately their in vivo role is not known. The minor proteins can, however, be placed with confidence in the assembled structures in vivo by making antibodies to them and using histochemical techniques to study them.

The advantage to this approach, in principle, is that it can be used to study cytoskeletal elements from complex tissues - like brain - where sheer volume and geometry make isolation of the individual organelles a difficult proposition. The disadvantage is that it relies upon in vitro interactions, which may give both false positive and false negative results. Indeed, in vitro assembly experiments failed to reveal any microtubule associated proteins in a yeast[20], although there are other possible explanations for the absence of co-assembling species in this situation.

WHAT'S GOOD OR BAD ABOUT THE YEAST CYTOSKELETON?

One of the basic features of the cytoskeleton is that it contains morphologically and biochemically conserved structures which occur in a wide variety of structural and, apparently, functional contexts. The microfilaments of the fibroblast look the same and contain many of the same components as the thin filaments of the sarcomere, but the two are packed into quite different arrays with respect to one another and to other structures. Similarly, microtubules of cilia, of spindle, of axon, and of interphase cytoplasm are all much the same by EM, and contain tubulins which are also much the same, but the microtubule organelles themselves are all quite different. Figuring out where this diversity comes from - where the vocabularly arises for specifying form and function - is one of the dominating problems in this area.

It seems very unlikely that yeast can satisfy curiosity on this subject. The repertoire of motile functions displayed by animal cells and assigned to cytoskeleton is much more restricted in yeast, which do not send out neurites or migrate across solid substrata, change their shape in response to contact with other cells or swim off in the direction of some chemoattractant. They have cell walls, so motility in the plane of the membrane - patching and capping phenomena, for example - have not yet been detected. This limitation is not an unmitigated drawback. Fewer functions may help sort out phenotypes. Too, the functions which yeast clearly do show, such as cell division or assignment of polarity within the cytoplasm, are ones that more familiar eukaryotic cells also demonstrate. Most important, there are clear functions already associated with yeast cytoskeletal elements. That is, it is quite clear that microtubules are absolutely essential for cell division[21] and for nuclear fusion during mating[22]. An understanding of the mechanism of just those events, at the level of detail presaged by similar analyses by biochemistry and genetics of metabolism or phage assembly, would more than justify the search.

There will be disappointments, and in fact there already has been one. To the dismay of cell biologists who have followed the phenomenology of cell division for a long time, the phenotype of actin mutations has been discouraging. These mutations do not appear to block cells in cell division, even at the step which in the budding yeasts should roughly correspond to cytokinesis[23]. That is counterintuitive, because the neck between the mother and daughter cell in yeast constricts in a fashion which suggests nothing so much as a contractile ring in a dividing animal cell, and there is good and well-known evidence - morphological as well as from drug and antibody interference experiments - that argues for a role for actin filaments in this event. The resolution of this possible paradox may come by examining analogous mutations in other yeasts - for example, Schizosaccharomyces pombe, which undergoes a more "animal-like" fission. Or it may be that in some details, results in yeast may not extend directly to other eukaryotes.

APPROACHES AND EXPECTATIONS

Yeast cytoskeleton is being attacked by a variety of procedures. All depend upon the transfer of information from other eukaryotic systems, either in the form of a method of analysis, or a probe. An example of the former is the identification of yeast microtubule associated proteins which we described above. It seems likely that other minor cytoskeletal components can also be identified in yeast using affinity techniques already employed elsewhere. There are also many sorts of probes available, both at the level of protein and nucleic acid, which should facilitate identification of intersting proteins. And then, as described in the preceding chapter in this volume, it should be a relatively straightforward matter to go from protein to mutation, and then from mutation to complementing mutation.

What a cell biologist can reasonably expect from this work is rather less clear. A probably realistic ambition is to have in hand a cytoskeletal protein, localized to a structure, with a clear phenotype associated with mutations in the relevant gene. For example, does a mutation in a microtubule associated protein result in failure of the tubulin to polymerize, or does it polymerize but not organize into spindles or extranuclear arrays, or does it organize but not function in cell division or nuclear fusion? A set of such results would be a major contribution to understanding how the cytoskeleton is organized for motility. One is not confident in predicting that cytoskeletal assembly will present a more complex replay of phage assembly, which has long since become accessible to detailed analysis of the sort promised here. The two systems are quite different, for example in the functions performed and in the reversibility of the steps. But the construction of an extensive genetics of the cytoskeleton seems the obvious next step in an attempt to bring a wealth of in vitro observation back to the cell itself.

ACKNOWLEDGEMENTS

Work in our laboratory is supported by grants from the National Cancer Institute and the American Cancer Society (to F.S.) and a Center Grant from the National Cancer Institute (to P. Sharp). Lorraine Pillus is a graduate student and was supported in part by a Johnson and Johnson Fellowship.

REFERENCES

1. "Organization of the Cytoplasm," (1982). Cold Spring Harbor Symposia on Quantitative Biology, Vol. XLVI, Cold Spring Harbor Laboratory, New York.
2. Luck DJL (1984). Genetic and biochemical dissection of the eucaryotic flagellum. J Cell Biol 98:789.
3. Matsudaira PT, Burgess DR (1979). Identification and organization of the components of the isolated microvillus cytoskeleton. J Cell Biol 83:667.
4. Gonella PA, Nachmias VT (1981). Platelet activation and microfilament bundling. J Cell Biol 89:146.
5. Yu J, Fischman D, Steck T (1973). Selective solubilization of proteins and phospholipids from red blood cell membranes by nonionic detergents. J Supramol Struct 2:233.
6. Brown S, Levinson W, Spucich JA (1976). Cytoskeletal elements of chick embryo fibroblasts revealed by detergent extraction. J Supramol Struct 5:119.
7. Osborn M, Weber K (1977). The display of microtubules in transformed cells. Cell 12:561.
8. Solomon F, Magendantz M, Salzman A (1979). Identification with cytoplasmic microtubules of one of the coassembling microtubule associated proteins. Cell 18:431.
9. Solomon F (1985). Direct identification of MAPs by selective extraction of cultured cells. Methods in Enzymology, in press.
10. Duerr D, Pallas D, Solomon F (1981). Molecular analysis of cytoplasmic microtubles in situ: identification of both widespread and specific proteins. Cell 24:203.
11. Pallas D, Solomon F (1982). Cytoplasmic microtubule associated proteins: phosphorylation at novel sites is correlated with their incorporation into assembled microtubules. Cell 30:407.
12. Zieve G, Solomon F (1982). Proteins specifically associated with the microtubules of the mammalian mitotic spindle. Cell 28:233.
13. Black M, Kurdyla JT (1983) Microtubule associated proteins of neurons. J Cell Biol 97:1020.
14. Drubin D, Kirschner M, Feinstein S (1984). Microtubule-associated protein induced by nerve growth factor during neurite outgrowth in PC12 cells. in Molecular Biology of the Cytoskeleton, Cold Spring Harbor Press, Cold Spring Harbor, New York.

15. Magendantz M, Solomon F (1985). Analyzing the components of microtubules: antibodies against chartins, associated proteins from cultured cells. Proc Natl Acad Sci USA in press.
16. Izant JG, Weatherbee JA, McIntosh JR (1983). A microtubule associated protein antigen unique to mitotic spindle microtubules in PtK_1 cells. J Cell Biol 96:424.
17. Pillus L, Solomon F. Submitted for publication.
18. Weeds A (1982). Actin-binding proteins - regulators of cell architecture and motility. Nature 296:811.
19. Kirschner M (1978). Microtubule assembly and nucleation. Int Rev Cytol 54:1.
20. Kilmartin JV (1981). Purification of yeast tubulin by self-assembly in vitro. Biochemistry 20:3629.
21. Thomas J, Neff N, Botstein D (1985). Isolation and characterization of mutations in the beta-tubulin gene of Saccharomyces cerevisiae. Genetics, in press.
22. Thomas J (1984). PhD. Thesis, M.I.T.
23. Novick P, Botstein D (1985). Phenotypic analysis of temperature-sensitive yeast actin mutants. Cell 40:405.

GENETICS OF THE YEAST CYTOSKELETON[*]

James H. Thomas, Peter Novick and David Botstein

Department of Biology, Massachusetts Institute of Technology, Cambridge, Massachusetts 02139

The eukaryotic cell-division cycle consists of an ordered series of events, some of which entail major morphological changes. The cell's cytoskeleton is a central element in these changes, and it has been clear for many years that an understanding of mitosis, for example, will involve understanding the way in which tubulin and its associated proteins interact in the assembly, function, and disassembly of the mitotic spindle. The cell cycle and the cytoskeleton have traditionally been studied by cell biologists and biochemists, usually in higher eukaryotic cells, by observation, by application of drugs that interfere with cell-cycle functions or cytoskeletal assemblies, and by isolation and characterization of the major protein species (tubulin, microtubule-associated proteins, actin, actin-binding proteins, etc.).

The cell-division cycle has also been the subject of genetic analysis, principally in the budding yeast *Saccharomyces cerevisiae* and the fission yeast *Schizosaccharomyces pombe*. Through the pioneering work of Hartwell (20), a class of conditional-lethal mutations of *S. cerevisiae* has been recognized that conditionally blocks the progress of the cell-division cycle at a particular point. Such *cdc* mutations define a large set of *CDC* genes, all of which affect progress of cell division; these genes can be ordered into dependent pathways of function by analysis of the phenotypes of double mutants.

---

[*] *This chapter is reprinted from "Molecular Biology of the Cytoskeleton", Cold Spring Harbor: Cold Spring Harbor Laboratory, p. 153-174, with permission of the authors and Cold Spring Harbor Laboratory.*

Recently, it has become clear that yeasts have actin and tubulins that are strikingly similar in primary structure to their analogs in higher cells: There is 92% identity in the amino acid sequences of actin from *S. cerevisiae* and chicken brain and 71% identity between the β-tubulins from the same two species. Furthermore, as described by Pringle *et al.* (this volume), the visualization by immunofluorescence of microtubules and microfilaments in yeast shows a strong similarity in the organization of actin and tubulin in yeasts and other eukaryotic cells.

The conservation of primary structure and the apparent similarity in supramolecular structure between yeast and other eukaryotes, combined with the well-known advantages of yeast as a system for study of genetics by both classical and recombinant DNA methods (for review, see ref. 2), suggest that a true molecular genetics of the cell cycle and cytoskeleton can be achieved using yeast as a model system. It has recently been shown that yeast contains a single essential actin gene (16, 33, 38) and a single essential β-tubulin gene (32), making analysis of any mutations in these genes simple relative to organisms in which there are multiple copies of these genes.

The goal of such a molecular genetic approach is the association of genes specifying proteins involved in the cytoskeleton and cell cycle with their function. The connection between gene, product, and function is best made through the isolation of mutations: The mutant gene can be associated with the mutant protein by a number of criteria, often by direct determination of both the DNA and amino acid sequence, whereas the connection between the gene product and its function can be made by examination of the mutant phenotype *in vivo* or in a suitable *in vitro* system.

Even in yeast, however, one still faces technical problems in associating genes, products, and functions essential for growth, since the mutations must be conditional to permit growth of the organism. Straightforward biochemical analysis is also complicated by the fact that very little is known about the proteins that make up the cell structure. Thus, the geneticist can identify candidate genes by observing mutant properties that suggest failure in cell architecture or cell cycle. The biochemist and cell biologist, on the other hand, can find proteins (like actin and tubulin) that are abundant in structures implicated in basic cell functions. The challenge to the yeast molecular geneticist is to find ways to bring these two lines of endeavor together so that the geneticist's genes can be associated with the cell biologist's structures and functions.

In principle, there are two general ways in which to proceed. The classical genetic route (plan A) begins with mutations, which define a gene and whose properties indicate failure in a particular cellular function. Using mutations, one can, in yeast, readily isolate the gene as a DNA fragment by using DNA transformation with gene libraries to complement the mutation. Given the isolated gene, one can analyze the gene as a physical entity and, using newly developed techniques (e.g., raising antibodies to fusion proteins [41] or to synthetic peptides determined from DNA sequences [42, 43]) proceed to find the gene product. In favorable cases then, one can learn, through plan A, something about the way in which this protein contributes to the cellular function. By this route, one eventually obtains all the elements: the gene, the function, and the product.

The second approach (plan B) begins with a cellular process, let us say mitosis. One identifies a gene product (e.g., a protein like tubulin) whose involvement in this cellular process is known. Then, using recombinant DNA methods, one isolates the gene using information provided by the protein (as in the case of insulin; 44). Once the gene is available as a DNA fragment, one can readily induce mutations in the gene by directly altering the DNA by chemical or biochemical means. In yeast, methods have been devised to replace the normal gene on the chromosome with the mutagenized DNA so that the effects of particular mutations can be studied. It should be clear that with plan B, one ends up with the same elements as with plan A, but in a different order: product, gene, and function.

It is vital to recognize that both routes of approach have in common the fusion of classical and modern genetics. It is not really a question of "new" and "old" but of finding a strategy using elements of both. Second, it should be noted that both plans A and B contain elements of modern recombinant DNA technology, but they are not all the same for both plans. Third, it should be remembered that both plans result in the achievement of the same goal: association of gene, product, and function. Finally, it is necessary to recognize that the most interesting functions in the cell are executed through the interaction of many genes and gene products so that the analysis of gene and protein interactions is a crucial element of any attempt to construct a true molecular picture of cell functions. The most successful approaches to such interactions in the cell are firmly based in classical genetics: the isolation and characterization of mutations in secondary genes that modify or "suppress" mutations in the primary genes of interest. These mutations in primary genes can be mutations obtained by either plan A or plan B and can be analyzed further by modifications of either plan.

Plan A and plan B also have important and complementary defects. The success of plan B is limited to proteins (and their genes) that have been identified by biochemical means. This means that they will strongly tend to be the abundant proteins and not necessarily the most interesting ones. Plan A overcomes this difficulty, since mutations do not respect the abundance of the product they affect. With plan A, however, to target specific functions is often difficult, and functions that mutate rarely to a conditional form may be difficult to identify. The β-tubulin gene is a good example of this problem: Despite the fact that β-tubulin mutants show a Cdc phenotype, none have been isolated among the hundreds of Cdc mutants.

The remainder of this paper describes recent progess in our laboratory in defining the molecular nature of the yeast cytoskeleton and its relationship to the cell-division cycle, using the complementary advantages of plans A and B.

Genetics of Yeast Actin

Background

The first evidence that yeast contains actin was the isolation of an actinlike protein from cell extracts based on its affinity for DNase (26; 45). The tight binding of actin to calf thymus DNase in a 1:1 stolchiometry, resulting in its loss of enzymatic activity, had been established previously (28). Water *et al.* (45) used DNase coupled to Sepharose to purify yeast actin by affinity chromatography. Elution required denaturing conditions, although gentler elution conditions have now been developed (47). Native yeast actin has more recently been purified by conventional techniques (18), using DNase inhibition as an assay. The purified protein shares many properties with rabbit muscle actin. Polymerization of yeast actin into 7-nm-thick filaments could be induced, and the filaments could be decorated with a proteolytic fragment of muscle myosin (HMM). Copolymerization of yeast and rabbit muscle actin has been demonstrated, although some differences were noted (19).

Two laboratories have reported the presence of a single actin gene in yeast (15; 33), in contrast to higher eukaryotes, which typically contain up to 15 actin genes. Using the highly conserved coding sequence of the actin gene, these two groups cloned the yeast actin gene by screening a plasmid library of yeast genomic DNA for hybridization to a *Dictyostelium discoideum* actin gene. The complete nucleotide sequence of the cloned yeast gene has been determined (16; 33). The 373 amino acid residues of the yeast actin protein are perfectly colinear with rabbit muscle actin and differ at only 44 residues. The yeast actin gene was found to contain a single intervening sequence of 309 bp, beginning after the fourth codon. The presence of this intron is not thought to be important for actin gene expression, since it can be precisely deleted with no effect on expression of the actin gene (27).

The biological function of actin in a simple nonmotile eukaryote such as yeast is unknown. However, its striking conservation during evolution suggests that yeast actin plays a role or roles similar to those of cytoplasmic actins in higher eukaryotes. Among the proposed functions are maintenance of a cytoskeleton, organization of cytoplasmic membrane proteins, and transport of material within cells.

Constructing Actin Mutants *In Vitro*

The construction and phenotypic characterization of actin mutants *in vitro* is an excellent example of plan B at work. The first actin mutant constructed was a null mutation generated by disruption of the chromosomal actin gene (38). A restriction fragment that was internal to the coding sequence of the actin gene was subcloned into a yeast integrating plasmid vector, YIp5. The resulting plasmid (called pBR111) carries the yeast selectable marker *URA3* and transforms *ura3-* yeast to Ura+ only by integration into the yeast genome by homologous recombination between the yeast sequences found on the plasmid and a yeast chromosomal locus. Thus, when this plasmid transforms yeast, it can integrate via either its *URA3* homology or its actin gene homology. Integration at the *URA3* locus has no effect on the function of the endogenous actin gene. However, Figure 1 shows that integration into the actin gene results in disruption of the coding sequence, leaving a direct repeat, each copy of which contains an incomplete actin gene.

A disruptive integration into the actin gene might result in a lethal mutation in a haploid yeast. However, if the loss of the actin gene function due to disruption is recessive, then an integrant into only one chromosome of a diploid should be viable but cause a recessive lethal mutation (associated with the disrupted actin gene). Six independent Ura3+ transformants of a

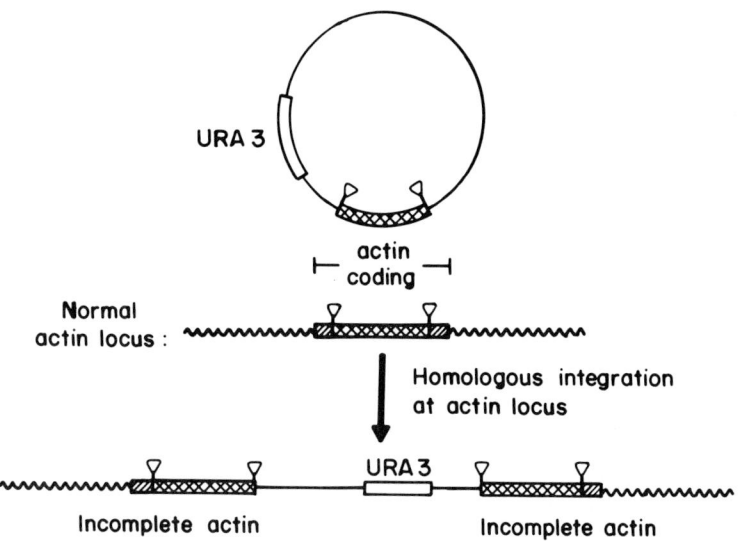

Figure 1
Disruption of the actin gene by integrative transformation. A fragment of the actin gene internal to the coding sequence subcloned on an integrating vector directs integration to the actin locus. This results in two incomplete actin sequences.

homozygous *ura3-* diploid strain were analyzed. Gel-transfer hybridization experiments demonstrated that four of these transformants resulted from integration at the actin locus, and the remaining two were integrated at the *URA3* locus (leaving the actin gene intact). Each of these diploids was sporulated, and the viability of the spores from the resulting tetrads was noted. All spores that had inherited the plasmid (marked by Ura3+) integrated at the actin locus were inviable, whereas those integrated at the *ura3* locus were viable. This indicated that the disruption of the actin gene was a recessive lethal event and that actin is required (at least) for growth of spores into a viable colony.

Constructing Conditional Mutants in the Actin Gene

Studying the function of actin by a genetic approach requires mutants conditional for actin function(s). Since we now know that the actin gene was essential for viability, we possessed a phenotype by which to recognize conditional actin mutants (conditional-lethality). Our problem was to devise a simple method to recognize recessive conditional-lethal mutations in an actin gene introduced into yeast by transformation. Simple introduction of the mutant actin gene on a replicating plasmid was not suitable, since the endogenous actin gene would mask all but dominant plasmid actin mutants. To achieve immediate expression of the *in-vitro*-mutagenized actin gene, a variation of the method of gene disruption was used. Rather than a plasmid carrying a DNA fragment entirely internal to the actin-coding region, a plasmid containing a larger DNA fragment with only one end lying within the coding sequence was used. The result of integration of such a plasmid into the actin locus is shown in Figure 2. It can be seen that a duplication of the actin locus results, only one copy of which remains functional. If the plasmid contained a mutation in the actin region and the integrative recombinant event occurred on the appropriate side of the actin mutation, it would result in the immediate expression of the mutant gene in yeast.

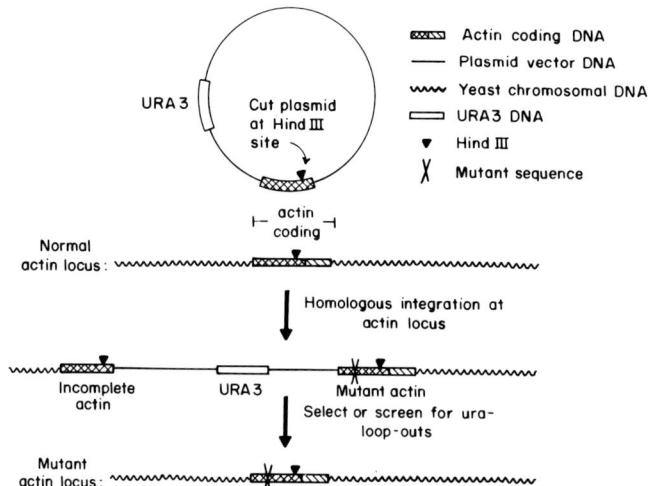

Figure 2
Construction of chromosomal actin mutants. A mutant actin sequence is transferred from a plasmid to the chromosomal actin locus by integration, followed by excision. The plasmid shown, carrying a portion of a mutant actin gene, is directed to integrate at the actin locus. This results in one copy deleted for the carboxyterminal coding sequence and a complete second copy that may contain a mutation. This structure allows expression of even recessive actin mutations, since a single functional actin gene is present in the transformant. After recognition of the mutation by its conditional-lethal phenotype, the plasmid sequences are excised by homologous recombination, leaving the actin mutation on an otherwise normal chromosome.

To ensure that the integration event occurred at the actin locus rather than at the *ura3* locus, the plasmid was cut at the unique *Hin*dIII restriction site located in the region of actin homology. The effect on yeast transformation of forming such a double-strand break is to stimulate greatly the frequency of transformation as well as to direct the integration exclusively to the chromosomal locus homologous to the region on the plasmid in which the double-strand break lies (34).

The mutagenesis of the actin gene *in vitro* could, in principle, be achieved in a variety of ways. In fact, the mutagenesis was achieved by RecA-mediated D-loop generation and treatment of the resulting single-strand region with the single-strand-specific mutagen sodium bisulfite (40). The details of this procedure are unimportant here and have in any case been largely supplanted by other methods that generate a wider variety of base changes (see, e.g., 39). A large number of mutant plasmids were generated and pooled together for transformation into yeast. Ura3+ yeast transformants were screened for conditional-lethality and several temperature-sensitive strains were identified. As a first test of the identity of these mutants, tetrad analysis was used to test the linkage of the temperature-sensitive phenotype to the actin locus (marked by the plasmid *URA3* gene). Only 10% of the mutations mapped at the actin locus. The other mutations were presumably due to background mutations unrelated to actin.

Three mutations were found that were closely linked to the actin gene. The plasmid sequences were removed from these three mutant strains, leaving only the mutant actin gene behind at an otherwise normal chromosomal locus. This was achieved by screening for spontaneous Ura- segregants that result from looping out of the plasmid sequences from the chromosome by homologous recombination between the two copies of the actin locus. Such segregants arise at a frequency of about 10^{-4}, making screening laborious. A simple positive selection for Ura3- now exists (F. Lacroute, pers. comm.). The site of recombination in the loop-out event can lie on either side of the actin mutant site. Thus, some Ura- segregants are *ts+* and others are *ts-*. This is another indication that the *ts-* lethal lies in the actin gene, since other mutations will not produce frequent *ts+* segregants during loop-out of the plasmid. Three actin temperature-sensitive mutants (*act1-1*, *act1-2*, and *act1-3*) were constructed using this protocol.

Phenotypes of Actin Mutants

In trying to understand the phenotype of these actin mutants, some features of the yeast cell-division cycle are important. *S. cerevisiae* divides by budding. Near the start of each cell cycle, a small round bud appears on the mother cell. This bud grows as the cell cycle progresses, remaining separated from the mother by a short neck region. As the bud enlarges, the nucleus migrates to the bud neck, elongates into the daughter cell, and divides. By this point, the bud is approaching the mother cell in size and cytokinesis and cell separation rapidly ensues, completing the cell cycle.

Several features characterized the neck separating the mother cell and the bud. A collar of chitin (a polymer of N-acetyl glucosamine) is formed in the cell wall surrounding the bud neck (7). This chitin ring can be visualized in the fluorescence microscope with the dye Calcofluor. Following cell division, the chitin ring remains associated with the mother cell and is stable for many generations, giving rise to a series of chitin rings (called bud scars) on "old" mothers cells. The plasma membrane of the bud neck is completely lined on the cytoplasmic side by rings on a 10-nm filament visible by thin-section electron microscopy (6). This internal collar disappears just prior to cytokinesis. The size of these filaments is inconsistent with that of either microfilaments (7 nm) or microtubules (30 nm). The rings are more likely to be equivalent to the intermdeiate filaments of higher eukaryotes.

The phenotype of the temperature-sensitive actin mutants has been studied primarily through the use of two fluorescence microscopic techniques for visualizing the intracellular distribution of actin in yeast. Adams and Pringle (1) have used rhodamine-tagged phallodin, a mushroom toxin known to interact with polymerized actin, whereas Kilmartin and Adams (24) have used affinity-purified anti-actin antibody and a fluorescently tagged secondary antibody. In either case, the cells are first fixed with formaldehyde. Rhodamine-phalloidin requires no further permeabilization of the cells, but for antibody staining, removal of the cell wall is necessary to allow penetration of the antibody. These techniques produce similar results. Our observations using the anti-actin antibody, kindly supplied by J. Kilmartin, are discussed below.

Through much of the cell cycle, the actin-staining pattern is asymmetric, as shown in Figure 3. The mother cell shows cables of actin directed toward the bud neck, and the bud shows brightly staining dots just below the cell surface. Few dots are seen in the mother cell and few cables are seen in the bud. This actin asymmetry persists until shortly before cytokinesis, at which time randomly directed cables and scattered dots are seen in both mother and bud. At the time of bud emergence, in addition to the dots and cables, a ring of actin appears to form around the bud neck only

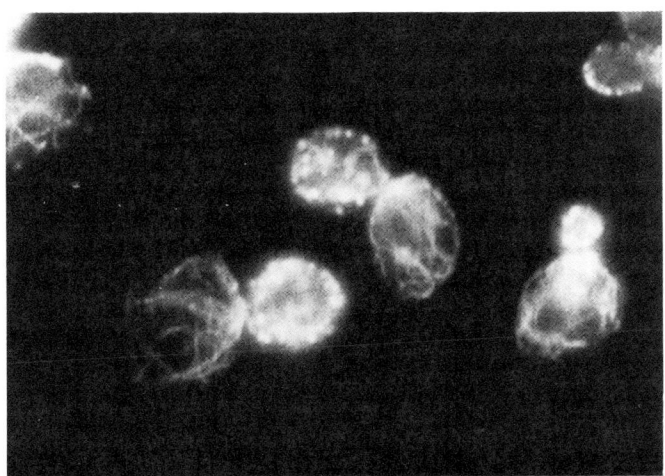

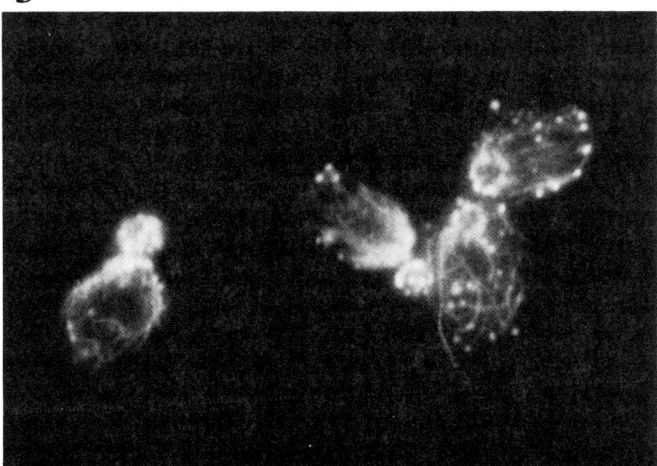

Figure 3
Immunofluorescent staining of actin in wild-type diploid yeast. Budded cells show cables in the mother cell and dots near the surface of the bud (*a*). A ring of actin dots forms just prior to bud emergence and persists as a collar around the neck of cells with a small bud (*b*).

on the mother side of the neck. This actin ring persists as a collar around the neck throughout the early portion of the cell cycle but disappears at later times. In contrast, the ring of 10-nm filaments seen in the electron microscope is found through the full length of the bud neck, supporting the conclusion that this ring is not composed of actin.

Several differences between the results presented above and the rhodamine-phalloidin staining pattern observed by Adams and Pringle (1) should be noted. The asymmetry of the actin dots seen with phalloidin is maintained only through the early part of bud growth, and a collar around the neck is seen at the time of cytokinesis. Whether these differences reflect the use of a different sample preparation, different staining reagents, or differences in the yeast strains used is presently unclear.

Mutations in actin might affect its localization in the cell. The pattern of actin staining in *act1-2* and *act1-3* mutants has been examined using anti-actin antibody. At permissive temperature, *act1-3* cells show only a few faint cables, and the surface dots are found in both the mother cell and the bud. After 2 hours at the restrictive temperature, no cables are seen, only randomly distributed dots. A somewhat different phenotype is observed for *act1-2*. At permissive temperature, dots are seen in both mother and bud; however, many heavy bars of actin are also seen, often crossing the cell diametrically. After 2 hours at the restrictive temperature, the bars have dissipated, a meshwork of fine filaments appears to fill the cell, and randomly distributed dots lie near the cell surface. The aberrant actin-staining pattern of the mutants even at their permissive termperature is consistent with their poor growth at this temperature, and presumably reflects a partial actin defect.

The actin mutants affect other cell processes as well. At permissive temperature, the mutants appear considerably larger and more rounded than wild-type cells. At the restrictive temperature, this trend becomes more pronounced, leading to very large and nearly spherical cells. The vacuole swells, filling much of the cytoplasm. The percentage of budded cells drops to about 30% after 2 hours; a wild-type exponential culture contains 50-60% budded cells. After 4 hours, cell lysis becomes prevalent. Both actin mutants show a pronounced osmotic sensitivity; at temperatures approaching the restrictive temperature, addition of 0.5-1 M sorbitol, KCl, or ethylene glycol to the medium abolishes growth. This phenomenon is not specific to these alleles, but rather reflects a reduced level of functioning actin, since diploid strains with one wild-type copy of actin and one disrupted copy also display osmotic sensitivity.

The actin mutants also affect chitin localization. At permissive temperature, chitin rings are seen, but their diameters are much larger than in wild type. After arrest for 2 hours at the restrictive temperature, the mutant cells are uniformly bright, suggesting delocalized chitin synthesis.

These observations suggest at least two possible roles for actin in yeast. Actin may play a role in bud emergence, as suggested by the actin collar formed at the base of the bud neck, the delocalized chitin staining, and the reduction in budded cells at restrictive temperature. Actin may also play a role in osmotic regulation of the cell, as suggested by the cell lysis, osmotic sensitivity, and swollen vacuoles of the actin mutants.

Suppressors of Actin Mutants

The elaborate cell-cycle-dependent pattern of actin localization must be controlled both temporally and spatially. Such control most likely involves proteins that regulate the site and extent of actin polymerization by means of direct or indirect physical interactin with actin. Candidates for such proteins have been identified in higher eukaryotes by *in vitro* polymerization assays (21). We hope to identify such proteins in yeast by genetic methods. The deleterious effects of a mutation in one protein may be compensated for by a mutation in a second protein that physically interacts with it (22; 31). Among such extragenic suppressors, some will simultaneously confer a new phenotype that may be informative.

To perform this sort of analysis for actin, pseudorevertants of a temperature-sensitive (*ts*) actin mutant were screened for cold sensitivity (*cs*) for growth. Restricting our attention to this class of mutants allowed simple genetic manipulation and phenotypic analysis of the suppressor mutants, since they have a phenotype of their own that can be assessed in the absence of any actin defect.

An *act1-3* strain was incubated on plates at 37°C (the restrictive temperature) to allow the growth of revertants. Colonies that arose were tested for cold sensitivity for growth at 17°C. Tetrad analysis of these strains identified those that had a single locus responsible for both the *ts* suppressor and the *cs*-lethal phenotypes. Eighteen revertants (50%) fell into this class; the others were the result of more than one mutation and were not pursued. These 18 mutants were arranged into complementation groups according to their *cs* phenotype. Three groups were defined, *sac1*, *sac2*, and *sac3* (suppressor of actin). The largest group, *sac1*, contains 13 independently isolated members, *sac2* has 3 members, and *sac3* has 2 members. All of these mutants were recessive for both cold sensitivity and suppression of temperature sensitivity. Suppressors of this sort are typically dominant in yeast (see, e.g., 30), suggesting that there is a special property of these actin suppressors that explains their recessivity. One model to explain this phenomenon is that *SAC*-gene products interact along the length of an actin polymer. If all copies of the *SAC* protein must be the suppressor type to function as a suppressor, then the mutant allele would show recessive suppression.

If the *SAC*-gene products physically interact with actin, two predictions can be made. First, *sac* mutants should exhibit allele specificity in suppression. In fact, one allele of each group has been shown to suppress *act1-3*, the mutations with which they were isolated, but not *act1-2*. The second prediction is that the *sac* mutants should show a phenotype at their restrictive temperature that is related to the phenotype of the original actin mutant in some way. Immunofluorescent staining with anti-actin antibody was performed on strains containing a mutant *sac* gene but wild-type actin. At the restrictive temperature, 14°C, *sac1* shows dots on the cell surface of both mother and bud and loss of actin cables. *sac2* also exhibits randomly localized dots as well as occasional heavy actin bars, whereas *sac3* contains randomly distributed dots and heavy bars extending the full length of budded cells. It is thus clear that mutations in the *SAC* genes can affect the localization of wild-type actin. The pattern of chitin deposition is also aberrant in the *sac* mutants; *sac2* and *sac3* show delocalized chitin synthesis at the restrictive temperature, whereas *sac1* exhibits bright patches of chitin at the tip of small buds or on the mother cell in addition to the normal collar around the neck. These results indirectly support the hypothesis that the *SAC*-gene products physically interact with actin. Future work will be directed toward identification and localization of the *SAC*-gene products.

Genetics of Yeast Tubulin

Background

Mitotic spindles consisting of microtubules were first observed in *Saccharomyces* in 1966 by electron microscopy (36). Extensive cytological observations of microtubules at all phases of the yeast life cycle have been made at the electron microscopic level (for a recent review, see ref. 3), and more recently by the use of monoclonal antibodies against tubulin as an immunofluorescent probe for microtubules (25; 1; 24). Yeast has a typical mitotic spindle that is organized by spindle-pole bodies (analogous to the higher eukaryotic centrosome) that are embedded in the nuclear envelope. All microtubules in yeast have one end closely associated with a spindle-pole body, radiating both into and out of the nucleus to form the spindle and the cytoplasmic microtubules, respectively. Duplication of the spindle-pole body is an important event in the yeast cell cycle that immediately follows "start" (commitment to a new cell cycle) and coincides with bud emergence (5). Observation of the state of the spindle-pole bodies and associated microtubules in mutants that affect the yeast cell-division cycle indicates that these structures are closely regulated during the cell cycle (4).

During karyogamy following conjugation in yeast, microtubules appear to play a key role. The spindle-pole body and asociated cytoplasmic microtubules undergo a series of characteristic changes that appear to orient and mediate migration of the two nuclei, resulting in fusion of their nuclear envelopes and spindle-pole bodies (4). Microtubules are also, of course, involved in meiosis, forming the meiotic spindle (29).

Tubulin has been purified from yeast by self-assembly *in vitro* (23). The purified protein coassembles with mammalian tubulin and appears very similar in physical properties. The gene for β-tubulin has been cloned, sequenced, and demonstrated to be present in one copy in yeast (32), in contrast to the many copies of the gene found in higher eukaryotes. The amino acid sequence of yeast β-tubulin protein, inferred from the gene sequence, is identical to chicken β-tubulin at 71% of its residues. Using the gene-disruption technique described above in detail for actin, Neff et al. (32) showed that this β-tubulin gene is essential for growth in yeast.

Mutations in the β-Tubulin Gene by Drug Resistance

At the time we set out to make mutants in the tubulin gene(s) of yeast, no clones of either α- or β-tubulin genes of yeast were available. We therefore chose to attempt to isolate such mutants by selecting resistance to the anti-mitotic drug benomyl (10). Some mutants resistant to benomyl in *Aspergillus nidulans* had been shown to lie in a β-tubulin gene (37). Reasoning that this might also be true in yeast, we isolated a large number of spontaneous benomyl-resistant (Ben^R) mutants. These mutants were shown to lie in a single gene$_R$ (J.H. Thomas et al., in prep.). Among the many Ben^R mutants, six also displayed cold sensitivity and three others displayed temperature sensitivity for growth. In crosses, these conditional-lethal phenotypes perfectly cosegregated with the Ben^R phenotype, and where two mutants shared a common recessive phenotype (Ben^R or conditional-lethality), they were shown to fail to complement.

The locus defined by these Ben^R mutants was identified as the gene encoding β-tubulin by use of a newly isolated clone of yeast β-tubulin (32). This clone was isolated by virtue of its DNA homology with a previously isolated chicken β-tubulin clone (9). It was positively identified as a gene for β-tubulin by sequence analysis. The connection between the clone of yeast β-tubulin and the Ben^R locus was made by two separate tests: a complementation test and a genetic linkage test. To test complementation, a recessive, cold-sensitive Ben^R mutant, *tub2-104*, was transformed with a centromere-containing vector (8; 2) carrying the entire wild-type β-tubulin gene. In every transformant, the wild-type (benomyl-sensitive, cold-resistant) phenotype was restored. Segregants of several transformants that had lost the plasmid were isolated, and the complementing activity always cosegregated with the plasmid. This demonstrated that the yeast insert in this plasmid can supply the functions lacking in the mutant. Another plasmid that contained only a part of the β-tubulin gene carried on a yeast-integrating vector was integrated into the chromosome at the β-tubulin locus and shown by tetrad analysis to be tightly linked to the Ben^R locus.

The gene-disruption method, described in detail for the actin gene (see Fig. 1), was also applied to the β-tubulin gene, with the same results: When the β-tubulin gene is disrupted by transformation, it creates a recessive lethal mutation that is linked to the gene disruption and the Ben^R locus. In this case, one further experiment was possible (since mutants in the β-tubulin gene existed) that demonstrated that the disrupted β-tubulin gene in the diploid was nonfunctional. The gene disruption was performed in a diploid heterozygous for a cold-sensitive Ben^R mutant that was recessive for both phenotypes ($tub2$-104). We found that when the β-tubulin gene disruption occurred in the wild-type homolog of this diploid, it resulted in the expression of both the cold-sensitivity and Ben^R phenotypes of the β-tubulin mutant gene on the other homolog. This confirmed that the DNA resulting in the gene disruption does indeed affect the same gene that is defined by the Ben^R mutations.

Phenotypes of β-Tubulin Mutants

Studies have suggested that microtubules are involved in a variety of cell processes in yeast, including chromosome segregation, bud emergence (3), and nuclear fusion during conjugation (5). Until recently, there has been no good experimental evidence addressing any of these β-tubulin mutants allowed us to ask directly what cellular defects are associated with a defect in microtubules.

If microtubules are required for progression through mitosis but at no other stage of the cell cycle, then β-tubulin mutants should show a cell-cycle-specific terminal arrest morphology (20). We have observed the arrest of several different β-tubulin mutant alleles and have found a classic cell-division-cycle defect in mitosis.

In the tightest mutant alleles, over 90% of the cells arrest as large-budded doublet cells. The nuclear DNA in these cells (as detected by fluorescence microscopy with a DNA-staining dye, DAPI) is undivided and usually remains unelongated and fails to migrate to the neck between the mother and bud. The mutant block is rapidly imposed, since cells arrest uniformly after only one cell-cycle time at the restrictive temperature. In general, the arrest closely resembles that observed with the anti-mitotic drugs methyl benzimidazole-2-yl-carbamate (446) or nocodazole (J.H. Thomas et al., unpubl.). This cell-cycle arrest supports the hypothesis that microtubules are involved in chromosome segregation, thus resulting in a block in mitosis. At the same time, it argues against a microtubule requirement for bud emergence, since this cell-cycle event is apparently unaffected in the mutants.

Using indirect immunofluorescent staining with antibodies directed against yeast tubulin (25; 24), we determined that the microtubules in a $tub2$-104 strain largely, but not entirely, disappear at the restrictive temperature. They are reduced to a dot or a faint short bar of staining, similar to the pattern of staining normally seen during part of the G_1 phase in wild type.

The mutant $tub2$-104 also shows a karyogamy defect during conjugation. When two permissively grown $tub2$-104 mutant strains were allowed to conjugate at their restrictive temperature, they underwent normal cytoplasmic fusion. However, if large-budded zygotes were recovered from such a conjugation and allowed to divide to form colonies at permissive temperature, over 90% of them segregated "cytoductants" (haploid cells of one parental genotype or the other), indicating that nuclear fusion had failed in these zygotes. When the same experiment was performed using a $tub2$-104 mutant crossed to a $TUB2$+ strain, the zygotes rarely formed cytoductants, indicating that the $tub2$-104 karyogamy defect can be compensated for by the wild-type gene product contributed by the other parent. Thus, tubulin has the properties of a "bilateral" karogamy deficiency (both parent cells must be defective; see ref. 13), in contrast to most other karogamy-defective mutants, which are unilateral (13; 11).

When the zygotes produced by these matings were fixed and their nuclear DNA and microtubules were viewed by fluorescence microscopy, the nature of the karyogamy defect was clear. In the *tub2-104* cross by *TUB2+* at the restrictive temperature (as well as in wild-type crosses at either temperature or a *tub2-104* by *tub2-104* cross at permissive temperature), cytoplasmic microtubules are normally elaborated and the nuclei are closely associated or fused at very early times after cell fusion. In the *tub2-104* cross by *tub2-104* at the restrictive temperature, there is a general failure to elaborate cytoplasmic microtubules and the nuclei remain well separated.

Suppressors of a β-Tubulin Mutant

Assembly and function of tubulin, like actin, must be controlled spatially and temporally to result in the complex pattern of changes seen during the cell cycle. We have used pseudoreversion as a means of genetically identifying proteins that interact with tubulin. The cold-sensitive mutant *tub2-104* was reverted to *cs+* and screened for those revertants that had simultaneously acquired temperature sensitivity for growth (*ts*). Subsequent crosses established that the cold-sensitive suppressor and *ts*-lethal phenotypes were tightly linked and extragenic (except one, which was shown to be an intragenic *ts* revertant). Twenty-six such mutants were classified into 16 complementation groups, of which 6 have more than one member. When examined for cell-cycle phenotype, mutants from one complementation group (six members) showed a phenotype similar to that of β-tubulin temperature-sensitive mutants, even in a *TUB2+* strain. This preliminary observation suggests that at least some of these mutants may define genes whose products are functionally related to tubulin.

Several of the suppressor mutants were tested for dominance of their suppressor phenotype. All were found to be recessive, supporting the argument that recessive suppressors may be common for proteins that form part of a highly polymerized structure (see above).

Map Position of the *ACT1* and *TUB2* Genes

The *ACT1* gene was mapped to the left arm of chromosome VI of yeast (12), using a new method involving the integration of a plasmid containing 2-μm sequences into the chromosome at the site of a cloned gene. When the same method was applied to the *TUB2* gene, identical results were obtained (J.H. Thomas et al., in prep.). Tetrad analysis of a cross between an actin locus marked by an integrated *URA3* gene and a *tub2* benomyl-resistant allele showed close linkage of the two genes (PD:NPD:TT, 22:0:0). Examination of the plasmid clones carrying each of these genes and gel-transfer hybridization experiments confirming the structure of the genomic locus demonstrated that *ACT1* and *TUB2* are immediately adjacent, with about 1 kb of DNA separating them. Analysis of transcription and DNA sequence studies of this region (14; 17; N.N. Neff and J.H. Thomas, unpubl.) indicate that the *ACT1* and *TUB2* genes are arranged in opposite orientation (Fig. 4) so that transcription is divergent. A small transcript of unknown function lies between the genes in the same orientation as *ACT1*.

Plan A Revisited

It is clear that plan B provides powerful tools for

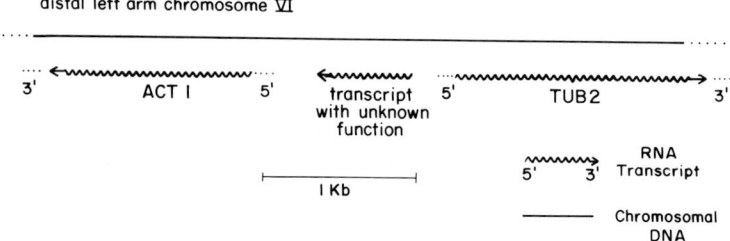

Figure 4
Arrangement of the actin and β-tubulin region of chromosome VI. The actin and β-tubulin messages are divergently transcribed and a third transcrip of unknown function lies between them. The precise ends of the transcripts are not known.

the genetic analysis of proteins already thought to be important in the function of the cytoskeleton. This leaves the geneticist at the mercy of cell biologists for the identification of interesting problems. One solution to this difficulty is plan A: the isolation of mutations in random genes that display "interesting" properties, such as a cell-division-cycle-specific arrest. One of the crucial decisions to make when using this approach is what constitutes an interesting phenotype. The identification of Cdc mutants, for instance, has proved very fruitful (35); however, no Cdc mutant has yet been shown to be a defect in a structural component of the cytoskeleton. We now know that actin mutants would never have been found by this criterion, and although mutants in β-tubulin should have shown a Cdc phenotype, none were isolated. Recently, we have found a mutation in a gene whose protein product is unknown, but whose properties are intriguing. Here, we discuss this mutation, *ndc1-1*, as an example of the application of plan A to the identification of new genes related to cytoskeletal function.

Recognition of *ndc1-1*

The mutation *ndc1-1* was originally isolated by D. Moir as part of a search through random *cs*-lethal mutants for those that displayed a Cdc arrest (30). The *ndc1-1* mutant was recognized as having a Cdc-related arrest, although its terminal arrest morphology was not uniform. The mutant was not pursued by Moir *et al.*, but recently we became interested in this mutant, since its nonpermissive arrest phenotype seemed related to the β-tubulin arrest.

We have characterized the terminal arrest phenotype of *ndc1-1* strains primarily by two methods: (1) Cells arrested in liquid culture can be fixed to allow observation of nuclear DNA, microtubules, or other structures by fluorescence microscopy, and (2) cells can be arrested on agar slabs, and the products of a single cell's lineage can be assessed. The first method does not permit assessment of the lineage of the cells being observed, and the second method does not permit the detection of nuclei and microtubules. When the two methods are applied together, they provide complementary information about the pattern of arrest.

When an asynchronous population of *ndc1-1* cells grown at the permissive temperature was shifted to the restrictive temperature on an agar slab and individual cells were followed by photomicroscopy, the result shown in Table 1 was seen. Initially, unbudded cells produced five to six cell bodies over several generation times. This suggests that either the arrest due to the *ndc1-1* defect is "leaky" (allowing cells to perform the mutant function slowly) or the cell cycle is not immediately arrested during the block. Also, the five or six cell bodies at the end of the arrest time might be part of one undivided cell or they might be due to several divided cells. Restrictive arrest in liquid medium distinguished these possibilities, as shown in Table 2.

Table 1
Mean Number of Cell Bodies at Times after Shift of *ndc1-1* to Restrictive Temperature

Cell morphology at time of shift	Generation Times		
	1.5	3.3	4.4
Unbudded	2.7	4.1	4.6
Small budded	4.2	6.2	75.
Large budded	5.5	8.1	9.0

Cells were grown in liquid at permissive temperature and were lightly sonicated to disrupt clumps prior to shift. A sample of cells was spotted onto a prechilled slab of nutrient agar on a microscope slide and covered with a coverslip. A wire mesh screen embedded in the agar allowed the same field to be photographed at time 0 and several subsequent times. Wild-type cells, when subjected to this protocol, grow normally for several generations, with the same generation time as in log phase liquid growth.

Very few multiply budded cells accumulated even at long arrest times, indicating that the cells are undergoing division prior to the production of a new bud. Furthermore, nuclear DNA division was completely blocked

Table 2
Percentage of Various Cell Types at Times after Shift of Wild-type or ndc1-1 Cells to Restrictive Temperature

Strain	No. of generation times after shift	Cell Types				
		unbudded		small budded	large budded	
		+ nuclear DNA	- nuclear DNA		1 region of nuclear DNA	2 regions of nuclear DNA
Wild type	2.0	60	0	23	17	
ndc1-1	0	64	0	26	2	8
ndc1-1	0.7	58	0	22	10	10
ndc1-1	1.5	21.5	0.5	20	51	7
ndc1-1	2.6	10	22	7	59	2
ndc1-1	3.3	5	40	8	47	0

Cells were grown in liquid medium at permissive temperature and shifted in early log phase to the restrictive temperature for ndc1-1 cells. At time points, cells were removed and prepared for observation by fixation and treatment as if for immunofluorescence (Kilmartin and Adams 1984). We found that this method gave DAPI staining superior to other techniques, which is important in the certain identification of aploid cells. A representative time point is shown for wild-type cells; these numbers do not change significantly at any time during the arrest and aploid cells are never seen. ndc1-1 cells grown at permissive temperature are also indistinguishable from wild-type.

in the *ndc1-1* arrest, since no cells with divided nuclear DNA were seen. Together, these results indicate that although DNA division is defective in the *ndc1-1*-arrested cells, the cell cycle progresses for several generations. As would be expected for such an arrest, "aploid" cells (lacking detectable nuclear DNA staining) accumulated during the arrest, eventually becoming the predominant cell type. These cells were always unbudded and never succeeded in progressing through another cell cycle. Thus, we can infer that the daughter of an *ndc1-1*-defective cell division that retains the nuclear DNA must rebud and continue to divide several times over the course of the arrest, giving rise to the several cells seen during arrest on solid medium.

We tested whether DNA replication occurs in the *ndc1-1*-arrested cells by asking if the nucleated cells that arise from the first division have duplicated their genome, becoming 2N diploids. We arrested haploid *ndc1-1* cells at the nonpermissive temperature for about 1.5 generation times, at which point nearly all of the large-budded cells have the characteristic undivided nuclear DNA. These cells were plated onto agar medium at permissive temperature, and large-budded cells were micromanipulated out. These cells were allowed to divide, and the viability and genotype of the resulting daughters were assessed. Most cells divided shortly after shift to permissive temperature, giving rise to one viable and one inviable cell (this cell never budded). By crossing the colony that arose from the viable cell and dissecting tetrads, we were able to show that they were precise or nearly precise 2N diploids (12/13 cases). The two cells produced by the division presumably correspond to the aploid daughter (never budded) and the daughter that contains all of the nuclear DNA (diploidized). Thus, DNA replication occurred normally in these cells, but the chromosomes all segregated to one daughter.

Since microtubules are involved in chromosome segregation, we fixed and stained arrested *ndc1-1* cells with antibody directed against tubulin (25), visualizing the microtubules by indirect immunofluorescence. To appreciate the aberrant patterns seen in these *ndc1-1* cells, it is necessary to be familiar with the normal cycle of nuclear division in yeast. Figure 5 shows a sketch of the events pertinent to this discussion. All microtubules in yeast have one end associated with

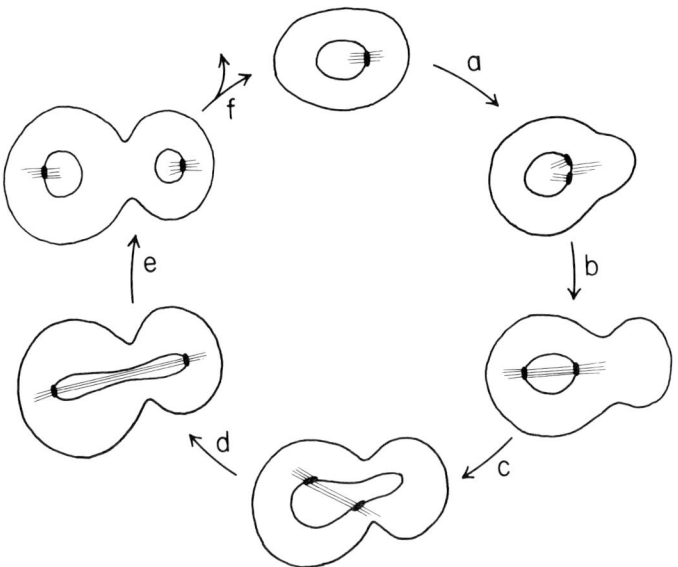

Figure 5
Schematic diagram of the yeast cell cycle, showing the major features of the nuclear division pathway, derived primarily from electron microscopic data (Byers 1981). Shown in each cell are the cell wall, the nuclear envelope, the spindle-pole body (embedded in the nuclear envelope), and the intra- and extranuclear microtubules radiating from the spindle-pole body. (a) One of the first steps in the cell cycle is the duplication of the spindle-pole body and the nearly simultaneous emergence of the bud. DNA replication initiates shortly thereafter. (b) As the bud grows, DNA replication is completed; the two spindle-pole bodies migrate to opposite poles of the nucleus, lining up roughly on the long axis of the cell; and a short intranuclear spindle forms between them. (c) The nucleus then migrates into the neck of the bud and begins elongation. (d) The spindle elongates rapidly until it extends nearly to the ends of the two cell bodies. The appropriate chromosomes remain closely associated with each spindle-pole body during this stage, resulting in their segregation. (e) The spindle breaks down from the middle toward the ends and the nuclear envelope divides. (f) Cytokinesis and cell separation ensue.

the spindle-pole body (or bodies), which remains embedded in the nuclear envelope throughout the cell cycle (in yeast, the nuclear envelope remains intact during mitosis). Some microtubules radiate into the nucleus to form the spindle, and others radiate out to form cytoplasmic microtubules. At the time of bud emergence, the single spindle-pole body duplicates and the two products migrate to opposite sides of the nearly spherical nuclear envelope, forming the spindle between them. At the same time, DNA replication occurs, followed by migration of the nucleus into the neck of the mother and bud. There, the spindle and the nucleus elongate through the neck of the bud, finally spanning the length of the two cell bodies. During this elongation, the nuclear DNA can be seen to separate into equal regions, which segregate to their respective poles. Finally, the elongated spindle breaks near the middle into half-spindles, and the nuclear envelope divides, followed shortly by cell division.

The appearance of typical *ndc1-1* cells after about 1.5 generations of arrest is shown in Figure 6. The microtubules remain intact and are roughly normal in number and general appearance. Nearly all of the large-budded cells have microtubule bundles radiating from a single point in each cell body. These structures appear similar to the half-spindles seen in normal cells just prior to cell division. In these cells, nuclear DNA is present in only one cell body. Judging from the appearance of the microtubules found in the aploid cell body, it appears that the spindle-pole body has divided and properly segregated, since the microtubules appear to originate from a single point and are morphologically normal in appearance. Also shown in Figure 6 is an aploid daughter cell, in which it appears that there is a spindle-pole body and its associated microtubules, but no nuclear DNA.

It seemed possible that cells defective for *ndc1-1* function proceed along an aberrant pathway not related to the normal cell-division cycle. To test this possibility we imposed a double block with *ndc1-1* and another treatment that normally results in a block at a specific stage of the cell cycle. If the step blocked by this second agent were absent from the aberrant *ndc1-1* pathway, then it would not affect the arrest of the *ndc1-1*-defective cells. If, however, the aberrant *ndc1-1* pathway is also dependent on the second block, then

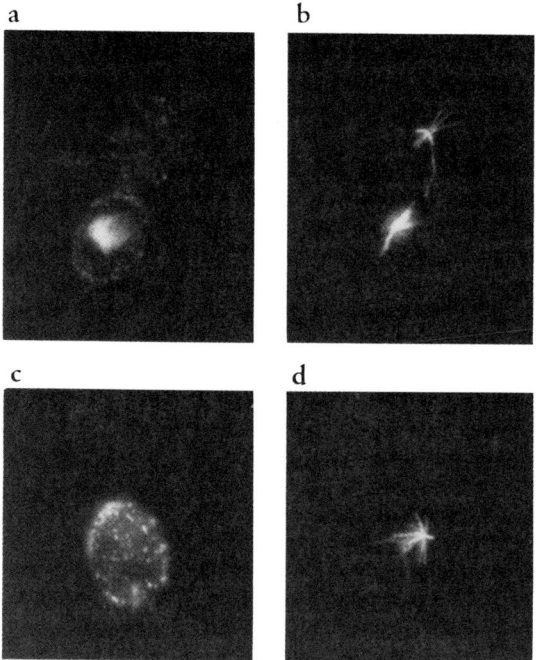

Figure 6
Pattern of DNA and microtubules found in *ndc1-1* mutants arrested at restrictive temperature for 1.5 generation times. a and b show the same large-budded cell stained with the DNA staining dye DAPI (a) and with monoclonal antibody against tubulin (b). The two stains can be viewed independently by the use of appropriate excitation and emission filters. The DAPI stain reveals the nuclear DNA (large bright area) clearly situated in only one cell body, whereas the microtubules are found in both cell bodies, apparently radiating from two properly segregated spindle-pole bodies. Small dots of DAPI stain distributed primarily on the periphery of the cell are mitochondrial DNA. (c,d) Analogous stains of an unbudded aploid cell from the same arrest. The DAPI-stained panel is overexposed to show the absence of nuclear DNA despite excellent staining of the mitochondrial DNA. The microtubules in this cell nonetheless appear normal, all appearing to radiate from a single site.

that block should be epistatic to *ndc1-1*. The two secondary blocks we employed were α factor, a yeast pheromone that blocks cells in G_1, and hydroxyurea, which blocks DNA synthesis. Epistasis of these blocks was tested both by morphological criteria (whether the double-blocked cells followed the pattern of *ndc1-1* arrest or showed the α factor or hydroxyurea pattern) and by assessing the ploidy of cells that were allowed to recover from such a double block. Both α factor and hydroxyurea were epistatic to *ndc1-1* for both the morphology and diploidization phenotypes, producing only arrested cells that were typical of the drug block alone and preventing the diploidization of the *ndc1-1* cells. Thus, the aberrant pathway followed by cells during the *ndc1-1* block still passes through these two cell-cycle steps. This supports the argument that *ndc1-1*-arrested cells replicate their DNA.

The behavior of the *ndc1-1* cells has important implications for analysis of the yeast cell-division cycle. First, it suggests that not all defects in a cell-cycle function result in a uniform arrest morphology; there may be many other mutants that affect cell-cycle functions but have not been recognized for this reason. Second, these results indicate that proper chromosome segregation is not required for progression through the cell cycle. In the parlance of Jarvik and Botstein (22), chromosome segregation in the *ndc1-1* mutant cells is not on a dependent pathway with any other function required for the cell-division cycle.

In conclusion, we have presented examples of application of the combined methods of classical cell biology, genetics, and the newly developed recombinant DNA methods to *S. cerevisiae*. It seems clear that continued application of these methods, by plan A and plan B, holds real promise for the understanding of the cell cycle and cytoskeleton at the molecular level.

REFERENCES
1. Adams, A.E.M. and J.R. Pringle. 1984. Localization of actin and tubulin in wild-type and morphogenetic mutants of *Saccharomyces cerevisiae*. J. Cell Biology 98: 934.
2. Botstein, D. and R.W. Davis. 1982. Principles and practice of recombinant DNA research with yeast. In: The molecular biology of the yeast

Saccharomyces. *Metabolism and gene expression* (ed. J.N. Strathern et al.), p. 607. Cold Spring Harbor Laboratory, Cold Spring Harbor, New York.

3. Byers, B. 1981. Cytology of the yeast life cycle. In: *The molecular biology of the yeast Saccharomyces. Life cycle and inheritance* (ed. J.N. Strathern et al.), p. 59. Cold Spring Harbor Laboratory, Cold Spring Harbor, New York.

4. Byers, B. and L. Goetsch. 1974. Duplication of spindle plaques and integration of the yeast cell cycle. Cold Spring Harbor Symp. Quant. Biol. 38: 123.

5. Byers, B. and L. Goetsch. 1975. The behavior of spindles and spindle plaques in the cell cycle and conjugation of *Saccharomyces cerevisiae*. J. Bacteriol. 124: 511.

6. Byers, B. and L. Goetsch. 1976. A highly ordered ring of membrane-associated filaments in budding yeast. J. Cell Biol. 69: 717.

7. Cabib, E. 1975. Molecular aspects of yeast morphogenesis. Annu. Rev. Microbiol. 29: 191.

8. Clarke, L. and J. Carbon. 1980. Isolation of a yeast centromere and construction of functional small circular chromosomes. Nature 287: 504.

9. Cleveland, D., M. Lopata, R. MacDonald, N. Cowan, W. Rutter, and M. Kirschner. 1980. Number and evolutionary conservation of α- and β-tubulin and cytoplasmic α- and γ-actin genes using specific cloned cDNA probes. Cell 20: 95.

10. Davidse, L. and W. Flach. 1977. Differential binding of methyl benzimidazol-2-yl carbamate to fungal tubulin as a mechanism of resistance to this antimitotic agent in strains of *Aspergillus niculans*. J. Cell Biol. 72: 174.

11. Dutcher, S.K. and L.H. Harwell. 1983. Genes that act before conjugation to prepare the *Saccharomyces* nucleus for caryogamy. Cell 33: 203.

12. Falco, S.C. and D. Botstein. 1983. A rapid chromosome-mapping method for cloned fragments of yeast DNA. Genetics 105: 857.

13. Fink, G.R. and J. Conde. 1977. Studies of *KAR1*, a gene required for nuclear fusion in yeast. In: *International cell biology 1976-77* (ed. B.R.

Brinkley and K.R. Porter), p. 414. Rockefeller University Press, New York.
14. Gallwitz, D. 1982. Construction of a yeast actin gene intron deletion mutant that is defective in splicing and leads to the accumulation of precursor RNA in transformed yeast cells. Proc. Natl. Acad. Sci. 79: 3493.
15. Gallwitz, D. and R. Seidel. 1980. Molecular cloning of the actin gene from yeast *Saccharomyces cerevisiae.* Nucleic Acids Res. 8: 1043.
16. Gallwitz, D. and I. Sures. 1980. Structure of a split yeast gene: Complete nucleotide sequence of the actin gene in *Saccharomyces cerevisiae.* Proc. Natl. Acad. Sci. 77: 2546.
17. Gallwitz, D., C. Donath, and C. Sander. 1983. A yeast gene encoding a protein homologous to the human *c-has/bas* proto-oncogene product. Nature 306: 704.
18. Greer, C. and R. Schekman. 1982a. Actin from *Saccharomyces cerevisiae.* Mol. Cell. Biol. 2: 1270.
19. Greer, C. and R. Schekman. 1982b. Calcium control of *Saccharomyces cerevisiae* actin assembly. Mol. Cell. Biol. 2: 1279.
20. Hartwell, L.H. 1974. *Saccharomyces cerevisiae* cell cycle. Bacteriol. Rev. 38: 164.
21. Hitchcock-DeGregori, S.E. 1980. Actin assembly. Nature 288: 437.
22. Jarvik, J. and D. Botstein. 1975. Conditional-lethal mutations that suppress genetic defects in morphogenesis by altering structural proteins. Proc. Natl. Acad. Sci. 72: 2738.
23. Kilmartin, J. 1981. Purification of yeast tubulin by self-assembly *in vitro.* Biochemistry 20: 3629.
24. Kilmartin, J. and A. Adams. 1984. Structural rearrangements of tubulin and actin during the cell cycle of the yeast *Saccharomyces.* J. Cell Biol. 98: 922.
25. Kilmartin, J., B. Wright, and C. Milstein. 1982. Rat monoclonal antitubulin antibodies derived by using a new nonsecreting rat cell line. J. Cell Biol. 93: 576.
26. Koteleansky, U., M. Gluckhova, M. Benjanian, A. Surguchov, and V. Smirnov. 1979. Isolation and characterization of actin-like protein from yeast

Saccharomyces cerevisiae. FEBS Lett. 102: 55.

27. Larson, G.P., K. Itakura, H. Ito, and J.J. Rossi. 1983. *Saccharomyces cerevisiae* actin-*Escherichia coli* lacZ gene fusions: Synthetic-oligonucleotide-mediated deletion of the 309 base pair intervening sequence in the actin gene. Gene 22: 31.
28. Lazarides, E. and U. Lindberg. 1974. Actin is the naturally occurring inhibitor of deoxyribonuclease I. Proc. Natl. Acad. Sci. 71: 4742.
29. Moens, P.B. and E. Rapoport. 1971. Spindles, spindle plaques, and meiosis in the yeast *Saccharomyces cerevisiae* (Hansen). J. Cell. Biol. 50: 344.
30. Moir, D., S. Stewart, B. Osmond, and D. Botstein. 1982. Cold-sensitive cell-division -cycle mutants of yeast: Isolation, properties, and pseudoreversion studies. Genetics 100: 547.
31. Morris, N.R., M.H. Lai, and C.E. Oakley. 1979. Identification of a gene of α-tubulin gene in *Aspergillus nidulans*. Cell 16: 437.
32. Neff, N.N., J.H. Thomas, P. Grisafi, and D. Botstein. 1983. Isolation of the β-tubulin gene from yeast and demonstration of its essential function *in vivo*. Cell 33: 211.
33. Ng, R. and J. Abelson. 1980. Isolation and sequence of the gene for actin in *Saccharomyces cerevisiae*. Proc. Natl. Acad. Sci. 77: 3912.
34. Orr-Weaver, T., J. Szostak, and R. Rothstein. 1981. Yeast transformation: A model system for the study of recombination. Proc. Natl. Acad. Sci. 78: 6354.
35. Pringle, J.R. and L.H. Hartwell. 1981. The *Saccharomyces cerevisiae* cell cycle. In: The molecular biology of the yeast *Saccharomyces*. Life cycle and inheritance (ed. J.N. Strathern *et al.*), p. 97. Cold Spring Harbor Laboratory, Cold Spring Harbor, New York.
36. Robinow, C.F. and J. Marak. 1966. A fiber apparatus in the nucleus of the yeast cell. J. Cell Biol. 29: 129.
37. Sheirr-Neiss, G., M. Lai, and N. Morris. 1978. Identification of a gene for tubulin in *Aspergillus nidulans*. Cell 15: 639.
38. Shortle, D., J. Haber, and D. Botstein. 1982a. Lethal disruption of the yeast actin gene by

integrative DNA transformation. Science 217: 371.
39. Shortle, D., P. Grisafi, S.J. Benkovic, and D. Botstein. 1982b. Gap misrepair mutagenesis: Efficient site-directed induction of transition, transversion, and frameshift mutations *in vitro*. Proc. Natl. Acad. Sci. 79: 1588.
40. Shortle, D., D. Koshland, G. Weinstock, and D. Botstein. 1980. Segment-directed mutagenesis: Construction *in vitro* of point mutations limited to a small predetermined region of a circular DNA molecule. Proc. Natl. AQcad. Sci. 77: 5375.
41. Shuman, H.A., T.J. Silhavy, and J.R. Beckwith. 1980. Labelling of proteins with β-galactosidase by gene fusion. J. Biol. Chem. 255: 168.
42. Sutcliffe, J.G., R.J. Miller, T.M. Shinnick, and F.E. Bloom. 1983a. Identifying the protein products of brain-specific genes with antibodies to chemically synthesized peptides. Cell 33: 671.
43. Sutcliffe, J.G., T.M. Shinnick, N. Green, and R.A. Lerner. 1983b. Antibodies that react with predetermined sites on proteins. Scinece 219: 660.
44. Villa-Komaroff, L.A., A. Efstratiadis, S. Broome, P. Lomedico, R. Tizard, S.P. Naker, W.L. Chick, and W. Gilbert. 1978. A bacterial clone synthesizing proinsulin. Proc. Natl. Acad. Sci. 75: 3727.
45. Water, D., J. Pringle, and L. Kleinsmith. 1980. Identification of an actin-like protein and of its messenger ribonucleic acid in *Saccharomyces cerevisiae*. J. Bacteriol. 144: 1143.
46. Wood, J.S. and L.H. Hartwell. 1982. A dependent pathway of gene functions leading to chromosome segregation in *S. cerevisiae*. J. Cell Biol. 94: 718.
47. Zechel, K. 1980. Dissociation of the DNase I-actin complex by formamide. Eur. J. Biochem. 110: 337.

II. CONTROL OF THE CELL CYCLE

CELLULAR MORPHOGENESIS IN THE YEAST CELL CYCLE[1]

J.R. Pringle, S.H. Lillie, A.E.M. Adams, C.W. Jacobs,
B.K. Haarer, K.G. Coleman, J.S. Robinson, L. Bloom,
and R.A. Preston

Division of Biological Sciences, The University of Michigan
Ann Arbor, Michigan 48109

ABSTRACT The cell-division cycle of the budding yeast Saccharomyces cerevisiae includes a number of processes that are morphogenetic in the sense that they generate cellular shape and spatial organization. Direct studies of known cytoskeletal elements such as microtubules and actin-containing filaments are beginning to reveal the composition of these elements and their roles in the morphogenetic processes of the cell cycle. In addition, a genetic approach has been undertaken that has the potential to elucidate further the composition and functions of the known cytoskeletal elements, to identify novel mechanisms involved in cellular morphogenesis, or both. This genetic approach exploits the exceptional susceptibility of yeast to both classical and molecular genetic manipulations.

INTRODUCTION

Among the most characteristic features of living organisms are their complex three-dimensional organization and the morphogenetic processes that generate this organ-

---

[1]This chapter is a revised and updated version of ref. 1. Our work has been supported by grants from the USPHS (GM-31006) and NSF (PCM 78-25607), the Biomedical Research Support Grant to The University of Michigan, USPHS fellowships and training-grant support to C.W.J., B.K.H., and K.G.C., and by The University of Michigan Center for Molecular Genetics.

ization (2). At the cellular level, the morphogenetic processes include the generation of cell shape and cell polarity, the asymmetric distribution of cell-surface components, the intracellular movement and positioning of organelles, and asymmetric cell division (2,3). The yeast cell-division cycle (4) appears to offer an outstanding opportunity to explore the mechanisms of such processes. The morphogenetic processes of the yeast cell cycle include (Fig. 1; refs. 4-7 and others cited therein): (a) the selection of a nonrandom site on the cell surface at which the single bud will emerge; (b) the formation of a ring of chitin (the "bud scar") in the largely nonchitinous cell wall at that site; (c) the localization of new cell-wall growth to the region bounded by the chitin ring, resulting in the appearance and selective growth of a bud; (d) the localization of secretion of other materials to the surface of the bud; (e) the localization of new cell-wall growth to the tip of the bud during much of the period of bud growth (cf. Fig. 4A, below); (f) the balancing of tip growth against periods of uniform (isotropic) growth of the bud cell wall, resulting in the normal ellipsoidal shape of the daughter cell; (g) the migration of the nucleus from a position within the mother cell into the neck connecting mother and bud; (h) cytokinesis and the formation of septal cell wall; and (i) the asymmetric distribution of materials, including regulatory elements (8), to mother and daughter cells at division.

Although these processes are relatively simple and, in their details, unique to yeast, they appear to involve the same general principles as morphogenetic processes in other types of eukaryotic cells, and they seem likely to employ similar mechanisms. Thus, in our studies, we hope to exploit the great experimental advantages of yeast (including its exceptional susceptibility to the manipulations of both classical and molecular genetics) to yield generally useful information about the mechanisms of eukaryotic cellular morphogenesis as well as about the mechanisms of eukaryotic cellular reproduction. Until recently, it seemed possible that the advantages offered by yeast for the study of cellular morphogenesis might be outweighed by the difficulties of performing adequate morphological studies on these small, tough-walled cells. However, the recent development of effective immunofluorescence procedures (7,9) to complement the effective procedures for electron microscopy (10) has mitigated this concern.

Studies of cellular morphogenesis in yeast might take either of two rather different approaches. First, we could

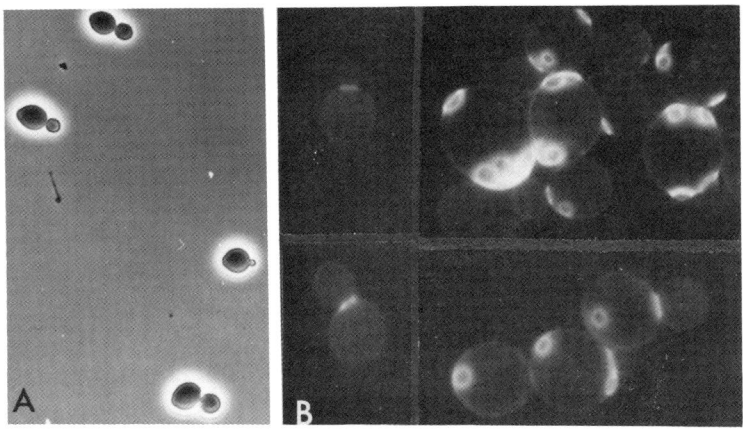

FIGURE 1. A, phase-contrast micrograph of wild-type S. cerevisiae cells, illustrating the ellipsoidal shape of the cells, the selective growth of the bud relative to the mother cell (four different cells are shown, but four successive views of the same cell would look much the same), and the nonrandomness of bud position (note that each bud has formed near a pole of the mother cell). B, fluorescence micrographs of wild-type cells stained with Calcofluor, a dye that binds specifically to chitin in the yeast cell wall (5,6). Note that each developing bud has a ring of chitin at its base. These chitin rings remain on the mother cells after division; their distribution illustrates further the nonrandomness of bud position. The bipolar pattern shown is typical of cells expressing both MATa and MATα information at the mating-type locus; such a cell also typically forms its first bud near the pole opposite to that at which it was attached to its own mother cell. In contrast, a cell expressing only MATa or MATα information typically forms its first and subsequent buds near the pole at which it was attached to its own mother cell; Calcofluor staining thus reveals a unipolar cluster of scars (not shown).

make the plausible hypothesis that known cytoskeletal elements such as microtubules and actin filaments are involved, and thus proceed to study such elements directly by morphological, biochemical, and genetic methods. Second, we could undertake a genetic approach that involves no a priori assumption that the processes of interest are controlled by the known cytoskeletal elements. In this chapter, we summarize briefly our progress with these two approaches.

MICROTUBULE AND ACTIN DISTRIBUTION IN RELATION TO BUD GROWTH

Yeast contains microtubules that are similar in morphology (Fig. 2A; refs. 10-12) and in the structures and general properties of their constituent tubulins (11-16) to microtubules from other eukaryotic cells. Intranuclear (spindle) and cytoplasmic microtubules emanate from the opposite faces of the spindle-pole bodies (SPBs), structures embedded in the nuclear envelope that appear to function as microtubule-organizing centers (10,17). Temporal and spatial correlations observed in the serial-section electron-microscopic studies by Byers and Goetsch (reviewed in ref. 10) suggested that the SPBs and the cytoplasmic microtubules (cMTs) were involved in the selection of the budding site and/or in directing secretory vesicles to that site and into the growing bud. First, in wild-type vegetative cells, zygotes, and various mutants (notably cdc4; see below), bud emergence and the early stages of bud growth always occurred in cells with duplicated but unseparated SPBs. In such cells, the double SPBs were always oriented toward the budding site, and the cMTs ran from the SPB into the bud, often seeming (at least superficially) to be associated with the secretory vesicles that were accumulated there. Second, centrifugation of newly formed zygotes altered both the typical position of the nuclei (and hence of the SPBs) within the zygotes and the typical location of the zygotes' first buds, but the orientation of the SPBs and cMTs toward the budding sites was retained. Third, involvement of the SPBs and cMTs in the directed movement of secretory vesicles was also suggested by their orientation toward the sites of localized cell-wall alterations in cells undergoing zygote formation.

The development of effective immunofluorescence procedures for yeast (7,9) allowed extension of these correlations in two ways. First, it became clear that cMTs continued to run from the nuclear envelope into the growing bud as the SPBs separated, the spindle formed, and the nucleus migrated into the neck (Fig. 2B; cf. also Fig. 2A). Second, it was observed that cdc4 mutant (4) cells (a) often had active cell-wall growth occurring at two or more sites simultaneously (see Fig. 4A, below) and (b) typically had multiple bundles of cMTs emanating from the one double SPB in each cell (10) and extending to the multiple bud tips (Fig. 2C). These observations appeared to strengthen the

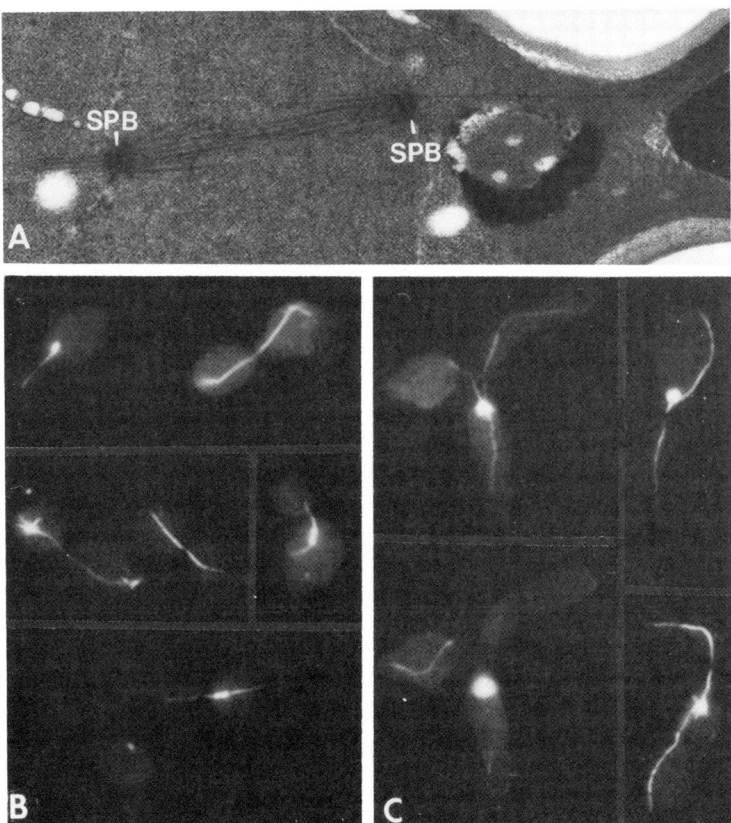

FIGURE 2. A, electron micrograph of a budding cell (the bud is on the right), illustrating spindle-pole bodies (SPBs), spindle microtubules (note that the nuclear envelope does not break down during mitosis in yeast), and cytoplasmic microtubules (cMTs; note that one set extends into the bud). B-C, immunofluorescence micrographs illustrating the distribution of microtubules, as revealed using a monoclonal antitubulin antibody (7,9). B, normal proliferating cells; note that cMTs extend into the bud in each cell in which a complete spindle exists within the mother cell. C, temperature-sensitive cdc4 cells (4) stained after several hours at 36°. At this restrictive temperature, the cells form multiple, abnormally elongated buds that fail to divide while the nuclear cycle is arrested prior to DNA synthesis and with the SPB duplicated but not separated. Three cells are shown; that on the left is shown at two levels of focus.

case for involvement of the SPB and cMTs in the selective growth of the bud and the polarization of secretion.

At the same time, observations on the intracellular localization of actin in relation to the sites of cell-wall growth suggested that the actin (and myosin?) system of the cells might be involved in these processes. Although it was clear that yeast contains substantial amounts of an actin that resembles other eukaryotic actins (18-21) and that the single actin gene per haploid cell is essential (22), electron microscopy had provided no good clues as to the localization and function of actin in the cells (see Fig. 4 of ref. 7). This problem was alleviated with the application of immunofluorescence and staining of actin with fluorochrome-labeled phallotoxins (7,9,23). Actin was seen to be disposed in cytoplasmic fibers and in numerous "dots" that appear to be cortical in location (Fig. 3A,B; also Fig. 4B, below). Although the exact nature of the actin dots is not known (sites of anchorage of actin fibers to the membrane? actin-coated vesicles?), their distribution and that of the actin fibers strongly suggest an involvement in localized growth of the cell wall. Specifically, during the cell cycle of wild-type cells, the dots cluster at sites of active cell-wall deposition, whereas the fibers tend to run along the long axis of the cell between mother and bud (Fig. 3A). Both the clustering of actin dots and the longitudinality of the actin fibers appear accentuated in morphogenetic mutants that show an exaggerated tip-growth of abnormally elongated buds (Figs. 3B,4B). Especially suggestive is the correlation between active cell-wall growth and clustering of actin dots at particular bud tips on cdc4 cells (Fig. 4). Surprisingly, the intracellular localization recently reported for yeast myosin (24) is quite distinct (a concentration exclusively in the neck region) from that of actin; thus, the mechanism of the putative involvement of actin in secretory-vesicle movement and cell-wall growth remains problematic.

EFFECTS ON MORPHOGENESIS OF MICROTUBULE-DISRUPTING DRUGS

Several groups (25,26) reported that treatment of yeast with the microtubule inhibitor methylbenzimidazole-2-yl-carbamate (MBC) arrested cell division, apparently by blocking nuclear division. The arrested cells terminated development as mother cells with large buds, suggesting that bud emergence and selective growth of the bud could occur in the absence of functional cMTs. However, it was not clear that the cMTs were really effectively disrupted by the drug. Indeed,

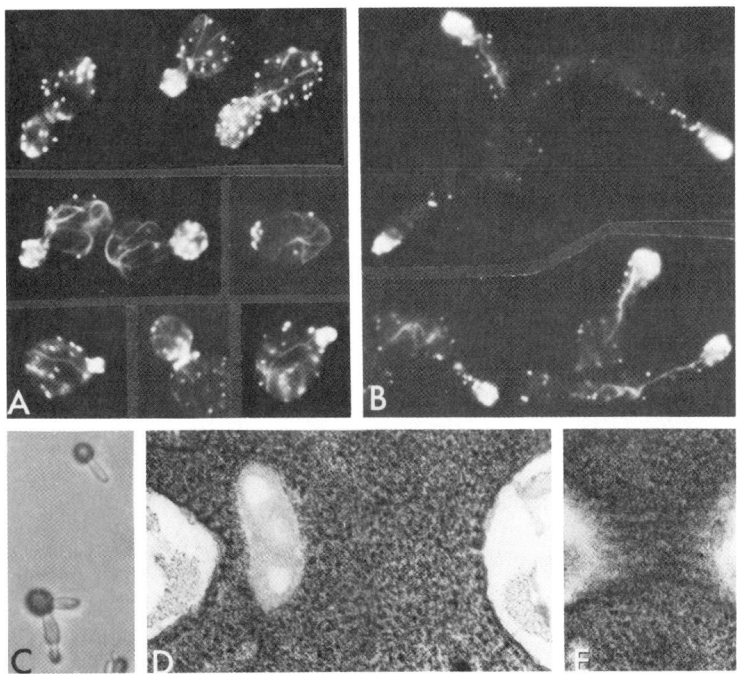

FIGURE 3. A-B, fluorescence micrographs illustrating the distribution of actin, as revealed by staining with rhodamine-conjugated phalloidin (7,9). A, wild-type cells. Note the clustering of actin "dots" over buds (particularly small buds), at a site of presumed incipient budding (9) on an unbudded cell, and in the neck region of a cell that may be beginning septum formation. B, temperature-sensitive cdc3 (upper cell) and cdc12 (lower cells) mutants after several hours growth at 36°. At this restrictive temperature, these mutants (as well as cdc10 and cdc11 mutants) form multiple, abnormally elongated buds (Fig. 3C) that fail to divide, while the nuclear cycle continues, yielding multinucleate cells (4). These mutants also fail to form normal chitin rings (unpublished results) and are defective (10) in forming the ring of 10-nm filaments [of unknown biochemical nature (10,24)] that normally lies immediately subjacent to the plasma membrane in the neck region (Fig. 3D,E; ref. 10). C, phase-contrast micrograph of cdc10 cells budding at restrictive temperature. D-E, electron micrographs showing transverse (D) and glancing (E) sections of the ring of 10-nm filaments.

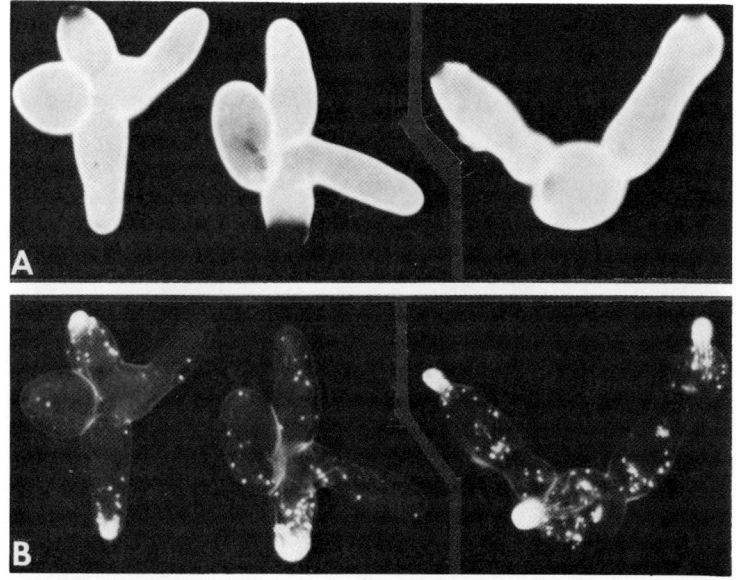

FIGURE 4. Fluorescence micrographs illustrating the correlation between the distribution of actin and the pattern of cell-wall deposition. A, cdc4 mutant cells whose surface mannan has been stained with fluorescein-conjugated concanavalin A to reveal (as dark patches) the sites of active cell-wall deposition (7). B, the same cells as in A, stained with rhodamine-conjugated phalloidin to reveal the distribution of actin.

examination of MBC-arrested cells using anti-tubulin immunofluorescence revealed that extended, tubulin-containing structures could persist in such cells even after several hours of exposure to the drug (27). Although the structures visualized looked abnormal and might well be nonfunctional, this was difficult to test; thus, the validity of inferences from such experiments about the functions of the cMTs was uncertain.

Fortunately, another inhibitor, nocodazole (28), provided less equivocal results (27). Examination of nocodazole-treated cells by anti-tubulin immunofluorescence revealed that microtubules began to be lost within a few minutes of exposure to the drug. By 30 min of exposure, most cells showed no localized staining, although a few cells showed small, residual spots of fluorescence (Fig. 5A); by 60 min, even the spots were gone from virtually all cells. Electron microscopy confirmed this rapid loss of both spindle and cytoplasmic microtubules. Corresponding to the rapid loss of microtubules was a rapid arrest of cell division: exposure of an asynchronous, exponentially growing population to nocodazole arrested division within about 40 min (residual increase in cell number about 25%). The arrested cells were uniformly large-budded (Fig. 5B). Similar arrested populations were obtained when populations enriched by centrifugation for small, unbudded cells were inoculated into fresh medium containing nocodazole. Moreover, such cells appeared to form both normal chitin rings and normal rings of 10-nm filaments (cf. Fig. 3D,E) at the bases of the buds that emerged in the presence of nocodazole. These data indicated that bud emergence, chitin-ring formation, 10-nm-filament-ring formation, and selective bud enlargement can all occur in the absence of cMTs. In addition, time-lapse observations revealed no detectable perturbation of the normal mating-type-dependent selection of budding sites (Fig. 1B) when exponentially growing cells were plated on nocodazole-containing medium. [The ~25% of the cells that divided in the presence of the drug (see above) were able to bud once more, allowing the locations of the new buds to be scored.] Finally, polarized deposition of new cell wall at the tips of buds on cdc4 cells (similar to that of Fig. 4A) could be observed in nocodazole-treated cells that appeared devoid of microtubules by immunofluorescence. Probably significantly, nocodazole was not observed to affect the intracellular distribution of actin in these experiments.

In contrast to their apparent noninvolvement in the processes of polarized growth and secretion, the cMTs may be involved in nuclear migration. Thus, the nuclei in nocodazole-arrested cells appeared to be randomly positioned within the mother cells [Fig. 5C; a similar experiment in which cells were double-stained with DAPI (to reveal nuclear DNA) and Calcofluor (to reveal bud scars and thus identify the mother cells) confirmed that the nuclei were indeed in the mother cells and not in the buds.] This impression of randomness was strengthened by determining the "nuclear-migration index" (defined as the shortest distance from any part of the nucleus to the mother-bud neck, divided by the length of the mother cell) for many individual nocodazole-arrested and control (exponentially growing) cells. In the control population, 93% of the uninucleate, budded cells had nuclear-migration indices ≤ 0.25, whereas in the arrested

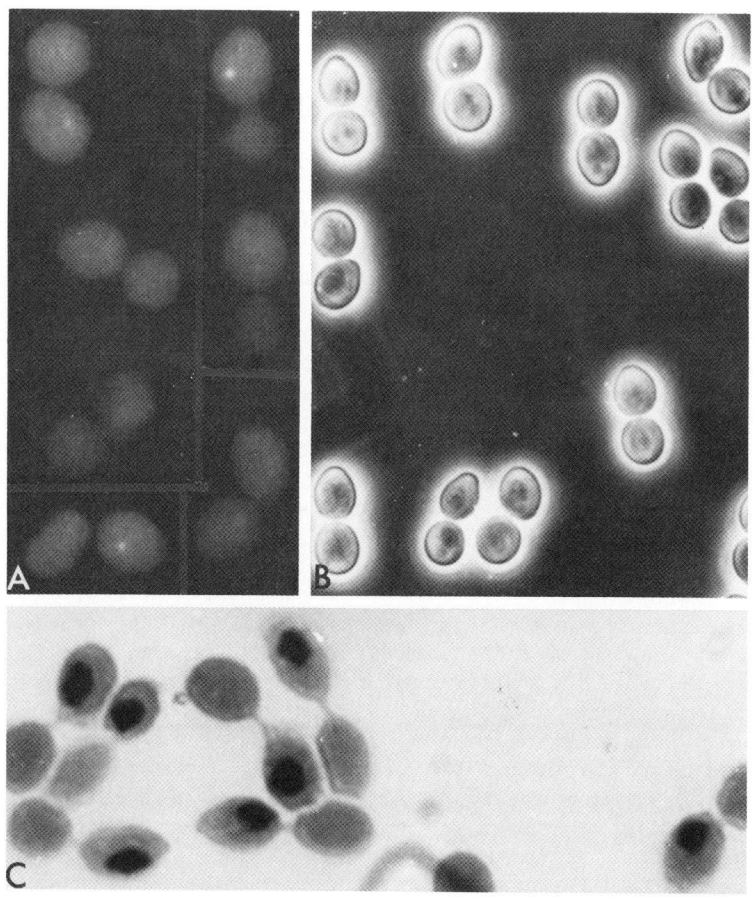

FIGURE 5. Effects of nocodazole on wild-type cells. 15 μg/ml nocodazole was added to a culture growing exponentially in rich medium at 23°. A, at 30 min, a sample was taken for anti-tubulin immunofluorescence (cf. Fig. 2B). B-C, at 3 h, samples were taken (B) for phase-contrast microscopy and (C) for bright-field microscopy after staining with Giemsa to reveal the positions of the nuclear DNA.

population only 52% of the cells had indices ≤0.25 and 27% had indices ≥0.4. These differences were accentuated by repeating the experiment with a cdc2 strain. When this temperature-sensitive mutant is incubated at restrictive temperature, the cells arrest with large buds and nuclei that have typically migrated to, but not into, the necks (4,29); in our experiment, 90% of the cells had nuclear-migration indices ≤ 0.1. In contrast, when the cells were exposed to nocodazole during the incubation at restrictive temperature, only 25% had indices ≤ 0.1. Interestingly, when nocodazole was added to cdc2 cells that had already arrested at restrictive temperature, the nuclei appeared to remain at the necks: 88% of the cells still had nuclear-migration indices ≤ 0.1 after 2 h of exposure to nocodazole at restrictive temperature.

Electron microscopic observations suggested that the cMTs might also be involved in the positioning of the SPBs, and hence in the orientation of the spindle. Thus, in nocodazole-treated cells that still had small buds, the double SPBs were often not oriented toward the budding site (Fig. 6), in striking contrast to the situation in control cells. These studies also indicated that SPB separation failed to occur in nocodazole-arrested cells; this failure, like that of nuclear division itself, might be a consequence of loss of function by spindle microtubules, by cMTs, or both.

It should be noted that these tentative conclusions about what the cMTs may do (in contrast to the conclusions above about what the cMTs do not do) are subject to a local version of the universal caveat about inhibitor experiments; i.e., disruption of a process by nocodazole does not necessarily result from the action of the drug on microtubules. However, the inference that nuclear migration may be dependent on the cMTs appears consistent with the role of microtubules in nuclear movement in the hyphal fungus Aspergillus (30), as well as with genetic evidence obtained in yeast (see below).

Studies by others of zygote formation have also suggested that the cMTs function in the movement and orientation of the nucleus rather than in the directed movement of secretory vesicles. In particular, both tubulin mutants (15; see also below) and wild-type cells treated with MBC (31) can undergo cell fusion, with its attendant localized reorganization of the cell surface, but fail to undergo nuclear migration or nuclear fusion (which normally begins by fusion of the apposed SPBs). Moreover, mutants isolated on the basis of a specific defect in nuclear fusion (in cells competent for cell fusion) show abnormalities of their cMTs (32).

GENETIC APPROACHES TO MORPHOGENESIS

Studies of morphological correlations, the biochemical

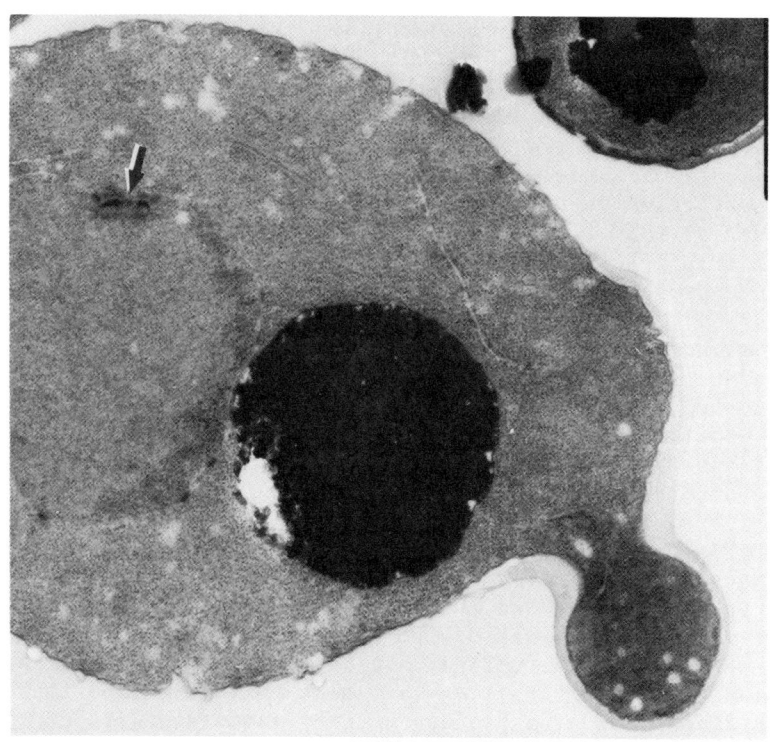

FIGURE 6. Electron micrograph illustrating the failure of the double SPB (arrow) to orient toward the site of a small bud in a wild-type cell treated with nocodazole (as in Fig. 5).

activities of known proteins, and the effects of inhibitors can be revealing (see above) but have significant limitations. These limitations are particularly acute as regards the establishment of causality relations and the identification of the (presumably) multiple components of the machinery responsible for the cellular processes of interest. Thus, it is crucial also to investigate these processes using the powerful methods of classical and molecular genetics, and it is principally because of its exceptional susceptibility to these methods (33-36) that yeast has become such an attractive object for studies of fundamental problems in cell biology. Two distinct and complementary approaches are possible for the genetic investigation of the morphogenetic processes of the yeast cell cycle. First, one can use either classical methods (14,15,34,37) or in vitro mutagenesis and gene-replacement techniques (14-16,22,23,33-36,38) to seek mutations in genes encoding known components of the cytoskeleton, such as actin and the tubulins. The phenotypes of these mutants should reveal the functions of the affected components, whereas the isolation and analysis of "pseudorevertants" of these mutants (carrying extragenic suppressors of the original mutations) should reveal interactions among the known cytoskeletal components and identify new components that interact with those previously known (15,34,39,40). Exciting progress has been made with this important approach during the past few years by D. Botstein and his co-workers (14-16,22,23,37,38); their conclusions about the functions of actin and of microtubules in yeast agree well with those reached from morphological and inhibitor studies (see above), and they have already identified numerous new genes whose products may interact with actin and tubulin. More progress using this general approach can be expected in the near future (24,41,42).

However, this first approach also has potentially serious limitations. These arise in part from the possibility that some important components of the cytoskeleton may not be sufficiently abundant or highly organized to have been recognized by biochemists and cell biologists or to serve as effective suppressors of defects in the abundant components. Moreover, there may be molecules or even whole mechanisms involved in morphogenesis that are connected only loosely, or not at all, to the presently known elements of the cytoskeleton (2,3,43). A particular possibility with some precedent is that the polarization of growth and secretion in yeast may depend on transcellular ion currents (44,45) rather than (or in addition to) the actions of cytoskeletal components. However, an even more intriguing possibility is that mechanisms are involved in morphogenesis that are as yet undreamt of in the philosophies of biochemists and cell biologists. Interest in these possibilities has been stimulated by the emerging molecular data (at both the protein and nucleic-acid levels) on the genetic complexity of eukaryotic cells and their constituent processes (for reviews, see refs. 46,47). For example, it is sobering to realize that a structure as (seemingly) simple as the Chlamydomonas flagellar axoneme may contain more than 280 polypeptides (48; R. Segal and D. Luck, pers. commun.), or that even in the lowly and simple yeast, we apparently know or can guess at the functions of only perhaps 30-40% of the more than

5,000 genes in the haploid genome (46,47). If such presently unrecognized molecules or mechanisms indeed exist, they will be difficult or impossible to identify by a genetic approach that works "outward" from mutations affecting the already known major components of the cytoskeleton.

Thus, we have deliberately chosen the alternative genetic approach, which attempts to work "inward" from the isolation of mutants to the identification of molecules involved in morphogenesis. This approach makes no a priori assumptions as to what structures or molecules are involved in the morphogenetic processes of the cell cycle and imposes no requirement that the genes to be studied encode abundant products. Thus, we begin by seeking mutants using aberrant morphogenesis as our criterion for isolation; in effect, this represents an attempt to let the organism identify for us the genes encoding significant components of its morphogenetic machinery. We then use molecular-genetic techniques to attempt identification of the products of these genes. We expect that information about the structures of the gene products and about their intracellular localizations will provide clues to guide the early stages of biochemical investigation of gene-product function (see below). Simultaneously, we analyze suppressors of the original morphogenetic mutations in the hopes (a) of identifying additional genes involved in the processes of interest and (b) of identifying interactions among the known and/or newly identified genes or their products (see below).

MORPHOGENETIC MUTANTS

We have focused to date on three classes of high-temperature-sensitive (ts) morphogenetic mutants. The outstanding characteristics of the prototype for one class, the cdc4 mutant, have been presented briefly in Figs. 2C and 4 (see ref. 4 for more detailed summaries of the properties of this and other cdc mutants). Mutations in two other known genes (CDC34 and an as-yet-unnamed gene defined by the JPT175 and JPTA1528 ts mutations recently isolated in our laboratory) produce phenotypes similar to that of cdc4 mutants. The outstanding characteristics of the prototypes for the second class of mutants (cdc3, cdc10, cdc11, cdc12) have been presented briefly in Fig. 3B,C. The prototype for the third class of mutants is the intensively studied cdc24 mutant (4, 5); the cdc42 and cdc43 mutants are similar (4,49). At restrictive temperature, these mutants continue growing but fail to form buds and exhibit an apparently complete loss of ability to localize secretion; thus, huge, round cells showing a nonpolarized deposition of new cell-wall material are formed (Fig. 7A). The CDC24 gene product, at least, also appears to be involved in the selection of budding sites, as one cdc24 mutant exhibits a loss of the normal pattern of budding during growth at permissive temperature, where other aspects of budding and cell-wall deposition appear normal (Fig. 7B). Presumably, the mutant gene product is not working quite normally even at the "permissive" temperature.

The original cdc4, cdc34, cdc3, cdc10, cdc11, cdc12, and cdc24 mutants were isolated by L. Hartwell and his collaborators by screening a collection of approximately 2,000 ts-lethal mutants for mutants that arrested at restrictive temperature with homogeneous terminal cell morphologies indicative of defects in specific steps of the cell cycle (4,50). More recently, we have used the same criterion to screen about 5,000 additional ts-lethal mutants in an attempt to identify more genes whose products are involved in the morphogenetic processes of the cell cycle. Numerous additional isolates defective in the known genes were ob-

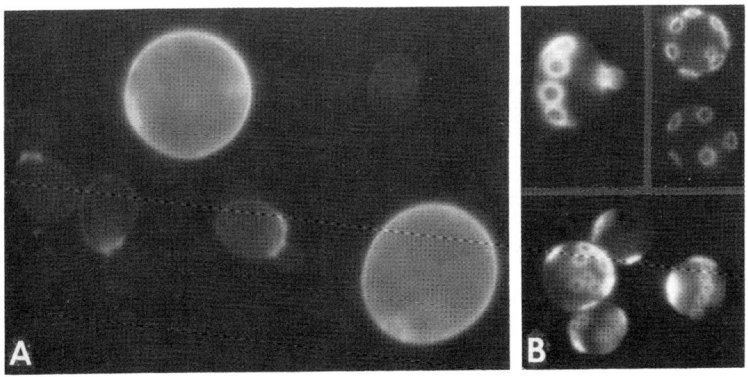

FIGURE 7. Fluorescence micrographs of Calcofluor-stained cells, illustrating some of the morphogenetic abnormalities caused by cdc42 and cdc24 mutations. A, cdc42 cells from a culture incubated 5 h at restrictive temperature (the two large, generally bright cells) were mixed after fixation with cells from the parent culture grown at permissive temperature. B, MATa and MATa/MATα cdc24-4 cells that had been grown at permissive temperature, illustrating the loss of the normal unipolar or bipolar budding pattern (cf. Fig. 1).

tained, but only three new genes (CDC42, CDC43, and that represented by JPT175 and JPTA1528) were identified. This might mean that nearly all of the genes of interest are now known, but it appears much more likely (40,46,47) to reflect the limitations of an approach based solely on ts-lethal mutations. In particular, it now seems likely both that many genes do not readily make ts mutations and that many genes cannot give rise to lethal mutations, either because the function encoded is not essential or because it is duplicated elsewhere in the genome. It is also possible that there are many genes whose mutants do not satisfy the criterion of a homogeneous terminal morphology at arrest, even though their products function in specific steps of the cell cycle (23,46). To avoid the apparent limitations of ts mutations, we have begun to search for additional morphogenetic mutants using morphological screening of cold-sensitive (cs)-lethal and suppressible-nonsense-lethal mutations. Results to date with the cs mutants are encouraging; the mutants producing multiple, abnormally elongated buds fall into six complementation groups. As expected (40), one of these is cdc11, but at least some of the others appear to be genes not previously identified using ts mutations.

SUPPRESSORS OF MORPHOGENETIC MUTATIONS

Pseudorevertants

Another approach to the identification of additional genes involved in the morphogenetic processes of the cell cycle is the analysis of "pseudorevertants" carrying extragenic suppressors of the available ts and cs morphogenetic mutations (15,34,39,40,48,51-53). This approach is attractive because it circumvents many, if not all, of the factors that appear to limit attempts to identify additional genes using ts- or cs-lethal cdc mutations (see above). In addition, whether the suppressor mutations prove to lie in newly identified genes or in previously known ones, the observation that a mutation in one gene can suppress a mutation in another suggests that the two genes or their products interact fairly directly in the cell. Such information should contribute to the development of molecular models for the integrated function of the various gene products in the morphogenetic processes. Given that the original ts and cs mutations are presumably mostly missense (54), the types of interactions likely to allow suppression should be mostly at the level of regulation or of protein-protein interaction.

For example, a mutation in a regulatory gene that increases
the amount of some other gene product should sometimes suppress a mutation that reduces the efficiency of function of
that gene product (cf. notes g and m of Table 1, below).
Alternatively, if two gene products interact physically in
the cell, a mutation in one should sometimes be able to compensate for the structure-destabilizing effects of a mutation in the other (34,39). Another interesting possibility
is that the suppressor mutation simply bypasses the need for
the gene product that is defective in the original mutant
[34,48,51 (and P. Nurse, pers. commun.),52,53].

To identify the suppressor-carrying strains, each apparent revertant to be analyzed is crossed to a wild-type
strain and tetrads are dissected. The appearance of segregants with the phenotype of the parental mutant strain (ts
or cs lethality; the appropriate aberrant cell morphology;
Fig. 8) demonstrates that the reversional event was neither
a true (original-site) reversion nor an intragenic secondsite alteration. The segregation ratios from this cross and
from a cross of the revertant by the original mutant then
indicate whether the suppression was due to a single linked
or unlinked suppressor mutation, to a more complex set of
mutations producing suppression, or to an increase in copy
number of the mutant gene resulting from aneuploidy. Additional genetic analyses allow the suppressors to be sorted
into genes and these genes to be tested for possible identity with those known previously as the sites of morphogenetic mutations. The results to date of our pseudoreversion
analyses are summarized in Table 1. It is clear that in
some cases (notably cdc3 and cdc10), simple extragenic suppressors are found easily and their analysis leads promptly
to comprehensible and useful results. In other cases (e.g.,
cdc12), the analyses are laborious, at least with the particular mutants used so far as parents.

It should be noted that our protocol differs from the
best-known paradigm for pseudoreversion studies (15,34,40),
which demands that each suppressor to be analyzed confers a
new temperature-conditional lethality (e.g., that a suppressor of a ts-lethal mutation be itself a cs-lethal mutation)
in addition to its suppression. This paradigmatic strategy
offers three significant advantages. First, it facilitates
complementation tests among the suppressor mutations and
allows such tests between the suppressor mutations and the
available ts or cs morphogenetic mutations. Second, it facilitates the cloning of the suppressor genes (by complementation of the ts or cs lethality by plasmids from libraries

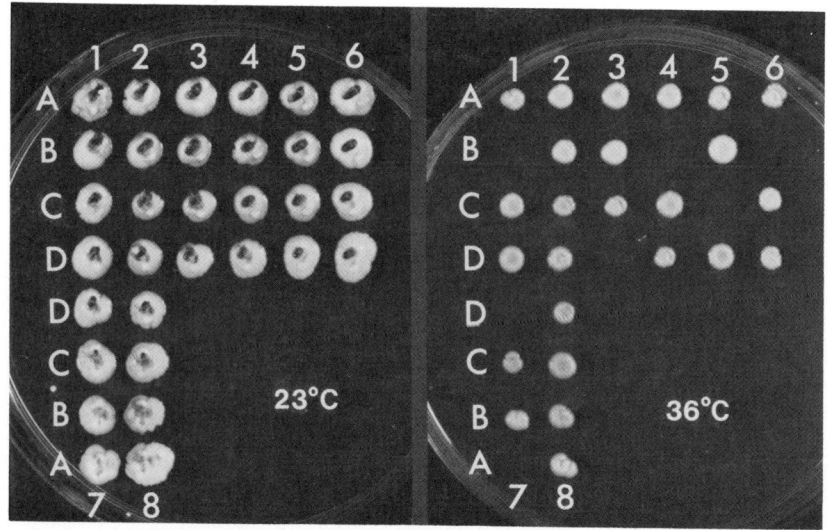

FIGURE 8. Segregrants obtained when an apparent revertant of a cdc10 strain was crossed by wild type. Numbers denote the tetrads, letters A-D the four spores from a given tetrad. The permissive-temperature (23°) master plate is on the left, the restrictive-temperature (36°) replica on the right. Dead cells scraped from the nongrowing colonies on the 36° plate had the aberrant cell morphology characteristic of cdc10 mutants. The segregation pattern obtained indicates that the original revertant carried a single extragenic suppressor mutation that was not linked to CDC10 itself. Note that all four segregants are growing about equally well in the two tetrads showing 4 Ts$^+$:0 Ts$^-$ segregations. Thus, with this suppressor, the suppressed cdc10 strains grow about as well as CDC10$^+$ strains.

of wild-type DNA). Third, and probably most important, the new mutant phenotype can provide good evidence at an early stage in the analyses that the suppressor gene is directly involved in the processes of interest, and thus (i) that the suppression under analysis is due to some interesting form of gene-gene interaction and not to something trivial (e.g., stabilization of the originally meta-unstable product of the mutant gene by a new mutation that alters slightly some non-specific aspect of the intracellular physiological milieu) and (ii) that further detailed investigation of the suppres-

TABLE 1
PSEUDOREVERSION ANALYSES OF MORPHOGENETIC MUTANTS[a]

Gene	No. of alleles studied[b]	No. of revertants analyzed[c]	No. of simple extragenic suppressors[d]	No. of suppressor genes[e]
CDC3	2	24	13	2? (12[f],1?)
CDC10	3[g]	52[h]	33	$\geq$1 (27[i])
CDC11	2	66	17[j]	$\geq$2 (2[k],1)
CDC12	2	75	1	1[l]
CDC24	1	27	6	1(6)
CDC42	1	31	15	4 (5,4,4,2)
X[m]	2	12	0	-

[a]All of these studies are still in progress; presented here is a tentative summary of results to date. Most revertants analyzed to date have been spontaneous, but we will also analyze mutagen-induced revertants of each mutant, in case these yield different spectra of suppressor mutations.

[b]Except for "bypass" suppressors (see text), the suppressors of interest are likely to be allele-specific (15,34,40); thus, the chances of identifying all genes that interact with a particular gene of interest are increased by analyzing suppressors obtained with several different mutant alleles of that gene.

[c]Each revertant that grew approximately as well as wild-type at the originally restrictive temperature was analyzed genetically. None of these revertants displayed a new temperature-conditional lethality opposite to that of the original mutant (see text).

[d]That is, cases in which a single, extragenic suppressor mutation appeared to acount for the original "reversion". The remaining revertants involved intragenic events, a pair of mutations required for suppression, etc. (See also notes g and m.) None of the suppressors conferred a new ts or cs lethality when separated from the original mutations.

[e]Based on linkage analyses plus, in some cases, the complementation patterns among recessive suppressors (see text). Numbers in parentheses are the numbers of independent isolates per suppressor gene defined by the analyses to date.

[f]This gene is tightly linked to CDC10. As, in addition, all cdc10 suppressors analyzed to date are tightly linked to CDC3 (note i), it seems clear that mutations in these genes can suppress each other.

[g]Six independently isolated ts cdc10 strains were used, but some of these appear to carry the same mutant cdc10 allele. One of the strains yielded spontaneous revertants at frequencies at least 100X greater than those obtained with other strains. In the several cases analyzed, these revertants appeared simply to have become disomic for the chromosome carrying the ts cdc10 allele. The frequency of reversion was comparable to the typical frequency of mitotic nondisjunction (55). Evidently, doubling the copy number of the mutant gene sufficed to allow growth at the formerly restrictive temperature. The other entries in this row refer to results with the other five cdc10 strains examined.

[h]150 additional spontaneous Ts+ revertants were collected and simply tested for cs lethality; no clearly cs revertants were observed.

[i]This gene is tightly linked to CDC3. See also note f.

[j]Upon separation from the original cdc11 mutation, two of these suppressors conferred abnormal cell morphologies reminiscent of that displayed by cdc11 cells themselves at restrictive temperature.

[k]This gene is $\sim$25 cM from CDC11 itself, and thus is not CDC3, CDC10,

or CDC12.
   [l]This suppressor gene is not linked to CDC3, CDC10, or CDC11.
   [m]This as-yet-unnamed gene is defined by the JPT175 and JPTA1528 ts mutants (see text). JPTA1528 yielded spontaneous revertants at a normal frequency, but all 12 analyzed so far have proved to be due to intragenic events. JPT175 yielded high-frequency revertants that (from preliminary evidence) are probably disomic for the chromosome carrying the mutant gene (cf. note g).

sor gene (see below) will be rewarding. For example, it was clearly encouraging when the suppressors of actin mutations were found to cause morphological abnormalities similar to those caused by the actin mutations themselves (15).

However, we have deliberately sacrificed these advantages in order to avoid overlooking potential suppressor genes that do not readily give rise to ts- or cs-lethal mutations. This seems important in light of the evidence (i) that such genes may be the majority of those in the genome (see above and refs. 46,47) and (ii) that the double constraint (suppression plus ts or cs lethality) may be particularly difficult to satisfy (notes c, d, and h, Table 1; refs. 34,40; P. Novick, J. Thomas, and D. Botstein, pers. commun.). We are encouraged by the following considerations to think that our approach is feasible despite its greater difficulties and risks. (a) Abundant evidence both from our our own studies (the suppression of cdc3 mutations by cdc10 mutations and vice versa: Table 1) and from others (e.g., refs. 15,34,39,40,51,52) shows that the genes identified by suppressor mutations often are directly involved in the processes of interest. Thus, although we may pursue some false leads, many of the genes identified as suppressor loci should be interesting enough to repay intensive investigation (see below). (b) We can probably maximize the fraction of suppressor genes that are interesting by demanding that suppression be efficient (i.e., that the suppressed mutant strain grow approximately as well as wild type; cf. Fig. 8 and note c, Table 1). (c) Although we do not demand that the suppressors we isolate confer a new ts or cs lethality, we expect that some will; in such cases, the advantages indicated above become available. Even when no ts or cs lethality is observed, we should sometimes observe morphological peculiarities of the CDC$^+$, suppressor-carrying segregants (e.g., abnormal bud shape or delocalized chitin deposition; cf. note j, Table 1) that will indicate that the suppressor gene's product is really involved in the morphogenetic processes of interest. (d) Most of the suppressors analyzed to

date have been recessive (i.e., diploids homozygous for the
cdc allele and heterozygous for the suppressor fail to grow
at the restrictive temperature.) This result was not expected (40), but now clearly parallels the results with suppressors of temperature-conditional actin and tubulin mutations (15). This recessiveness is both an interesting clue
to the mechanisms of the suppression (15) and an experimental convenience that mitigates considerably the difficulties
of our protocol vis-à-vis the paradigmatic one. In particular, it allows complementation analyses among the suppressors (diploids homozygous for the cdc mutation and heterozygous for each of two independently isolated suppressors are
tested for growth at the restrictive temperature) to supplement linkage analyses in assigning the suppressors to genes.
Moreover, the recessiveness of the suppressors should facilitate the cloning of the suppressor genes (i.e., libraries
of wild-type DNA in suitable plasmid vectors can be screened
for sequences that can convert a suppressed mutant strain
back to its original temperature sensitivity). (e) The fact
that the suppressors analyzed in each of our studies to date
have fallen into modest numbers of genes (Table 1; cf. refs.
15,40,51,52) both suggests that these genes are worth studying further by molecular methods and makes such study seem
a manageable goal.

The classical genetic analyses of suppressors and their
associated phenotypes may provide clues to the actual roles
of the suppressor genes' products and the nature of the interactions leading to suppression (see above). However, full
elucidation of these matters will require detailed molecular
investigation of the suppressor genes. Such investigations
will presumably involve both in vitro mutagenesis and gene
replacement (33-36,38) to generate a variety of mutations in
the suppressor genes, as well as identification and characterization (see below) of the suppressor genes' products.
(For example, if the products of CDC3 and CDC10 both localize to the same part of the cell, it will suggest that their
interaction is at the protein-protein level, whereas if one
localizes to the nucleus, it will suggest that it is involved in regulation of expression of the other.)

Plasmid Cross-Complementation

Recently, both we (56; note e of Table 2, below) and
others (57-59) have observed that not infrequently a cellular mutation in a particular gene can be "complemented" not
only by plasmids carrying that gene but also by plasmids

carrying some other gene. Presumably this "plasmid cross-complementation" depends on overexpression of the heterologous gene (resulting from its increased copy number and/or from enhanced transcription of the plasmid-borne copy), as the normal chromosomal copy of this gene does not suffice for suppression. Regardless of the specific mechanisms, these observations suggest a general and novel method to seek new genes involved in some process of interest. This method has essentially the same attractions as the pseudorevertant approach; i.e., it avoids the apparent limitations of traditional methods of seeking cell-cycle genes, and it provides immediate information about intergenic interactions. Moreover, once a new gene of interest is identified, it is already cloned, thus facilitating its further investigation by in vitro mutagenesis and gene replacement.

IDENTIFICATION AND CHARACTERIZATION OF GENE PRODUCTS

We hope eventually to understand morphogenesis in yeast at the molecular level. Realization of this hope will depend upon our ability to identify and elucidate the functions of the products of the genes defined by analysis of morphogenetic mutations and their suppressors. Fortunately, many of the techniques needed to move systematically from mutant to identification of the gene product have become available during the past few years. Although elucidation of gene-product function remains challenging, it is at least now possible to get in position for a serious assault on the problem. We have been working along these lines on several of the genes of interest; our progress to date is summarized in Table 2 (see also refs. 56,60).

Our hopes of elucidating gene-product functions depend largely on two approaches. First, we will attempt to infer properties of the gene products from the sequences of the coding regions. Although neither methods for inferring protein secondary and tertiary structure from amino-acid sequence data nor methods for inferring protein function from secondary and tertiary structure are yet very good (61,62), these methods can be expected to improve. Even now there is a good chance of recognizing an integral-membrane protein by virtue of its amino-terminal signal sequence (63,64) and/or its membrane-spanning domains (65,66). Such a result would not be particularly surprising for the gene products of interest to us, and would have clear implications for subsequent experimentation. Moreover, computer-assisted comparison of sequence data for the genes of interest to the ever-

TABLE 2
MOLECULAR ANALYSES OF GENES INVOLVED IN MORPHOGENESIS[a]

	Gene						
Step	CDC3	CDC10	CDC11	CDC12	CDC24	CDC42	CDC43
Yeast transformation[b]	+	+	+	.+	+	+	+
Yeast → E. coli → yeast[b]	+	+	+	+	+	+	+
Southern hybridization[c]	+	+	+	+	+	+	-
Fragment size (kb)[d]	2.4	4.3	2.3	1.4	2.7	4.0	-
Transcript localization[d]	+	+	+	+	+	-	-
Gene identification[e]	M	M	I/M	I/M	I/D	-	-
Sequence determination	-	-	-	-	-	-	-
Gene-fusion construction[f]	-	+	-	+	+	-	-
Fusion-protein isolation[f]	-	+	-	+	+?	-	-
Antibody production	-	-	-	+[g]	-	-	-

[a]+ implies that a particular step has been accomplished for that gene; - implies that this step has not yet been accomplished.

[b]Libraries of wild-type yeast DNA in "shuttle vectors" [plasmids capable of replicating, and being selected for, both in yeast and in E. coli (33)] are screened for sequences that complement a ts mutation in the gene of interest. Instability of the transformed phenotype during growth under nonselective conditions confirms that the transforming information is plasmid borne. Plasmid DNA is then recovered from the yeast transformant into E. coli for amplification; efficient retransformation of yeast then confirms that a particular E. coli clone contains the desired plasmid.

[c]After restriction mapping of the plasmids, Southern blots of chromosomal DNA are hybridized to appropriate plasmid-derived probes to check for possible rearrangements during cloning (none have been detected) and to see if the genes appear to be single copy (all have so appeared).

[d]Localization of the transcribed regions responsible for the cdc-complementing activities has been done using various combinations of subcloning, Northern blotting, R-loop analysis, and marker-rescue and gene-disruption (33,35) experiments. This step is crucial because of the high density of transcribed regions in yeast (typically one per 2.5 kb of genomic DNA). Indicated for each gene are the size of the smallest fragment yet obtained that retains complementing activity and whether the relevant transcribed region has been localized at least to some extent within that fragment.

[e]Evidence that the plasmids complementing particular cdc mutations actually contain those CDC genes is obtained by showing that the plasmids integrate by homologous recombination at the proper map positions (I), by gene-disruption experiments (D), or by marker-rescue experiments (M). This step is crucial because sometimes genes other than that harboring the original mutation can "complement" that mutation when borne on plasmids (see text). In our own studies, we have observed both complementation of cdc11 mutations by CDC12-containing plasmids and apparent complementation of cdc3 mutations by CDC10-containing plasmids (56).

[f]Of the several possible methods for proceeding from a cloned, identified gene to a suitable immunogen (see text), we have been working so far primarily with the pORF and pATH vectors for producing fusion proteins in E. coli (Fig. 9).

[g]Both ompF-CDC12-beta-galactosidase and trpE-CDC12 fusion proteins (see Fig. 9) have been injected into rabbits, who have been duly "boosted" and bled. We are just beginning to characterize the antisera obtained.

growing libraries of sequence data from known genes (67,68) may well reveal clues to the function(s) of the protein of interest (e.g., refs. 69-71). Note that the conservation of structure within protein domains with specific functions (e.g., ATP-binding sites) means that useful information may be obtained from such a search even in the likely event that no known protein is highly homologous, overall, to the gene product of interest (67,70,72,73).

Second, we are attempting to generate antisera specific for the gene products of interest. The several methods available for obtaining suitable immunogens include overproduction of the normal polypeptides in yeast or E. coli, production of any of several types of fusion proteins containing antigenic regions of the protein of interest, or synthesis of oligopeptide haptens based on nucleotide-sequence information. We expect that flexibility in choosing methods will be necessary to adapt to such exigencies of particular cases as the presence of inconvenient introns, instability of normal or fusion polypeptides, or difficulties in raising antisera of sufficiently high titer. However, it seems that in general the method of choice will be the construction of gene fusions leading to the synthesis of fusion proteins in E. coli. A variety of suitable vectors is available (74-77). In successful cases such as CDC12 (Fig. 9), fusion proteins are obtained in good yield that contain long segments of the proteins of interest. Polyclonal antisera raised against these proteins are likely to contain antibodies specific for multiple determinants in the protein of interest, thus maximizing the chances of having a detectable signal in subsequent immunolocalization experiments. Moreover, the fusion proteins are easily purified on the basis of their high molecular weights, insolubility, and/or enzymatic activity, and are contaminated, if at all, by E. coli proteins that are unlikely to elicit antibodies that will cause confusion by cross-reacting with yeast proteins.

The antisera obtained will be used initially to investigate the intracellular localizations of the gene products, using immunofluorescence (7,9,23,78), immuno-electron microscopy (6,79-82), cell fractionation (13,17,83-86), or a combination of these techniques (13,86). Such localization data seem quite likely to be informative with the gene products we are studying, as their involvement in cellular morphogenesis makes it likely that they will localize to relevant cellular structures such as elements of the cytoskeleton, the cell cortex, or the nucleus. Clearly, any of these results would be suggestive both as to possible functions of

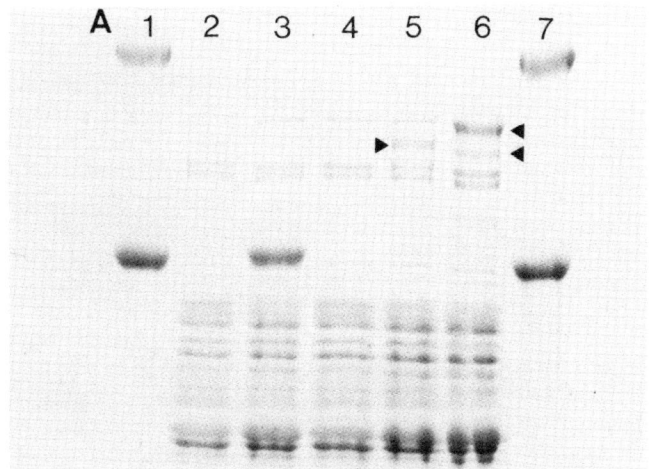
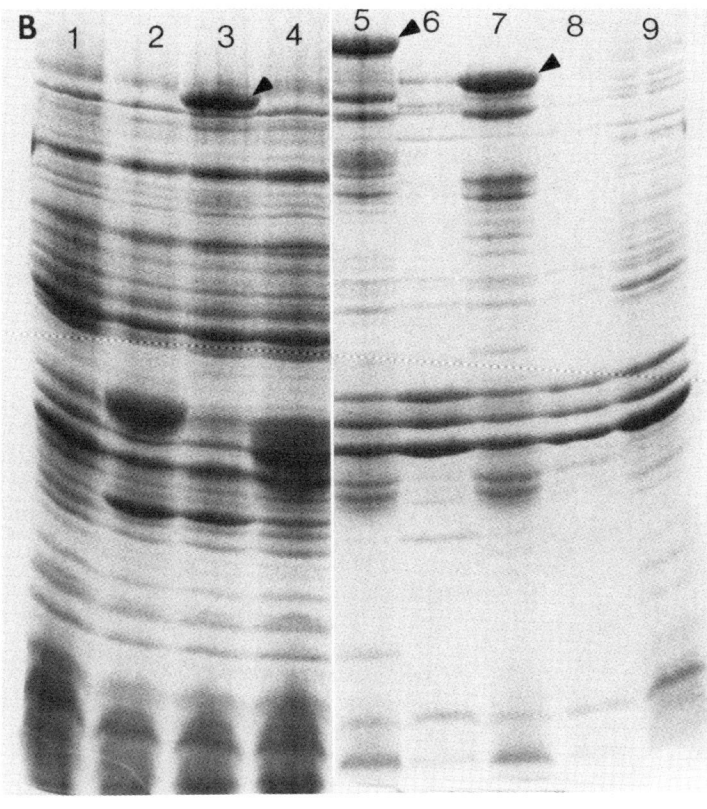

FIGURE 9. Proteins produced by E. coli strains with or without plasmids encoding fusions of the yeast CDC12 gene to E. coli genes. Proteins were separated by sodium-dodecyl-sulfate polyacrylamide-gel electrophoresis and stained with Coomassie blue. A, production of fusion proteins using the pORF1 vector (74). This plasmid encodes an out-of-frame fusion between the 5' terminus of the E. coli ompF gene (which encodes a major outer-membrane protein) and all but the 5' terminus of the E. coli lacZ gene (which encodes beta-galactosidase). A deletion or insertion that puts the lacZ gene fragment into frame with the ompF gene fragment leads to synthesis of a fusion protein possessing beta-galactosidase activity. Lanes 1 and 7, molecular-weight standards myosin (205,000) and beta-galactosidase (116,000). Lanes 2-6, total cellular proteins from strain TK1046 harboring no plasmid (lane 2), plasmid pORF1 (lane 4), a plasmid in which ompF and lacZ have been put into frame by making a small deletion in the polylinker that connects them in pORF1 (lane 3; note the fusion protein with a molecular weight slightly greater than that of normal beta-galactosidase), or plasmids encoding ompF-CDC12-lacZ "tribrid" fusion proteins (lanes 5 and 6; the fusion proteins are indicated by the arrowheads; the lower-molecular-weight polypeptide in lane 6 is probably a degradation product of the higher-molecular-weight polypeptide). B, production of fusion proteins using the pATH vector (75; T. Koerner and A. Tzagoloff, pers. commun.). This plasmid contains the regulatory regions of the E. coli trp operon and a truncated trpE coding region. Insertion of a properly oriented, in-frame fragment of the CDC12 coding region results in production of a trpE-CDC12 fusion protein under trp control. Such fusion proteins tend to precipitate and can be recovered in an "insoluble fraction" (75). Shown are total cellular proteins (lanes 1-4) or insoluble proteins (lanes 5-9) produced by E. coli strain HB101 (with or without plasmid) after induction of the trp operon. Lanes 1 and 9, no plasmid. Lanes 2 and 8, pATH plasmid with no insert of yeast DNA. Lanes 3 and 7, pATH plasmid with an insert of CDC12 DNA. Lanes 4 and 6, pATH plasmid with the same insert as for lanes 3 and 7, but in the opposite orientation. Lane 5, pATH plasmid with a longer insert of CDC12 DNA. The fusion proteins are denoted by arrowheads.

the gene product and as to the most promising directions for subsequent investigations. Once the clues provided by localization studies and by sequence analyses (see above) have suggested promising lines of biochemical investigation, the antibodies can be used further to facilitate purification of the gene products (87-89) as a step in such investigation.

SUMMARY AND CONCLUSIONS

It seems clear that known cytoskeletal elements such as microtubules, actin, and the 10-nm filaments of the neck region are involved in the morphogenetic processes of the cell cycle. Thus, direct study of these elements (and of other known elements such as myosin, spectrin, calmodulin, etc.) by morphological, biochemical and genetic methods should contribute to understanding of the morphogenetic processes. However, it also seems likely that these known elements function as parts of complex systems, not all components of which may be revealed easily by direct study of the known elements. In addition, it is possible that the morphogenetic processes also depend on more novel mechanisms. Thus, it seems crucial also to study the processes of interest using a genetic approach that involves no assumptions as to what the important mechanisms are. Fortunately, the techniques necessary to make such an approach practical have been developing rapidly during the past few years. Although acquisition of a detailed molecular understanding of morphogenesis in yeast will doubtless require much hard work and ingenuity, it at least appears to be a realistic goal.

ACKNOWLEDGEMENTS

Other past and present members of our laboratory (Chris Evans, Barbara Sloat, Rich Longnecker, Almuth Tschunko, Anne Stapleton, Paul Shiels, Paul Oeller, Mike Kelley, Stuart Ketcham, Susan White, and Doug Johnson) have also made significant contributions to the developing story presented here. We also thank John Kilmartin, Breck Byers, T. Wieland, David Kaback, Kerry Bloom, and many other colleagues for their gifts of materials and advice, and David Botstein, Jim Thomas, Peter Novick, Peter Schatz, and Mark Rose for so freely sharing results with us prior to publication.

REFERENCES

1. Pringle JR, Coleman K, Adams A, Lillie S, Haarer B, Jacobs C, Robinson J, Evans C (1984). Cellular morphogenesis in the yeast cell cycle. In Borisy GG, Cleveland DW, Murphy DB (eds): "Molecular Biology of the Cytoske-

leton", Cold Spring Harbor: Cold Spring Harbor Laboratory, p 193.
2. Malacinski GM, Bryant SV (eds) (1984). "Pattern Formation: A Primer in Developmental Biology". New York: Macmillan.
3. McIntosh JR (ed) (1983). "Spatial Organization of Eukaryotic Cells (Modern Cell Biology, Vol. 2)". New York: Alan R. Liss.
4. Pringle JR, Hartwell LH (1981). The Saccharomyces cerevisiae cell cycle. In Strathern JN, Jones EW, Broach JR (eds): "The Molecular Biology of the Yeast Saccharomyces: Life Cycle and Inheritance", Cold Spring Harbor: Cold Spring Harbor Laboratory, p 97.
5. Sloat BF, Adams A, Pringle JR (1981). Roles of the CDC24 gene product in cellular morphogenesis during the Saccharomyces cerevisiae cell cycle. J Cell Biol 89:395.
6. Roberts RL, Bowers B, Slater ML, Cabib E (1983). Chitin synthesis and localization in cell division cycle mutants of Saccharomyces cerevisiae. Mol Cell Biol 3:922.
7. Adams AEM, Pringle JR (1984). Relationship of actin and tubulin distribution to bud growth in wild-type and morphogenetic-mutant Saccharomyces cerevisiae. J Cell Biol 98:934.
8. Nasmyth K (1983). Molecular analysis of a cell lineage. Nature 302:670.
9. Kilmartin JV, Adams AEM (1984). Structural rearrangements of tubulin and actin during the cell cycle of the yeast Saccharomyces. J Cell Biol 98:922.
10. Byers B (1981). Cytology of the yeast life cycle. In Strathern JN, Jones EW, Broach JR, (eds): "The Molecular Biology of the Yeast Saccharomyces: Life Cycle and Inheritance", Cold Spring Harbor: Cold Spring Harbor Laboratory, p 59.
11. King SM, Hyams JS (1982). The mitotic spindle of Saccharomyces cerevisiae: assembly, structure and function. Micron 13:93.
12. Kilmartin JV (1981). Purification of yeast tubulin by self-assembly in vitro. Biochemistry 20:3629.
13. Kilmartin JV, Wright B, Milstein C (1982). Rat monoclonal antitubulin antibodies derived by using a new nonsecreting rat cell line. J Cell Biol 93:576.
14. Neff NF, Thomas JH, Grisafi P, Botstein D (1983). Isolation of the beta-tubulin gene from yeast and demonstration of its essential function in vivo. Cell 33:211.
15. Thomas JH, Novick P, Botstein D (1984). Genetics of the yeast cytoskeleton. In Borisy GG, Cleveland DW,

Murphy DG (eds): "Molecular Biology of the Cytoskeleton", Cold Spring Harbor: Cold Spring Harbor Laboratory, p 153.
16. Schatz PJ, Botstein D (1985). Two functional alpha-tubulin genes in Saccharomyces cerevisiae. J Cell Biochem 9C:111.
17. Kilmartin JV (1984). Microtubules and actin filaments during the yeast cell cycle. In Borisy GG, Cleveland DW, Murphy DG (eds): "Molecular Biology of the Cytoskeleton", Cold Spring Harbor: Cold Spring Harbor Laboratory, p 185.
18. Water RD, Pringle JR, Kleinsmith LJ (1980). Identification of an actin-like protein and of its messenger ribonucleic acid in Sacharomyces cerevisiae. J Bacteriol 144:1143.
19. Greer C, Schekman R (1982). Actin from Saccharomyces cerevisiae. Mol Cell Biol 2:1270.
20. Ng R, Abelson J (1980). Isolation and sequence of the gene for actin in Saccharomyces cerevisiae. Proc Natl Acad Sci USA 77:3912.
21. Gallwitz D (1982). Construction of a yeast actin gene intron deletion mutant that is defective in splicing and leads to the accumulation of precursor RNA in transformed yeast cells. Proc Natl Acad Sci USA 79:3493.
22. Shortle D, Haber JE, Botstein D (1982). Lethal disruption of the yeast actin gene by integrative DNA transformation. Science 217:371.
23. Novick P, Botstein D (1985). Phenotypic analysis of temperature-sensitive yeast actin mutants. Cell 40:405.
24. Watts FZ, Miller DM, Orr E (1985). Identification of myosin heavy chain in Saccharomyces cerevisiae. Nature 316:83.
25. Quinlan RA, Pogson CI, Gull K (1980). The influence of the microtubule inhibitor, methyl benzimidazol-2-yl-carbamate (MBC) on nuclear division and the cell cycle in Saccharomyces cerevisiae. J Cell Sci 46:341.
26. Wood JS, Hartwell LH (1982). A dependent pathway of gene functions leading to chromosome segregation in Saccharomyces cerevisiae. J Cell Biol 94:718.
27. Jacobs CW, Adams AEM, Szaniszlo P, Pringle JR (1985). Functions of microtubules during the Saccharomyces cerevisiae cell cycle. Submitted for publication.
28. Hoebeke J, Van Nijen G, De Brabander M (1976). Interaction of nocodazole (R17934), a new antitumoral drug, with rat brain tubulin. Biochem Biophys Res Commun 69:319.
29. Culotti J, Hartwell LH (1971). Genetic control of the cell division cycle in yeast. III. Seven genes control-

ling nuclear division. Exp Cell Res 67:389.
30. Oakley BR, Morris NR (1980). Nuclear movement is beta-tubulin-dependent in Aspergillus nidulans. Cell 19:255.
31. Delgado MA, Conde J (1984). Benomyl prevents nuclear fusion in Saccharomyces cerevisiae. Mol Gen Genet 193:188.
32. Rose MD, Fink GR (1985). Nuclear fusion in yeast. J Cell Biochem 9C:116.
33. Botstein D, Davis RW (1982). Principles and practice of recombinant DNA research with yeast. In Strathern JN, Jones EW, Broach JR (eds): "The Molecular Biology of the Yeast Saccharomyces. Metabolism and Gene Expression", Cold Spring Harbor: Cold Spring Harbor Laboratory, p 607.
34. Botstein D, Maurer R (1982). Genetic approaches to the analysis of microbial development. Annu Rev Genet 16:61.
35. Rothstein RJ (1983). One-step gene disruption in yeast. Meth Enzymol 101:202.
36. Struhl K (1983). The new yeast genetics. Nature 305:391.
37. Thomas JH, Neff N, Botstein D (1985). Isolation and characterization of mutations in the beta-tubulin gene of Saccharomyces cerevisiae. Genetics, in press.
38. Shortle D, Novick P, Botstein D (1984). Construction and genetic characterization of temperature-sensitive mutant alleles of the yeast actin gene. Proc Natl Acad Sci USA 81:4889.
39. Morris NR, Lai MH, Oakley CE (1979). Identification of a gene for alpha-tubulin in Aspergillus nidulans. Cell 16:437.
40. Moir D, Stewart SE, Osmond BC, Botstein D (1982). Cold-sensitive cell-division-cycle mutants of yeast: Isolation, properties and pseudoreversion studies. Genetics 100:547.
41. Troock, DC, Tyler JM (1984). Isolation of a spectrin-like protein from Saccharomyces cerevisiae. Abst Twelfth Internat Conf Yeast Genet Mol Biol, p 276.
42. Davis TN, Thorner J (1985). This volume.
43. Porter KR (ed) (1984). "The Cytoplasmic Matrix and the Integration of Cellular Function". J Cell Biol 99: No 1, Part 2.
44. Nuccitelli R (1983). Transcellular ion currents: Signals and effectors of cell polarity. In McIntosh JR (ed): "Modern Cell Biology, Vol. 2: Spatial Organization of Eukaryotic Cells", New York: Alan R Liss, p 451.
45. Kropf DL, Caldwell JH, Gow NAR, Harold FM (1984). Transcellular ion currents in the water mold Achlya. Amino acid proton symport as a mechanism of current entry. J Cell Biol 99:486.

46. Pringle JR (1981). The genetic approach to the study of the cell cycle. In Zimmerman AM, Forer A (eds): "Mitosis/Cytokinesis", New York: Academic Press, p 3.
47. Kaback DB, Oeller P, Steensma HY, Hirschman J, Ruezinsky D, Coleman KG, Pringle JR (1984). Temperature-sensitive lethal mutations on yeast chromosome I appear to define only a small number of genes. Genetics 108:67.
48. Luck DJL (1984). Genetic and biochemical dissection of the eucaryotic flagellum. J Cell Biol 98:789.
49. Adams AEM, Evans CT, Sloat BF, Pringle JR. CDC42 and CDC43, two new genes involved in budding and the polarization of secretion in the yeast Saccharomyces cerevisiae. In preparation.
50. Hartwell LH, Mortimer RK, Culotti J, Culotti M (1973). Genetic control of the cell division cycle in yeast: V. Genetic analysis of cdc mutants. Genetics 74:267.
51. Fantes PA (1981). Isolation of cell size mutants of a fission yeast by a new selective method. J Bacteriol 146:746.
52. Matsumoto K, Uno I, Ishikawa T (1985). Genetic analysis of the role of cAMP in yeast. Yeast 1:15
53. Tatchell K, Robinson LC, Breitenbach M (1985). RAS2 of Saccharomyces cerevisiae is required for gluconeogenic growth and proper response to nutrient limitation. Proc Natl Acad Sci USA 82:3785.
54. Pringle JR (1975). Induction, selection, and experimental uses of temperature-sensitive and other conditional mutants of yeast. Meth Cell Biol 12:233.
55. Hartwell LH, Smith D (1985). Altered fidelity of mitotic chromosome transmission in cell cycle mutants of S. cerevisiae. Genetics 110:381.
56. Lillie SH, Haarer B, Bloom L, Pringle JR (1986). Isolation and characterization of the Saccharomyces cerevisiae CDC11 and CDC12 genes. In preparation.
57. MacKay VL (1983). Cloning of yeast STE genes in 2 micron vectors. Meth Enzymol 101:325.
58. Kuo C-L, Campbell JL (1983). Cloning of Saccharomyces cerevisiae DNA replication genes: Isolation of the CDC8 gene and two genes that compensate for the cdc8-1 mutation. Mol Cell Biol 3:1730.
59. Patterson M, Field S, Venning B, Rosamond J (1984). Characterization of cloned genomic fragments which complement cdc mutants of Saccharomyces cerevisiae. Abst Twelfth Internat Conf Yeast Genet Mol Biol, p 251.
60. Coleman KG, Steensma HY, Kaback DB, Pringle JR (1986).

Molecular cloning of chromosome I DNA from Saccharomyces cerevisiae: isolation and characterization of the CDC24 gene and adjacent regions of the chromosome. Submitted for publication.
61. Chothia C (1984). Principles that determine the structure of proteins. Annu Rev Biochem 53:537.
62. Kabsch W, Sander C (1984). On the use of sequence homologies to predict protein structure: identical pentapeptides can have completely different conformations. Proc Natl Acad Sci USA 81:1075.
63. Schekman R, Novick P (1982). The secretory process and yeast cell-surface assembly. In Strathern JN, Jones EW, Broach JR (eds): "The Molecular Biology of the Yeast Saccharomyces. Metabolism and Gene Expression", Cold Spring Harbor: Cold Spring Harbor Laboratory, p 361.
64. Tschopp J, Esmon PC, Schekman R (1984). Defective plasma membrane assembly in yeast secretory mutants. J Bacteriol 160:966.
65. Kyte J, Doolittle RF (1982). A simple method for displaying the hydropathic character of a protein. J Mol Biol 157:105.
66. Eisenberg D (1984). Three-dimensional structure of membrane and surface proteins. Annu Rev Biochem 53:595.
67. Doolittle RF (1981). Similar amino acid sequences: chance or common ancestry? Science 214:149.
68. Lipman DJ, Pearson WR (1985). Rapid and sensitive protein similarity searches. Science 227:1435.
69. Waterfield, MD, Scrace GT, Whittle N, Stroobant P, Johnsson A, Wasteson A, Westermark B, Helden C-H, Huang JS, Deuel TF (1983). Platelet-derived growth factor is structurally related to the putative transforming protein $p28^{sis}$ of simian sarcoma virus. Nature 304:35.
70. Nurse P. (1985). Cell cycle control genes in yeast. Trends Genet 1:51.
71. Reed SI, Hadwiger JA, Lorincz AT (1985). Protein kinase activity associated with the product of the yeast cell division cycle gene CDC28. Proc Natl Acad Sci USA 82:4055.
72. Gilbert W (1985). Genes-in-pieces revisited. Science 228:823.
73. Lochrie MA, Hurley JB, Simon MI (1985). Sequence of the alpha subunit of photoreceptor G protein: Homologies between transducin, ras, and elongation factors. Science 228:96.
74. Weinstock GM, ap Rhys C, Berman ML, Hampar B, Jackson D, Silhavy TJ, Weisemann J, Zweig M (1983). Open reading

frame expression vectors: A general method for antigen production in Escherichia coli using protein fusions to beta-galactosidase. Proc Natl Acad Sci USA 80:4432.
75. Spindler KR, Rosser DSE, Berk AJ (1984). Analysis of adenovirus transforming proteins from early regions 1A and 1B with antisera to inducible fusion antigens produced in Escherichia coli. J Virol 49:132.
76. Reed SI (1982). Preparation of product-specific anti sera by gene fusion: antibodies specific for the product of the yeast cell-division-cycle gene CDC28. Gene 20:255.
77. Rüther U, Müller-Hill B (1983). Easy identification of cDNA clones. EMBO J 2:1791.
78. Hall MN, Hereford L, Herskowitz I (1984). Targeting of E. coli beta-galactosidase to the nucleus in yeast. Cell 36:1057.
79. Horisberger M, Vonlanthen M (1977). Location of mannan and chitin on thin sections of budding yeasts with gold markers. Arch Microbiol 115:1.
80. Geuze HJ, Slot JW, Van der Ley PA, Scheffer RC, Griffith JM (1981). Use of colloidal gold particles in double-labeling immunoelectron microscopy of ultrathin frozen tissue sections. J Cell Biol 89:653.
81. Roth J, Berger EG (1982). Immunocytochemical localization of galactosyltransferase in HeLa cells: Codistribution with thiamine pyrophosphatase in trans-Golgi cisternae. J Cell Biol 93:223.
82. Griffiths G, Simons K, Warren G, Tokuyasu KT (1983). Immunoelectron microscopy using thin, frozen sections: Application to studies of the intracellular transport of Semliki Forest virus spike glycoproteins. Meth Enzymol 96:466.
83. De Duve C, Beaufay H (1981). A short history of tissue fractionation. J Cell Biol 91:293s.
84. Oliver DB, Beckwith J (1982). Regulation of a membrane component required for protein secretion in Escherichia coli. Cell 30:311.
85. Schekman R (1982). Biochemical markers for yeast organelles. In Strathern JN, Jones EW, Broach JR (eds): "Molecular Biology of the Yeast Saccharomyces. Metabolism and Gene Expression", Cold Spring Harbor: Cold Spring Harbor Laboratory, p 651.
86. Kilmartin JV, Fogg J (1982). Partial purification of yeast spindle pole bodies. In Cappuccinelli P, Morris NR (eds): "Microtubules in Microorganisms", New York: Marcel Dekker, p 157.
87. MacSween JM, Eastwood SL (1981). Recovery of antigen

from staphylococcal protein A-antibody adsorbents. Meth Enzymol 73:459.
88. Burke DJ, Ward S (1983). Identification of a large multigene family encoding the major sperm protein of <u>Caenorhabditis elegans</u>. J Mol Biol 171:1.
89. Langone JJ, Van Vunakis H (eds) (1983). "Immunochemical Techniques, Part E". Meth Enzymol 92.

ADENYLATE CYCLASE AND GTP-BINDING PROTEINS IN YEAST AND VERTEBRATE CELLS[1]

Henry R. Bourne, Gerald F. Casperson,
Naomi Walker, Kathleen Sullivan,
and Dan C. Medynski

Departments of Pharmacology and Medicine,
University of California,
San Francisco, CA 94143

ABSTRACT The ras proteins of yeast and mammals exhibit striking structural and functional homologies to a vertebrate family of GTP-binding proteins that includes G_s, G_i, and retinal transducin. Comparison of the amino acid sequences of transducin subunits with those of the ras proteins leads to strong predictions regarding function of ras structural domains and suggests an evolutionary relationship among all the GTP-binding signal transducers.

INTRODUCTION

Other chapters in this volume detail recent advances in our understanding of the role of GTP-binding membrane proteins and cAMP in yeast cell biology. Elegant genetic and biochemical studies by Matsumoto, Uno, and their colleagues (1-4) established that cAMP acts as a positive regulator of yeast cell proliferation and a negative regulator of sporulation. Wigler and his colleagues found (5) that the yeast RAS genes, which are close homologs of the vertebrate ras oncogenes (6,7), encode membrane proteins that are required for GTP-dependent synthesis of cAMP by adenylate cyclase. Thus the RAS proteins of yeast, like their vertebrate

---

[1]This work was supported by NIH grants GM 28310 and GM 27800.

counterparts, promote cell proliferation.

Aside from their importance for understanding the control of cell proliferation, these findings point to striking homologies between mechanisms of transmembrane signaling by GTP-binding proteins in yeast and vertebrate cells. Many vertebrate membranes contain one or more of a family of oligomeric GTP-binding proteins that transduce a diverse series of extracellular signals into regulation of adenylate cyclase and other effector enzymes. This family includes three well characterized proteins: (a) G_s, which mediates hormonal stimulation of adenylate cyclase (8); (b) G_i, which mediates hormonal inhibition of adenylate cyclase (8) and (in some cells) activation of phosphoinositide metabolism and mobilization of cellular calcium (9); (c) transducin, a retinal protein that mediates stimulation of cGMP phosphodiesterase (PDE) by photoexcited rhodopsin (10). In addition, the 21-kDa ras polypeptide gene products (termed p21) promote cell proliferation in vertebrate cells but appear not to stimulate adenylate cyclase.

Thus GTP-binding proteins represent a strategy devised by evolution to solve the fundamental biological problem of transducing information across lipid membranes. Because of the accessibility of yeast to genetic manipulation, the discovery that GTP regulates yeast adenylate cyclase via GTP-binding proteins homologous to vertebrate ras proteins presents an exciting prospect for future exploration of molecular mechanisms that are shared by all the GTP-binding signal transducers.

In this chapter we will first describe work from this and other laboratories on regulation of adenylate cyclase in yeast. We will then compare three families of cognate GTP-binding signal transducers -- the oligomeric G/transducin family and p21, in vertebrates, and the RAS proteins of yeast. Structural homologies and differences among these proteins raise interesting questions regarding their function and evolution.

CONTROL OF ADENYLATE CYCLASE BY GUANINE NUCLEOTIDES AND RAS GENE PRODUCTS

Casperson et al. (11) first reported that guanine nucleotides regulate adenylate cyclase in membranes of S. cerevisiae. Just as in vertebrate membranes, guanosine triphosphates stimulate the yeast enzyme, while GDPβS, a stable GDP analog, blocks the stimulation. Previous studies of the yeast enzyme did not reveal regulation by guanine nucleotides, in part because assays were performed in the presence of Mn^{2+}. The enzyme's activity in yeast is most sensitive to stimulation by GTP in the presence of Mg^{2+}, rather than Mn^{2+}. Presumably this is because the catalytic unit (C) of the yeast enzyme, like that of vertebrates (12), requires stimulation by a G protein to utilize Mg-ATP as substrate, but shows no such requirement with Mn-ATP as substrate.

Activity of yeast adenylate cyclase, measured with Mn-ATP as substrate, is thermally stable, while GTP-dependent activity (with Mg-ATP as substrate) rapidly disappears upon incubation of membranes at 30° (11). This observation suggested that the yeast enzyme contained at least two components, one that is thermally stable (presumably C) and one labile component, which is required for GTP-dependent activity in the presence of Mg^{2+}. Stabilization of the GTP-dependent activity by Gpp(NH)p suggested that the labile component is a GTP-binding protein (11).

Regulation of the yeast and vertebrate enzymes differ, in that several agents that activate cAMP synthesis in vertebrate membranes fail to do so in yeast. These include fluoride ion and cholera toxin, both of which act on vertebrate G_s (13), and forskolin, which directly stimulates vertebrate C (14).

Yeast genes encoding both the G and C components have now been cloned. The G component genes, RAS1 and RAS2, were cloned by virtue of their ability to hybridize with the Ha-ras oncogene (6,7). Insertional activation of both RAS genes (but not of either gene alone) produces cells that cannot progress through G_1 (15). The C component gene was cloned (16) by complementation of the cyr1-1 mutation, which also prevents cells from progressing through G_1 (2).

Compelling evidence (5) indicates that the RAS genes encode proteins that can stimulate C in a GTP-dependent fashion: Products of both genes bind and hydrolyze GTP (17,18). Insertional inactivation of RAS2 causes decreased cellular cAMP and decreased GTP-dependent adenylate cyclase activity in membranes. Yeast strains in which both RAS1 and RAS2 genes are inactivated can proliferate in the presence of the bcy1 mutation, which causes persistent activation of the cAMP-dependent protein kinase (1). These doubly RAS-deficient cells contain extremely low cellular cAMP, and their membranes lack detectable GTP-dependent adenylate cyclase activity. Strains bearing a point mutation in RAS2 that substitutes valine for glycine at position 19 exhibit marked increases in cellular cAMP and membrane adenylate cyclase activity measured in the presence of Mg^{2+}; this amino acid substitution at the equivalent position in the homologous c-Ha-ras protein reduces its GTPase activity and increases its capacity for oncogenic transformation (19). As expected, none of these mutations affects C activity measured with Mn-ATP as substrate.

Finally, a membrane mixing experiment (5) convincingly rules out the possibility that RAS proteins indirectly confer GTP-sensitivity on C: cyr1-1 membranes (C^-, putatively G^+) are mixed in the presence of detergent with membranes from $RAS1^-/RAS2^-$ cells (C^+, putatively G^-); guanine nucleotide-dependent adenylate cyclase activity, undetectable in either type of membrane alone, is reconstituted in the mixture. This biochemical complementation experiment makes it extremely likely that the protein products of the yeast RAS genes are G proteins that directly interact with and stimulate C.

The fact that either RAS gene is sufficient to support cAMP synthesis and traversal through G_1 explains why none of the complementation groups among the cAMP auxotrophs selected by Matsumoto and his co-workers (1) was associated with a lesion in a G protein gene. It is not clear, however, why S. cerevisiae should enjoy the luxury of two G protein genes. Either gene alone allows the cells to traverse G_1, and each presumably can support GTP-dependent adenylate cyclase. Only one difference between $RAS1^-$ and $RAS2^-$ strains has so far been reported: Homozygous $RAS2^-$ diploid cells, but not homozygous $RAS1^-$ diploids, sporulate in rich medium (5). The ability to sporulate in rich medium, mimicking the phenotype of homozygous cyr1-1 diploids, suggests that the RAS2 protein's activity is more critical for cAMP synthesis. Perhaps the RAS1 gene product primarily regulates an effector enzyme other than C, although it can support the GTP-dependent activity of C (in the absence of RAS2) well enough to allow cells to traverse G_1.

Biochemical characterization of vertebrate adenylate cyclase has progressed much more rapidly for G_s than for C, in part because vertebrate C is a much more labile activity. The similar regulation of the vertebrate and yeast enzymes suggests that isolation and characterization of the gene encoding yeast C may open avenues to elucidation of the structure and function of this important effector enzyme. We have reported (16) molecular cloning of genomic DNA containing the yeast C gene (CYR1), accomplished by complementation of the cyr1-1 mutation. The isolated DNA restored adenylate cyclase activity to cyr1-1 mutants and directed integration at the CYR1 locus. Although we have not yet sequenced the CYR1 gene, this work has produced two interesting experimental observations:

1. Wild type strains transformed with CYR1 DNA on the high copy number vector YEp24 contain four- to six-fold more adenylate cyclase activity than strains carrying the plasmid with no insert. The increased activity is observed not only with Mn-ATP as substrate (as would be expected from increased expression of the C gene), but also when the enzyme is stimulated by guanine nucleotides with Mg-ATP as substrate. This result suggests that expression of C, rather than that of other polypeptide components, including the RAS gene products, limits total adenylate cyclase activity in S. cerevisiae.

2. CYR1-containing plasmids also complement the temperature-sensitive growth defect of cells containing the cell division cycle mutation cdc35-1, which confers a phenotype under restrictive conditions similar to that of cyr1-1 (20,21) and maps to the same locus (1,22). Furthermore, cdc35-1 cam mutants, which contain mutations that enable them to take up cAMP from the medium (23), grow at the restrictive temperature in the presence of exogenous cAMP. These observations support the view that CDC35 and CYR1 are allelic, and confirm the hypothesis that cAMP synthesis is required for cells to pass through the "start" position of the cell division cycle.

DETECTOR-EFFECTOR COUPLING BY GTP-BINDING PROTEINS

While information concerning the interaction between yeast RAS proteins and yeast C is just beginning to emerge, biochemical understanding of the coupling of the vertebrate G proteins and transducin is more complete. Each member of the G protein/transducin family is a heterotrimer of α, β, and γ subunits. Fig. 1 depicts a generalized model of the molecular interactions by which these proteins transduce signals between detector (D) and effector (E) components. Subunits of the G protein are denoted α, β or γ. In this model, based upon that proposed for transducin (24,25) and studies of the G proteins (8), the G protein's α subunit shuttles between GTP- and GDP-bound states. The GTP-bound state activates E (e.g., C or PDE), while the GDP-bound state binds to D* (activated hormone receptor or photorhodopsin). The intrinsic GTPase activity of the active state of α (α*-GTP) terminates its ability to activate E.

Accordingly, light and specific hormones maintain activation of their respective coupling proteins (and effector enzymes) only by catalyzing replacement of GDP by GTP at the guanine nucleotide binding site.

In this scheme, the detector's interaction with α could not occur without βγ.[2] Free βγ binds to α-GDP, and the

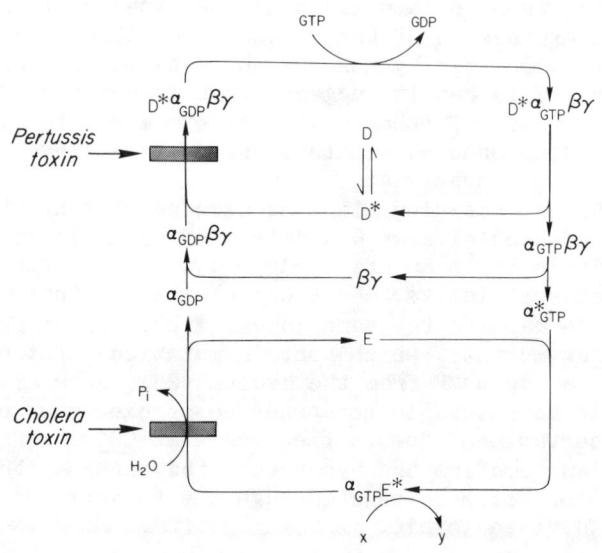

FIGURE 1. Model of signal transduction by G proteins and transducin, illustrating interactions among subunits of the G protein (α, βγ), guanine nucleotides, detector component (D), and effector component (E). Bars indicate sites of blockade by covalent modifications of α catalyzed by bacterial toxins. See text for details.

resulting αβγ complex has a high affinity for the activated detector component, D*. Guanine nucleotides and D* heterotropically displace one another from binding to αβγ.

---

[2] In both the G-protein and transducin systems, the β and γ subunits bind together tightly upon dissociation from α, and act functionally as a single subunit, hereafter referred to as βγ.

Consequently, GTP replaces GDP at the guanine nucleotide binding site of α, with concomitant dissociation of $\beta\gamma$ from D*. GTP stabilizes a conformational change in α (to α*) that reduces its affinity for $\beta\gamma$, so that free α*-GTP is released to activate E once again. Note that in this scheme the changing affinities of α for $\beta\gamma$ determine the directionality of the process: α*-GTP and α-GDP have, respectively, low and high affinities for $\beta\gamma$. Furthermore, $\beta\gamma$ is absolutely required for binding of α-GDP to D*, and therefore for regeneration of α*-GTP. Using radiolabeled and functionally active α and $\beta\gamma$ subunits of transducin, Fung and coworkers showed that neither α nor $\beta\gamma$ alone binds to photorhodopsin, but that the two together bind stoichiometrically to the photoreceptor in the absence of GTP (24,25).

Evidence for homology (8,10,26) among the G proteins and transducin includes their similar subunit structures, similar amino acid compositions of their respective subunits, and -- in their α subunits -- the occurrence of guanine nucleotide binding sites and sites covalently modified by bacterial toxins. Cholera toxin catalyzes ADP-ribosylation of the α subunit of G_s (27); the ADP-ribosylation stabilizes the active GTP-bound form of G_s, increasing stimulation of the effector enzyme, adenylate cyclase. Pertussis toxin catalyzes ADP-ribosylation of the α subunit of G_i (28); this covalent modification stabilizes the inactive (GDP-bound) form of G_i and interrupts signal transduction from hormone receptors to the effector enzymes of G_i, adenylate cyclase and phospholipase C (for review, see (8)). Each of the bacterial toxins catalyzes ADP-ribosylation at a different site on transducin's α subunit, with functional effects that exactly parallel its effects on G_s or G_i, respectively (29,30).

The role of vertebrate <u>ras</u> proteins is much less well understood than that of transducin and the G proteins. Indeed, p21-<u>ras</u> proteins cannot even be formally identified as signal transducing proteins, inasmuch as specific detector and effector elements coupled by these proteins have not yet been identified. Three analogies with the G proteins, however, strongly suggest that p21-<u>ras</u> must transduce signals across plasma membranes: 1. p21 proteins bind and hydrolyze GTP. 2. p21 proteins are located on the cytoplasmic face of the plasma membrane.[3] 3. Mutations that

---

[3]Indeed, mutations that prevent binding of Ha-<u>ras</u> to the membrane decrease its capacity for malignant transformation (31).

substitute almost any amino acid for the glycine at position 12 of the p21 proteins produce a protein with reduced GTPase activity and enhanced capacity for malignant transformation (19); this suggests that the GTP-bound form of p21 is active in promoting cell proliferation.

A fourth analogy makes an even stronger case for the p21 proteins as signal transducers. This is the role of the yeast RAS proteins as GTP-dependent stimulatory components of adenylate cyclase in yeast (5). Although adenylate cyclase is the effector stimulated by yeast RAS, no detector component has yet been identified; the simplest guess is that such D proteins would serve to detect critical nutrient molecules (or their metabolites) whose presence or absence regulates transition through the "start" position in G_1 of the yeast cell cycle. Finally, the c-Ha-ras gene can substitute for the yeast RAS2 gene in supporting transition of yeast cells through "start" (15). This finding strongly implies that the p21 protein can stimulate yeast adenylate cyclase, making it quite likely that the p21 protein similarly regulates an effector enzyme in vertebrate cells.

Taken together, these observations raise several critical questions:

1. Do the two classes of ras proteins transduce signals by a molecular mechanism identical to that of the transducers of light and hormonal signals? More specifically, do the ras proteins interact with a component similar to $\beta\gamma$?

2. What structural domains of the ras proteins interact with detector and effector components and with guanine nucleotides?

3. What is the evolutionary relationship between the ras proteins and the G proteins and transducin?

The isolation of cDNAs encoding the α and γ subunits of transducin (32-34) allows deduction of their primary structure. Comparisons of the amino acid sequences of transducin subunits and the ras proteins provides strong hints at answers to some of these questions.

STRUCTURAL COMPARISON OF RETINAL TRANSDUCIN SUBUNITS AND RAS PROTEINS

Functional characterization ((25) and B.K.-K. Fung, personal communication) of polypeptide domains of transducin's α and γ subunits (T_α and T_γ, respectively), combined with knowledge of their amino acid sequences, may shed some light on the functional roles of structural domains in the ras proteins.

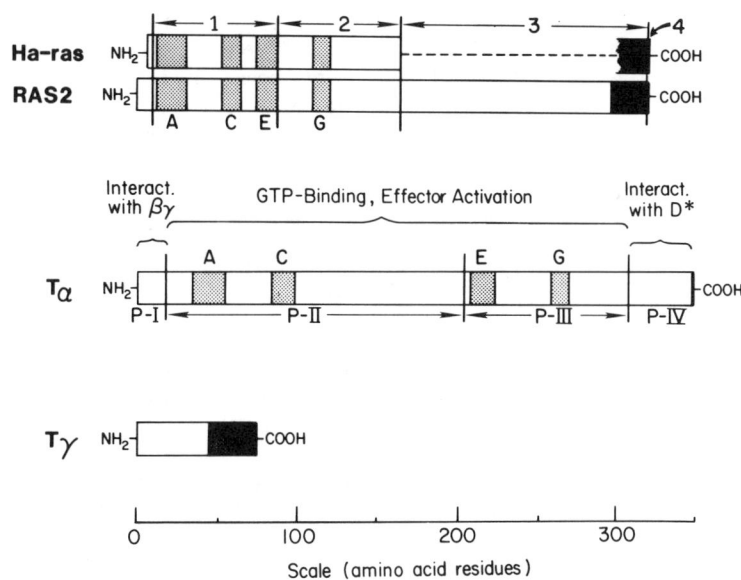

FIGURE 2. Schematic diagram of c-Ha-ras, yeast RAS2, T_α, and T_γ polypeptides. Domains 1-4 of amino acid sequence homology between the vertebrate and yeast ras proteins (top) and tryptic peptides P-I through P-IV of T_α (middle) are separated by vertical lines. Cross-hatched areas in the ras proteins and T_α (denoted A, C, E, and G) indicate regions of amino acid sequence that are homologous to regions of the ras proteins and the elongation factors and probably contribute to GTP binding (see Table 1). The filled area at the carboxy terminus of each polypeptide indicates a conserved tetrapeptide sequence that contains a cysteine required for oncogenic activity of H-ras. Heavily cross-hatched areas near the carboxy termini of Ha-ras, RAS2, and T_γ indicate regions of limited but significant amino acid sequence homology. Domains of T_α postulated to interact with other components of the phototransduction system are indicated above the polypeptide. See text for details.

The amino acid sequences of the yeast and vertebrate ras proteins can be conveniently divided into four domains, each of which shows a different degree of homology within the ras protein family. Beginning at the amino terminus (see Fig. 2), these are:

Domain 1 (amino acid residues 3-79 of c-Ha-ras, c-Ki-ras, and N-ras, and residues 10-88 of yeast RAS1 and RAS2) shows the highest degree of sequence homology: 100% homology within the vertebrate p21 proteins, 90% between RAS1 and RAS2, and approximately 85% between p21 and either yeast protein. This domain contains several sites at which point mutations increase oncogenic capacity of the p21 family, most notably gly^{12} (reviewed in (19)). This site corresponds to gly^{19} in the yeast proteins, where substitution of a valine residue in RAS2 enhances RAS-dependent adenylate cyclase activity (5).

In comparison to the first domain, domain 2 (residues 80-159 in p21 and 87-167 in the yeast proteins) shows slightly decreased homology, within the p21 family (80%), between RAS1 and RAS2 (80%), and in comparisons of any p21 protein to either yeast protein (50%).

The next domain, domain 3, is termed the "variable domain" (15) because it shows the largest variation in amino acid sequence. In addition, this domain is very much shorter (26 residues) in the p21 proteins than in the yeast RAS proteins (137 residues in RAS1 and 150 in RAS2), accounting for most of the difference in size between the vertebrate and yeast proteins.

Domain 4 comprises the four residues at the carboxy terminus of all the proteins. Here all the proteins show a similar sequence: -cys-hydrophobic-hydrophobic-X. In Ha-ras, mutations within this domain (including a point mutation that replaces the cysteine with a serine residue) prevent a post-translational modification (probably acylation) of the protein that is necessary for its attachment to the plasma membrane; these mutations also destroy the protein's capacity for malignant transformation (31,35).

Halliday (36) recently described four regions of conserved amino acid sequence that are common to bacterial elongation factors G and Tu and to the mammalian and yeast ras proteins (see Table 1). Biochemical studies of the elongation factors, including photoaffinity labeling with radioactive GTP derivatives and covalent modification of specific amino acid residues (summarized in (32,36)), suggest that all four regions contribute to binding or hydrolysis of GTP. The four regions, which Halliday termed A, C, E, and G, occur in the same order in the elongation factors and the ras proteins, where they appear in homologous domains 1 and 2. Region A contains gly^{12} of mammalian ras

TABLE 1
HOMOLOGIES IN GTP-BINDING DOMAINS[a]

		A (19 residues)	Homology
T_α	31:	K L L L L G A G E S G K S T I V K Q M	
c-Ha-<u>ras</u>1	5:	K L V V V G A G G V G K S A L T I Q L	15 (9)
RAS1	12:	K I V V V G G G G V G K S A L T I Q F	15 (7)
RAS2	12:	K L V V V G G G G V G K S A L T I Q L	15 (8)
EF-Tu	14:	N V G T I G H V D H G K T T L T A A I	10 (4)

		C (13 residues)	
T_α	80:	I V R A M T T L N I Q Y G	
c-Ha-<u>ras</u>1	55:	I L D T A G Q E - - E Y S	7 (2)
RAS1	62:	I L D T A G Q E - - E Y S	7 (2)
RAS2	62:	I L D T A G Q E - - E Y S	7 (2)
EF-Tu	79:	H V D C P G H A - - D Y V	4 (2)

		E (14 residues)	
T_α	203:	I H C F E G V T C I I F I A	
c-Ha-<u>ras</u>1	72:	M R T G E G F L C V F A I N	7 (4)
RAS1	79:	M R T G E G F L L V Y S V T	7 (2)
RAS2	79:	M R T G E G F L L V Y S I T	7 (3)
EF-Tu	95:	A A Q M D G A I L V V A A T	5 (1)

		G (11 residues)	
T_α	259:	S I V L F L N K K D V	
c-Ha-<u>ras</u>1	110:	P M V L V G N K C D L	8 (5)
RAS1	117:	P V V V V G N K L D L	8 (4)
RAS2	117:	P I V V V G N K S D L	8 (5)
EF-Tu	129:	Y I I V P L N K C D M	8 (5)

[a] Amino acid sequences from conserved regions of human (c-Ha-<u>ras</u>1) and yeast (RAS1) <u>ras</u> proteins and bacterial elongation factor EF-Tu are aligned with corresponding regions of T_α. Regions of exact homology or conservative (37) substitutions between T_α and other proteins are included within boxes. Numbers on the right indicate the number of conservative substitutions and, in parentheses, the number of identities, both with respect to T_α. Sources for sequences are summarized in (32,36). Single letter abbreviations of amino acid residues are as follows: A, ala; C, cys; D, asp; E, glu; F, phe; G, gly; H, his; I, ile; K, lys; M, met; N, asn; P, pro; Q, gln; R, arg; S, ser; T, thr; V, val; W, trp; Y, tyr. The following Dayhoff conservative categories were used: C; S,T,P,A,G; N,D,E,Q; H,R,K; M,I,L,V; F,Y,W.

and gly^{19} of RAS1 and RAS2, the sites at which mutations enhance effector-stimulating activities of both sets of proteins (Table 1).

Amino acid sequences of transducin's α and γ subunits contain regions of limited but significant homology to regions of the ras proteins. Four regions of the α subunit (32) correspond to the conserved GTP-binding regions of ras proteins and elongation factors; the four regions occur in the same order in all these proteins (Table 1, Fig. 2). Note that region A of T_α contains at position 38 a glycine residue corresponding to the critical glycine residues at positions 12 or 19 of p21 or yeast RAS. We have discerned only one other region of amino acid sequence homology between T_α and the ras proteins, the carboxy terminal tetrapeptide of T_α, -cys-gly-leu-phe. This sequence resembles the consensus sequence of the carboxy terminal tetrapeptide of the ras proteins (see above).

Surprisingly, the carboxy terminal half of T_γ shows homology with the amino acid sequence encoded by the last exon of c-Ha-ras (34). Comparison of the two sequences reveals that residues at 8 of 34 positions are identical, while an additional 6 positions show conservative[4] substitutions. Note that this region of c-Ha-ras includes domains 3 and 4, the "variable domain" and the carboxy terminal tetrapeptide. The carboxy terminal tetrapeptide of T_γ (-cys-val-ile-ser) shows strong homology to the fourth domains of the p21 family (-cys-val-leu-ser in c-Ha-ras) and yeast RAS (-cys-ile-ile-ser in RAS2).

While these structural homologies are intriguing, we would very much like to identify the functions of structural domains in transducin subunits and ras proteins. Biochemical studies employing tryptic fragments of native T_α point the way to identifying structural correlates of some of these functions. Limited digestion with trypsin cleaves native T_α at three sites (24,25,38) to produce four peptides termed, in order from the amino terminus, P-I through P-IV (Fig. 2). Functional studies (25) of these four tryptic cleavage products have suggested roles for each in mediating the reversible interactions of T_α with $T_{\beta\gamma}$, photorhodopsin, GTP and PDE.

Trypsin treatment rapidly removes the P-I polypeptide from the amino terminus of T_α; the remaining 38-kDa

---

[4]By the Dayhoff criteria (37) summarized in the legend to Table 1.

polypeptide cannot exchange bound guanine nucleotide or hydrolyze GTP when reconstituted with photorhodopsin and intact $\beta\gamma$ (25). Fung's laboratory has recently shown that these functional defects result from inability of the 38-kDa polypeptide to interact with $\beta\gamma$.[5] Thus all or part of the 18-residue P-I peptide is required for interaction with $T_{\beta\gamma}$. Note that the vertebrate and yeast <u>ras</u> proteins lack the amino terminal 19-26 residues, corresponding to P-I, that precede the A region of homology in T_α. This suggests that these proteins will not be found to interact with components that correspond to $T_{\beta\gamma}$.

While the amino terminus of T_α is critical for binding to $T_{\beta\gamma}$, the polypeptide's extreme carboxy terminal sequence appears to regulate binding of holo-transducin ($\alpha\beta\gamma$) to the signal detector, photorhodopsin. This conclusion is based on experiments with pertussis toxin. This toxin catalyzes ADP-ribosylation of an amino acid near the carboxy terminus of T_α;[6] this covalent modification stabilizes binding of the α subunit to $\beta\gamma$ but markedly reduces the affinity of $\alpha\beta\gamma$ for photorhodopsin and thereby "uncouples" transducin from its detector component (30).

The specific domain (or domains) of T_α that interacts with the effector enzyme, PDE, has not been identified. Like the GTP-binding domains, however, the effector interaction site apparently does not require the 18 amino acid residues at the amino terminus (P-I), or the 40 residues at the carboxy terminus (P-IV). Prolonged trypsin treatment of T_α bound to a stable GTP analog produces a truncated polypeptide that has lost both P-I and P-IV. The truncated molecule (the PII-PIII polypeptide, comprising amino acid residues 19-310) is still capable of stimulating retinal PDE, although it cannot bind to $\beta\gamma$ or to photorhodopsin, and

---

[5] T_α serves as a substrate for ADP-ribosylation by pertussis toxin only in the presence of $T_{\beta\gamma}$. Even in the presence of $T_{\beta\gamma}$, however, the 38-kDa polypeptide is not ADP-ribosylated by pertussis toxin (B.K.-K. Fung, personal communication), despite the fact that the site ADP-ribosylated by the toxin is located at the carboxy terminus of the molecule (39).

[6] While ADP-ribose is attached to a site within the last five amino acids of the polypeptide, the specific residue has not been definitively identified (32,38,39). For a variety of reasons we suspect that the amino acid is cys-347, the fourth residue from the carboxy terminus (32).

therefore cannot exchange bound guanine nucleotide in a light and rhodopsin-dependent fashion (25).

The PII-PIII polypeptide retains tightly bound guanine nucleotide (25), as would be predicted from the fact that it contains all four Halliday regions of T_d (see Fig. 2). Indeed, bound Gpp(NH)p prevents cleavage of the native polypeptide at the tryptic cleavage site between P-II and P-III (25), which just precedes the third Halliday region (Fig. 2).

SPECULATION AND PREDICTIONS

The confluence of new information derived from studies of the ras oncogenes, yeast genetics, hormone action and retinal phototransduction presents exciting prospects for investigators in all these fields. Evidence summarized in this chapter allows a series of predictions for results of future research. Some of these predictions are general, some quite specific; all are speculative.

Detector and Effector Elements Associated with ras Proteins

Almost certainly, the vertebrate ras proteins will prove to serve as transducers of external chemical signals into intracellular messengers that regulate cell proliferation. The signal detector elements coupled to the ras proteins will probably turn out to include cell surface receptors for growth factors. To date, none of the growth factor receptors has been definitively identified as a detector protein coupled to the ras proteins. Instead, several known growth factors are currently thought to act by stimulating a tyrosine protein kinase activity intrinsic to their cell surface receptors. Kamata and Feramisco (40) recently reported that two of these growth factors, insulin and epidermal growth factor, enhance binding of guanine nucleotides by p21-ras proteins in isolated membranes. If receptors for these two ligands are coupled to p21-ras, the relation between this coupling mechanism and their ability to phosphorylate tyrosine will have to be determined.

Although detector elements associated with the yeast RAS proteins have not been identified, further genetic and biochemical investigation will probably identify receptor proteins that detect nutritional signals. Characterization of these proteins may provide clues that will help in identifying homologous detector elements in mammalian cells.

The effector components coupled to p21-ras are at present equally mysterious. Toda et al. (5) suggested that p21-ras may act by stimulating adenylate cyclase, based upon the analogy with yeast RAS and the observation that the c-Ha-ras gene can complement deficiency of the yeast RAS proteins. This seems unlikely. Aside from the fact that cAMP is not a universal stimulator of proliferation in mammalian cells, transformation with the Ki-ras oncogene actually reduces cAMP content of transformed cells (41). It appears more likely that evolution has preserved the proliferation-promoting function of the ras proteins but has done so by coupling them to different effectors. The possibility that p21-ras and cellular homologs of other oncogenes interact with phosphoinositide metabolism (9) remains to be explored.

Functional and Structural Domains of ras Molecules

Based on the amino acid sequence homologies and functional studies of tryptic fragments of T_α (summarized in Fig. 2), we can predict which domains of the ras proteins will be found to interact with other components of the signal transduction cascade. These predictions include:

Interaction with guanine nucleotide. The four Halliday regions of amino acid sequence homology in both p21-ras and the yeast RAS proteins will be found to contribute to formation of a binding pocket for GTP. As suggested by mutational substitution of a critical glycine in in region A (5,19), this region will be found to mediate or control GTP hydrolysis in all the GTP-binding proteins.

Interaction with $\beta\gamma$. The initial 18-20 residues of T_α will be found to contribute to binding transducin's $\beta\gamma$ component. Because such an amino terminal sequence is lacking in the ras proteins, the action of these proteins will be found to not to require a $\beta\gamma$ component.

Interaction with the detector component. In the case of transducin, both the extreme carboxy terminus of the α subunit and associated $\beta\gamma$ are required for association with the signal detector, photorhodopsin. If the prediction that ras proteins do not associate with a $\beta\gamma$ component is correct, then we must determine which ras domain (or domains) performs the detector-binding function of $\beta\gamma$. As noted above, domains 3 and 4 of the ras proteins -- the "variable" region and the conserved tetrapeptide sequence at the carboxy terminus -- exhibit sequence homology with the carboxy termini of both T_α (in the terminal tetrapeptide) and T_γ (weak but significant homology of the last 34 residues of T_γ to the peptide segment encoded by the last exon of c-Ha-ras). Accordingly, we predict that these carboxy terminal domains of the p21-ras and yeast RAS proteins will prove to interact with their respective detector components. Although domain 3 shows considerable variation in amino acid sequence among the different ras proteins, these sequences are conserved between species for each of the three known mammalian ras genes, suggesting that they are subject to evolutionary pressure. Such pressure may preserve specificity of interaction with different detector components on the cell surface.

Interaction with the effector component. Interaction of T_α with its effector enzyme requires neither $\beta\gamma$ nor the terminal 40 amino acids of the α subunit. Accordingly, we predict that the corresponding homologous region of the ras proteins (the carboxy terminus and at least a portion of the variable domain) is not required for interaction with the putative effector.

Evolution of GTP-binding signal transducers

In addition to the obvious functional homologies, we have noted homologies between amino acid sequences of transducin's α and γ subunits, on the one hand, and those of p21-ras and the yeast RAS proteins, on the other. The apparent sequence homology is stronger for the amino terminal GTP-binding domains than for the carboxy terminal region of the ras proteins and the carboxy terminus of T_γ. The latter homology is probably real, however, as sugésted by the 44% homology between the DNA base sequences of exon 4 of the c-Ha-ras1 proto-oncogene and the corresponding segment of the cDNA encoding T_γ (34).

Accordingly, it appears likely that both the ras and G/transducin protein families derive from a common precursor in evolution. Such a precursor protein could have evolved directly into the ras proteins of yeast, Drosophila, and vertebrates. Duplication of the precursor gene could have given rise to portions of the G/transducin subunit genes, one duplicated gene evolving to encode α subunits and the other to encode γ. Cloning of genes encoding α, β, and γ subunits of G_s and G_i and of genes that encode cognate signal transducing proteins in invertebrates may confirm or contradict this speculative scenario.

REFERENCES

1. Matsumoto K, Uno I, Oshima Y, Ishikawa T (1982). Isolation and characterization of yeast mutants deficient in adenylate cyclase and cAMP-dependent protein kinase. Proc Natl Acad Sci USA 79:2355.
2. Matsumoto K, Uno I, Ishikawa T (1983). Initiation of meiosis in yeast mutants defective in adenylate cyclase and cyclic AMP-dependent protein kinase. Cell 32:417.
3. Matsumoto K, Uno I, Ishikawa T (1983). Control of cell division in Saccharomyces cerevisiae mutants defective in adenylate cyclase and cAMP-dependent protein kinase. Exp Cell Res 146:151.
4. Matsumoto K, Uno I, Ishikawa T (1984). Identification of the structural gene and nonsense alleles for adenylate cyclase in Saccharomyces cerevisiae. J Bacteriol 157:277.
5. Toda T, Uno I, Ishikawa T, Powers S, Kataoka T, Broek D, Cameron S, Broach J, Matsumoto K, Wigler M (1985). In yeast, RAS proteins are controlling elements of adenylate cyclase. Cell 40:27.
6. DeFeo-Jones D, Scolnick EM, Koller R, Dhar R (1983). ras-Related gene sequences identified and isolated from Saccharomyces cerevisiae. Nature 306:707.
7. Powers S, Kataoka T, Fasano O, Goldfarb M, Strathern J, Broach J, Wigler M (1984). Genes in S. cerevisiae encoding proteins with domains homologous to the mammalian ras proteins. Cell 36:607.
8. Gilman AG (1984). G proteins and dual control of adenylate cyclase. Cell 36:577.
9. Berridge MJ, Irvine RF (1984). Inositol trisphosphate, a novel second messenger in signal transduction. Nature 312:315.
10. Stryer L (1983). Transducin and the cyclic GMP phosphodiesterase: Amplifier proteins in vision. Cold Spring Harbor Symp Quant Biol 48:841.
11. Casperson GF, Walker N, Brasier AR, Bourne HR (1983). A guanine nucleotide sensitive adenylate cyclase in the yeast Saccharomyces cerevisiae. J Biol Chem 258:7911.
12. Ross EM, Howlett AC, Ferguson KM, Gilman AG (1978). Reconstitution of hormone-sensitive adenylate cyclase activity with resolved components of the enzyme. J Biol Chem 253:6401.
13. Ross EM, Gilman AG (1980). Biochemical properties of

hormone-sensitive adenylate cyclase. Ann Rev Biochem 49:533.
14. Seamon K, Daly JW (1981). Activation of adenylate cyclase by the diterpene forskolin does not require the guanine nucleotide regulatory protein. J Biol Chem 257:9799.
15. Kataoka T, Powers S, Cameron S, Fasano O, Goldfarb M, Broach M, Wigler M (1985). Functional homology of mammalian and yeast RAS genes. Cell 40:19.
16. Casperson GF, Walker N, Bourne HR (1985). Isolation of the gene encoding adenylate cyclase in Saccharomyces cerevisiae. Proc Natl Acad Sci USA (in press).
17. Tamanoi F, Walsh M, Kataoka T, Wigler M (1984). A product of yeast RAS2 gene is a guanine nucleotide binding protein. Proc Natl Acad Sci USA 81:6924.
18. Temeles GL, Gibbs JB, D'Alonzo JS, Sigal IS, Scolnick EM (1985). Yeast and mammalian ras proteins have conserved biochemical properties. Nature 313:700.
19. McGrath JP, Capon DJ, Goeddel DV, Levinson AD (1984). Comparative biochemical properties of normal and activated human ras p21 protein. Nature 310:644.
20. Pringle JR, Hartwell LH (1981). The Saccharomyces cerevisiae cell cycle. In: Strathern J, Jones EW, Broach JR, eds. The Molecular Biology of the Yeast Saccharomyces - Life Cycle and Inheritance. Cold Spring Harbor, New York, Cold Spring Harbor Laboratory, p 97.
21. Shilo V, Simchen G, Shilo B (1978). Initiation of merosin in cell cycle initiation mutants of Saccharomyces cerevisiae. Exp Cell Res 112:241.
22. Boutelet F, Hilger F (1980). A mapping study on fourteen centromere-linked temperature sensitive mutations. In: 10th International Conference on Yeast Genetics & Molecular Biology. Louvain-la-Neuve, Belgium, p 177.
23. Matsumoto K, Uno I, Toh-E A, Ishikawa T, Oshima Y (1982). Cyclic AMP may not be involved in catabolite repression in Saccharomyces cerevisiae: Evidence from mutants capable of utilizing it as an adenine source. J Bacteriol 150:277.
24. Fung BK-K (1983). Characterization of transducin from bovine retinal rod outer segments. I. Separation and reconstitution of the subunits. J Biol Chem 258:10495.
25. Fung BK-K, Nash CR (1983). Characterization of transducin from bovine retinal rod outer segments. II. Evidence for distinct binding sites and conformational

changes revealed by limited proteolysis with trypsin. J Biol Chem 258:10503.
26. Manning DR, Gilman AG (1983). The regulatory components of adenylate cyclase and transducin. A family of structurally homologous guanine nucleotide binding proteins. J Biol Chem 258:7059.
27. Cassel D, Selinger Z (1977). Mechanism of adenylate cyclase activation by cholera toxin: inhibition of GTP hydrolysis at the regulatory site. Proc Natl Acad Sci USA 74:3307.
28. Bokoch GM, Katada T, Northup JK, Hewlett EL, Gilman AG (1983). Identification of the predominant substrate for ADP-ribosylation by islet-activating protein. J Biol Chem 258:2072.
29. Abood ME, Hurley JB, Pappone M-C, Bourne HR, Stryer L (1982). Functional homology between signal-coupling proteins: Cholera toxin inactivates the GTPase activity of transducin. J Biol Chem 257:10540.
30. Van Dop C, Yamanaka G, Steinberg F, Sekura RD, Manclark CR, Stryer L, Bourne HR (1984). ADP-ribosylation of transducin by pertussis toxin blocks the light-stimulated hydrolysis of GTP and cGMP in retinal photoreceptors. J Biol Chem 259:23.
31. Willumsen BM, Norris K, Papageorge AG, Hubbert NL, Lowy DR (1984). Harvey murine sarcoma virus p21 ras protein: biological and biochemical significance of the cysteine nearest the carboxy terminus. EMBO J 3:2581.
32. Medynski DC, Sullivan K, Smith D, Van Dop C, Chang FH, Fung BKK, Seeburg PH, Bourne HR (1985). Amino acid sequence of the alpha subunit of transducin deduced from the cDNA sequence. Proc Natl Acad Sci USA (in press).
33. Van Dop C, Medynski D, Sullivan K, Wu AM, Fung BK-K, Bourne HR (1984). Partial cDNA sequence of the gamma subunit of transducin. Biochem Biophys Res Comm 124:250.
34. Hurley JB, Fong HKW, Teplow DB, Dreyer WJ, Simon MI (1984). Isolation and characterization of a cDNA clone for the γ subunit of bovine retinal transducin. Proc Natl Acad Sci USA 81:6948.
35. Willumsen BM, Christensen A, Hubbert NL, Papageorge AG, Lovy DR (1984). The p21 ras C-terminus is required for transformation and membrane association. Nature 310:583.
36. Halliday K (1984). Regional homology in GTP-binding

proto-oncogene products and elongation factors. J Cyclic Nucl Res 9:435.
37. Dayhoff MO, Schwartz RM, Orcutt BC (1978). A model of evolutionary change in proteins. Atlas of Protein Sequence and Structure 5:345.
38. Hurley JB, Simon MI, Teplow DB, Robishaw JD, Gilman AG (1984). Homologies between signal transducing G proteins and ras gene products. Science 226:860.
39. Manning DR, Fraser BA, Kahn RA, Gilman AG (1984). ADP-ribosylation of transducin by islet-activating protein. Identification of asparagine as the site of ADP-ribosylation. J Biol Chem 259:749.
40. Kamata T, Feramisco JR (1984). Epidermal growth factor stimulates guanine nucleotide binding activity and phosphorylation of ras oncongene proteins. Nature 310:147.
41. Carchman RA, Johnson GS, Pastan I, Scolnick EM (1974). Studies on the levels of cyclic AMP in cells transformed by wild-type and temperature-sensitive Kirsten sarcoma virus. Cell 1:59.

ROLE OF CYCLIC AMP IN CELL DIVISION

Kunihiro Matsumoto[1], Isao Uno*, and
Tatsuo Ishikawa*

Department of Industrial Chemistry, Tottori University,
Tottori-shi, Tottori 680, Japan; *Institute of Applied
Microbiology, University of Tokyo, Bunkyo-ku,
Tokyo 113, Japan

ABSTRACT Three groups of cAMP-requiring mutants of yeast, cyr1, cyr2 and CYR3, were isolated. The cyr1 mutation caused the deficiency of adenylate cyclase activity. The cyr2 and CYR3 mutants were altered in the catalytic and regulatory subunits of cAMP-dependent protein kinase, respectively. These mutations were suppressed by the bcy1 mutation, which caused production of high level of cAMP-independent protein kinase. The cyr2 mutation was suppressed by the ppd1 mutation, which was defective in phosphoprotein phosphatase. The CYR3 mutation was suppressed by the pde1 and IAC mutations; pde1 was deficient in high Km phosphodiesterase, and IAC caused the production of a high level of cAMP. The cAMP-requiring mutants were arrested at the G1 phase of the mitotic cell cycle in the absence of cAMP. The temperature-sensitive cyr1, cyr2 and CYR3 mutations permitted the initiation of meiosis, but resulted in the frequent production of two-spored asci. In diploids homozygous for the bcy1 and ppd1 mutations, no premeiotic DNA replication occurred, and no spores were formed.

---

[1]Present address: DNAX Research Institute of Molecular and Cellular Biology, Palo Alto, California 94304, U. S. A.

These data suggest that cAMP works as a positive effector at the start of yeast mitosis via the activation of cAMP-dependent protein kinase, but that it does as a negative effector at the start of meiosis.

INTRODUCTION

Cyclic AMP has been shown to play important roles in the control of various metabolic processes in eukaryotes as well as prokaryotes (1). Biochemical studies have identified at least some of the enzymes and regulatory proteins involved in the cAMP metabolism. However, a detailed understanding of the mechanisms involved in the cAMP action at the cellular level has not yet been attained particularly in eukaryotes. We have been studying yeast mutants which are altered in cAMP metabolism, in expectation that the powerful genetic approaches possible in yeast will enable us to understand cAMP function in this organism. Specific biochemical changes caused by the mutation can then be correlated altered cellular functions such as mitosis and meiosis. In this report, we discuss the major progress that has been made in genetic analyses of cAMP-requiring mutants and their suppressors, and studies of cAMP functions in yeast cells.

ISOLATION AND CHARACTERIZATION OF CYCLIC AMP-REQUIRING MUTANTS AND THEIR SUPPRESSORS

The cAMP-requiring Mutants.

Cyclic AMP is synthesized from ATP by adenylate cyclase, and degraded to 5'-AMP by phosphodiesterase. In eukaryotic cells, it is known that cAMP binds with the regulatory subunit of protein kinase and activates the enzyme. The catalytic subunit of this enzyme phosphorylates cellular proteins, and phosphorylated proteins may exert various biological functions. Therefore, if we induce mutations at any one of these reactions, we may be able to obtain mutants related to the cAMP metabolism. Since we found that adenine-requiring mutants of yeast were unable to take up cAMP, mutants which are able to utilize cAMP as the adenine source were isolated from an ade6 ade10 double mutant (2). The mutant phenotype was proven to be due to the simultaneous occurrence of triple mutations designated as cam1, cam2 and cam3. To isolate cAMP-requiring mutants, cells of the cam mutant strain which can take up cAMP as an adenine source were subjected to ethylmethane sulfonate mutagenesis and the nystatin selection procedure (3). Each of the mutants obtained was crossed to the parental cam strain and the resulting diploids were tested for their cAMP-requirement. Most diploids thus constructed grew without supplementing cAMP, indicating that these mutations are recessive to the wild-type counterparts. Complementation tests among all recessive mutants showed the presence of two distinct genes. We call these cyr1 and cyr2. One other mutant group was dominant and called CYR3.

The cyr1 mutants had no detectable adenylate cyclase activity. Among these mutants, temperature-sensitve cyr1 alleles which grew at $25^\circ C$ but not at $35^\circ C$ in the absence of cAMP have been obtained (Table 1). We obtained several evidences to show that the cyr1 mutants carry lesions in the structural gene for adenylate cyclase (4). First, adenylate cyclase obtained from the temperature-sensitive mutants was thermolabile. To exclude a possibility that the alteration of plasma membrane causes the thermolability of mutant enzyme, adenylate cyclase was solubilized by Luburol, but the mutant enzyme was still thermolabile. Second, three cyr1 mutants which carry the nonsense mutation susceptible to UAA ochre suppressors were isolated. This suggests that the CYR1 gene product is the enzyme protein. Third, adenylate cyclase activity observed in tetraploid strains carrying different dosages of the CYR1 gene was proportional to the dosage of the active CYR1 gene. Finally, the CYR1 gene has been cloned by complementation of tsm0185 (cdc35) and cyr1 mutations (5). Adenylate cyclase activity of products from the cloned CYR1 gene in yeast and Escherichia coli cells was confirmed (6).

The cyr2 and CYR3 mutants required cAMP for growth at higher temperature. No significant differences were found in the levels of adenylate cyclase and phosphodiesterase activities compared between these mutants and wild types (Table 1), but abnormality in cAMP-dependent protein kinase was observed in these mutants. The cAMP-dependent protein kinase activity of these mutant cells showed greater K_a value for activation by cAMP at $35^\circ C$ than that of the wild-type cells. The protein kinase levels of cyr2 cells assayed at $35^\circ C$ was very low, and the DEAE-Sephacel elution profile of the cyr2 enzyme was markedly different from that observed for the wild-type enzyme (7). The free catalytic subunit of the cyr2 mutant enzyme showed extremely decreased affinity for ATP and was more thermolabile compared with that of the wild-type enzyme. On the other hand, the electrophoretic mobility of regulatory subunit of the CYR3 enzyme was clearly different from that of the wild-type enzyme (8). These results indicated that the cyr2 and CYR3 mutants were altered in the catalytic and regulatory subunits of cAMP-dependent protein kinase, respectively.

TABLE 1
COMPARISON OF CYCLIC AMP LEVELS, ADENYLATE CYCLASE AND PHOSPHODIESTERASE ACTIVITIES IN WILD-TYPE AND MUTANT STRAINS

Strain	Growth condition	Adenylate cyclase[a]	phosphodiesterase[a]	cAMP level[a]
+	$25°C$	11.4	75	1.0
	$35°C$	19.2	92	1.4
cyr1-2	$27°C$	0.3	157	0.4
	$35°C$+cAMP	0.0	92	-[b]
cyr2	$27°C$	9.6	69	1.0
	$35°C$+cAMP	12.0	92	-
CYR3	$27°C$	7.8	63	2.1
	$35°C$+cAMP	10.4	70	-
bcy1	$30°C$	10.9	118	1.9
ppd1	$27°C$	15.3	119	2.0
pde1	$27°C$	7.7	5	6.7
IAC	$27°C$	22.4	52	10.1

[a] Yeast cells were grown with or without cAMP (1.0 mM). Enzyme activities were expressed as units per mg protein. Cyclic AMP levels were expressed as pmol per mg protein.
[b] Cyclic AMP levels could not be measured because the culture medium contained cAMP.

The mutant cells may produce a normal level of cAMP, but growth of these mutant cells will be limited without supplementing cAMP, because a large amount of cAMP is required for the activation of the mutant enzymes, judging from the activation constant of the mutant enzymes.

Suppressors of cAMP-requiring Mutations.

The most suppressors of the primary mutation circumvent the primary genetic lesion indirectly by substituting for the function of the product of the first gene, or providing cytoplasmic conditions which affect the level of the product of the first mutation. Such suppressors for cAMP-requiring mutations were isolated as revertants from the cAMP-requiring mutants, and characterized by genetic and biochemical analyses.

By analyzing revertants obtained from cyr1 mutants, we have found a suppressor mutation, bcy1, that bypassed the cyr1 defect without restoring the adenylate cyclase activity (3). Chromatographic analyses indicated that the bcy1 mutants had extremely low levels of cAMP-binding activity and cAMP-dependent protein kinase but produced a high level of cAMP-independent protein kinase. It seems likely that the bcy1 mutation results in the production of free catalytic subunits of this enzyme because of the lack of the regulatory subunit or an altered enzyme complex which is active in the absence of cAMP. The bcy1 mutation was recessive and suppressed not only cry1 but also cyr2 and CYR3 and ras1 ras2. The bcy1 mutant cells carrying these mutations were able to grow in the absence of cAMP, because the cells always have active cAMP-independent protein kinase.

The ppd1 mutant was isolated as a suppressor of the cyr2 mutation which caused alteration of the catalytic subunit of cAMP-dependent protein kinase (9). Three peaks of phosphoprotein phosphatase activity were identified by DEAE-Sephacel chromatography of crude extracts of wild-type strain. The ppd1 mutant was deficient in one of the peaks of phosphoprotein phosphatase. Since it has been found that the ppd1 mutation resulted in the decreased dephosphorylation of cellular proteins which have been phosphorylated by cAMP-dependent protein kinase, the suppressor activity of the ppd1 mutation may be exerted by the accumulation of certain levels of phosphorylated proteins.

Since the CYR3 mutant required a high level of cAMP for activation of cAMP-dependent protein kinase, a mutation that results in the overproduction of cAMP may suppress the

CYR3 phenotype (Table 1). One of such type of mutations was pde1 which was deficient in high Km phosphodiesterase (10). The other type was dominant IAC which had high adenylate cyclase activity but normal level of phosphodiesterase (8). The pde1 and IAC cells accumulated several times over the intracellular cAMP found in wild-type cells (Table 1). The cAMP levels accumulated in pde1 CYR3 and IAC CYR3 double mutant cells may be enough to activate altered cAMP-dependent protein kinase of the CYR3 strain.

The Relationship between cAMP-requiring Mutations and Suppressors.

Table 2 summarizes the relationship between the cAMP-requiring mutations and their suppressors isolated to date.

TABLE 2
THE RELATIONSHIP BETWEEN CYCLIC AMP-REQUIRING MUTATIONS AND THEIR SUPPRESSORS

Mutation	Suppressor				Defect[a]
	bcy1	ppd1	pde1	IAC	
cyr1	+	−	−	−	AC(C)
ras1 ras2	+	−	?	−	AC(R)
cyr2	+	+	+	+	PK(C)
CYR3	+	−	+	+	PK(R)
Defect[a]	PK(R)	PPase	PDE	AC(C)	

[a]AC, adenylate cyclase; PK, protein kinase; PPase, phosphoprotein phosphatase; PDE, phosphodiesterase; (C), catalytic subunit; (R), regulatory subunit.

CYR1 and IAC may be the gene for the catalytic subunit of adenylate cyclase, and RAS1 and RAS2 may be genes for regulatory proteins for the same enzyme (11). CYR2 is the gene for the catalytic subunit of protein kinase, and CYR3 and bcy1 may be genes related with the regulatory subunit of protein kinase. PPD1 is the gene for one of phosphoprotein phosphatases, and pde1 may be a gene for one of phosphodiesterases. The bcy1 mutation suppressed all these cAMP-requiring mutations, but ppd1, pde1 and IAC showed no suppressor activity in some combinations. The mutations that we have isolated to date corresponded most enzymes involved in cAMP metabolism.

ROLES OF CYCLIC AMP IN CELLULAR FUNCTIONS

Cyclic AMP is involved in mitosis.

Examination of the proportion of unbudded cells, terminal nuclear phenotype and DNA content of nuclei indicated that the cyr1 mutants were arrested at the G1 phase of the cell cycle in the absence of cAMP (3). The temperature-sensitive cyr1 mutant cells were a kind of conditional start mutants and arrested before the mating pheromone-sensitive step at the restrictive temperature. The phenotypes of the cyr1 mutants mimicked those of nutritionally limited cells which are arrested at a step within the class-II region of the G1 phase (12). The bcy1 and ppd1 mutant cells were able to bypass the class-II arrest caused by the cyr1 mutation or nutritional limitation, and continued bud emergence for multiple cycles attaining multiple buds. From these data we concluded that cAMP works as a positive effector at the start of a yeast cell cycle via activation of cAMP-dependent protein kinase.

Cyclic AMP is involved in Conjugation.

The different classes of G1 phase mutations have been distinguished by the level of conjugational competence (13). The cdc28 mutant which is one of well known class-I start mutant is known to show low mating efficiency (14). In contrast the temperature-sensitive cyr1 cells did not retain the capacity to conjugate at a restrictive temperature (12). Based on the criteria to distinguish two classes of start mutants, the cyr1 mutants fall in the class-II group. The bcy1 and bcy1 cyr1 mutants were able to conjugate, indicating that the bcy1 mutation can suppress the class-II arrest leading to inability of conjugation caused by the cyr1 mutation and nutritional limitation. From these data, we concluded that cAMP works as a positive effector for conjugation of yeast cells.

Cyclic AMP Works as a Negative Effector on the Initiation of Meiosis.

The temperature-sensitive cyr1, cyr2 and CYR3 mutations permitted the initiation of meiosis at the restrictive temperature.

TABLE 3
SPORULATION OF CYCLIC AMP MUTANTS

Diploid strain	Sporulation		Frequency of each ascus type per total asci (%)		
	Temperature	Efficiency (%)	1-Spored	2-Spored	3,4-Spored
+/+	25	63	1	15	84
	33.5	45	3	25	72
cyr1-2/cyr1-2	25	44	12	46	42
	33.5	28	21	60	19
cyr2/cyr2	25	61	5	24	72
	33.5	55	5	49	39
CYR3/CYR3	25	34	4	27	70
	33.5	19	11	45	44
bcy1/bcy1	30	0	0	0	0
ppd1/ppd1	30	0	0	0	0

Unlike the wild-type cells, these mutant cells were capable of initiating meiosis even in nutrient growth media (15).

This unique feature of the cAMP-requiring mutants suggests that these mutations relate to the choice between mitotic and meiotic processes. Since cyr1, cyr2 and CYR3 homozygous diploid cells were able to sporulate at the restrictive temperature (Table 3), it is suggested that the initiation of meiosis occurs at the start step in the mitotic G1 phase. In diploids homozygous for the bcy1 or ppd1 mutation, no premeiotic DNA replication and commitment to intragenic recombination occurred, and no spores were formed (Table 3). From these data we concluded that the initiation of meiosis may be dependent upon the repression of cAMP production and the inactivation of cAMP-dependent protein kinase.

Cyclic AMP is Involved in Sporulation Events.

The cyr1, cyr2 and CYR3 mutations permitted the initiation of meiosis, but resulted in the frequent production of two-spored asci at the restrictive temperature (Table 3). Fluorescent microscopy of the mutant diploids incubated at the restrictive temperature revealed that four meiotic products are formed but only two nuclei are enclosed in each of the two spores after meiosis (16). Dyad analysis of asci indicated that meiotic products were randomly included into ascospores. About half of sporulating mutant diploid cells showed normal meiosis I producing two normal spindle pole bodies, but the other half exhibited abnormal meiosis I producing one normal spindle pole body with inner and outer plaques, and one defective spindle pole body without outer plaque. At meiosis II, a part of cells contained a pair of normal spindle pole bodies with both plaques and prospore wall. Two spores formed in this type of cells should contain sister nuclei. The other part of the cells contained pairs of spindle pole bodies; normal ones with both plaques and prospore wall, and abnormal ones without outer plaque and prospore wall. Two spores formed in this type of cells should contain nonsister nuclei. Thus, the enclosure of the products of meiosis in two spores of mutant diploid strains appears to be random with respect to the distribution of haploid genomes. The cAMP levels of wild-type and mutant diploid cells incubated at the permissive temperature were first decreased in presporulation medium and then increased by incubation in sporulation medium, but those in mutant diploid cells incubated at the restrictive temperature remained at an extremely low level in sporulation medium (16). These evidences indicated the connection between the availability of cAMP and the ability to assemble an outer plaque during sporulation. From these data we concluded that cAMP is required for the enclosure of all meiotic products in prospore walls.

DISCUSSION

Identification of the mutant products in cAMP-requiring mutants and the suppressor mutants was able to demonstrate that the cAMP levels are regulated by adenylate cyclase and phosphodiesterase activities, and cAMP-dependent protein kinase is an essential mediator of certain actions of cAMP in yeast cells. The mutant studies revealed that cAMP-dependent phosphorylation is involved in the G1 phase of the cell cycle, conjugation, and the post-meiotic stage of sporulation, but inhibition of cAMP-dependent protein phosphorylation is required to induce meiotic division. Although we have demonstrated the participation of cAMP in mitosis, conjugation and sporulation, the exact roles of cAMP-dependent phosphorylation in these processes are virtually unknown. Further work will be focused on the identification of the nature of the proteins which are phosphorylated by cAMP-dependent protein kinase in these processes.

REFERENCES

1. Robison GA, Butcher RW, Sutherland EW (1971). "Cyclic AMP." New York: Academic Press.
2. Matsumoto K, Uno I, Toh-e A, Ishikawa T, Oshima Y (1982). Cyclic AMP may not be involved in catabolite repression in Saccharomyces cerevisiae: Evidence from mutants capable of utilizing it as an adenine source. J Bacteriol 150:277.
3. Matsumoto K, Uno I, Oshima Y, Ishikawa T (1982). Isolation and characterization of yeast mutants deficient in adenylate cyclase and cAMP-dependent protein kinase. Proc Natl Acad Sci USA 79:2355.
4. Matsumoto K, Uno I, Ishikawa T (1984). Identification of the structural gene and nonsense alleles for adenylate cyclase in Saccharomyces cerevisiae. J Bacteriol 157:277.

5. Masson P, Jacquemin JM, Culot M (1984). Molecular cloning of the tsm0185 gene responsible for adenylate cyclase activity in Saccharomyces cerevisiae. Ann Microbiol (Inst Pasteur) 135:343.
6. Uno I, Matsumoto K, Ishikawa T (1985). The function of the CYR1 and RAS genes in yeast. This symposium.
7. Uno I, Matsumoto K, Adachi K, Ishikawa T (1984). Characterization of cyclic AMP-requiring yeast mutants altered in the catalytic subunit of protein kinase. J Biol Chem 259:12508.
8. Uno I, Matsumoto K, Ishikawa T (1983). Characterization of cyclic AMP-requiring yeast mutants altered in the regulatory subunit of protein kinase. J Biol Chem 257:14110.
9. Matsumoto K, Uno I, Kato K, Ishikawa T (1985). Isolation and characterization of a phosphoprotein phosphatase-deficient mutant in yeast. Yeast, in press.
10. Uno I, Matsumoto K, Ishikawa T (1983). Characterization of a cyclic nucleotide phosphodiesterase-deficient mutant in yeast. J Biol Chem 258:3539.
11. Toda T, Uno I, Ishikawa T, Powers S, Kataoka T, Broek D, Cameron S, Broach J, Matsumoto K, Wigler M (1985). In yeast, RAS proteins are controlling elements of adenylate cyclase. Cell 40:27.
12. Matsumoto K, Uno I, Ishikawa T (1983). Control of cell division in Saccharomyces cerevisiae mutants defective in adenylate cyclase and cAMP-dependent protein kinase. Exp Cell Res 146:151.
13. Reed SI (1980). The selection of S. cerevisiae mutants defective in the start event of cell division. Genetics 95:561.
14. Reid BJ, Hartwell LH (1977). Regulation of mating in the cell cycle of Saccharomyces cerevisiae. J Cell Biol 75:355.
15. Matsumoto K, Uno I, Ishikawa T (1983). Initiation of meiosis in yeast mutants defective in adenylate cyclase and cyclic AMP-dependent protein kinase. Cell 32:417.
16. Uno I, Matumoto K, Hirata A, Ishikawa T (1985). Outer plaque assembly and spore encapsulation are defective during sporulation of adenylate cyclase-deficient mutants of Saccharomyces cerevisiae. J Cell Biol, in press.

FUNCTIONS OF THE CYR1 AND RAS GENES IN YEAST

Isao Uno, Kunihiro Matsumoto*[1], and
Tatsuo Ishikawa

Institute of Applied Microbiology, University of Tokyo, Bunkyo-ku, Tokyo 113, Japan; *Department of Industrial Chemistry, Tottori University, Tottori-shi, Tottori 680, Japan

ABSTRACT The relationship between the cyr1 and ras mutations of Saccharomyces cerevisiae in production of cAMP was studied. Cloned CYR1 and RAS2 genes were expressed in S. cerevisiae and Escherichia coli cells. The products of the CYR1 and RAS2 genes reconstituted GTP-dependent adenylate cyclase in vivo and in vitro. The results suggest that yeast GTP-dependent adenylate cyclase is composed of catalytic and regulatory proteins encoded by the CYR1 and RAS2 genes, respectively.

INTRODUCTION

Regulatory roles of cAMP in yeast, Saccharomyces cerevisiae, have been studied by isolation of mutants requiring cAMP and their suppressors (1). Identification of the altered products in these mutant cells was able to show that the cAMP levels are regulated by adenylate cyclase and phosphodiesterase activities, and cAMP-dependent protein kinase is an essential mediator of cAMP actions in yeast cells (1-4). Among these mutants obtained, the cyr1 mutants possessed no detectable level of adenylate cyclase activity (1,5), and the growth of cells carrying the cyr1 mutation was arrested at the G1 phase of cell cycle in the

---

[1]Present address: DNAX Research Institute of Molecular and Cellular Biology, Palo Alto, California 94304

absence of cAMP (6). The G1 arrest by the cyr1 mutation
was bypassed by the bcy1 mutation. The bcy1 mutant had no
regulatory subunit of cAMP-dependent protein kinase, but
produced a high level of cAMP-independent one (1). Thus,
we were able to show that cAMP is an essential factor for
yeast cells to proceed through the cell cycle via activation of protein kinase.

It has been recently found that the bcy1 mutation can
suppress the lethality exerted by the ras1 ras2 mutations,
and that the phenotype of RAS2^{val19}, a RAS2 allele with a
missense mutation, resembles that of the bcy1 cells (7).
The ras genes are a highly conserved family of genes first
discovered as the oncogenes of rat sarcoma virus (8).
Mutant ras genes encoding altered proteins are found in
many tumor cells and are capable of the morphological and
tumorigenic tranformation of mouse cells (9-12). The yeast
contains two closely related, but distinct genes, RAS1 and
RAS2. These genes encode proteins that are highly homologous to the mammalian ras proteins (13-15). Neither RAS1
nor RAS2 is by itself an essential gene, but ras1 ras2
mutant cells were unable to survive (16-18) like the cyr1
mutants.

It has been reported that yeast adenylate cyclase is
composed of at least two protein components; one mediates
the catalytic activity of adenylate cyclase, and the other
is responsible for its regulation by guanine nucleotides
(19). There are evidences to show that the cyr1 gene encodes the catalytic subunit of yeast adenylate cyclase (5),
and the RAS gene product is GTP-binding protein (20) and
modulates adenylate cyclase activity (7). In this report,
we tried to characterize the CYR1 and RAS2 gene products by
cloning these genes in S. cerevisiae and Escherichia coli.

THE RELATIONSHIP BETWEEN CYR1 AND RAS MUTATIONS

The cyr1 mutant did not grow in the absence of cAMP,
but the cyr1 bcy1 double mutant was able to grow on minimal
medium (Table 1). Upon encountering nutritional conditions
that are unsatisfactory for growth, prototrophic yeast
cells are arrested in cell division at its onset in the G1
phase of the cell cycle (6). The cyr1 mutant cells
mimicked those of nutritionally limited cells which are
arrested at a step of the G1 phase. The bcy1 mutant cells
did not respond to nutritional limitation and were able to
bypass the G1 arrest caused by the nutritional limitation

(21). The bcy1 mutation suppressed the cyr1 mutation and bypassed the need for cAMP in the cyr1 mutant cells (1). By crossing ras1 bcy1 with ras2 or ras2 bcy1 with ras1, we were able to isolate ras1 ras2 bcy1 (7).

TABLE 1
GROWTH AND SPORULATION PHENOTYPES OF
cyr1, bcy1 AND ras MUTANTS

Strain	Growth on minimal medium	G1 arrest by nutitional limitation[a]	Sporulation[b]
+	+	+	+
cyr1	–		+
bcy1	+	–	–
cyr1 bcy1	+	–	–
ras1 ras2 bcy1	+	–	–
RAS2^{val19}	+	–	–

[a] +, Cells were arrested at G1 phase by nutritional starvation; –, not arrested.
[b] +, Each homozygous diploid cells sporulated in sporulation medium; –, did not sporulate. The diploid cyr1/cyr1 cells sporulated even in rich medium (21).

Unlike the wild-type strain, bcy1 and RAS2^{val19} cells were not arrested at the G1 phase by nutritional limitation (Table 1). Wild-type diploid cells sporulated well in sporulation medium, but not in rich medium, but the homozygous cyr1/cyr1 and ras2/ras2 diploid cells sporulated in rich medium. In contrast, diploid cells containing bcy1 or RAS2^{val19} failed to sporulate even in sporulation medium (Table 1).

The cyr1-2 mutant which produced thermolabile adenylate cyclase and a low level of cAMP at the permissive temperature showed a low level of trehalase activation and accumulated a large amount of trehalose (22). The bcy1 and

$RAS2^{val19}$ mutants produced high levels of active trehalase and accumulated small amounts of trehalose (7,22). All these experimental data indicate striking similarities between cyr1 and ras1 ras2, and between bcy1 and $RAS2^{val19}$, suggesting that the effector pathways of both CYR1 and RAS genes relate with the cAMP-cascade reaction. It is possible to consider that the yeast RAS gene products modulate adenylate cyclase activity. In fact, the yeast RAS2 gene product, like the mammalian ras proteins, is a GTP-binding protein (20), and several kinds of GTP-binding protein modulate adenylate cyclase of mammalian cells. Moreover, the yeast adenylate cyclase is modulated by GTP (5,19).

To test whether RAS proteins modulate adenylate cyclase, intracellular cAMP levels of yeast mutants carrying mutant RAS genes were examined.

TABLE 2
CYCLIC AMP LEVELS OF YEAST MUTANTS AND TRANSFORMANTS CARRYING CYR1

Relevant genotype	cAMP level (pmol/mg protein)
+	2.1
cyr1	$-^b$
bcy1	2.0
cyr1 bcy1	0.0
ras1	1.8
ras2	0.3
ras1 ras2 bcy1	0.0
$RAS2^{val19}$	7.6
cyr1/pCEY710[a]	4.7
cyr1/pHM9[a]	6.5

[a] The cyr1 cells carrying plasmids, pCEY710 or pHM9.
[b] cAMP level could not be measured because the culture medium contained cAMP.

As shown in Table 2, the ras1 RAS2 strain had only slightly depressed cAMP level, while the RAS1 ras2 strain had a low level of cAMP comparing to the RAS1 RAS2 strains. In the cyr1 bcy1 and ras1 ras2 bcy1 strains, cAMP levels were negligible. Cyclic AMP levels of RAS2^{val19} strains were increased about 4-fold, comparing with the RAS1 RAS2 strains. These results indicate that the RAS2 gene is the major determinant of cAMP level, while the RAS1 gene may only be a monor determinant.

CLONING AND EXPRESSION OF CYR1 AND RAS2 GENES

Plasmid pCEY710 was originally cloned by the ability to complement the tsm0185 mutation, a temperature-sensitive cell division cycle mutation (23). When pCEY710 plasmid was introduced into yeast cyr1 cells, which required cAMP for growth, transformant cells (cyr1/pCEY710) could produce cAMP and grow without cAMP (Table 2). This result indicates that pCEY710 carries the CYR1 gene. Further transformants carrying pHM9 (cyr1/pHM9), which contains 3.1-Kb Bgl II fragment from pCEY710, also produced cAMP (Table 2).
To examine whether the CYR1 gene could be expressed in E. coli cells, pHM10 was constructed by recloning the CYR1 gene into pACY184, which is compatible with pBR322, and introduced by transformation into E. coli Δcya mutant strain (CA8306), which has deletion in the structural gene of adenylate cyclase. Production of cAMP was not observed in transformant cells (Δcya/pHM10) carrying the CYR1 gene.
The RAS2 and RAS2^{val19} genes were cloned into E. coli expression vector carrying lacUV5 promoter, whose promoter activity was cAMP-independent, by substituting the RAS2 or RAS2^{val19} gene for IFN-γ gene of pGIF5, yielding placRAS2 or placRAS2^{val19}, respectively. Transformant cells carrying placRAS2 or placRAS2^{val19} could not produce cAMP (Table 3). Two plasmids, pHM10 and either placRAS2 or placRAS2^{val19}, were introduced sequentially by transformation into the Δcya strain of E. coli. Transformant cells carrying pHM10 and placRAS2 produced significant amount of cAMP, and those carrying pHM10 and placRAS2^{val19} produced larger amount of cAMP (Table 3). To confirm the cAMP accumulation in these transformants, β-galactosidase induction by the addition of IPTG was examined. Induction of β-galactosidase by IPTG could be clearly observed in transformants carrying pHM10 and placRAS2 or placRAS2^{val19},

but not in other transformants. The results indicate that yeast CYR1 and RAS2 genes are expressed in E. coli cells.

TABLE 3
CYCLIC AMP LEVELS OF E. COLI TRANSFORMANTS CARRYING CYR1 AND RAS2

Strain or transformant	cAMP level (pmol/mg protein)
CYA	24.1
Δcya	0.0
Δcya/pHM10	0.0
Δcya/placRAS2	0.0
Δcya/placRAS2^{val19}	0.0
Δcya/pHM10,placRAS2	1.5
Δcya/pHM10,placRAS2^{val19}	4.6

RECONSTITUTION OF GTP-DEPENDENT ADENYLATE CYCLASE IN VITRO

The properties of adenylate cyclase of membrane fractions prepared from strains carrying the cyr1 or ras mutations were examined. Membrane fractions from the wild-type cells had high adenylate cyclase activity, when assayed in the presence of manganese ions, but low activity in the presence of magnesium ions (Table 4). Adenylate cyclase activity from the wild-type cells assayed with magnesium ions was stimulated significantly by the addition of Gpp(NH)p, a nonhydrolyzable GTP analogue (Table 4). All yeast strains except cyr1 had equivalent amount of the catalytic subunit of adenylate cyclase which was detected by the assay in the presence of manganese ions. Membrane fractions from ras1 ras2 bcy1 lacked GTP-stimulated adenylate cyclase activity present in membrane fractions from the wild-type cells, and membrane fractions from RAS2^{val19} strain had elevated levels of apparently GTP-independent adenylate cyclase activity (Table 4). Mixing the membrane fraction from rasl ras2 bcyl with that from cyrl reconstituted a GTP-dependent adenylate cyclase. These results suggest that the yeast RAS2 protein is a GTP-binding protein which regulates adenylate cyclase.

TABLE 4
ADENYLATE CYCLASE ACTIVITY IN MEMBRANE FRACTIONS
OF YEAST cyrl AND ras MUTANTS

Strain	Adenylate cyclase activity (units/mg protein)		
	Mn^{2+}	Mg^{2+}	Mg^{2+}+Gpp(NH)p
+	65.0	3.0	16.0
cyrl	0.0	0.0	0.0
rasl ras2 bcyl	62.1	0.4	0.4
RAS2^{val19}	71.0	17.1	16.7
cyrl + rasl ras2 bcyl[a]	64.2	2.9	14.3

[a] Mixture of membrane fractions from cyrl and rasl ras2 bcyl.

Crude extracts of E. coli transformants carrying pHM6 possessed adenylate cyclase activity in the presence of manganese ions, but no stimulation of activity by the addition of magnesium ions and Gpp(NH)p was observed (Table 5). Crude extracts of E. coli transformants carrying either RAS2 or RAS2^{val19} had no adenylate cyclase activity in the presence of manganese ions or magnesium ions (Table 5). When crude extracts of transformant carrying pHM6 were mixed with those of transformant carrying placRAS2, adenylate cyclase activity was stimulated in the presence of magnesium ions by the addition of Gpp(NH)p (Table 5). GTP-independent activity was observed by mixing crude extract of transformant carrying placRAS2^{val19}, and the activity was at the partially activated level in the absence of Gpp(NH)p. These results indicated that the products of the CYR1, RAS2 and RAS2^{val19} genes made in E. coli were functional, and that the RAS proteins interact directly with the catalytic subunit of adenylate cyclase.

TABLE 5
ADENYLATE CYCLASE ACTIVITY OF E. COLI TRANSFORMANTS
CARRYING CYR1 and RAS2 GENES

Strain or transformant	Adenylate cyclase activity (units/mg protein)		
	Mn^{2+}	Mg^{2+}	Mg^{2+}+Gpp(NH)p
Δcya	0.0	0.0	0.0
Δcya/pHM6	2.2	0.1	0.1
Δcya/placRAS2	0.0	0.0	0.0
Δcya/placRAS2^{val19}	0.0	0.0	0.0
Δcya/pHM6 + Δcya/placRAS2 [a]	4.6	0.4	1.1
Δcya/pHM6 + Δcya/placRAS2^{val19} [a]	4.2	0.7	1.0

[a] Mixture of crude cell extracts from two transformants.

DISCUSSION

We showed that yeast CYR1 and RAS2 genes could be expressed in E. coli cells, and these gene products could compose GTP-dependent adenylate cyclase in vivo and in vitro. The CYR1 product alone was not active to produce cAMP in E. coli cells, although it exhibited adenylate cyclase activity in the presence of manganese ions. The RAS2 gene could be expressed in E. coli cells, and the product of this gene was a protein which has the molecular weight of approximately 39,000 and no adenylate cyclase activity. Both CYR1 and RAS2 products were required to constitute GTP-dependent adenylate cyclase which was active in the presence of magnesium ions and to produce cAMP in E. coli cells. The RAS2^{val19} mutation caused the increase in the levels of GTP-independent adenylate cyclase activity and cAMP in the presence of CYR1 product. If the yeast RAS2^{val19} product has altered GTP hydrolyzing activity, it will activate the CYR1 product constitutively. All these results indicate that in yeast cells GTP-dependent adenylate cyclase was regulated by two subunits, catalytic subunit encoded by the CYR1 gene, and GTP-binding regulatory subunit encoded by the RAS2 gene.

ACKNOWLEDGMENTS

We thank our colleagues, H. Mitsuzawa, K. Tanaka, T. Oshima and K. Kato for invaluable help in this work.

REFERENCES

1. Matsumoto K, Uno I, Oshima Y, Ishikawa T (1982). Isolation and characterization of yeast mutants deficient in adenylate cyclase and cAMP-dependent protein kinase. Proc Natl Acad Sci USA 79:2355.
2. Uno I, Matsumoto K, Ishikawa T (1983). Characterization of a cyclic nucleotide phosphodiesterase-deficient mutant in yeast. J Biol Chem 258:3539.
3. Uno I, Matsumoto K, Adachi K, Ishikawa T (1984). Characterization of cyclic AMP-requiring yeast mutants altered in the catalytic subunit of protein kinase. J Biol Chem 259:12508.
4. Uno I, Matsumoto K, Ishikawa T (1982). Characterization of cyclic AMP-requiring yeast mutants altered in the regulatory subunit of protein kinase. J Biol Chem 257:14110.
5. Matsumoto K, Uno I, Ishikawa T (1984). Identification of the structural gene and nonsense alleles for adenylate cyclase in Saccharomyces cerevisiae. J Bacteriol 157:277.
6. Matsumoto K, Uno I, Ishikawa T (1983). Control of cell division in Saccharomyces cerevisiae mutants defective in adenylate cyclase and cAMP-dependent protein kinase. Exp Cell Res 146:151.

7. Toda T, Uno I, Ishikawa T, Powers S, Kataoka T, Broek D, Cameron S, Broach J, Matsumoto K, Wigler M (1985). In yeast, RAS proteins are controlling elements of adenylate cyclase. Cell 40:27.
8. Ellis RW, DeFeo D, Shih TY, Gonda MA, Young HA, Tsuchida N, Lowy DR, Scolnick EM (1981). The p21 src genes of Harvey and Kirsten sarcoma viruses originate from divergent members of a family of normal vertebrate genes. Nature 292:506.
9. Capon DJ, Seeburg PH, McGrath JP, Hayflick JS, Edman U, Levinson AD, Goeddel DV (1983). Activation of Ki-ras2 gene in human colon and lung carcinoma by different point mutations. Nature 304:507.
10. Shimizu K, Birnbaum D, Ruley MA, Fasano O, Suard Y, Edlund L, Taparowsky E, Goldfarb M, Wigler M (1983). Structure of the K-ras gene of the human lung carcinoma cell line Calu-1. Nature 304:497.
11. Tabin CJ, Bradley SM, Bargmann CI, Weinberg RA, Papageorge AG, Scolnick EM, Dhar R, Lowy DR, Chang EH (1982). Mechanism of activation of a human oncogene. Nature 300:143.
12. Taparowsky E, Shimizu K, Goldfarb M, Wigler M (1983). Structure and activation of the human N-ras gene. Cell 34:581.
13. Defeo-Jones D, Scolnick EM, Koller R, Dhar R (1983). ras-Related gene sequences identified and isolated from Saccharomyces cerevisiae. Nature 306:707.
14. Dhar R, Nieto A, Koller R, Defeo-Jones D, Scolnick EM (1984). Nucleotide sequence of two ras^H related-genes isolated from the yeast Saccharomyces cerevisiae. Nucleic Acids Res 12:3611.
15. Powers S, Kataoka T, Fasano O, Goldfarb M, Strathern J, Broach J, Wigler M (1984). Genes in S. cerevisiae encoding proteins with domains homologous to the mammalian ras proteins. Cell 36:607.
16. Kataoka T, Powers S, Cameron S, Fasano O, Goldfarb M, Broach J, Wigler M (1985). Functional homology of mammalian and yeast RAS genes. Cell 40:19.
17. Kataoka T, Powers S, McGill C, Fasano O, Strathern J, Broach J, Wigler M (1984). Genetic analysis of yeast RAS1 and RAS2 genes. Cell 37:437.
18. Tatchell K, Chaleff DT, Defeo-Jones D, Scolnick EM (1984). Requirement of either of a pair of ras-related genes of Saccharomyces cerevisiae for spore viability. Nature 309:523.
19. Casperson GF, Walker N, Brasier AR, Bourne HR (1983).

A guanine nucleotide-sensitive adenylate cyclase in the yeast Saccharomyces cerevisiae. J Biol Chem 258:7911.
20. Tamanoi F, Walsh M, Kataoka T, Wigler M (1984). A product of yeast RAS2 gene is a guanine nucleotide binding protein. Proc Natl Acad Sci USA 81:6924.
21. Matsumoto K, Uno I, Ishikawa T (1983). Initiation of meiosis in yeast mutants defective in adenylate cyclase and cyclic AMP-dependent protein kinase. Cell 32:417.
22. Uno I, Matsumoto K, Adachi K, Ishikawa T (1983). Genetic and biochemical evidence that trehalase is a substrate of cAMP-dependent protein kinase in yeast. J Biol Chem 258:10867.
23. Masson P, Jacquemin JM, Culot M (1984). Molecular cloning of the tsm0185 gene responsible for adenylate cyclase activity in Saccharomyces cerevisiae. Ann Microbiol (Inst Pasteur) 135:343.

Biochemistry of Yeast RAS1 and RAS2 proteins

Asao Fujiyama*, Nasrollah Samiy, Madan Rao**
and Fuyuhiko Tamanoi*

Cold Spring Harbor Laboratory
P.O. Box 100, Cold Spring Harbor, NY 11724

ABSTRACT

The bacterially produced yeast RAS2 protein binds GDP and Gpp(NH)p with the Kd of 47 nM and 87 nM, respectively. In addition, it exhibits an activity to hydrolyze GTP. RAS1 protein also has similar activities. After incubation at 4°C, a stable complex between the RAS2 protein and GTP can be isolated. Subsequent incubation at 37°C results in the conversion of the bound GTP to GDP. Addition of GTP to RAS2-GDP complex results in the displacement of GDP by GTP. Thus, the RAS2 protein is capable of shuttling between a GTP bound form and a GDP bound form. The mutant RAS2 protein, containing an alteration at the 19th amino acid (glycine to valine), purified after its expression in E. coli binds GDP and Gpp(NH)p with a Kd similar to that of the wild-type protein. However the GTPase activity of the mutant protein is significantly decreased. The shuttling does not take place efficiently with the mutant because of its lowered GTPase activity.

To analyze the biosynthesis of the RAS proteins in yeast, we have overexpressed intact RAS1 and RAS2 proteins. The primary translation products are found in the cytosol. These precursor molecules appear to undergo a processing event which results in a 1,000 dalton decrease in the apparent molecular weight as seen on SDS-polyacrylamide gels. The faster migrating molecules are found both in cytosol and membrane fractions. Only the molecules found in the membranes can be labeled with [^{3}H] palmitic acid. The attached fatty acids can be released from labeled RAS proteins by alkaline hydrolysis, indicating ester bond between protein and fatty acid molecules.

*Present address: Dept. of Biochemistry and Molecular Biology, the University of Chicago, Chicago , Illinois 60637
**Present address: Genetics Program, state Univ. of New york at Stony Brook, Stony Brook, NY 11794

INTRODUCTION

The *ras* genes, originally discovered as oncogenes of Harvey and Kirsten murine sarcoma viruses, have been identified as the transforming genes of a variety of human tumors (reviewed in ref. 1). Biochemical analysis has shown that the *ras* proteins, $p21^{ras}$, bind guanine nucleotides (2, 3). In addition to the guanine nucleotide binding activity, bacterially produced *ras* proteins exhibit a GTPase activity which appears to be an intrinsic property of the $p21^{ras}$. The GTPase activity is impaired in the oncogenic $p21^{ras}$ where the amino acid at the 12th or 59th residues is altered (4-9). The $p21^{ras}$ is first synthesized as a soluble precursor, then is converted to a form which migrates faster on SDS-polyacrylamide gels (10). $p21^{ras}$ is also reported to be bound with palmitic acid via an ester bond (11). Recent studies have suggesed that the site of the modification may be the cysteine residue which lies near the COOH-terminus, and the mutant virus which produce non-modified proteins lacks transforming activity (12, 13).

Recently homologues of *ras* genes were found in yeast. Two such homologues, *RAS1* and *RAS2*, code for proteins which show more than 90% homology to the $p21^{ras}$ at the NH_2-terminus (14, 15). Both *RAS1* and *RAS2* proteins exhibit guanine nucleotide binding activity (16, 17). Recent genetic and *in vitro* studies indicate that the *RAS* genes are involved in the regulation of cAMP production (18, 19). To fascilitate biochemical and structural characterization of the yeast *RAS* proteins, we have expressed the cloned *RAS* genes in *E. coli* and purified a large quantities of the proteins. The purified proteins exhibit a high level of guanine nucleotide binding activity as well as a GTPase activity (20). A model for the action of the *RAS* protein will be discussed. We also report the pathway of the biosynthesis of the yeast *RAS* proteins. A precursor of the *RAS* proteins appear to be synthesized in the cytosol. Shortly after synthesis, the precursor molecules are processed, and translocated to the membranes. The majority of the *RAS* proteins in the membranes are associated with fatty acid moieties.

EXPRESSION OF YEAST *RAS2* GENE IN *E. COLI.*

In order to characterize biochemical activities of the yeast *RAS2* protein it was necessary to obtain a large amount of purified protein. For this purpose the HincII fragment of the *RAS2* gene was cloned into a pUC8 plasmid. The resulting construct, pUC8-*RAS2*, directs the synthesis of a fusion protein which contains twenty extra amino acids added onto the NH_2-terminus of the *RAS2* protein (Figure 1).

Expression of the *RAS2* gene was examined by the appearance of GDP binding activity immunoprecipitable by the monoclonal antibody raised against mammalian *ras* protein. As seen in Table I, crude extracts prepared from *E. coli* cells carrying the

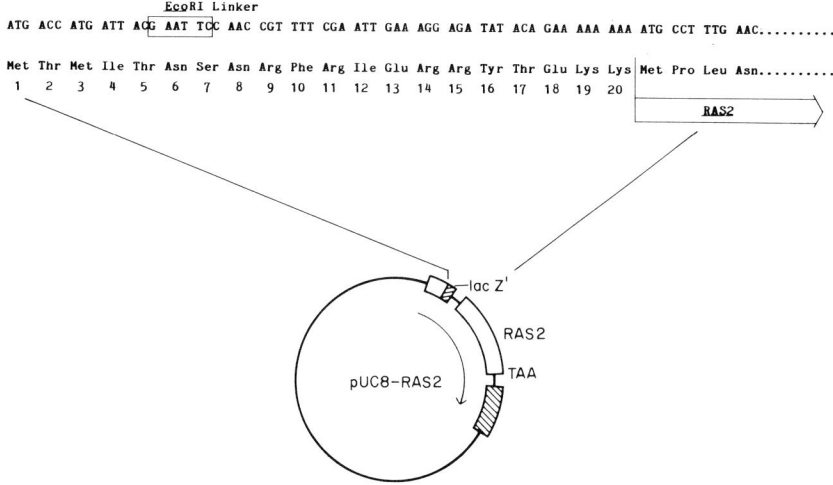

Figure 1. The RAS coding sequence present in pUC8-RAS2. HincII fragments containing wild-type and mutant RAS2 genes were kindly provided by Drs. Tohru Kataoka and Michael Wigler (Cold Spring Harbor Laboratory). EcoRI linker (GGAATTCC) was added to the HincII fragment and the fragment was ligated to pUC8 DNA digested with EcoRI. Amino acid positions 1-7 are derived from b-galactosidase coding region in pUC8. Amino acid positions 8-20 are derived from the sequence upstream of ATG of the RAS2 gene. The RAS2 gene is shown by an arrow.

Table I. Guanine nucleotide binding activity of RAS2 protein expressed in E. coli.

Extracts	[^{3}H]GDP bound pmol
None	0.08
pUC8-RAS2 no. 2 (50lg)	10.85
pUC8-RAS2 no. 5 (50lg)	0.11

To prepare crude extracts to examine GDP binding activity, E. coli JM103 cells carrying the pUC8-RAS2 construct were grown at 37°C in SOB media (33) containing 2mM IPTG and 100 lg/ml ampicillin to late log phase. The cells were collected, resuspended in Buffer A (10mM Tris HCl (pH 7.4), 50mM NaCl, 0.1mM DTT, 0.1mM EDTA, 1mM PMSF) + 5mM MgCl$_2$ + 1% Triton X-100 and disrupted by sonication. GDP binding activity was assayed by the incubation with 1.41M [^{3}H] GDP (7.8Ci/mmole, NEN) in a 100l1 reaction mixture containing 10mM Tris-HCl (pH 7.4), 50mM NaCl, 0.1mM EDTA, 0.1mM DTT, 0.5mM PMSF, 0.1% Triton X-100 and 5mM MgCl$_2$. After the incubation at 37°C, GDP binding was detected either by immunoprecipitation using a monoclonal antibody Y13-259 as previously described (16). One unit of activity is defined as the amount of protein that binds 1 pmol of [^{3}H]GDP in 60 min at 37°C under the above assay condition. The amount of total protein present in the extracts used is indicated in the parenthesis.

pUC8-RAS2 exhibited a high level of immunoprecipitable GDP binding activity. Control extracts did not show any GDP binding activity.

CHARACTERIZATION OF BACTERIALLY PRODUCED RAS2 PROTEIN.

Using conventional methods, the RAS2 fusion protein was purified to 70% homogeniety as judged by SDS-polyacrylamide gel electrophoresis (SDS-PAGE) (Figure 2). SDS-PAGE analysis showed a major band with a relative mobility of 43K daltons, which is the expected molecular weight of bacterially synthesized fusion RAS2 protein (Figure 2). The RAS2 protein eluted from a Sephacryl S-300 column with an apparent molecular weight of 46K. Thus the protein behaves as a monomer.

The purified RAS2 protein exhibited maximum guanine nucleotide binding activity at 45 °C. Very little binding was observed at 0 °C. The pH optimum was 7.6 and the optimum Mg^{2+} concentration was approximately 0.5mM. Some binding (33% of maximum level) was observed even in the presence of EDTA. Figure 3A shows the binding of [^{3}H]GDP to the RAS2 protein as a function of the GDP concentration. Saturation was reached at about 0.4lM. A Scatchard plot of the data (Figure 3A) shown in Figure 3B is consistent with single-site binding kinetics, which gives a calculated Kd of 47 nM. A similar experiment using a non-hydrolyzable GTP analogue, Gpp(NH)p, gave the Kd of 87 nM. This indicates that the RAS2 protein binds GDP and GTP with the affinity of the same order of magnitude. The $p21^{ras}$ binds GDP and GTP with the Kd of 60 nM, and 11 ~ 25 nM, respectively (21, 22). The Kd values of the yeast RAS proteins are also comparable to those obtained with other GTP binding proteins; Gs and Gi (23, 24). These situations are in contrast to other GTP binding proteins which exhibit different affinity for GTP and GDP. Elongation factor EF-Tu binds GTP with the Kd of 200 ~ 300 nM (25). But, the affinity of GDP for the EF-Tu is two orders of magnitude lower than that seen for GTP (25, 26). A difference in the binding affinity of GDP and GTP is also observed with eukaryotic translation initiation factor, eIF-2 (27).

GTPASE ACTIVITY OF THE RAS2 PROTEIN.

The purified RAS2 protein exhibited GTPase activity. As shown in table II, the RAS2 preparation was treated with the monoclonal antibody and the immune-complexes were removed by protein A-Sepharose. This treatment resulted in the loss of 81% of the GTPase activity which was comparable to the loss of the GDP binding activity (90%). The products of hydrolysis are GDP and inorganic phosphate. Properties of the GTPase activity were similar to those of the GDP binding activity. The hydrolysis is specific to guanine nucleotides since no hydrolysis of ATP, CTP, dATP, dCTP, and dTTP was observed. However, the RAS2 protein

PURIFICATION OF RAS2 PROTEIN

FRACTION	PROTEIN	TOTAL ACTIVITY	SPECIFIC ACTIVITY	FOLD PURIFICATION
	mg	UNITS	UNITS/mg	
CRUDE EXTRACT	187	35.4X10³	189	1
HIGH SPEED SUPERNATANT	128	19.6X10³	153	0.8
PHOSPHOCELLULOSE	2.7	10.9X10³	4098	22
SEPHADEX G-150	0.3*	1.6X10³*	5611	30

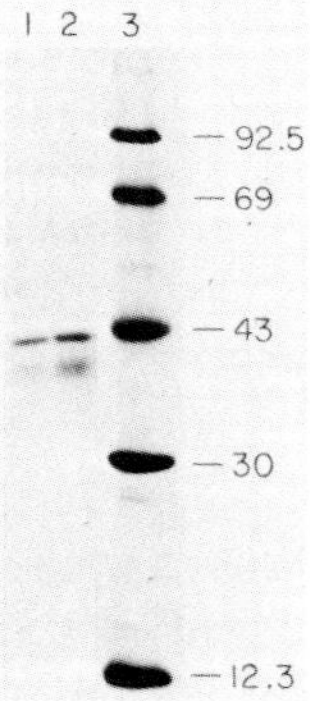

Lanes 1, 2, 3 — size markers: 92.5, 69, 43, 30, 12.3

Figure 2. Purification of RAS2 Protein. E. coli JM103 cells carrying pUC8-RAS2 no. 2 were grown at 37°C in the presence of IPTG. 3g of the cells were suspended in 6ml Buffer A + 5mM $MgCl_2$ + 1% Triton X-100, and broken by sonication. The crude extract was loaded onto phosphocellulose and eleted with a linear gradient of 0 to 0.4M NaCl in Buffer A + 5 mM $MgCl_2$. Portion of the pooled peak fractions was loaded onto Sephadex G-150 or Sephacryl S-300. The RAS2 proteins were eluted with Buffer G. The purified proteins were kept at -20°C after the addition of glycerol to 50%. *Values for only peak fractions are shown. GDP binding activity was determined as described in Table I. Amount of protein was determined by the method of Lowry (34). Coomassie blue staining of the final preparation is shown in the right hand panel. Lanes 1 and 2: Peak fractions of Sephadex G150. Lane 3: Size markers, numbers indicate M_r x 10^{-3}, phosphorylase B (92.5), albumin (69), ovalbumin (43), carbonic anhydrase (30) and cytochrome C (12.3).

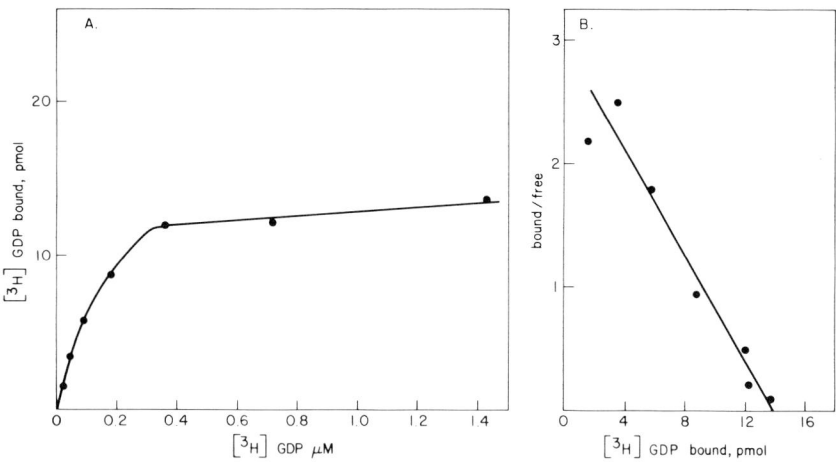

Figure 3 (A). GDP binding to RAS2 protein as a function of the GDP concentration. The incubation was for 1 hr and essentially the same binding was obtained with 2 hr incubation. (B) Scatchard plot.

Table II. Removal of GTPase by Immunoprecipitation

Exp.		GTP hydrolyzed pmol	GDP bound pmol
1.	-antibody (sup1)	0.43	5.58
2.	+antibody (sup2)	0.08	0.55
3.	sup2 + RAS2	0.42	ND
4.	RAS2	0.36	ND

Purified RAS2 protein in Buffer G was mixed with monoclonal Y13-259. After rotating in the cold room for 2 hr, rabbit anti-rat IgG was added and the solution was sat on ice for 1 hr. Protein A-Sepharose was added and rotated for 40 min in the cold room. After spinning, the supernatant was recovered (sup2). The same treatment was carried out without the addition of the monoclonal antibody (sup1). Exp1,2: 4 ll and 20 ll aliquots were used to assay GTPase and GDP binding activity, respectively, as described below except that 100 lM ATP was included in the reaction mixture for GTPase. Exp3: 4 ll of sup2 was mixed with RAS2 protein and assayed for GTPase activity. Exp4: GTPase activity of RAS2 protein. ND = not done. GTPase assay was carried out in a 10ll reaction mixture containing 50mM Tris-HCl (pH 7.6), 25mM KCl, 5mM $MgCl_2$, 15mM $(NH_4)_2SO_4$, 0.24lM c-^{32}P-GTP (15Ci/mmol, Amersham) and various amounts of RAS2 proteins. After incubation at 37°C, 200ll of 4% activated charcoal in 0.7M $HClO_4$, 25 mM KH_2PO_4 was added to the reaction mixture. After vigorous mixing, the tubes were centrifuged and 105 ll aliquot was counted in 5ml Aquasol.

does not distinguish ribose and deoxyribose since dGTP was also hydrolyzed. The rate of the hydrolysis of GTP by RAS2 protein is calculated to be 60mmol/min/mol of protein at 37°C. This is comparable to the rate by known GTP binding proteins. For example; Ta of transducin, approximately 50 mmol/min/mol of protein at 25°C (28), Gs, less than 30 mmol/min/mol of protein at 37°C (24). These values are small in comparison to those of typical metabolic enzymes.

CONVERSION FROM GTP BOUND FORM TO GDP BOUND FORM.
Striking similarities in the requirements for the GDP binding and GTP hydrolysis activities prompted us to test whether these activities are coupled. First, the RAS2 protein was incubated with α-^{32}P-GTP at low temperature in order to bind GTP under the condition where GTP hydrolysis is minimized (Fig. 4). Then, the reaction mixture was applied to a column of Sephadex G-50. As seen in Figure 4A, a small peak of ^{32}P was detected at the position where excluded proteins elute. The radioactivity at this position was trapped on a nitrocellulose filter indicating that the GTP is bound to the RAS2 protein (data not shown). The radioactivity in the complex was in the form of GTP as revealed by thin layer chromatography (Figure 4B). Incubation of this complex at 37°C resulted in the conversion of almost all of GTP to GDP (Figure 4B). The resulting GDP was still complexed to the RAS2 protein, since 70% of the radioactivity could still be trapped by nitrocellulose filters at the end of incubation. Thus, these results show that the RAS2 protein binds GTP and the bound GTP is hydrolyzed to GDP.

EXCHANGE OF BOUND GDP WITH FREE GTP.
Hydrolysis of the bound GTP results in the production of RAS2 protein with a bound GDP. In order to determine whether this GDP can be exchanged with fresh GTP, the following experiment was carried out. First, the RAS2 protein was incubated with [^{3}H]GDP. When the binding approached a plateau level, α-^{32}P-GTP was added at a final concentration equal to that of [^{3}H]GDP and incubated further. As can be seen in Figure 5, a decrease of [^{3}H] radioactivity was observed after the addition of GTP whereas a gradual increase was seen when the GTP was not added. Concomitant with the ^{3}H release, incorporation of ^{32}P radioactivity into a nitrocellulose bound form was observed. This result indicates that the bound GDP can be exchanged with GTP by the addition of GTP to the RAS2-GDP complex. Hydrolysis of GTP is negligible (less than 3.5% of the input GTP) under the condition used.

MUTANT RAS2 PROTEIN.
Yeast cells carrying a mutant RAS2 gene which encodes for valine at position 19 rather than glycine exhibit a drastically altered

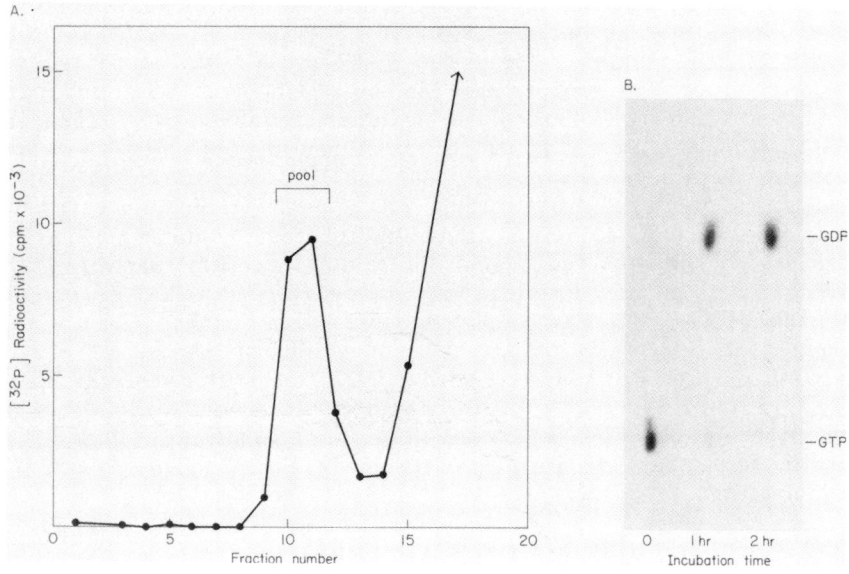

Figure 4. Isolation of RAS2-GTP Complex by Sephadex G50. A: Purified RAS2 protein was incubated with a-^{32}P-GTP in 180ll reaction mixture as for GDP binding. The incubation was carried out at 4°C for 1 hr and then loaded onto 5ml column of Sephadex G50. Elution was carried out with Buffer A + 5mM MgCl$_2$. 200ll fractions were collected and an aliquot was counted in Aquasol. B: 10ll of fraction 11 of the Sephadex G50 column was spotted onto PEI cellulose before and after 1hr or 2 hr incubation at 37°C. The PEI cellulose was developed with 1M HCOOH/1M LiCl and autoradiographed.

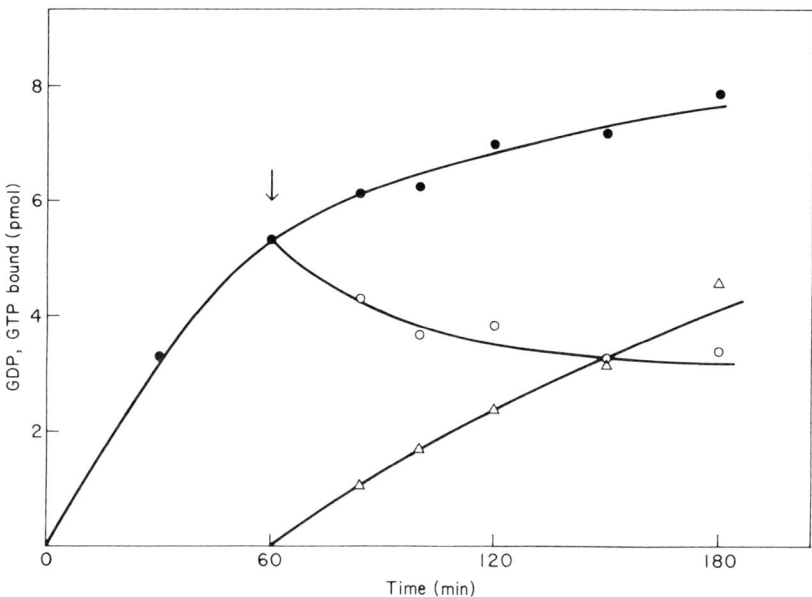

Figure 5. Exchange of Bound GDP with GTP. GDP binding was carried out using the purified RAS2 protein and 1.4 1M [^{3}H]GDP. After the incubation at 37°C for 1 hr, the reaction mixture was divided into two portions and a-^{32}P-GTP was added to one portion (a) with a final concentration of 1.4 1M (shown by an arrow). No addition was made to the other portion (b). After further incubation at 37°C, aliquots were removed at certain time intervals and filter binding assay was used to detect the amount of the bound [^{3}H]GDP and [^{32}P]GTP. o----o ^{3}H in the portion (a) bound to nitrocellulose filter. ---- ^{32}P in the portion (a) bound to nitrocellulose filter. o----o ^{3}H in the portion (b) bound to nitrocellulose filter. Note that the total amount of bound GDP and GTP in the reaction (a) roughly equals the amount of bound GDP in the reaction (b).

physiology (29). To investigate whether any biochemical activity of RAS2 protein is affected by the mutation, the mutant RAS2 gene was cloned into the pUC8 plasmid. Extracts of E. coli cells expressing the mutant gene exhibited high level of GDP binding activity comparable to that observed with the extracts expressing the wild-type RAS2 gene. Using the GDP binding activity as an assay, the mutant RAS2 fusion protein was purified. The mutant RAS2 protein exhibited GDP binding activity similar to that seen for the wild-type protein (Figure 6A). The Kd value for the binding of GDP was 57 nM. In addition, GDP bound to the mutant protein could be exchanged with free GTP with an efficiency comparable to that seen with the wild-type protein (data not shown). However, a significant difference was observed in their GTPase activities. As shown in Figure 6B, the GTPase activity of the valine mutant protein was decreased to about 16% of that of the wild-type protein. The rate of GTP hydrolysis is 60 mmol/min/mol of protein for the wild type RAS2 and 9.6 mmol/min/mol of protein for the valine mutant. A similar decrease in GTPase activity was observed for another mutant protein containing alteration from alanine to threonine at the 66th amino acid (20). Temeles et al. (30) expressed NH_2-terminal portion of yeast RAS1 protein in E. coli. The polypeptide exhibited a GTPase activity that was reduced in variants containing amino acid alterations at the 66th or 68th residues.

A MODEL FOR THE ACTION OF RAS PROTEINS.
Based on the enzymatic activities, a model for the action of RAS2 protein is proposed as shown in Figure 7. We propose that the RAS2 protein exists in two different forms; one complexed with GTP (RAS2-GTP) and the other complexed with GDP (RAS2-GDP). The RAS2-GTP can be converted to the RAS2-GDP by its intrinsic GTPase activity. The RAS2-GDP can be converted back to the RAS2-GTP by its exchange reaction. Thus the RAS2 protein can shuttle between a GTP-RAS2 form and a GDP-RAS2 form. The shuttling between the two forms appears to be an important aspect of the function of the protein. We further speculate that there is a difference in the structure and activity between the RAS2-GTP and the RAS2-GDP.

The model proposed in Figure 7 indicates that there are three stages during which the action of RAS2 protein can be regulated. One stage is during the binding of GTP to the protein. It may be possible to envision a factor affecting the conformation of the RAS2 protein which alters the affinity of the protein to GTP and thus the functions of the RAS2 protein. Another stage could be the conversion from GTP-RAS2 to GDP-RAS2 by the intrinsic GTPase activity. This is evident in the case of the mutant RAS2 protein which contains an alteration from glycine to valine at the 19th amino acid. Since the mutant protein shows much reduced GTPase activity, the mutant protein is most likely to be locked in

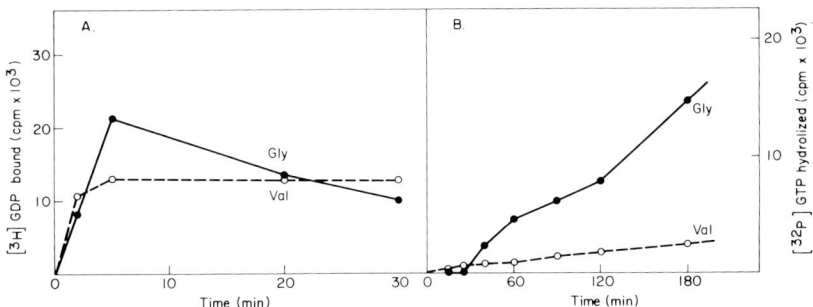

Figure 6 (A). GDP binding activity of wild-type RAS2 (Gly) and mutant RAS2 (Val) proteins. (B). GTPase activity of wild-type and mutant RAS2 proteins. Wild-type (Gly) and mutant (Val) RAS2 proteins were incubated at 37°C with c-^{32}P-GTP and the release of inorganic phosphate was determined. The amount of proteins used was adjusted so that both proteins give identical level of GDP binding. ^{32}P released with wild-type protein after 180 min incubation corresponds to 39% of total GTP in the reaction mixture.

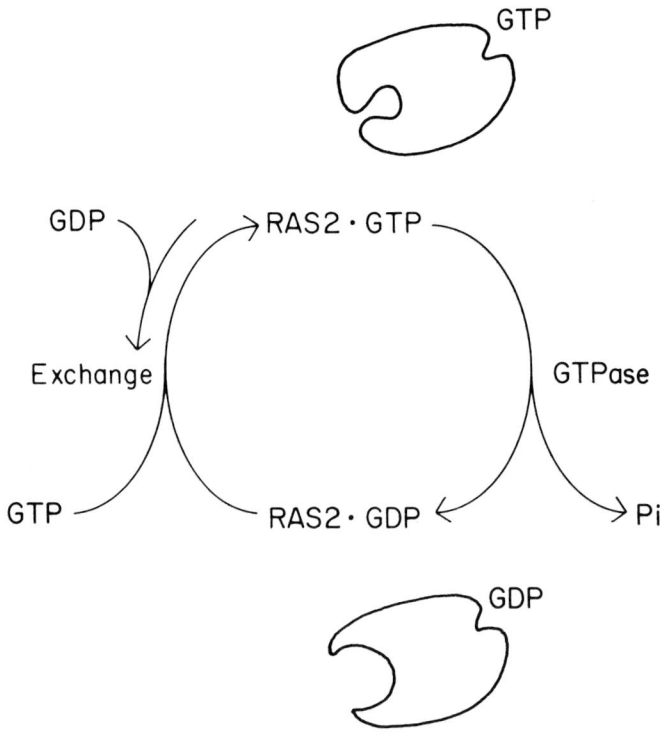

Figure 7. A Model for the Action of RAS2 Protein.

the RAS2-GTP form. Yeast cells expressing the mutant protein exhibits dramatically altered physiology which includes reduced sporulation efficiency and reduced survivability upon nutrient starvation (18). Finally, the activity of RAS2 protein can be influenced during GDP/GTP exchange reaction. This exchange reaction may depend on the intracellular pool of GTP relative to GDP. Since the pool of GTP is generally much higher than that of GDP (31), it is possible that the RAS2-GDP is quickly converted to the RAS2-GTP.

How regulation of the action of RAS2 protein and hence regulation of adenylate cyclase in yeast take place would become clearer by biochemical analyses of interactions between these proteins. The model proposed here could presumably be applied to RAS1 protein, since the protein exhibits GTP binding and GTPase activities similar to those observed with the RAS2 protein (unpublished results)(30).

BIOSYNTHESIS OF THE RAS PROTEINS IN YEAST.

To characterize RAS proteins produced in yeast, we used the plasmids, YEp51-RAS1 or YEp51-RAS2 (see Figure 8), which overexpress intact RAS proteins in yeast. As shown in Fig. 9, both RAS1 and RAS2 proteins synthesized in a cell-free translation system migrated more slowly on SDS-polyacrylamide gel than those isolated from yeast after a 20 min labeling with [^{35}S] methionine. The apparent molecular weights of the primary translation products were M_r37,000 and M_r41,000 for RAS1 and RAS2, respectively. On the other hand, the products of yeast showed M_r36,000 for RAS1 and M_r40,000 for RAS2, suggesting that RAS proteins undergo some processing or modification in yeast cells. The 1,000 dalton decrease in the molecular weight is roughly equal to the loss of nine amino acid residues from the proteins. To examine the exact feature of the processing event in yeast cells, we carried out pulse-chase experiment. Cells carrying YEp51-RAS2 were labeled with [^{35}S] methionine and fractionated into membrane and cytosol fractions. As can be seen in Fig. 10, the band of M_r41,000 protein was detected in the cytosol fraction of the 1 min labeled cells (lanes 2), and diminished after chase (lane 6). We also found that the majority of the labeled RAS2 protein has a relative mobility of M_r40,000 even after a very brief labeling shorter than 1 min (unpublished results) (16). These results indicate that the processing proceeds very rapidly in cytosol. Three other bands of about M_r30,000, were also detected in the cytosol as reported by Tamanoi et al. (16). However, these bands do not appear in the cells labeled with [^{35}S] cysteine (data not shown). Because the cysteine residues in the RAS2 protein are located only in the COOH-terminal region (15), these molecules might be products of pre-mature termination during translation process.

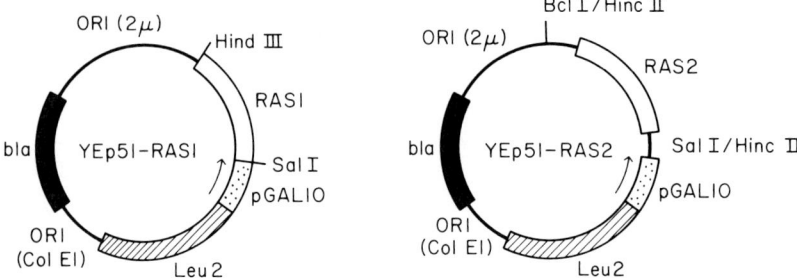

Figure 8. Construction of the expression plasmids for yeast. S.cerevisiae JR25-2A (leu2, ura3, trp1, his3, can1) carrying expression plasmid was grown at 31°C to 0.7-1.1x 10^7 cells/ml in a synthetic medium containing 4% galactose (16).

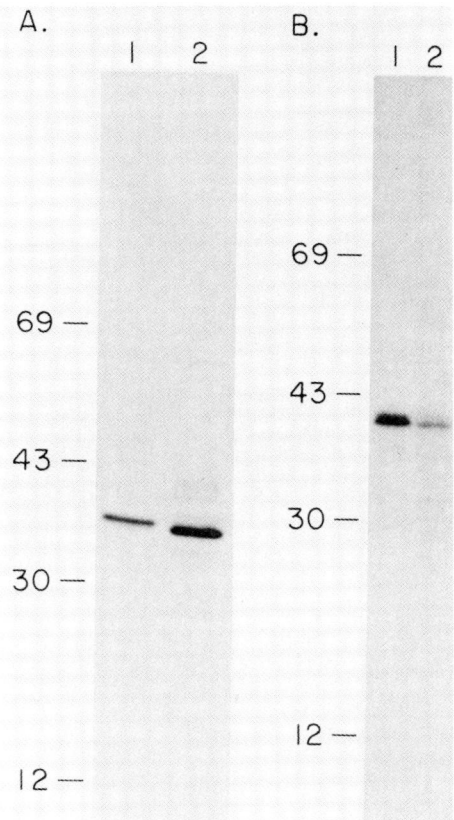

Figure 9. SDS-polyarylamide gel analysis of RAS proteins. (A). Detection of RAS1 proteins synthesized in vitro (lane 1) and in yeast cells (lane2). Poly(A) RNA was extracted from the yeast cells carrying YEp51-RAS1 plasmid. The mRNA was incubated with rabbit reticulocyte lysate and [^{35}S] methionine at 30 °C for 60 min. Yeast cells containing YEp51-RAS1 were labeled with [^{35}S] methionine for 20 min at 31 °C. (B). RAS2 protein synthesized in vitro (lane 1) and in yeast cells. Poly(A)RNA was isolated from the yeast cells containing YEp51-RAS2 (16) and translated as in A. Yeast cells (YEp51-RAS2) were labeled with [^{35}S] methionine for 20 min at 31 °C. RAS proteins were precipitated with the monoclonal antibody Y13-259 and separated in a 12.5% SDS-polyacrylamide gel. Bands of the radioactive protein were visualized by fluorography using Amplify (Amersham) and Kodak-XAR5 films. Size markers ($M_r \times 10^{-3}$): albumin (69), ovalbumin (43), carbonic anhydrase (30), and cytochrome C (12).

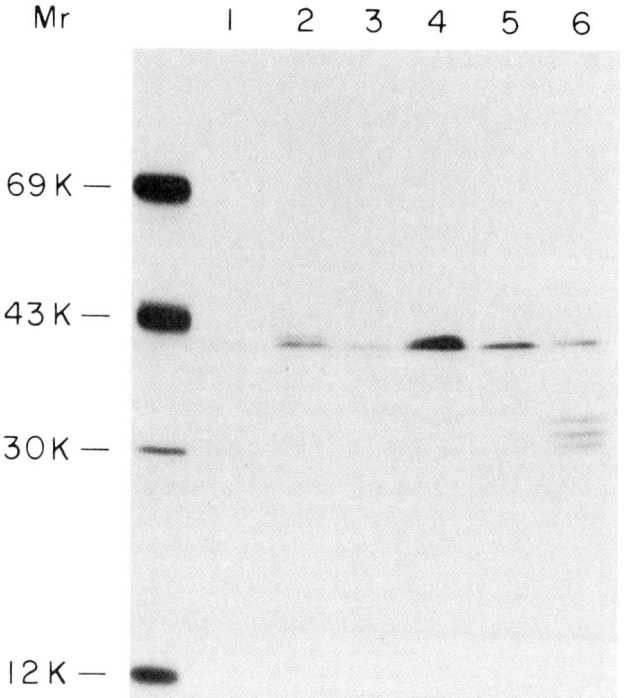

Figure 10. Pulse labeling and subcellular fractionation of the yeast cells carrying YEp51-RAS2. Cells were grown in 10 ml of the synthetic medium and pulse labeled with [^{35}S] methionine for 1 min (lanes 1 and 2) or 2 min (lanes 3 and 4). For chase experiment, 1 min labeled cells were further incubated in the presence of 20 ug/ml cold methionine for 90 min (lanes 5 and 6). Labeled cells were mixed with unlabeled cells (50 ml culture) and membrane extracts (lanes 1,3, and 5) and soluble fractions (lanes 2,4, and 6) were prepared according to the method of Willsky (35). Proteins were separated by 15% SDS-polyacrylamide gel and the fluorogram was taken as described in the legend to Fig. 9.

MODIFICATION BY FATTY ACID AND ASSOCIATION TO MEMBRANES.

The processed molecules were found in the membrane fractions as well as in the cytosol, and the radioactivity of the molecules in the membranes accumulated after long labeling or chase (Figure 10). This suggests that the processed RAS2 proteins in cytosol are transported to membranes at relatively slow rate. We first tested whether the molecules in the membranes are chemically the same with those in the cytosol. Since many membraneous proteins including $p21^{ras}$ and $p60^{src}$ have been reported to be modified by fatty acid attachment (11, 32), we labeled yeast cells with [^{3}H] palmitic acid follwed by subcellular fractionation. Results shown in Figure 11 clearly demonstrate that only the RAS2 proteins in the membrane fraction can be labeled with [^{3}H] palmitic acid. Virtually no RAS2 proteins in cytosol were labeled with the radioactive fatty acid. Therefore, the processed RAS2 proteins in the cytosol apparently undergo a modification by fatty acid attachment. Both the modified and the unmodified RAS proteins have same electrophoretic mobility on SDS-polyacrylamide gel. Bound fatty acids could be released by alkaline treatment, indicating ester bond between the protein molecules (Table III). The fatty acids resulted by the hydrolysis were a mixture of palmitic acid and myristic acid and other minor components. RAS1 proteins also undergo similar processing and modification events (data not shown).

From the lines of evidence described above, we propose a pathway of maturation of yeast RAS proteins as shown in Figure 12. The RAS proteins are first synthesized as a higher molecular weight precursor in cytosol. The precursor molecules then undergo a processing event to a faster-migrating form as seen on SDS-polyacrylamide gels. Since the COOH-terminal cysteine residue remains in the processed molecule (data not shown), the processing does not involve the removal of the COOH-terminal region. Because only the proteins in the membrane fractions were found to be labeled with [^{3}H] palmitic acid, it appears that the transport of the RAS proteins to the membrane is coupled with fatty acid attachment. Many possible functions of the attached fatty acids have been proposed (reviewed in ref. 32). In the case of the RAS proteins, unmodified proteins produced in E.coli as well as proteins which lack the COOH-terminal portion exhibit guanine nucleotide binding and GTPase activity. Thus, the modification by fatty acid may not be necessary for these biochemical activities of the RAS proteins, although further detailed analyses are needed to conclude this point. According to a hydrophobicity analysis based on the predicted amino acid sequence, the RAS proteins do not contain typical hydrophobic domain (unpublished data)(15), nonetheless, the RAS proteins appear to be membrane associated. It is possible that the fatty acid moiety gives hydrophobicity to the COOH-terminus of the RAS proteins, and helps the proteins to settle into the membranes to function.

Table III. Distribution of radioactivity of the [^{3}H] palmitic acid labeled RAS proteins after hydrolysis

	RAS1*	RAS2*
after hydrolysis by mild alkali**	60 %	70 %
after hydrolysis by 6M HCl***	40 %	30 %

*: ^{3}H radioactivities in the RAS1 or RAS2 proteins extracted from SDS-gels were 1,248 cpm and 1,330 cpm, respectively.

**: Most of the radioactivities were recovered as a mixure of free acids and methyl esters of palmitic acid and myristic acid from the extracts of alkaline hydrolyzate.

***: Most of the radioactivities released by acid hydrolysis were not palmitic acid or myristic acid.

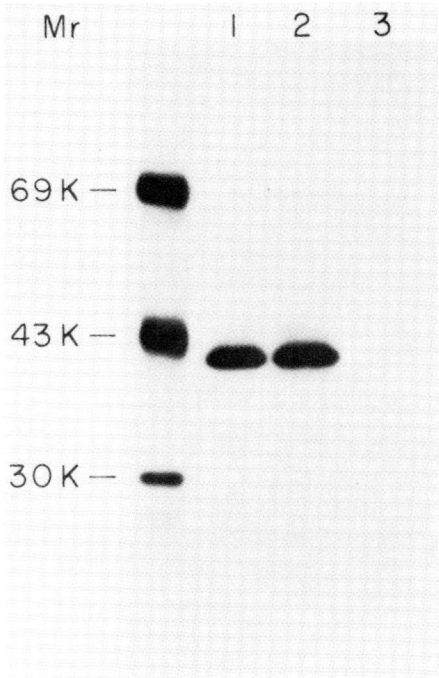

Figure 11. Distribution of the RAS2 protein labeled with [^{3}H] palmitic acid. 100 ml culture of yeast cells carrying YEp51-RAS2 were labeled by [^{3}H] palmitic acid for 30 min at 31°C. RAS proteins were precipitated with the monoclonal antibody Y13-259 from the extracts of 15,000 x g (lane 1) and 105,000 x g (lane 2) peletts, and the cytosol fraction (lane 3). SDS-PAGE and fluorography were performed as described in the legend to Fig. 9.

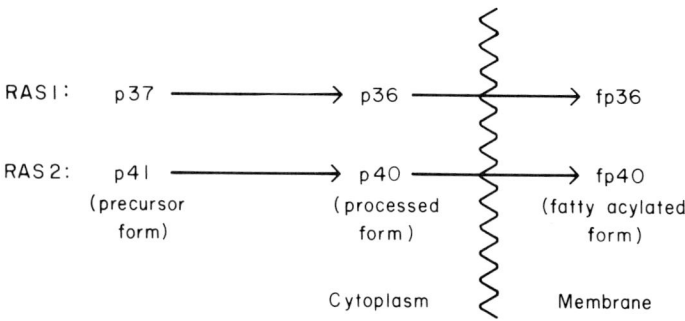

(1) Processed form without fatty acylation could be detected in yeast cells.
(2) Cysteine labeling suggests that the processing does not involve C terminus.

Figure 12. Proposed pathway of processing and modification for yeast RAS proteins.

ACKNOWLEDGEMENTS
We thnk Dr. Daniel Broek for reading the manuscript. We also thank Barbara Weinkauf and Louisa Dalessandro for preparing the manuscript. A.F. is on leave from Institute for Molecular and Cellular Biology, Osaka University, Japan. This work was supported by a Cancer Center Grant from the National Cancer Institute.

REFERENCES

1. Cooper, G.M. (1982) Science 217, 801-806.
2. Scolnick, E.M., Papageorge, A.G. and Shih, T.Y. (1979) Proc. Natl. Acad. Sci. USA 76, 5355-5359.
3. Shih, T.Y., Papageorge, A.G., Stokes, P.E., Weeks, M.O. and Scolnick, E.M. (1980) Nature 287, 686-691.
4. Lautenberg, J.A., Ulsh, L., Shih, T.Y. and Papas, T.S. (1983) Science 221, 858-860.
5. Stein, R.B., Robinson, P.S., and Scolnick, E.M. (1984) J. Virol. 50, 343-351
6. McGrath, J.P., Capon, D.J., Goeddel, D.V. and Levinson, A.D. (1984) Nature 310, 644-649.
7. Sweet, R.W., Yokoyama, S., Kamata, T., Feramisco, J.R., Rosenberg, M. and Gross, M. (1984) Nature 311, 273-275.
8. Lacal, J.C., Santos, E., Notario, V., Barbacid, M., Yamazaki, S., Kung, H-F, Seamans, C., McAndrew, S. and Crowl, R. (1984) Proc. Natl. Acad. Sci. USA 81, 5305-5309.
9. Gibbs, J.B., Sigal, I.S., Poe, M. and Scolnick, E.M. (1984) Proc. Natl. Acad. Sci. USA 81, 5704-5708.
10. Shih, T.Y., Weeks, M.O., Gruss, P., Dhar, R., Oroszlan, S., and Scolnick, E.M. (1982) J. Virol. 42, 253-261
11. Sefton, B.M., Trowbridge, I.S. and Cooper, J.A. and Scolnick, E.M. (1982) Cell 31, 465-474.
12. Willumsen, B., Christensen, A., Hubbert, N.L., Papageorge, A.G., and Lowy, D. (1984) Nature 310, 583-586
13. Weeks, M.O., Hager, G.L., Lowe, R., and Scolnick, E.M. (1985) J. Virol. 54, 586-597
14. DeFeo-Jones, D., Scolnick, E.M., Koller, R. and Dhar, R. (1983) Nature 306, 707-709.
15. Powers, S., Kataoka, T., Fasano, O., Goldfarb, M., Strathern, J., Broach, J. and Wigler, M. (1984) Cell 36, 607-612.
16. Tamanoi, F., Walsh, M., Kataoka, T. and Wigler, M. (1984) Proc.. Natl. Acad. Sci. USA 81, 6924-6928.
17. Temeles, G., DeFeo-Jones, D., Tatchell, K., Ellinger, M.S. and Scolnick, E.M. (1984) Mol. Cell. Biol. 4, 2298-2305.
18. Toda, T., Uno, I., Ishikawa, T., Powers, S., Kataoka, T., Broek, D., Cameron, S., Broach, J., Matsumoto, K. and Wigler, M. (1985) Cell 40, 27-36.
19. Broek, D., Samiy, N., Fasano, O., Fujiyama, A., Tamanoi, F., Northup, j., and Wigler M. (1985) Cell, in press
20. Tamanoi, F., Samiy, N., Rao, M. and Walsh, M. (1985) In Cancer Cells III: Growth Factors and Transformation; J. Feramisco, B. Ozanne and C. Stiles, eds, (Cold Spring Harbor Laboratory, Cold Spring Harbor, New York). pp251-256
21. Manne, V., Yamazaki, S. and Kung, H-F. (1984) Proc. Natl. Acad. Sci. USA 81, 6953-6957.
22. Finkel, T., Der, C.J. and Cooper, G.M. (1984) Cell 37, 151-158.

23. Bokoch, G.M., Katada, T., Northup, J.K., Ui, M. and Gilman, A.G. (1984) J. Biol. Chem. 259, 3560-3567.
24. Northup, J.K., Smigel, M.D. and Gilman, A.G. (1982) J. Biol. Chem. 257, 11416-11423.
25. Lucas-Lenard, J. and Lipmann, F. (1971) Annu. Rev. Biochem. 40, 409-448.
26. Kaziro, Y. (1978) Bioch. Bioph. Acta. 505, 95-127.
27. Panniers, R. and Henshaw, E.C. (1983) J. Biol. Chem. 258, 7928-7934.
28. Fung, B. K-K (1983) J. Biol. Chem. 258, 10495-10502.
29. Kataoka, T., Powers, S., McGill, C., Fasano, O., Strathern, J., Broach, J. and Wigler, M. (1984) Cell 37, 437-445.
30. Temeles, G., Gibbs, J.B., D'Alonzo, J.S., Sigal, I.S. and Scolnick, E.M. (1985) Nature 313, 700-703.
31. DeAbreu, R.A., vanBaal, J.M., Bakkeren, J.A.J.M., DeBruyn, C.H.M.M. and Schretlen, E.D.A.M. (1982) J. Chromato. 227, 45-52.
32. Schmidt, M.F.G. (1983) Curr. Top. Microbiol. Immunol. 102, 101-129
33. Hanahan, D. (1983) J. Mol. Biol. 166, 557-580.
34. Lowry, O.H., Rosebrough, N.J., Farr, A.L. and Randall, R.J. (1951) J. Biol. Chem. 193, 265-275.
35. Willsky, G.R. (1979) J. Biol. Chem. 254, 3326-3332

GENETICS OF SPINDLE POLE BODY REGULATION

Peter Baum, Loretta Goetsch, and Breck Byers

Department of Genetics, SK-50, University of Washington
Seattle, Washington 98195

ABSTRACT Microtubule organization in Saccharomyces cerevisiae depends on the behavior of the spindle pole body, which normally undergoes duplication late in the G1 phase of the cell division cycle. Mutations in two genes, cdc31 and esp1, disrupt the regulation of this process. Sequence analysis of CDC31, which is specifically required for the duplication to occur, reveals probable Ca^{++}-binding sites in the gene product, suggesting a role for Ca^{++}-fluxes in control of cell cycle events. By contrast, ESP1 is found to act as a negative regulator of spindle pole body duplication.

INTRODUCTION

The faithful reproduction of eukaryotic cells requires precise reorganization of complex microtubule arrays with each cycle of division. As the cell enters mitosis, the cytoskeleton, which defines the form of the interphase cell, must be disassembled and replaced by a bipolar spindle capable of separating the chromosomes with astounding fidelity. While reorganization of these microtubule arrays may depend in part on biochemical modifications of the tubulins and other proteins arrayed along the length of the fibers, critical regulation must also reside in the organizing centers[1]. In a mammalian cell, for example, material surrounding the pair of centrioles determines the distribution of microtubules radiating from the cytocentrum[2]. Entry into mitosis depends on duplication of these centers and modification of their properties to generate the precise distribution required for chromosome separation. Although the centrioles are cytologically distinct and thereby provide a visual guide to changes in the surrounding material, the organizing centers themselves are indistinct and difficult to observe when not marked in this manner, as is the case in the cells of higher plants. By contrast, the organizing centers of yeasts, fungi, and some lower plants constitute the spindle pole bodies, which are readily observed by electron microscopy.

Replication of Spindle Pole Bodies

The spindle pole bodies (SPB's) of Saccharomyces cerevisiae remain distinctive throughout the cell cycle, enabling one to observe the pivotal role they play in a variety of cellular processes [3]. These organelles function not only in formation of the mitotic spindle but also serve crucial roles in karyogamy, where they mediate the initiation of nuclear fusion within the zygotes[4], and in the formation of spore walls[5]. Accordingly, it is essential that the behavior of the SPB's be highly integrated with other cellular functions. In the cell division cycle, they must duplicate once, and only once, at the appropriate stage of the cycle. The first manifestation of duplication detectable by electron microscopy is the formation of a "satellite" adjacent to the extant SPB during the latter portion of G1 (Figure 1, solid arrows). Later, near the time of entry into S-phase, the satellite-bearing single SPB is replaced by a duplicated pair of SPB's which appear to arise by the maturation of the satellite to become one member of the pair. These SPB's then separate from one another to form the poles of the spindle. The terminal phenotypes of various temperature-sensitive cdc mutants resemble these stages in the SPB duplication pathway[6]: Arrested "Start" mutants possess either an unmodified SPB or one with a satellite. Mutants in two genes (cdc4 and cdc34) arrest in late G1 with a duplicated pair of SPB's, whereas several mutants defective in nuclear division have a complete spindle. Most of these arrests of SPB behavior appear to result from global regulation of cell cycle specific processes, for other features of the cell cycle (e.g., DNA replication and budding) are arrested in states roughly corresponding to similar stages of the ongoing cycle. The predominance of indirect influences on spindle development among the cdc mutants is exemplified by several DNA-defective mutants (such as cdc2, cdc8, cdc9, and cdc21), all of which become arrested with a short spindle following a primary failure of DNA metabolism.

By contrast, the primary defect of one mutation -- cdc31 -- appears to be more directly related to SPB function[7]. This mutation differs uniquely from other mutations of the cell division cycle in that it uncouples SPB duplication from other aspects of cell division. When a cell mutant in cdc31 is transferred to the restrictive temperature during vegetative growth, budding and chromosomal replication continue but the SPB fails to undergo duplication (Figure 1, dashed arrows); the ensuing aberrant mitosis upon a monopolar spindle results in polyploidization[8]. Although the SPB has failed to double, its dimensions are increased to an extent suggesting that the principal constituents of the daughter SPB had been formed but failed to become assembled into a separate entity. Indeed, it is possible that the sole function of CDC31 is in SPB duplication, for when diploids homozygous

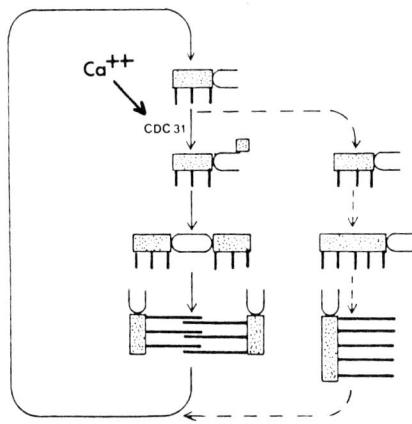

FIGURE 1. Possible mechanism of alteration in the pathway of spindle pole body (SPB) duplication by failure of CDC31 function. Normally the SPB (stippled rectangle at top of diagram) gains a satellite (small stippled square) at the distal margin of its half-bridge, leading next to formation of a daughter SPB and finally to formation of the opposite spindle pole. No satellite forms in cdc31, whereupon the single SPB simply expands, bearing a greater number of microtubules (thick lines) and forming a monopolar spindle. Structure of the derived CDC31 product (this report) implies its regulation by Ca^{++}.

for cdc31 are sporulated under restrictive conditions, asci containing two diploid spores are formed. These result from failure of SPB duplication in meiosis II and thereby appear to reflect the same specific requirement for this gene function.

Negative Regulation by ESP1

Reasoning that polyploidization of this sort might serve to identify other mutants defective in SPB regulation, we screened for mutants causing diploidization after a brief incubation at elevated temperature. This search led to the isolation of a mutation, termed esp1 for extra spindle pole bodies, which causes yet another aberrant pattern of SPB behavior. When growing cells are shifted to the restrictive temperature, they continue to increase their number of SPB's incessantly despite failure of other mitotic processes to proceed. Electron microscopy reveals that many cells come to contain large numbers of SPB's (as many as eight have been found in a single nucleus). SPB's become interconnected by spindle microtubules, resulting in formation of several multipolar spindles. Each SPB appears to be normal in its structure and dimensions; moreover, each retains the ability to mediate the polymerization of microtubules. The apparent defect is one of co-ordination -- specifically, in the regulation of SPB number.

As outlined earlier, cells mutant in other cdc genes become arrested in a stage-specific manner with one or two SPB's, as would be appropriate for the stage of arrest. We have sought to determine whether the apparent retention of control over SPB number at the cdc arrests is mediated through the action of ESP1 by constructing strains mutant both in esp1 and in one of various cdc genes. Further, we have tested the ability of the mating pheromone, α-factor, to exert its normal mode of control over the cell cycle in esp1 mutants. The results to date (Figure 2) are that several of the cdc - esp1 double mutants formed extra spindle pole bodies, revealing that the characteristic arrests of these cdc mutants depend upon negative regulation of SPB number under the control of ESP1. On the other hand, the pheromone-mediated arrest and at least one other cdc arrest (that by cdc28) occurred normally; that is, they were epistatic to esp1, suggesting that the arrest of SPB duplication in these cases results from another mode of regulation (unless these arrests were

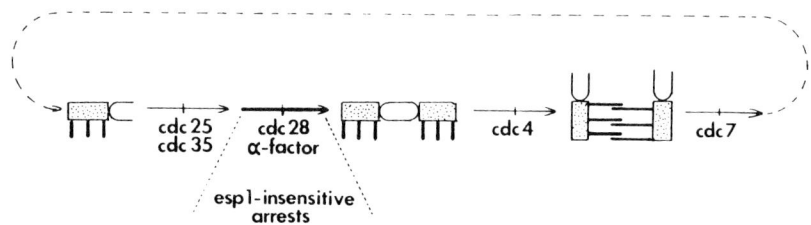

FIGURE 2. Epistatic relationships between cdc mutations which cause G1 arrests and esp1. Only the esp1 - insensitive arrests fail to result in the formation of extra spindle pole bodies, suggesting a second mode of regulation in this phase of the cell cycle. (The indicated order of cdc25/cdc35 and cdc28/α-factor steps has not been well established.)

simply more permissive of function by the esp1 allele). We have obtained a molecular clone of ESP1, with which we hope to determine whether the differences detected here are related to cell-cycle-specific changes in gene expression.

Calcium-ion Binding Pockets in the CDC31 Product

Analysis of the function controlled by CDC31 itself is also being pursued by characterization of the cloned gene, and DNA sequence analysis has already provided insight into the mechanism of gene action. The amino acid sequence was derived from the nucleotide sequence of the genomic clone and compared with other proteins whose sequences are known. This comparison revealed that the CDC31 gene product displays significant homology to calmodulin and several other members of the eukaryotic Ca^{++}-binding family of proteins (Figure 3). In addition, consideration of the criteria for cation binding specified by the EF-hand rule of Kretsinger and Barry[9] suggests the presence of at least two effective Ca^{++}-binding pockets in this protein (also see Thorner, this volume).

The probability that the CDC31 product is a Ca^{++}-binding protein leads us to the working hypothesis that the regulation of its action with respect to SPB duplication depends on the generation of a flux of calcium ions at a specific stage of the cell cycle (Figure 1). This

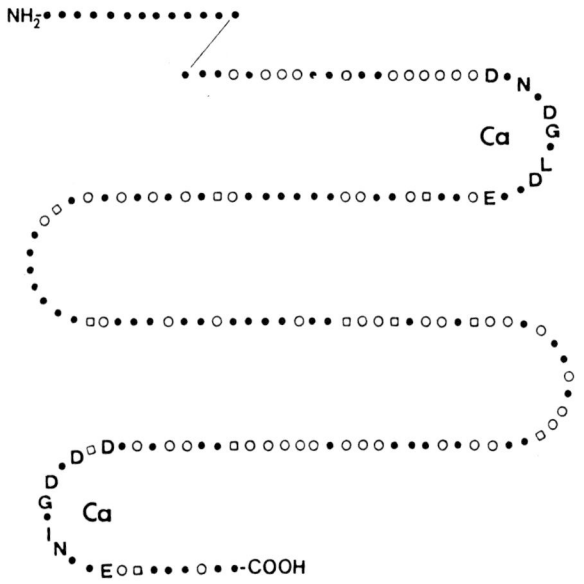

FIGURE 3. Representation of the derived CDC31 product indicating its homology with calmodulin. Exact alignment of the carboxyl termini leaves an extension at the amino terminus of the CDC31 product but reveals numerous identities (o) and conserved differences (□) throughout the rest of the protein. The positions of four calcium-ion binding sites in calmodulin are represented by the curved regions; the corresponding first and fourth sites in the CDC31 product contain the indicated residues crucial for binding, according to the EF-hand rule.

ion flux might be expected to co-ordinate cellular morphogenesis by controlling other processes appropriate to the same stage. Calcium transients have previously been implicated in the transition from quiescence to cell division in a number of systems including replated mammalian cells in tissue culture[10] and the mitotic activation of fertilized eggs of sea urchins[11] and amphibians[12]. One report also documents a role for calcium ions in the continuance of division under regulation of maturation promoting factor and cytostolic factor in Xenopus embryos[13]. The recent finding that yeast cells may contain an activity similar to maturation promoting factor suggests relevance to the present system[14]. There do, indeed, appear to be high levels of calcium ions maintained in the vacuoles of yeast by the action of a Ca^{++}/H^{+} antiport pump[15]. This reservoir might well serve as the source of the suggested fluxes into other cellular compartments.

Identification of the CDC31 product as a member of the family of Ca^{++}-binding proteins not only implicates calcium ions in its regulation but also provides a hint as to its mode of action. Other members of the family (calmodulin, troponin C, and myosin light chain) function by interacting with one or more other proteins to activate enzymatic activity in the latter[16,17]. We are therefore led to speculate that the CDC31 product acts in an analogous manner, mediating the action of another protein to which it binds. The nature of such enzymatic activity remains obscure but the specificity of the cdc31 phenotype to failure of SPB duplication suggests that the activity may occur in the vicinity of the SPB. In particular, we note from previous work that no satellite is formed adjacent to the existing SPB (as diagrammed in Figure 1) in the cycle during which SPB duplication fails[7]. This raises several likely possibilities for the putative CDC31-controlled function: One is that this function is required to generate the satellite from which the daughter SPB develops. Another is that it modifies the parental SPB, which thereby becomes unable to compete with the locus of the incipient satellite for assembly of nascent SPB materials. Investigation of these possibilities, as well as of the unbridled duplication of SPB's in espl, will require further genetic analysis and utilization of the cloned genes to probe the behavior of their gene products.

ACKNOWLEDGMENTS

This research was supported by the National Institutes of Health (GM18541).

REFERENCES

1. McIntosh, J.R. (1983) Modern Cell Biol. 2, 115-142.
2. Robbins, E., Jentzsch, G. and Micali, A. (1968) J. Cell Biol. 36, 329-339.
3. Byers, B., (1981) in The Molecular Biology of the Yeast Saccharomyces: Life Cycle and Inheritance, pp. 59-96.
4. Byers, B. and Goetsch, L. (1975) J. Bact. 124, 511-523.
5. Moens, P.B. and Rapport, E. (1971) J. Cell Biol. 50, 344-361.
6. Pringle, J.R. and Hartwell, L.H. (1981) in The Molecular Biology of the Yeast Saccharomyces: Life Cycle and Inheritance, pp. 97-142.
7. Byers, B. (1981) in Molecular Genetics in Yeast, Alfred Benzon Symp. 16, pp. 119-131.
8. Schild, D., Ananthaswarmy, A.N. and Mortimer, R.K. (1981) Genetics 97, 551-562.
9. Kretsinger, R.H. and Barry, C.D. (1975) Biochem. Biophys. Acta 405, 40-52
10. Hesketh, T.R., Moore, J.P., Morris, J.D.H., Taylor, M.V., Rogers, J., Smith, G. A. and Metcalfe, J.C. (1985) Nature 313, 481-484.
11. Poenie, M., Alderton, J., Tsieu, R.Y. and Steinhardt, R.A. (1985) Nature 315, 147-149.
12. Busa, W.B. and Nuccitelli, R. (1985) J. Cell Biol. 100, 1325-1329.
13. Newport, J.W. and Kirschner, M.W. (1984) Cell 37, 731-742.
14. Weintraub, H., Buscaglia, M., Ferrez, S., Weiller, A. Boulet, F., Fabre, F. and Baulieu, E.E. (1982) C. R. Acad. Sci. (Paris) Ser. III 295, 787-790.
15. Ohsumi, Y. and Anraku, Y. (1983) J. Biol. Chem. 258, 5614-5617.
16. Klee, C.B., Crouch, T.H. and P.G. Richman (1980) Ann. Rev. Biochem. 49, 489-515.
17. Adelstein, R.S. and Eisenberg, E. (1980) Ann. Rev. Biochem. 49, 921-956.

GENE EXPRESSION DURING SPORULATION

Stephen Kurtz, Janice Rossi and Susan Lindquist

Department of Molecular Genetics and Cell Biology
The University of Chicago
Chicago, Illinois 60637

ABSTRACT. A dramatic series of changes in gene expression occur during the course of sporulation in <u>Saccharomyces cerevisiae</u>. Some of these are a response to the nutrient-deficient conditions required to induce sporulation since they are observed in both sporulation-competent and sporulation-incompetent cells. Others occur only in sporulating cells and appear to play a role in the major morphogenic events of ascospore development.

In <u>Saccharomyces cerevisiae</u>, sporulation is a developmental process that is restricted, in most genetic backgrounds, to diploid a/α cells. These cells, when deprived of essential nutrients, undergo meiosis and form haploid spores that are specialized to withstand environmental extremes. Many of the genetic, morphological and biochemical aspects of sporulation have been described (reviewed in 1,2), but changes in gene expression which accompany or effect spore formation are only now being elucidated. This review examines recent progress on gene expression in sporulating cells.

The identification of gene products induced in sporulating cells is an essential first step in the study of developmental gene regulation in yeast. It is necessary to distinguish those that are specific to sporulation from those that are part of a more general cellular response to starvation. This is accomplished by comparing gene products isolated from a/α diploids undergoing sporulation with gene products isolated from asporogenous diploids (a/a and α/α) cultured under identical conditions. Those found only in a/α cells are considered to be sporulation-specific. Using this criterion, many investigators have examined gene expression during sporulation.

Sporulation-Specific Enzyme Activities

Several proteins, identified on the basis of their enzymatic activities, are expressed in a sporulation-specific manner. One of these, β-1,4-glucosidase, is a glycogenolytic enzyme that is present late in sporulation, during a period of extensive glycogen degradation (3). An enzyme required for glucosamine synthesis, 2-amino-2-deoxy-D-glucose-6-phosphate-ketolisomerase, appears at the tetranucleate stage of meiosis in sporulating cells and contributes to the maturation of the outer spore wall (4). Yet another enzymatic activity that appears to be sporulation-specific is 1,3-β-glucanase; however, its role in spore formation has not yet been established (5). The activities of certain enzymes, e.g., proteases A,B and C and ribonuclease, are elevated during sporulation but are not restricted to sporulating cells. They may be responsible for the macromolecule degradation that provides new substates for the synthesis of sporulation specific structures (reviewed in 1).

Gene Expression--Early In Vivo Studies

Until recently, attempts to obtain a more general view of gene expression in sporulating cells had been largely unsuccessful. An early problem was the impermeability of sporulating yeast cells to radiolabelled precursors. As cells sporulate the pH of the medium increases, inhibiting the activity of amino acid transferases. The problem was overcome simply by buffering the medium at pH 5.5 to 7 (6-9). Using this method, many changes in the protein composition of sporulating cells were identified by the incorporation of radiolabeled amino acids. Surprisingly, the same changes were observed in asporagenous diploids. That is, they were not sporulation-specific.

Another attempt to circumvent the impermeability problem relied on the reutilization of isotopes introduced during presporulation growth (10). This study identified several sporulation-specific changes in protein composition but the interpretation of these changes was problematic. It was not possible to distinguish between those due to de novo protein synthesis and those due to post-translational modification of pre-existing proteins. Thus, although protein synthesis was known to be required for sporulation (11), the enormous differences in structure between sporulating and non-sporulating cells could not be specifically associated with the expression of new proteins.

Gene Expression--Recent Studies

Recently, two different approaches have provided substantial insights into sporulation gene expression. In one, cDNAs were used to screen yeast genomic libraries for sequences preferentially expressed during sporulation (12,13). The cDNA probes, prepared from the messages of sporulating a/α cells and asporogenous a/a cells, identified about 14 genes that are induced only in sporulating a/α cells. The restriction endonuclease maps of these clones indicate that different genes have been isolated in different laboratories (14). The analysis of some of these genes confirms that they are transcribed in a specific temporal manner during sporulation. Furthermore, a mutation produced in one of these genes results in loss of the ability to sporulate (14). Although the protein products of most of these genes have yet to be identified, they will provide an important basis for further study.

We have taken a different approach to the analysis of sporulation specific gene expression. We isolated RNAs from cultures of sporulating cells and determined their protein coding capacities by translation in cell-free lysates. To identify changes unique to sporulation RNAs were also isolated from isogenic non-sporulating a/a and α/α diploids which were placed in the same induction medium. The results demonstrate a dramatic series of changes in gene expression during sporulation (15). Some of these are unique to sporulating cells; others are not.

A variety of different strains of S. cerevisiae were used in this study. The results presented here were obtained with the diploid AP3, a strain that sporulates in a rapid and nearly synchronous manner and achieves high levels of ascus formation. Similar results were obtained with other strains except that the timing of events varied.

Cells, preadjusted to respiratory metabolism, were collected in mid-logarithmic growth and resuspended in nitrogen-deficient sporulation medium. Using these culture conditions, tetranucleate cells, indicating the completion of meiosis, were first detected at eight hours. Within 24 hours, 90% of the cells had completed sporulation.

At various intervals during sporulation, total cellular RNAs were extracted and translated in wheat germ lysates. Figure 1 displays the translation products encoded by these RNAs. In this experiment, each translation reaction contained an identical quantity of total yeast cellular RNA at a concentration that was well below saturation for the lysate. Thus, the changing intensities of individual protein bands during this time course reflect the changing abundance of their messenger RNAs.

The initial response to the change in medium was a rapid increase in the concentration of RNAs encoding a very broad range of polypeptides. RNAs isolated 2 hours after transfer to nitrogen-deficient medium directed twice as much incorporation into protein as RNAs from vegetative cells. Most of the induced species were already present during vegetative growth but at lower concentrations.

Another change in the message population became apparent 4 to 6 hours after transfer to sporulation medium. Namely, a number of messages characteristic of vegetative growth were sharply diminished. A few examples, detectable at the level of exposure shown in Figure 1, are marked with dots (◉).

Superimposed upon the general increase in translatable RNAs, there was a dramatic induction of messenger RNA for a 26 kDa protein. This message was barely detectable in vegetative cells but within 6 hours after transfer to nitrogen-free medium it was the most intensely translated species in the in vitro reaction. Its protein product comigrated on SDS-polyacrylamide gels with the 26 kDa yeast heat-shock protein, hsp26. (An in vitro translation reaction with RNAs isolated from heat-shocked cells is displayed on the far right of Figure 1.) Message for a polypeptide of 84 kDa, which comigrated with another of the major heat-shock proteins, was also induced at this time. That these RNAs were messages for heat-shock proteins was demonstrated by hybridizing electrophoretically separated RNAs to DNA probes for the heat shock genes (Figure 2).

Later in sporulation two new sets of mRNAs were coordinately induced. In vitro translation products of these RNAs are marked with dots (●) in Figure 1. They first appeared after 6 hours and encoded polypeptides of 17, 20, 25, 31, 38, 50, 65 and 68.5 kDa. These RNAs were maximally induced between 8 and 14 hours and then began to disappear. After 16 hours, the second set of RNAs, encoding 21.5 and 34 kDa polypeptides, was induced. These RNAs continued to accumulate late into sporulation and were retained in the mature ascus.

Non-sporulating diploid strains were grown in an identical manner and transferred to nitrogen-deficient sporulation medium. These strains are isogenic with AP3 except at the mating type locus where they are homozygous for a or α. Initially, both the a/a and α/α strains responded to the change in medium in a manner similar to the a/α diploid. That is, many RNAs showed an enhanced abundance and several RNAs characteristic of vegetative growth were diminished over the first few hours (data not shown). Again, most prominent among the induced species were RNAs encoding polypeptides that comigrate with hsp26 and hsp84.

The hsp26 and hsp84 RNAs became abundant in a/a and α/α

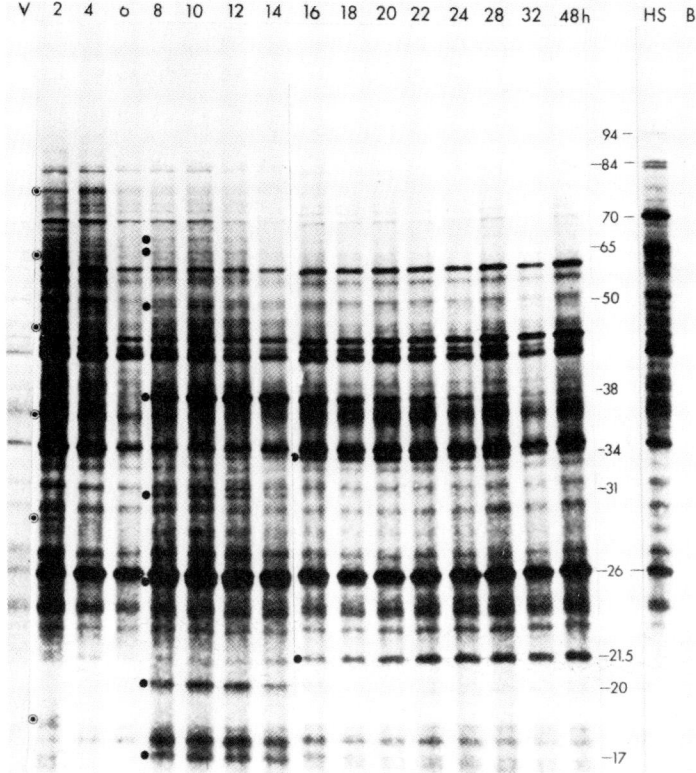

Figure 1.--The changing message composition of sporulating cells. RNAs were extracted from aliquots of a continuous culture of diploid AP3 (a/α) at various times after transfer to nitrogen-deficient medium (times indicated in hours, lanes 2-48 h). The RNAs were translated in a wheat germ lysate and the products separated on SDS polyacrylamide gels. Each reaction was normalized to contain the same concentration of ribosomal RNA. Lane V, translation products of vegetative cell RNAs; lane HS, translation products or RNAs from cells heat shocked at 39°C for 1 h; lane B, translation reaction in the absence of exogenous RNAs (blank). Apparent molecular weights (x 10^3) are indicated. Dots (◉) to the right of lane V indicate polypeptides repressed during sporulation. Dots (●) to the right of lane 6 h and 14 h indicate polypeptides induced during sporulation.

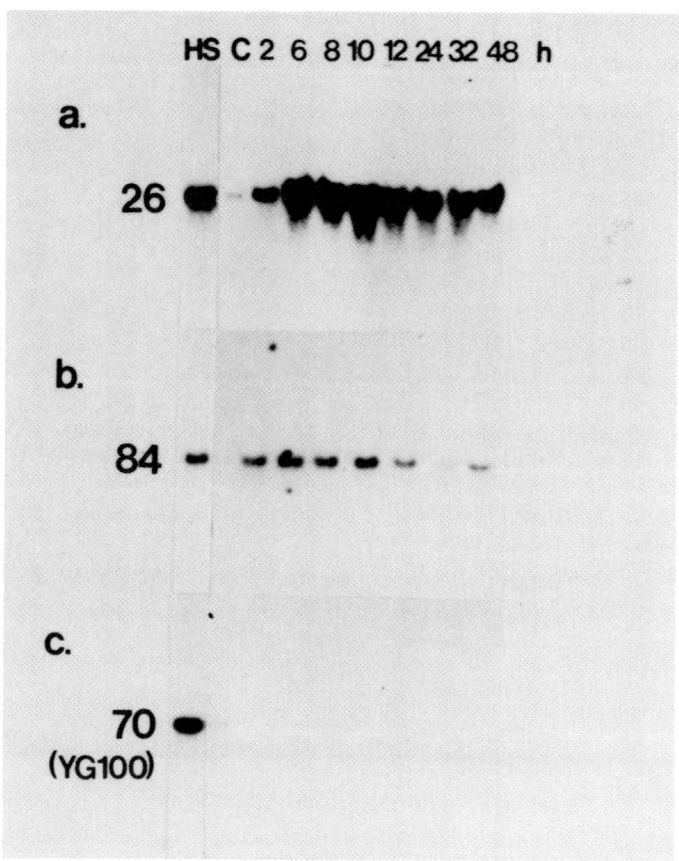

Figure 2.—Expression of heat shock genes during sporulation. For each time point, 2 ug of total RNAs were separated on 1.2% agarose gels containing 5 mM methylmercury hydroxide, transferred to nitrocellulose, and hybridized with ^{32}P labelled DNA probes for the yeast hsp26 (panel A), hsp84 (panel B), and hsp70 (panel C) genes. Lane designations: HS, heat shock RNAs extracted from midlogarthmic cells after a 1 hour incubation at 39°C; C, control RNAs extracted from midlogarthmic cells grown at 30°C; 2-48 h, RNAs extracted from cells at various times after transfer to sporulation medium. (Note: the 32 h time point was omitted from panels B and C.)

cells at about the same time they became abundant in sporulating cells. These RNAs were also produced in both haploid and diploid cells during the transition from log to stationary phase growth. These inductions differed from typical stress responses (such as those caused by exposure to high temperatures, ethanol, and sodium arsenite) in a significant way. Namely, messages for hsp70 were not induced. Instead, proteins related to hsp70, with apparent molecular weights of 70 and 71 kDa, were produced (17). Clearly, these changes in gene expression are not related to the major morphogenic events of spore formation. They appear to be involved in a more general aspect of cellular differentiation. A remarkably similar pattern of heat shock gene expression has been reported to occur in <u>Drosophila</u> during puparium formation and oogenesis (16,17).

A dramatic difference between the sporulating diploid and its non-sporulating derivatives occurred after eight hours in sporulation medium. Unlike the sporulating diploid, neither the a/a nor the α/α cells produced the coordinately regulated sets of messages for the 17, 20, 25, 31, 38, 50, 65, and 68.5 kDa polypeptides and for the 21.5 and 34 kDa polypeptides. Therefore, the induction of these two sets of RNAs is unique to sporulating cells.

Characterization of Sporulation-Specific Messages

Several observations suggest that the first set of sporulation-specific RNAs are involved in spore wall assembly. First, the intensity of their translation <u>in vitro</u> suggests that the RNAs are very abundant and most likely encode structural proteins. Second, the messages appear at the tetranucleate stage of meiosis, are detectable for a few hours and are then degraded. Electron microscopy combined with biochemical analysis indicates that their appearance is coincident with the initiation of spore wall synthesis and their disappearance is coincident with its completion (18).

A third piece of evidence is provided by the process of spore wall assembly itself. The yeast spore wall is formed by the deposition of material within a double membrane structure surrounding the nascent spore (reviewed in 19, Figure 3). Components of the wall, complex carbohydrates and mannoproteins, must therefore cross this membrane and may use signal sequences in a manner similar to secreted proteins. The translation behavior of the first set of sporulation RNAs <u>in vitro</u> suggests that their translation is associated with membranes. In cell-free lysates that contain high levels of SRPs, these messages are translated much less efficiently than the other messages in the

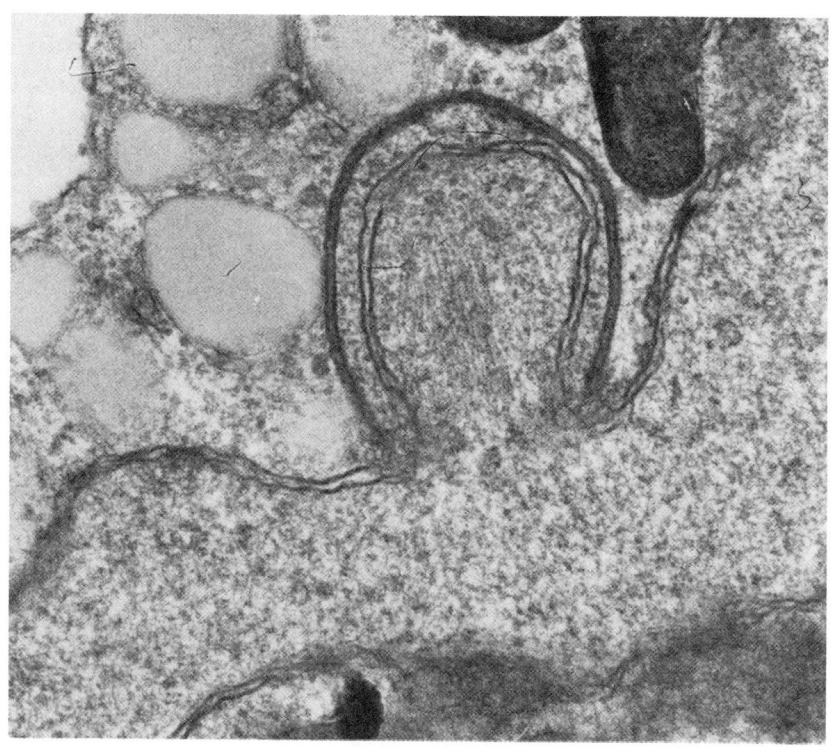

Figure 3.—A double membrane engulfs the emerging spore nucleus. Thin section of strain AP3 at 8 hours in sporulation. Magnification X 50,000.

population. They are translated with greater efficiency when the in vitro reaction is supplemented with microsomal membranes (18).

The second set of sporulation-specific RNAs, encoding the 34 and 21.5 kDa proteins, are induced when the first set of sporulation-specific RNAs is being degraded. They continue to accumulate late into sporulation and are maintained in a stable form in mature asci. Their accumulation indicates that the spore genome is transcriptionally active late in development.

The increasing concentrations of the second set of RNAs suggests that their polypeptides are required in significant amounts. The functions encoded by these RNAs are unknown, however, their presence in mature spores and their rapid degradation during germination (18) suggest that they are either involved in establishing the final stages of dormancy or stored for use during the initial stages of spore outgrowth.

The two sets of sporulation specific RNAs are induced at very different times in development. To determine whether they were present in the same or different subcellular compartments, we fractionated sporulating cells to obtain ascal cytoplasm and immature spores (18). RNAs were extracted from each fraction and analyzed by translation in vitro. The distribution of RNAs in these two fractions was dramatically different. The first set of RNAs was found only in the ascal cytoplasm. In contrast, the immature spore contained a broad array of RNAs not found in the ascal cytoplasm, including the second set of sporulation-specific RNAs. This distribution suggests that the developing ascus is differentiated into functionally distinct compartments, each responsible for particular aspects of spore formation.

A question arising from these results is why had the proteins produced by sporulation-specific messages in vitro not been detected in earlier in vivo studies? It is not simply that our strains produce more messages than others. We also have been unable to detect the proteins by labeling cells in vivo. There are two possible explanations. The first is that the RNAs simply do not code for proteins synthesized in the developing ascus. While this must remain a formal possibility, we consider it extremely unlikely that yeast cells would produce abundant, highly translatable RNAs at the time of spore wall synthesis and degrade them immediately thereafter without using them. Furthermore, we know that other RNAs are translated at this time. Antibodies specific to the 26 kDa heat-shock protein indicate its message is translated to produce one of the most abundant proteins in sporulating cells (17).

We believe the most likely explanation for the failure to detect these proteins in vivo is that they are immediately processed into a form which is either insoluble in sample buffer or unable to enter the gel matrix. This might be expected for proteins that are associated with or crosslinked into the carbohydrate complex of the cell wall. Co-translational modification of proteins secreted across cellular membranes is common in yeast and in other organisms (21). The nature of the spore wall, specialized to produce a structurally resilient, impermeable barrier to the environment, makes it highly likely that such protein modifications would occur during its synthesis.

Another question is how do the genes selected by differential cDNA hybridizations in other laboratories relate to the messages detected by in vitro translation in our laboratory? In collaboration with J. Segall we have recently demonstrated by in vitro translation of hybrid selected messenger RNAs that the clone designated p275 (13) codes for the 38 kDa polypeptide and that the clones designated p18 and p84 code for the 50 kDa polypeptide (F. Rosenberg, J. Segall, and S. Lindquist, unpublished results). Work is now concentrated on determining the specific functions of the proteins and the mechanisms by which they are regulated.

REFERENCES

1. Esposito RE, Klapholz S (1981). Meiosis and Ascospore Development. In Strathern JN, Jones EW, Broach JR (eds): "The Molecular Biology of the Yeast Saccharomyces," Cold Spring Harbor, New York: Cold Spring Harbor Laboratory, p. 211.
2. Dawes IW (1983). Genetic control and gene expression during meiosis and sporulation in S. cerevisiae. In Spencer JFT, Spencer DM, Smith ARW (eds): "Yeast Genetics", New York: Springer-Verlag, p.29.
3. Colonna WJ and Magee PT (1978). Glycogenolytic enzymes in sporulating yeast. J. Bacteriol 134, 844.
4. Ballou CE, Maitra SK, Walker JW, Whelan WL (1977). Developmental defects associated with glucosamine auxotrophy in Saccharomyces cerevisiae. Proc. Natl. Acad. Sci. USA 74, 4357.
5. delRay F, Santos T, Garcia-Acha I, Nombela C (1980). Synthesis of β-glucanases during sporulation in S. cerevisiae. J. Bacteriol 143, 621.
6. Hopper AK, Magee PT, Welch SK, Friedman M, Hall BD (1974). Macromolecule synthesis and breakdown in relation to sporulation and meiosis in yeast. J Bacteriol 119, 619.
7. Petersen JG, Kielland-Brandt MC, Nilson-Tillgren T (1979). Protein patterns of yeast during sporulation. Carlsberg Res Commun 44: 149.

8. Trew BJ, Friesen JD, Moens PB (1979). Two-dimensional protein patterns during growth and sporulation in Saccharomyces cerevisiae. J Bacteriol 138, 60.
9. Kraig E, Haber JE (1980). Messenger ribonucleic acid and protein metabolism during sporulation of Saccharomyces cerevisiae. J Bacteriol 144, 1098.
10. Wright JF, Dawes JW (1979). Sporulation-specific protein changes in yeast. FEBS Lett 104, 183.
11. Magee PT, Hopper AK (1974). Protein synthesis in relation to sporulation and meiosis in yeast. J Bacteriol 119, 592.
12. Clancy MJ, Buten-Magee B, Straight DJ, Kennedy AL, Partridge RM, Magee PT (1983). Isolation of genes expressed preferentially during sporulation in the yeast Saccharomyces cerevisiae. Proc. Natl. Acad. Sci. USA 80, 3000.
13. Percival-Smith AP, Segall J (1984). Isolation of DNA sequences preferentially expressed during sporulation in Saccharomyces cerevisiae. Mol Cell Biol 4, 142.
14. Segall J, personal communication.
15. Kurtz S, Lindquist S (1984). Changing patterns of gene expression during sporulation in yeast. Proc Natl Acad Sci (USA) 81, 7323.
16. Zimmerman JL, Petri WL, Meselson M (1983). Accumulation of specific subsets of D. melanogaster heat shock mRNAs in normal development without heat shock. Cell 32, 1161.
17. Kurtz S, Rossi J, Petko L, Lindquist S (1986). An ancient developmental induction: same heat shock proteins induced in Saccharomyces sporulation and Drosophila oogenesis. Science, in press.
18. Kurtz S, Lindquist S (1985) manuscript submitted.
19. Byers B (1981). Cytology of the yeast life cycle. In Strathern JN, Jones EW, Broach JR (eds): "The Molecular Biology of the Yeast Saccharomyces," Cold Spring Harbor, New York: Cold Spring Harbor Laboratory, p. 59.
20. Walter P, Gilmore R, Blobel G (1984). Protein translocation across the endoplasmic reticulum. Cell 38, 5.

III. DNA REPLICATION AND CHROMOSOME STRUCTURE

YEAST CHROMOSOMAL DNA REPLICATION

J. L. Campbell, L. M. Johnson, A.Y.S. Jong, M. Budd, K. Sweder and F. Srienc

Divisions of Biology and Chemistry
California Institute of Technology
Pasadena, California 91125

ABSTRACT This paper presents a summary of current strategies to dissect the yeast DNA replication apparatus. We report the isolation of the genes encoding DNA polymerase I and SSB-1. We demonstrate that a 19bp sequence in ARS1 is sufficient for autonomous replication and provide evidence for a protein in extracts that binds the ARS consensus sequence.

INTRODUCTION

For the past several years, we and others have been using in vitro replication systems in yeast in order to identify the proteins and DNA sequences involved in eukaryotic replication. Several in vitro replication systems have been developed (for review see 1) and several sets of mutants that probably have defects in replication have been described (2-7). There are two classical methods of utilizing in vitro replication systems to identify proteins involved in DNA synthesis. First, in vitro complementation assays employ extracts of mutants containing an inactive replication protein. The mutant extracts are complemented by extracts of wild-type cells that contain the normal replication protein. Second, resolution and reconstitution uses enzyme fractionation to identify proteins involved in replication and then reconstitutes the system by combining appropriate enzyme fractions. Here we will give specific examples of how we are employing both approaches to dissect the yeast replication apparatus.

RESULTS

The Role of the CDC8 Protein in Yeast.

Synthesis in several in vitro replication systems has been shown to be thermolabile in extracts prepared from cdc8 mutant strains (8-12). Since the nucleoside triphosphates are provided in the in vitro systems, the requirement for CDC8 in vitro was interpreted to mean that the CDC8 gene product is required directly for DNA synthesis and not for the production of the nucleotide precursors of DNA synthesis. Sclafani and Fangman (13), however, showed that Herpes simplex thymidine kinase could complement cdc8 defects. They further showed that cdc8 mutants were deficient in thymidylate kinase which normally catalyzes the following reaction:

$$dTMP + ATP \rightleftharpoons dTDP + ADP.$$

To establish that this activity was really the product of the CDC8 gene, we purified thymidylate kinase from an overproducing strain and determined 18 amino acids of the N-terminal amino acid sequence (14-16). This sequence was identical to that derived from the sequence we had determined for the cloned CDC8 gene (14-16).

The demonstration that CDC8 encoded thymidylate kinase was surprising because a 37,000 Da single-strand binding protein had been found to complement cdc8 extracts in vitro and was thus proposed to be the product of the CDC8 gene (17). This 37,000 Da protein is probably the same protein as the SSB-1 protein described by LaBonne and Dumas (18). In order to see if the above proteins were related to thymidylate kinase we purified several SSB proteins from yeast (data not shown). The proteins we purified have molecular weights of 20,000, 31,000, 45,000 and 50,000 and will be referred to as p20, p31, p45 and p50. p45 corresponds to SSB-1 and therefore to the protein that was claimed to be the product of CDC8 (17) based on the amino acid composition of p45 being 85% homologous to SSB-1 (17, 18 and Table 1) and on the fact that p45 also specifically stimulates yeast DNA polymerase I (see below). However, it is unlikely that p45 is related to the product of the CDC8 gene, which we have shown is thymidylate kinase, for several reasons. The molecular weight of the CDC8 gene product from DNA sequence is 24,792 and thus cannot encode p45. In contrast, thymidylate kinase migrates as a 25,000 Da species on polyacrylamide gels containing SDS, corresponding to the size calculated from DNA sequence. Furthermore, the amino acid composition of thymidylate kinase is not similar to that of the p45 SSB (Table 1). We have also determined the amino acid sequence of 53 residues of this SSB (p45) (data not shown) but failed to locate this sequence in the open reading frame defined by the CDC8 gene. Finally, we have determined the map position of the cloned SSB-1 gene (see below) by orthogonal pulse electrophoresis (19) and concluded that the gene does not reside on chromosome X as does CDC8 (20).

TABLE 1. AMINO ACID COMPOSITION OF THYMIDYLATE KINASE AND SSB-45,000.

AMINO ACID RESIDUE	AMINO ACID ANALYSIS OF dTMP KINASE	AMINO ACID ANALYSIS OF SSB - 45,000
ALA	9	23
ARG	9	17
ASP+ASN	28	43
CYS	N.D.	N.D.
GLU+GLN	25	55
GLY	16	43
HIS	4	9
ILE	12	27
LEU	25	12
LYS	24	27
MET	N.D.	5
PHE	9	32
PRO	3	15
SER	13	14
THR	14	21
TRP	N.D.	N.D.
TYR	N.D.	3
VAL	8	24

Thus, the in vitro complementation assay probably reflects non-specific stimulation of these extracts by the 37,000 Da protein and not specific complementation. The in vitro defect in cdc8 extracts may be due to limiting thymidine triphosphate due to phosphatase activity, or it may reveal a requirement for thymidylate kinase in the actual complex at the replication fork (see refs. 13 -15 for discussion). A number of models propose that ribonucleotide reductase and other enzymes involved in the synthesis of nucleotides directly participate in replication at the replication fork.

Resolution and Reconstitution - DNA Polymerase I.

Initiation of replication on naked DNA templates during in vitro replication can be divided for the purposes of discussion into a series of discrete steps--a sequence-specific event at the origin of replication followed by non-sequence specific steps that lead to the initiation of the replication fork. In bacteriophage lambda, for instance, the O protein recognizes four sites in the origin region and binds there (21). This step is thought to somehow alter the configuration of the DNA allowing prepriming proteins to bind, followed by initiation of primer synthesis by primase and elongation of the leading strand by DNA polymerase. Fork movement is facilitated by a helicase, which by analogy with phages T7 and T4, acts on the lagging strand (21). We propose this as a working model for replication in yeast which we wish to reconstitute from purified proteins.

Let us look at the non-sequence specific proteins first. The first goal is to identify the DNA polymerase involved. Yeast contains three DNA polymerases. DNA polymerase I is large and composed of several subunits (1,22). Recently it has been shown to co-purify with a DNA primase activity, suggesting that it might be involved in DNA replication (22-24). It is similar to DNA polymerase from higher cells in its subunit composition and its sensitivity to aphidicolin. DNA polymerase II is more prokaryotic in nature and co-purifies with a 3'-exonuclease activity, although it is also inhibited by aphidicolin. A third activity is primarily associated with the mitochondria. These properties suggested but did not prove that the DNA polymerase involved in replication was polymerase I. In order to investigate which DNA polymerase is active at the replication fork we have isolated the gene for DNA polymerase I.

The DNA polymerase I gene was isolated (manuscript in preparation) using rabbit polyclonal antibodies (25) to screen a lambda gt11 expression library (26) according to the method of Young and Davis (27). The identity of the polymerase I gene was verified by three methods: 1) crossreaction of antibodies (affinity-purified) using peptides synthesized from lambda gt11 clones with DNA polymerase I; 2) overproduction of polymerase activity in yeast when the gene is contained on a multicopy vector; 3) synthesis of catalytically active yeast DNA polymerase I in E. coli cells containing the appropriate clone.

The latter experiment is shown in Table 2. The yeast DNA polymerase I gene was inserted downstream of the phage T7 RNA polymerase promoter in the vector pT7-1 (28). The recombinant plasmid, pJLC33, was introduced into a strain carrying a plasmid encoding the T7 RNA polymerase. Induction of T7 RNA polymerase allows transcription of the yeast polymerase gene from the T7 promoter (28). For these studies we assumed that appropriate signals near the beginning of the DNA polymerase gene would initiate translation of the protein.

Extracts of the plasmid-containing cells were made and assayed for DNA polymerase activity. To distinguish between the E. coli polymerases and the yeast polymerase, we used specific inhibitors and antibody to yeast DNA polymerase I. As shown in Table 2, pT7-1 cells give 41 pmol incorporation, the 100% control value. Cells containing pJLC33, carrying the yeast DNA polymerase gene, have 166% more polymerase activity than the parental vector pT7-1. The addition of dideoxy-TTP, a specific inhibitor of E. coli DNA polymerases, does not inhibit the yeast DNA polymerase (line 2, Table 2). Aphidicolin and anti-yeast DNA polymerase I antibody, conversely, inhibit the yeast polymerase and leave the E. coli polymerase active (89% and 93% in lines 4 and 5, respectively). Thus, this is unequivocally the gene for yeast DNA polymerase I and we have designated the gene POL1.

TABLE 2. DNA POLYMERASE ACTIVITY IN PLASMID-CONTAINING E.COLI CELLS.

Inhibitor[2]			% DNA Polymerase Activity[1,3]	
ddTTP	aph	α-PolI	pT7-1	pJLC33
−	−	−	100	266
+	−	−	22	164
+	+	−	24	20
−	+	−	89	87
−	−	+	93	94

[1] DNA polymerase activity determined for 5 μg crude extract made from JM101 containing plasmids pGP1-2 and either pT7-1 or pJLC33.

[2] ddTTP = dideoxy-TTP
aph = aphidicolin
α-PolI = anti yeast DNA polymerase I antibody

[3] 100% activity = 41 pmoles total nucleotide incorporated in 1 hr at 35°C.

Gene disruption. The chromosomal yeast polymerase I gene was disrupted using a plasmid containing an internal 1500 bp fragment of the POL1 gene inserted into pBR322. The plasmid pJC22-P2 has a 4.1 kb LEU2 fragment inserted into the PstI site of pBR322 and the plasmid pJC22-P11 has LEU2 inserted into the PstI site of the polymerase gene. A diploid strain was transformed with pJC22-P2 or pJC22-P11 and stable leucine prototrophs were selected. Southern blot analysis was used to show that integration had occurred at POL1 in the transformants to be studied further. Dissection of sporulated transformants of pJC22-P2 yielded the following segregation: 6 spores, 2:2 viable:inviable and 2 spores 1:3 viable:inviable. The transformants of pJC22-P11 gave the following segregation: 7 spores, 2:2 viable:inviable, 3 spores, 1:3 viable:inviable, and 1 spore 3:1 viable:inviable. LEU$^+$ segregated with the inviable spores. These results are consistent with the POL1 gene being a single copy, essential gene.

Finally, examination of inviable spores revealed a terminal phenotype typical of cell-division cycle mutations having a defect in replication (Fig. 1). Arrest is at medial nuclear division (29).

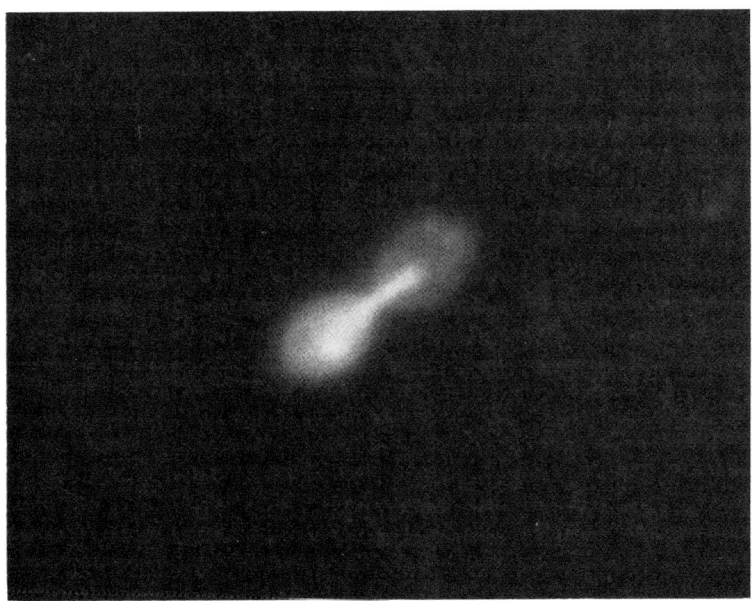

FIGURE 1. DAPI stained cells containing a disrupted POL1 gene. Tetrads were germinated after 1 week on sporulation plates, fixed with methanolic:acetic acid (3:1), stained with DAPI and observed under a Zeiss microscope.

Yeast SSB Proteins.

One additional protein required in all prokaryotic replication systems is a single-strand DNA binding protein. While several such proteins have been isolated in eukaryotes and some even stimulate their homologous DNA polymerases, a direct role for SSB proteins in eukaryotic replication remains to be established. Yeast offers an ideal opportunity to take a genetic approach to the question of the role of such proteins. There are several reports describing yeast proteins that bind single-stranded DNA. Chang et al. (30) have isolated a 37,000 Da protein that stimulates DNA polymerase I, called protein C, which binds ssDNA preferentially to dsDNA, though dsDNA is also bound. Recently LaBonne and Dumas (18) have isolated, on the basis of binding to ssDNA, another SSB (SSB-1) of M_r = 40,000. Finally, there is the SSB described by Arendes et al. (17). All of these proteins have similar molecular weights, DNA binding properties and all stimulate DNA polymerase I. However, the relationship of these species is not clear.

Our studies are aimed at systematically identifying the SSB proteins of yeast, as defined by preferential binding to ssDNA, with the intention of establishing through genetic, as well as biochemical means, their roles in DNA metabolism. As shown in Fig. 3, we began by purifying 3 SSB proteins chosen because of their relative abundance and their relatively high affinity for ssDNA. Two of the proteins are designated SSB-1 and SSB-2 based on the order of elution from ssDNA and on the fact that SSB-1 may be the same protein that LaBonne and Dumas designated SSB-1 (18). SSB-m is found only in mitochondrial subcellular fractions, and we have not investigated it further, since we are interested now in nuclear DNA synthesis. The M_r of SSB-1, SSB-2 and SSB-m are 45,000, 50,000, and 20,000, respectively.

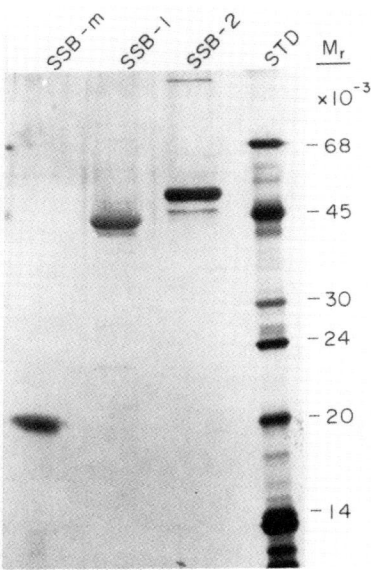

FIGURE 2. Purification of single-stranded DNA binding proteins from yeast. Purification procedures are described elsewhere (Jong and Campbell, in preparation). Each lane contains 3 ug of protein. Electrophoresis in an SDS-polyacrylamide gel and silver staining were by standard techniques. The three DNA binding proteins SSB-m, SSB-1 and SSB-2 are indicated on the top of the panel. In the right lane, STD, are standard proteins from Sigma Chemical Company. Bovine albumin (66,800), egg albumin (45,000), carbonic anhydrase (29,000), trypsinogen (24,000), trypsin inhibitor (20,100) and lactalbumin (14,200).

<u>Cloning of the gene for SSB-1</u>. We now wish to elucidate the roles of these proteins by isolating the genes and carrying out studies like those begun with DNA polymerase I. As an initial effort to assess the role of SSB-1 and SSB-2 in yeast, we have isolated the gene for

SSB-1. Rabbit polyclonal antibodies were prepared against SSB-1 and SSB-2; and the two proteins were shown to be unrelated immunologically. Since SSB-1 was found to stimulate DNA polymerase I (data not shown), we isolated this gene first. A lambda gt11 library was screened using affinity purified polyclonal antibody. Positives were used to identify homologous clones in a library prepared in the high copy number plasmid YEp24. YEp24 clones containing the SSB-1 gene, designated SSB1, were identified as those that showed overproduction of the SSB in yeast upon protein blot analysis of extracts of cells containing the cloned gene (Fig. 3). Thus, the starting material for genetic analysis is now available.

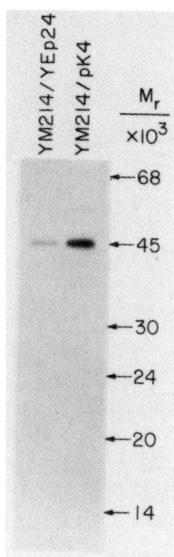

FIGURE 3. Western analysis of overproduction of SSB-1.

Strains containing the vector YEp24 (left lane) or pk4, which is YEp24 carrying SSB1, were grown to logarithmic phase. Extracts were prepared and proteins separated by electrophoresis in a polyacrylamide gel containing SDS. Proteins were transferred to nitrocellulose and probed with affinity purified antibody to SSB-1. Antibody antigen complexes were visualized with ^{125}I-protein A.

Role of Specific Sequences in DNA Replication

While progress has been made in studying the non-sequence specific proteins involved in replication, we have no knowledge of the sequence-specific proteins. In order to define these activities, it is first essential to have an idea of the sequences themselves. ARS elements, autonomously replicating sequence elements (31), are good candidates for origins of replication, for reasons discussed at length elsewhere (32). However, there is no direct evidence that yeast chromosomes initiate replication at the same site during every cell-cycle. Alternative roles for ARS elements are transcriptional enhancer or nuclear localization sequences. If the proteins that interact with these sequences can be defined, then mutants can be isolated and their phenotype should confirm the role of ARS elements in the cell.

According to data published previously (31, 32), ARS1 consists of about 300bp divided into 3 domains—A, B and C. Domain A is of interest because it contains the consensus sequence identified by others (for review see 1). Deletion and point mutations in this consensus sequence inactivate ARS elements (32, 33), demonstrating in a negative manner, its essential role. We have added positive evidence to this by subcloning a 25bp EcoR1-BglII fragment of YRpsb25 (32), which contains the 14bp consensus and five additional base pairs from ARS1 plus 6 nucleotides derived from the EcoR1 linker, into a plasmid containing a centromere. The fragment was inserted into YCpG2 (Fig. 5), which contains the 3.55 kb EcoR1-BamH1 CEN4 fragment (34). This CEN4 fragment does not contain ARS4, a weak ARS shown by Stinchcomb et al. (34) to be closely linked to CEN4. Surprisingly, YCpG2 contains only 19bp of ARS1 and transforms yeast at the same frequency as YCpG2 containing the complete ARS1. The transformants with the smaller insert grow slowly, however, with a generation time in liquid medium of 15h compared to 2.3h for the complete ARS1 segments or 3.5h for a plasmid containing about 100bp flanking the consensus on each side, i.e. with small deletions of domains B and C. What is striking is that only 19bp is sufficient to signal that a piece of DNA is to replicate in yeast. Even in prokaryotes much larger sequences are thought to be required.

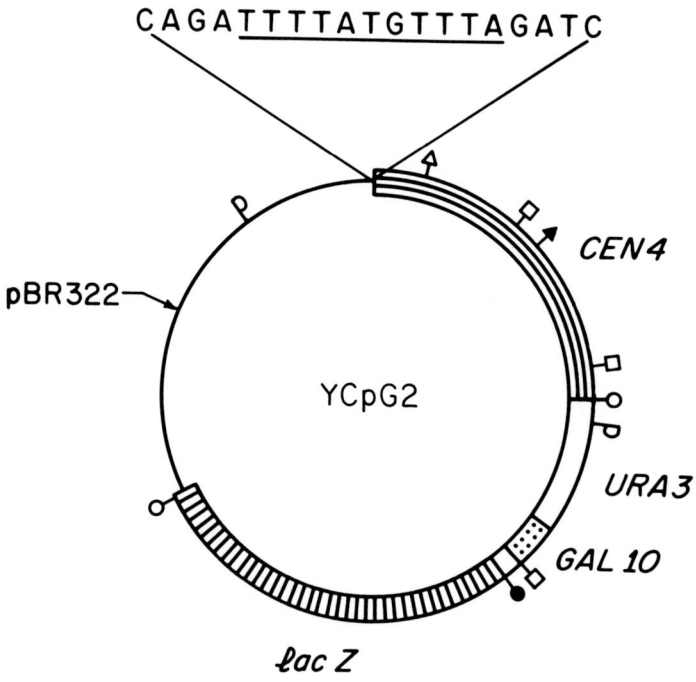

FIGURE 4. Minimal sequence for ARS1 function. The insert shows the 19bp fragment that allows autonomous replication of the YCpG2 vector (35).

This result, along with recent success by people working on transcription factors (36) suggested that we might be able to purify a protein important for ARS function on the basis of differential binding to DNA containing the consensus sequence. We chose DNase I "footprinting" as an assay (37), and measured protection of an EcoRl-HindIII DNA fragment (242bp) from plasmid H103 (32), since this fragment contains the consensus sequence in a convenient position. A protection pattern is shown in Fig. 5. Fractions from a DNA cellulose column when added at a critical concentration cause a shift in the banding pattern around the consensus sequence. While a large area is protected, these patterns are similar to those observed for binding of dnaA protein at the pBR322 origin of replication (38). The protein fraction is about 100-fold enriched over extracts.

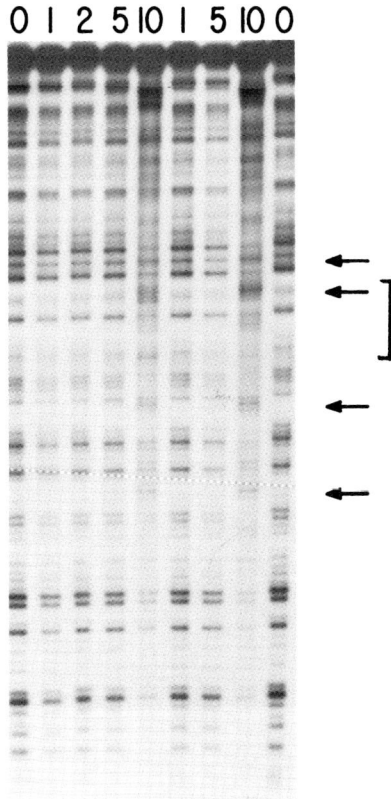

FIGURE 5. Footprinting of the ARS1 Consensus Sequence. Footprinting was carried out according to Galas and Schmitz (37) using fractions from a DNA cellulose column. Left and right lanes are

controls without added protein. Lanes labeled 1, 2, 5 and 10 have increasing amounts of neighboring fractions off the column. Arrows point to enhanced cutting sites. The consensus sequence is bracketed.

DISCUSSION

Our studies have come to focus on reconstitution as the best approach to working out the mechanism of replication in yeast. In order to insure that we are carrying out reconstitution studies with bona fide replication proteins, we have entered into a program of reverse genetics. For instance, we have described the cloning of the yeast DNA polymerase I gene and the use of the gene to prepare a deletion mutant whose phenotype shows that yeast DNA polymerase I is essential for DNA replication. The yeast POLI gene will be useful for determining the DNA sequence and for comparison to other polymerases. The crystal structure of E. coli DNA polymerase is known and therefore one may be able to obtain structural information about the yeast protein from primary sequence comparisons. The sequence may also reveal signals important for nuclear localization and cell cycle regulation. The clone will also be used to make an overproducer for biochemical studies on the protein. Finally, ts mutants will be useful for genetic analyses and for studying the residual polymerases--polymerase II and the mitochondrial polymerase--in the cell.

The single-strand binding proteins of eukaryotes have yet to be associated with any specific role in DNA metabolism. While a number of species that bind to ssDNA have more avidly than to dsDNA have been identified biochemically, lack of genetic approaches leave the question of their functions open. Recent reports verify that many proteins whose functions are known and that are not, therefore, expected to play a role in DNA metabolism, bind strongly to ssDNA cellulose (39). Therefore, the fact that a protein binds ssDNA does not implicate the protein in replication. Several proteins that bind ssDNA, however, can be expected to be involved in replication, because of additional properties demonstrated biochemically. The calf thymus UPI (M_r 24,000) and the mouse HD-1, all stimulate their respective DNA polymerase enzymes on single-stranded templates. Recent sequence analysis suggests remarkable amino acid sequence conservation between the mouse and the calf enzymes (39). Another class of binding protein, from Novikoff hepatoma cells, apparently stimulates DNA polymerase and not polymerase (40). A third class of enzyme has been shown to be HMG1 (41). Finally, the 72,000 M_r adenovirus ssDNA binding protein has been shown to be required for DNA replication both in vivo and in vitro (42). No mitochondrial SSBs have been previously reported. One interesting observation about these proteins is that many of them bind ssRNA as well as DNA. In fact, SSB proteins with molecular weights between 30 and 40,000 have been identified with core hnRNP proteins (43) and antibodies to the calf thymus SSB proteins (M_r 24,000) crossreact with the hnRNP proteins (44).

Yeast offers an ideal system in which to define the role of multiple SSB proteins through genetic analysis. The spectrum of proteins that we have identified in yeast that bind to ssDNA cellulose includes major species of M_r = 50,000, 45,000, 31,000, 22,500, 20,000 and some smaller proteins (data not shown). We have reported here, the purification of three of these to homogeneity--SSB-1, M_r = 45,000; SSB-2, M_r = 50,000, and the mitochondrial SSB, M_r = 20,000. Antibodies have been prepared against SSB-1 and SSB-2 and the gene for SSB-1 has been isolated in preparation for an investigation of the genetic role of this protein. SSB-2 will be looked at in turn. Availability of the cloned SSB-1 gene will allow us to compare its sequence to SSB proteins of higher cells, to make overproducers for biochemical studies and to make ts mutants for genetic analysis.

Yeast p31 and p22.5 SSB proteins are also of interest. While p31 binds to RNA and has been implicated as a component of ribonucleoprotein particles (John Abelson, personal communication), this does not rule out an additional role in DNA binding. p22.5 has not been studied further, but in view of the similarity in molecular weight to SSB proteins of higher cells, this should also be looked at. A 20,000 M_r SSB from *Ustilago maydis* has been described that stimulates DNA polymerase from the same source, but whether it is mitochondrial or nuclear has not been investigated (45).

Finally, in this work we have discussed an additional chapter in the quest for the function of ARS sequences in yeast, in particular ARS1. We have demonstrated that 19bp containing the consensus sequence is sufficient to allow a centromere-containing plasmid to replicate in yeast. Furthermore, we have partially purified a protein that binds the latter consensus sequence, which can now be investigated for its role in replication or transcription.

Several results suggest that the initiation of DNA replication in eukaryotes is relatively simple and occurs by binding of a sequence specific protein followed by the action of primase and polymerase. In the SV40 system, T antigen is clearly required for initiation of replication (46). Neither our in vitro replication system nor the recently described SV40 in vitro replication system is inhibited by -amanitin, suggesting that transcription is not required for initiation on naked DNA templates. Furthermore, polymerase and primase alone have been shown to make starts at the SV40 origin (47) in vitro similar to those observed by Hay and De Pamphilis in vivo (48). It is therefore possible that most of the proteins required for initiation of DNA replication in yeast are now known. Reconstitution studies on ssDNA with the core catalytic subunit of polymerase that we have cloned and with SSB should allow identification of other subunits of the polymerase holoenzyme. Synthesis in the presence of primase (23,24) should allow identification of any prepriming proteins that may be essential. Finally, synthesis on substrates that mimic replication forks (49) should allow identification of a eukaryotic helicase activity. These proteins, in conjunction with the ARS specific binding protein, as described here, may constitute an entire replication system. Much biochemistry remains to be carried out to prove this hypothesis, but combining the purification of putative replication proteins with the type of genetic approaches illustrated in this work in yeast should provide important guidelines and facilitate rapid progress in understanding replication.

The interesting questions to be pursued in the future are whether all 400 initiation sites for replication in the yeast genome use the same proteins. Do they initiate with a specific temporal pattern during S phase? What prevents reinitiation within an activated replicon during a single cell cycle? With the basic components of replication identified, we should now be able to address the question of how replication is integrated with other cellular processes such as transcription, cell growth and division.

ACKNOWLEDGMENTS

We would like to acknowledge the participation of M. P. Snyder, Stanford University, in the original screening for the POL1 gene. This work was supported by grants from the American Cancer Society, March of Dimes and National Institutes of Health.

REFERENCES

1. Campbell JL (1983). Yeast DNA replication in vitro and in vivo. In Setlow J and Hollaender A (eds): "Genetic Engineering," New York: Plenum Press, p 109.
2. Hartwell LH, Mortimer RK, Culotti J and Culotti M (1973). Genetics 74:267.
3. Johnston LH and Thomas AP (1982). Mol Gen Genet 187:42.
4. Dumas LB, Lussky JP, McFarland EJ and Shampay J (1982). Mol Gen Genet 187:42.
5. Kuo, Cl, Huang, Nh and Campbell JL (1983). Proc Natl Acad Sci USA 80:6469.
6. Hieter P, Mann C, Snyder M and Davis RW (1985). Cell 40:381.
7. Maine GT, Sinha P and Tye Bk (1984). Genetics 106:365.
8. Hereford LM and Hartwell LH (1971). Nature 234:171.
9. Jazwinski SM and Edelman GM (1979). Proc Natl Acad Sci USA 76:1223.
10. Kojo H, Greenberg BD and Sugino A (1981). Proc Natl Acad Sci USA 78:7261.
11. Kuo Cl and Campbell JL (1982) Proc Natl Acad Sci USA 79:4243.
12. Celniker S and Campbell JL (1982). Cell 31:563.
13. Sclafani RA and Fangman WL (1984). Proc Natl Acad Sci USA 81:5821.
14. Jong AYS, Kuo Cl and Campbell JL (1984). J Biol Chem 259:11052.
15. Kuo Cl and Campbell JL (1983). Mol Cell Biol 3:1730.
16. Jong AYS and Campbell JL (1984). J Biol Chem 259:14394.

17. Arendes J, Kim KC and Sugino A (1983). Proc Natl Acad Sci USA 80:673.
18. La Bonne S and Dumas LB (1983). Biochemistry 22:3214.
19. Schwartz DC and Cantor CR (1984). Cell 37:67.
20. Pringle JR and Hartwell LH (1981). In Strathern JN, Jones EW, Broach JR (eds): "The Molecular Biology of the Yeast Saccharomyces," New York: Cold Spring Harbor Laboratory, p 97.
21. McMacken R, personal communication.
22. Badaracco G, Capucci L, Plevani P and Chang LMS (1983). J Biol Chem 258:10720.
23. Plevani P, Badaracco G, Augl C and Chang LMS (1984). J Biol Chem 259:7532.
24. Singh H and Dumas LB (1984). J Biol Chem 259:7936.
25. Chang LMS (1977). J Biol Chem 252:1873.
26. Snyder M and Davis RW (1985). Screening lambda gt11 expression libraries with antibody probes. In Springer T (ed): "Hybridomas in Biotechnology and Medicine," New York: Plenum Press.
27. Young RA and Davis RW (1984). Science 222:778.
28. Tabor S and Richardson CC (1985). Proc Nat Acad Sci USA 82:1074.
29. Culotti J and Hartwell LH (1971). Exptl Cell Res 67:389.
30. Chang LMS, Lurie K and Plevani P (1978). Cold Spring Harbor Symp Quant Biol 43:587.
31. Stinchcomb DT, Struhl and Davis RW (1979). Nature 282:39.
32. Celniker SE, Sweder KS, Srienc F, Bailey JE and Campbell JL (1984). Molec Cell Biol 4:2455.
33. Kearsey S (1984). Cell 37:299.
34. Stinchcomb DT, Mann C and Davis RW (1982). J Mol Biol 158:157.
35. Srienc F, Bailey JE and Campbell JL (1985). Molec Cell Biol 5:in press.
36. Bram RJ and Kornberg RD (1985). Proc Natl Acad Sci USA 82:43.
37. Galas DJ and Schmitz A (1978). Nuc Acids Res 5:3157.
38. Fuller RS, Funnell BE and Kornberg A (1984). Cell 38:889.
39. Williams KR, Stone KL, LoPresti MB, Merrill BM and Planck SR (1985). Proc Natl Acad Sci USA 82:in press.
40. Koerner TJ and Meyer RR (1983). J Biol Chem 258:3126.

41. Bonne C, Sautiere P, Duguet M and de Recondo AM (1982). J Biol Chem 257:2722.
42. Challberg MD and Kelley TJ (1982). Ann Rev Biochem 51:901.
43. Beyer AL, Christensen ME, Walker BW and LeSturgeon LM (1977). Cell 11:127.
44. Valentini O, Biamonti G, Pandolfo M, Morandi C and Riva S (1985). Nuc Acids Res 13:337.
45. Banks GR and Spanos A (1975). J Mol Biol 93:63.
46. Li JJ and Kelly TJ (1984). Proc Natl Acad Sci USA 81:6973.
47. Tseng BY and Ahlem CN (1985). Fed Proc 44:1428.
48. Hay RT and DePamphilis ML (1982). Cell 28:767.
49. Tabor S and Richardson CC (1985). Proc Natl Acad Sci USA 82:1074.

REGULATION OF DNA REPLICATION INITIATION IN YEAST

Pratima Sinha[1], Clarence Chan[2], Gregory Maine[3]
Steve Passmore[2], DaMing Ren[4] and Bik-Kwoon Tye[2]

[1] IMTECH, 1383 Sector 33-C, Chandigarh 160031, INDIA

[2] Section of Biochemistry, Molecular and Cell Biology
Wing Hall, Cornell University, Ithaca, NY 14853

[3] BioTechnica International, Inc. 85 Bolton Street
Cambridge MA 02140

[4] Fudan University, Shanghai, People's Republic of China

ABSTRACT We previously reported the isolation of a series of mutants defective in the maintenance of minichromosomes (Mcm⁻)[1]. The properties of some of these mutants suggest that their defects may be targeted at the ARSs (autonomously replicating sequences), the putative chromosome replication origins. Two mutants, Mcm9 and Mcm46, which affect the function of only a subgroup of ARSs, have been characterized. Inspection of the nucleotide sequences of the ARSs that are active and inactive in these mutants suggests that the specificity site for these mutant gene products may lie within a region of 5-6 nucleotides, 10 base pairs from the ARS consensus sequence. Based on the properties of these mutants and their presumptive target sites, a model for the mode of DNA replication initiation in yeast is proposed.

INTRODUCTION

The eukaryotic genome contains a large number of replication origins. There is good evidence that the initiation of DNA replication at these origins is developmentally as well as temporally regulated. The work of Spradling and coworkers[2,3] showed that in Drosophila, the chorion genes in the embryo are amplified by the multiple reinitiation of the replication origins near these genes. Blumenthal et al. also showed that different sets of replication origins may be used in embryonic and somatic cells in Drosophila[4]. The most convincing evidence for the temporal regulation of replication initiation was demonstrated in yeast by Fangman and coworkers[5]. They showed by density transfer and quantitative hybridization experiments that different regions of the genome are replicated at specific times in the S phase of the cell cycle. Newlon and Burke[6] estimated that there are approximately 400 replication origins in the yeast genome. Although yeast is not a suitable system to study the developmental control of replication initiation, it is an appropriate system to examine the mechanism for the temporal initiation of replication origins.

In order to isolate yeast mutants that are defective in replication initiation, it is important to be able to predict the phenotype of such a mutant. One can imagine three possible modes of replication initiation in eukaryotes:

1. Each replication origin is initiated by a specific enzyme or enzyme complex. In this case, at least 400 genes would be required just for replication initiation in yeast.

2. All replication origins are initiated by a single enzyme or enzyme complex. This may seem too simple for the fine tuning expected for the temporal control of initiation.

3. There are classes of replication origins. A particular class of replication origins will be initiated by a particular enzyme or enzyme complex.

The phenotype of an initiation defective mutant depends very much on the mode of replication initiation used in the organism. In the first case, a mutant with an altered initiator protein would result in the inactivity of one of the replication origins. Such a mutant may not have any detectable defect in growth or in chromosome replication. In fact, we have shown that removal of an ARS sequence from the yeast genome has no observable change in the growth of the cell. In the second case, a mutant with an altered initiation protein would have a cell cycle block phenotype which has an execution point at the beginning of S phase. In the third case, an initiation mutant may or may not have an observable phenotype. If neighboring active replicons can replicate through silent replicons, then the mutant may not show any defect in the replication of the chromosomes. On the other hand, if the cell requires that all replication origins on a chromosome to be initiated in order to complete replication of the chromosome, then one may expect to see a cell cylce block occurring in S or late S phase. Thus, it is difficult to predict the phenotype of an initiation mutant without knowing the precise mechanisms used for replication initiation in the cell. In fact, two of the cdc (cell division cycle) mutants that have arrests during S or early S phase, and hence are good candidates for DNA replication mutants, were identified to be metabolic pathway mutants[7,8].

The difficulties mentioned above motivated us to look for mutants with a defect in a specific function rather a general phenotype. We designed a screening procedure that would allow us to look for chromosomal mutants defective in the activation of ARSs. If ARSs are the sites of chromosomal replication origins, then these studies would be informative in learning about the DNA replication initiation mechanism in yeast.

RESULTS

Minichromosome maintenance defective (Mcm⁻) mutants.

We constructed two circular minichromosomes, YCp1 and YCp131C, each of which contains an ARS, a centromere and two selectable markers, URA3 and LEU2 (Figure 1). These two minichromosmes differ only in their ARSs. Minichromosome YCp1 contains ARS1[9], which is present in a single copy in the yeast genome. Minichromosome YCp131C contains ARS131C, which is a repetitive ARS element found in most yeast telomeres[10]. Yeast

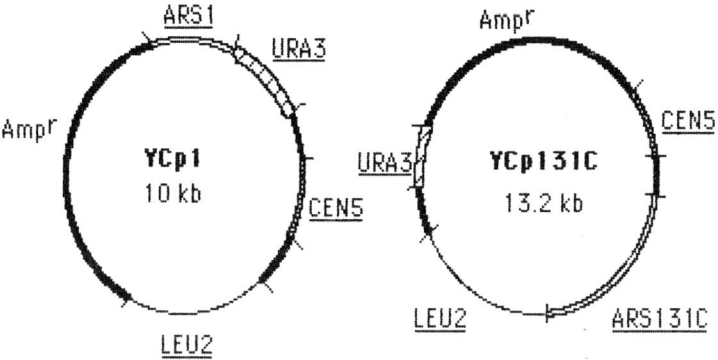

FIGURE 1 Minichromosomes YCp1 and YCp131C.

cells were transformed with either of these minichromosomes, mutagenized and screened for mutants that were unable to stably maintain these minichromosomes. The rationale for this screening procedure is that there is only one ARS element on each minichromosome but multiple ARSs are present on the natural chromosomes. Thus, the stability of these minichromosomes would be more sensitive to a defect in the initiation at ARS than the natural chromosomes. Forty Mcm⁻ mutants were isolated, comprising 16 complementation groups[1]. These mutants can be divided into two phenotypic classes, the specific mutants which affect only a subset of ARSs, and the nonspecific mutants which affect all ARSs tested (over 20 ARSs were tested).

Table 1 illustrates examples of each of these classes. Different minichromosomes were transformed into the mutants and their stability in the mutants were tested. The Mcm1 mutant is a nonspecific mutant since all minichromosomes tested were unstable in this mutant regardless of which ARS or centromere was present on the minichromosome. In contrast, the Mcm9 mutant is a specific mutant since only YCp1 and YCp131 were unstable but YCp121 and YCp131C were quite stable in this mutant. The mutant Mcm46 is temperature conditional for specificity. At the permissive temperature, it showed specificity for certain ARSs. At

TABLE 1

STABILITIES OF MINICHROMOSOMES IN SPECIFIC AND NONSPECIFIC MCM⁻ MUTANTS

Strain	YCp				
	1	121	131	131C	1-131C
Mcm1	<0.5	<0.5	<0.5	<0.5	<0.5
Mcm9	<0.5	54	<0.5	39	57
Mcm46(23C)	<0.3	67	<0.3	60	89
(35C)	<0.3	<0.2	<0.3	<0.5	12
Wild-type(25)	75	79	62	86	99
(35)	80	95	70	92	72

Stabilities are expressed as the percentage of plasmid-carrying cells in the cell population after 20 generations of growth in YEPD.

the nonpermissive temperature, it lost its specificity. Surprisingly, all of the specific mutants (a total of seven in four complementation groups) had the same specificity. This same pattern of specificity was observed when other centromeres were substituted for CEN5 on the minichromosomes[1]. When two ARSs which behaved differently in the specific mutants were placed on the same minichromosome, the active ARS (ARS131C) was epistatic over the inactive ARS (ARS1) (Table 1). This suggests that the defect in these specific mutants may be targeted at the ARS.

It should be pointed out that none of the Mcm$^-$ mutants isolated affected only one single ARS. Furthermore, all of the specific mutants were isolated from the procedure using the minichromosome YCp1. All of the mutants isolated from the procedure using minichromosome YCp131C were nonspecific mutants[1]. This suggests that there are some fundamental differences between ARS1 and ARS131C.

The Mcm46 mutant is temperature-conditional for ARS specificity.

The temperature-conditional specificity of the Mcm46 mutant suggests that the specificity groups in the ARSs may simply reflect the differential affinity of the mutant enzyme for the different ARSs. If so, the differential action of the mutant protein should be reflected in the rate of loss of each of the minichromosome from the mutant at various temperatures. We compared the rate of loss of YCp1 and YCp131C at different temperatures in the Mcm46 mutant and the wild-type strain. A log-phase culture of each strain, containing either YCp1 or YCp131C, growing at 23C in selective medium, was diluted into nonselective medium at 23C, 30C or 35C. At various times, the percentage of plasmid-bearing cells was determined in aliquots. As shown in figure 2, the rate of loss of YCp1 and YCp131C in the Mcm46 mutant varies as a function of temperature. This is consistent with the notion that the binding of the altered mutant protein to these two ARSs is dependent on temperature.

The instability of YCp1 and YCp131C can be explained by either a chromosome loss event such as nonreplication, or nondisjunction of the chromosome during mitosis. These two events can be distinguished if one follows the percent of plasmid-bearing cells in the culture and at the same time measures the total number of plasmids in the cell population. If instability of the plasmid is due to chromosome loss, then one would expect that while the percent of plasmid-bearing cells decreases during cell growth, the copy

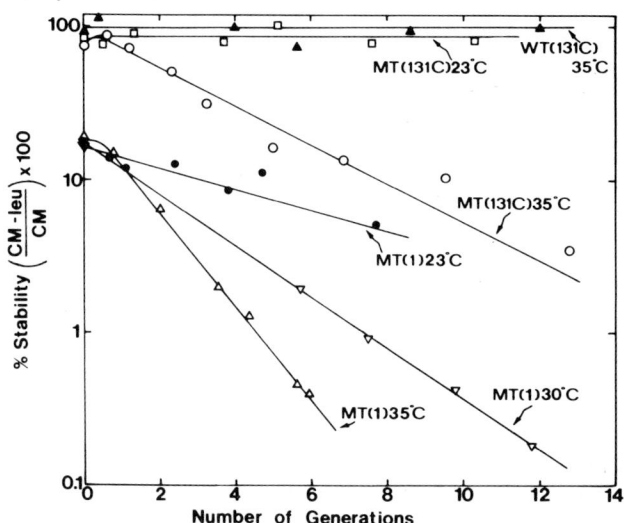

FIGURE 2. Kinetics of the loss of the minichromosomes YCp1 and YCp131C in the mutant Mcm46 (MT) and wild type (WT) at various temperatures. The curves for the loss of YCp 131C and YCp1 in the wild type were the same at all temperatures as that shown for YCp131C at 35C. The number in parenthesis indicates the minichromosomes under study.

number of minichromsome per plasmid-bearing cell would remain at a constant of approximately one. On the other hand, if instability is due to nondisjunction, then one would expect that the decrease in the percent of minichromosome-bearing cells would be associated with an increase in the copy number of minichromosome per plasmid-bearing cell. This experiment was carried out using YCp131C and the results are shown in Table 2. It appears that the instability of YCp131C is due to loss of the minichromosome. One can argue that the copy number of YCp131C is low because of the presence of the centromere on the minichromosome[11]. We repeated the above experiment using the TRP1ARS1 plasmid which does not contain a centromere. This plasmid is very stable and is maintained in about 100 copies per cell[12]. Its copy number in the mutant Mcm46 decreased from 100 to about 1 per cell when the transformant was allowed to propagate in YEPD at 35C (P. Sinha, unpublished results).

TABLE 2
STABILITY AND COPY NUMBER OF YCp131C IN THE MCM46 MUTANT

Temp.	Number of generations	percent of cells with plasmids	copy per plasmid carrying cell
25C	0	91	1.1
25C	4.5	60	1.1
25C	9.0	62	1.0
25C	12.6	59	1.1
35C	0.7	83	1.0
35C	4.7	48	0.8
35C	9.0	12	1.2
35C	12.2	3	N.D.

N.D. = Not determined
copy number of plasmid was determined by quantitaive hybridization and measured by densitometry.

Similar experiments were also performed using linear plasmids that contained ARS1 or ARS131C. We can conclude from these experiments that the defect in the Mcm46 strain affects circular and linear plasmids alike. Furthermore, the specificity exhibited by this mutant for the ARSs is the same whether the ARSs are present on linear or circular plasmids.

We showed that larger circular or linear chromosomes that contained multiple ARSs were unstable in this mutant at high temperature but stable at low temperature. Preliminary experiments indicate that the natural chromosomes may be lost at an increased rate in this mutant than in wild-type strains. These results suggest that even if there is a large number of ARSs present on a chromosome, if an insufficient number of ARSs are initiated to allow complete replication, it may result in the loss of the chromosome. This may explain the slower growth rate of this mutant at the nonpermissive temperature.

The Mcm9 mutant has pleiotrophic phenotypes.

The mcm9 mutation, besides affecting the stability of a subset of minichromosomes, also affects the mating ability of the cell when the mutation is present in an α cell. The Mcm⁻ and α-specific sterile phenotypes are caused by a single mutation since these two phenotypes corevert. The mcm9 mutation has been mapped at chromosome XIII, 2cM from LYS7 (G., Maine, unpublished results). So far, we are unaware of any known ste mutation that maps at this same location.

A gene which complements both the Mcm⁻ and the Ste⁻ phenotype has been cloned (G.Maine, unpublished results). This gene when it is integrated into the genome by homologous recombination maps at the same location as the mcm9 mutation. A transformant of the Mcm9 mutant by integration of this cloned gene is able to stabilize YCp1. Thus, this cloned gene corresponds to the MCM9 gene. The complementing activity of the MCM9 gene has been localized within a 500 base pair DNA fragment. We were able to show that the MCM9 gene is an essential gene by removal of one copy of the MCM9 genes from a diploid. Sporulation of this diploid yielded tetrads which gave 2 live and 2 dead spores. Thus, the mcm9 mutant allele is a leaky allele of an essential gene.

Determination of the nucleotide sequence of this DNA fragment reveals a single open reading frame that can encode a protein of 143 aminio acids. There is a cluster of basic residues close to the N-terminal and a cluster of acidic residues near the C-terminal of this protein (S.Passmore, unpublished results). A computer search for homology among the known protein sequences identified the E. coli sigma factor[13] to have some homology to the predicted MCM9 gene product. This homology lies within a stretch of 23 amino acids in the acidic region of the MCM9 gene product between residues 98 and 120 (Figure 3). If one considers the homology

```
            159
       Asp Asp Asp Glu Asp Glu Asp Glu Glu Asp Gly Asp Asp Asp
   Pro Asp Asp Glu Glu Glu Asp Glu Glu Glu Asp Gly Asp Asp Asp
   Leu  98

                                   181
   Ser Ala Asp Asp Asp Asn Ser Ile Asp
   Asp Asp Asp Asp Asp Asp Gly Asn Asp
                                   120
```

Top sequence = E. coli sigma factor (residues 159 – 181)
Bottom sequence = MCM9 gene product (residues 98 – 120)

FIGURE 3. Homology between the MCM9 gene product and the E. coli sigma factor.

that is invariant between the two protein sequences, then 14 out of 23 residues (position 98 to 120) or 13 out of 19 residues (position 98 to 116) are identical. If one considers that glutamic and aspartic acid are structurally and functional equivalent, then 18 out of 23 or 17 out of 19 of the residues are equivalent within this region. The biological significance of this homology is unknown. Comparison of the nucleotide sequences of the MCM9 gene and the mcm9 mutant allele indicates that the mcm9 allele contains a single base change which resulted in the substitution of leucine for proline at residue 97. This mutation is immediately in front of the acidic region which shows homology to the E. coli sigma factor, suggesting that the acidic region may be important for the activity of the MCM9 gene product. It is amusing to note that another protein which shows homology to the MCM9 gene product is the initiator protein of the E. coli plasmid, PSC101[14]. The homology between these two proteins is only about 50% and it resides within a 19 amino acids region distinct from the acidic region.

It is unclear how the Mcm⁻ and the α-Ste⁻ phenotypes are related. There are precedents which suggest that the ARS function may be associated with the repression of gene activity in yeast. Hereford and Osley[15] showed that there is an ARS located at the 3'-ends of both copies of the H2B histone genes in yeast. They were unable to separate the ARS element from the regulatory region responsible for the cell cycle regulation of the expression of the H2A and H2B genes. It is suggestive, but not proven, that the ARS function and the regulatory activity are coupled. A second example is the silent alleles, HMLα and HMRa, of the mating type locus MAT[16]. Again, the regulatory sites responsible for the suppression of the the silent alleles appear to be coincident with the ARS elements located on either side of each of the silent alleles. We do not think that the sterile phenotype observed in the MATα mcm9 strain is due to the expression of the silent a allele. We constructed a MATα mcm9 strain in which the silent a allele was replaced by a silent α allele. This strain still exhibited a Ste⁻ phenotype. One plausible explanation for the dual phenotypes exhibited by the Mcm9 mutant is to propose that the MCM9 gene product acts as an initiator protein for a family of ARSs. One of the ARSs may be located next to an α-specific gene whose expression is dependent on the activation of the ARS. Further investigations by identifying second site suppressors that are Ste⁺ but still Mcm⁻ may allow us to identify such a target site for the MCM9 gene product.

The target sites for the products of the MCM9 and MCM46 genes.

The ARS consensus sequence. Although the Mcm46 and the Mcm9 mutants have rather different properties and phenoytpes, they affect the same set of ARSs. In fact, all of the specific Mcm⁻ mutants share the same specificity for ARSs[1]. Thus, according to the specificity of these mutants, one can divide the ARSs into two classes, those that are active and those that are inactive in the specific mutants. We would like to know how these two classes of ARSs differ from each other. Comparison of the nucleotide sequence of ARSs determined by several groups revealed an (A+T)-rich consensus sequence of 11 base pairs[17,18]. Deletion analysis of some of these ARSs showed that the consensus sequence was essential but not sufficient for ARS activity[19,20] (C.Chan, unpublished results). However, most ARSs are very different in their nucleotide sequences and lack significant secondary structures such as direct or inverted repeats. Other features common to ARSs have not been identified. We have recently determined the nucleotide sequence of ARS120, 131S, 137, 245, and 249[21]. Comparison of these additional sequences with the available sequences in the literature allowed us to extend the consensus sequence to 13 base pairs (Figure 4). It was not obvious to us from the nucleotide sequence of these ARSs what are the features that might account for their differential activities in the specific mutants.

The homologous ARSs within the telomeric X sequences. We previously reported the isolation of a family of homologous ARSs that are located at the telomeres of yeast chromosomes[10] (Figure 5). The ARSs that are located within the highly conserved repeated Y' sequences are almost identical in their nucleotide sequences and are inactive in the specific mutants[1]. In contrast, the ARSs that are located within the repetitive telomeric X sequences are less conserved in their sequences. They also behave quite differently in the specific mutants[1]. Some of these class X ARSs are active while others are inactive in the specific mutants. By identifying the differences in the nucleotide sequences of these class X ARSs, it should be possible to identify the features that contribute to their different properties.

Initiation of DNA Replication in Yeast

ARS	Sequence		
H0	AAAAGTAAAA	TTTAATATTTT	GGATGAAAAA
F82	TCATTGTATG	TTTTATGTTTT	GTCTGGAAAA
1	ATTTTACAGA	TTTTATGTTTA	GATCTTTTAT
3	TTTAAT	ACTTATATTTA	AATATTGAAA
2	GGAAATATAA	ATTTATATTTA	GTAATACGCA
2	CTAAATATAA	ATTTATATTTC	CTGCGACCAT
HML, left	TTTTTGATTT	TTTTATGTTTT	TTTAAAACAT
HML, right	TTGTAATGAT	TTTTATATTTT	
HMR, left	TAATAACATT	TTTTATATTTA	GGTATTAAAT
HMR, right	GACCTCATTA	ATTAATATTTA	TTAATACCTT
2μ	CCTCTACATT	TTTTATGTTTA	TCTCTAGTAT
rDNA	GATTTGATT	GTTTATGTTTA	TCTCTAGTAT
rDNA	AAGGTATATT	TTGTATGTTTT	GTATGTTCCC
120	GATAGAATAT	TTTTATGTTTA	GGTGATTTTA
131S	ACTACCTTTA	TTTTATGTTTA	CTTTTTATAG
137	GGTCAATTCA	ATTAATGTTTT	GAGAAGATTG
245	TTAGCTTTCT	TTTTATGAGTT	GTCACCAATT
249	GACTTATTTA	ATTTATGTTTT	GTAAAGAATA

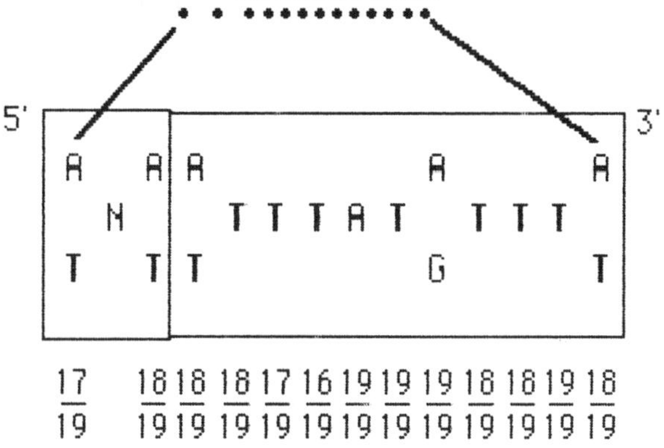

```
5'                                                      3'
    A   A A       A                A
        N     T T T A T        T T T
    T   T T                G            T

    17  18 18 18 17 16 19 19 19 18 18 19 18
    19  19 19 19 19 19 19 19 19 19 19 19 19
```

ARS CONSENSUS SEQUENCE

FIGURE 4. The ARS consensus sequence.
Note: not all sequences above have been shown by deletion analysis to be essential for ARS activity.

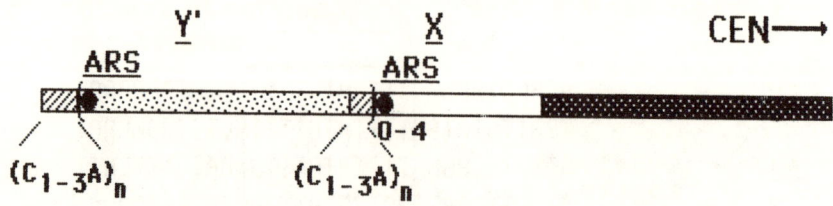

FIGURE 5. Organization of DNA sequences at yeast telomeres[10,22].

The specificity site. The effect of five class X ARSs on the stablity of minichromosomes in the Mcm46 and Mcm9 mutants was examined[1]. Three of them, ARS120, ARS131N, and ARS131J were active but ARS131A, and ARS131 were inactive in these mutants (Figure 6). Deletion analysis performed on some of these ARS sequences suggests that a region of about 40 base pairs is essential for autonomous replication in these ARSs (C.Chan, unpublished results). A comparison of the nucleotide sequence of this 40 base-pair region in these class X ARSs indicate that they all include the consensus sequence (Figure 6). ARS120 and ARS131N, which behave similarly in the Mcm46 and Mcm9 mutants have an identical nucleotide sequence within this region. ARS131J which differs from these two ARSs in five nucleotide changes is also active in the specific mutants. Thus, these five nucleotide changes probably do not contribute to the differential behavior of ARSs in the specific mutants. ARS131A and ARS131 are inactive in the specific mutants. In these two ARSs, there are nucleotide changes that are different from the three active class X ARSs. In fact, there is a single base deletion in this region in ARS131A. These differences lie within a region of six base pairs, ten base pairs from the consensus sequence. Thus, by inspection of the nucleotide sequences of these five class X ARSs, it seems likely that the specificity site recognized by the MCM9 gene product probably includes this six base-pair region. Further investigation by site specific mutagenesis in this 6 base-pair region should determine the significance contributed by these dissimilarities to the properties of these ARSs.

Telomeric X ARSs	Function in specific mutants
ARS120	+
ARS131N	+
ARS131J	+
ARS131A	–
ARS131	–

```
120   TAGAA TATTT TTATG TTTAG GTGAT TTTAG TGGTG ATTTT
131N  ----- --+-- ----- ---+- ----- ---+- ----- -----
131J  -G-TT A-+-- ----- ---+- ----- ---G- ----- -----
131A  ----- --+-- ----- ---+- ----- ---GA --A-◇ -----
131   ----- --+-- ----- ---+- ----- ---GA -A-C- -----
           CONSENSUS              SPECIFICITY
           SEQUENCE                    ?
```

FIGURE 6. Comparison of the nucleotide sequence of five class X ARSs.
☐ = single base deletion
-- = identity

SUMMARY

As a summary, we would like to present a working model (Figure 7) which is drawn from the concepts that were discussed in the introduction and which is consistent with the data presented here. If ARSs are in fact chromosomal replication origins, then the results presented suggest several features in the mechanism for DNA replication initiation in yeast.

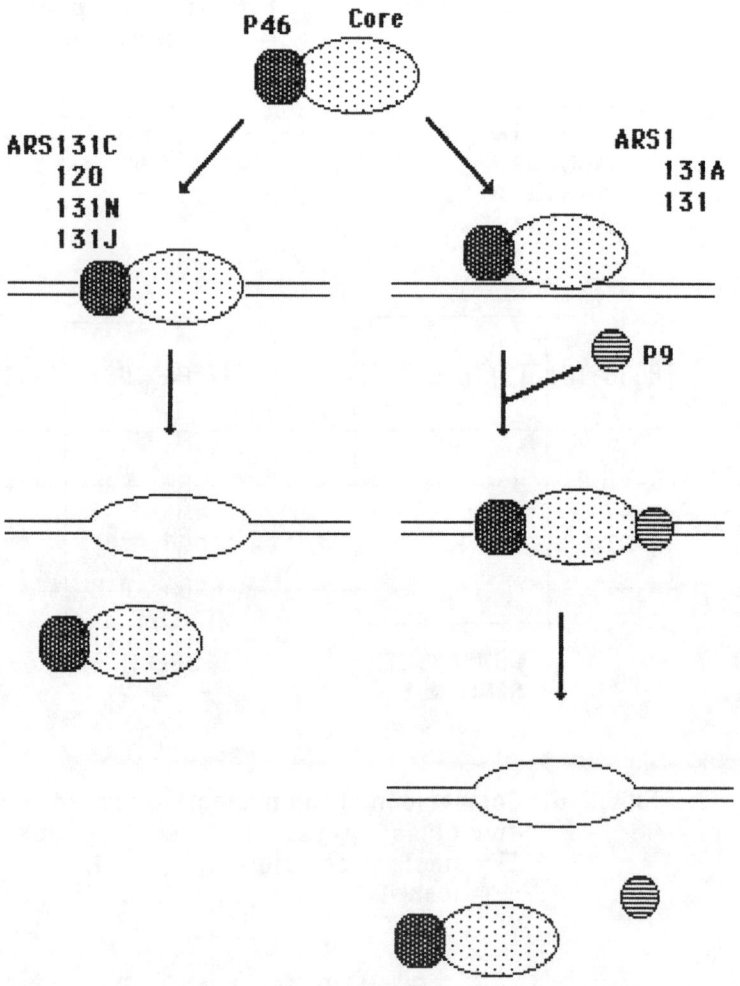

FIGURE 7. Model for replication initiation at ARSs.

First, there are classes of replication origins which serve as substrates for the binding of initiator proteins. This idea is consistent with our initial observation that all of the specific Mcm⁻ mutants that affected only a subset of the ARSs came from the isolation procedure using the minichromosome YCp1 which contains ARS1. In constrast, all of the mutants (30 in all) obtained from the isolation procedure using YCp131C were nonspecific and they affected all ARSs. The different outcomes from these two isolation procedures suggest that ARS1 and ARS131C are fundamentally different. ARS1 appears to be a more complex site which requires the activation by proteins, such as the MCM9 gene product, that are not required in the activation of ARS131C. Conversely, ARS131C is a simpler site since whatever that is required for the activation of ARS131C is also required for ARS1. Thus, from the specificity of the specific Mcm⁻ mutants, we can discern two classes of ARSs, the simpler ARSs and the more complex ARSs. The simpler ARSs are simpler not only in terms of their primary sequence requirement for the binding of initiation complex, they also require a simpler enzyme complex for initation activation. ARS131C, 120, 131N, and 131J belong to this class of replication origins. They can be activated by a simple enzyme complex shown as the core enzyme which includes P46, the product of the MCM46 gene. ARS1, 131A, and 131 belong to the more complex class of replication origins. They require additional accessory proteins such as P9, the product of MCM9 gene, for their activation.

Alternatively, we can think of the two classes of replication origins to be different only in their affinity to the replication initiation complex represented as the core enzyme and P46 in figure 7. ARS131C, 120, 131N and 131J have a high affinity for this enzyme complex but ARS1, 131A and 131 are bound only weakly by this enzyme complex and require the protein P9 to secure tight binding. Thus, in the absence of P9, only the simpler class of ARSs are initiated. In the Mcm46 mutant, in which P46 has an altered affinity for binding to ARSs, at low temperature P46 can still bind to ARS131C, 120 etc. with reasonable affinity to allow initiation but not to ARS1, 131A etc.. At high temperature, its affinity for its substrates is significantly altered so that it can no longer bind to any of the ARSs. In fact, the simplest model for temporal regulation of DNA replication initiation without imposing a large number of initiator proteins is to imagine that there are classes of replication origins that show differential affinities for the initiator protein(s). The change in concentration of the intiator protein(s) during S phase would result in the temporal initiation of these classes of replication origins.

Indeed, the idea of differential affinity of regulatory sites to their regulator proteins is not new. We can think of the regulation of replication initiation as analogous to the regulation of gene expression. As there are promotors of different strengths for the binding of the RNA polymerase, there may be replication origins that have different affinity for the binding of the initiation complex.

ACKNOWLEDGEMENTS

We thank Dr. Larraine Symington for critical reading of the manuscript. This work is supported by Public Health Service grants, GM26941 and AI14980, from the National Institutes of Health. GM and SP are predoctoral fellows supported by the National Research Service Award GM07273 awarded to the Section of Biochemistry, Molecular and Cell Biology.

REFERENCES

1. Maine GT, Sinha P, Tye B-K (1984). Mutants of S. cerevisiae defective in the maintenance of minichromosomes. Genetics 106:365.
2. Spradling AC (1981). The organization and amplification of two chromosomal domains containing Drosophila chorion genes. Cell 27:193.
3. Spradling AC, Mahowald AP (1981). A chromosome inversion alters the pattern of specific DNA replication in Drosophila follicle cells. Cell 27:203.
4. Blumenthal AB, Kriegstein HJ, Hogness DS (1973). The units of DNA replication in Drosophila melanogaster chromosomes. Cold Spring Harbor Symp. Quant. Biol. 38:205.
5. Fangman WL, Hice RH, Chlebowicz-Sledziewska (1983). ARS replication during the yeast S phase. Cell 32:831.
6. Newlon CS, Burke W (1980). Replication of small chromosomal DNAs in yeast. In ICN-UCLA Symposia: "Mechanistic Studies on DNA Replication and Genetic Recombination," New York: Alan R. Liss, Vol.XIX.
7. Sclafani RA, Fangman WL (1984). Yeast gene CDC8 encodes thymidylate kinase and is complemented by herpes thymidine kinase gene TK. Proc. Natl. Acad. Sci. USA. 81:5821.
8. Bisson L, Thorner J (1977). Thymidine 5'-monophosphate-requiring mutants of Saccharomyces cerevisiae are deficient in thymidylate synthetase. J. Bacteriol. 132:44.
9. Stinchcomb DT, Struhl K, Davis RW (1979). Isolation and characterization of a yeast chromosomal replicator. Nature 282:39.
10. Chan CSM, Tye B-K (1983). Organization of DNA sequences and replication origins at yeast telomeres. Cell 33:563.
11. Clarke L, Carbon J (1980) Isolation of a yeast centromere

and construction of functional small circular chromosomes. Nature 287:504.
12. Zakian VA, Scott JF (1982). Construction, replication, and chromatin structure of TRP1 RI circle, a multiple-copy synthetic plasmid derived from Saccharomyces cerevisiae chromosomal DNA. Mol. Cell Biol. 2:221.
13. Burton Z, Burgess RR, Lin J, Moore D, Holder S, Gross CA (1981). The nucleotide sequence of the cloned rpoD gene for the RNA polymerase sigma subunit from E. coli K12. Nucleic Acids Res. 9:2889.
14. Vocke C, Bastia D (1983). Primary structure of the essential replicon of the plasmid pSC101. Proc. Natl. Acad. Sci. USA. 80:6557.
15. Osley MA, Hereford L (1982). Identification of a sequence responsible for periodic synthesis of yeast histone 2A mRNA. Proc. Natl. Acad. Sci. USA 79:7689.
16. Abraham J, Feldman J, Nasmyth KA, Strathern JN, Klar AJS, Broach JR, Hicks JB (1983). Sites required for position-effect regulation of mating-type information in yeast. Cold Spring Harbor Symp. Quant. Biol. 47:989.
17. Stinchcomb DT, Mann c, Selker E, Davis RW (1981). DNA sequences that allow the replication and segregation of yeast chromosomes. ICN-UCLA Symp. on Mol. Cell. Biol. 22:473.
18. Broach JR, Li Y-Y, Feldman J, Jayaram M, Abraham J, Nasmyth KA, Hicks JB (1983). Location and sequence analysis of yeast origins of DNA replication. Cold Spring Harbor Symp. Quant. Biol. 47:1165.
19. Celniker SE, Sweder K, Srienc F, Bailey JE, Campbell JL (1984). Deletion mutations affecting autonomously replicating sequence ARS1 of Saccharomyces cerevisiae. Mol. Cell. Biol. 4:2455.
20. Kearsey S (1984). Structural requirements for the function of a yeast chromosomal replicator. Cell 37:299.
21. Chan CSM, Tye B-K (1980). Autonomously replicating sequences in Saccharomyces cerevisiae. Proc. Natl. Acad. Sci. USA. 77:6329.
22. Walmsley RW, Chan CSM, Tye B-K, Petes TD (1984). Unusual DNA sequences associated with the ends of yeast chromosomes. Nature 310:157.

STRUCTURE AND ORGANIZATION OF YEAST CHROMOSOME III[1]

Carol S. Newlon,[a,b] R.P. Green[a], K.J. Hardeman[a,b], K.E. Kim[a], L.R. Lipchitz[a], T.G. Palzkill[a,b], S. Synn[a] and S.T. Woody[a]

[a]Department of Biology, University of Iowa
Iowa City, Iowa 52242

[b]Department of Microbiology, UMDNJ-New Jersey
Medical School, Newark, N.J. 07103

ABSTRACT We have cloned and prepared a restriction map of a 200 kb circular derivative of chromosome III from Saccharomyces cerevisiae which contains all sequences from the HML locus on the left arm to MAT locus on the right arm. Analysis of each cloned fragment for autonomously replicating sequences (ARS's) has revealed the presence of at least twelve ARS elements on the ring chromosome. The number of ARS elements and their distribution are broadly consistent with the expected distribution of replication origins.
 The relative efficiencies of four chromosome III ARS elements have been measured using a colony color assay to distinguish 1:0 and 2:0 segregations of the ARS-bearing plasmids. The chromosome III ARS elements tested vary at least 6-fold in efficiency.
 To determine whether ARS elements are essential for chromosome stability, we are systematically deleting ARS's from wild type, linear chromosome III (350 kb) and two circular derivatives, one 200 kb and the other 66 kb. Deletion of two single ARS elements from the 66 kb derivative destabilizes it significantly while deletion of the same ARS elements individually from the normal chromosome has no detectable effect.

1. This work was supported by N.I.H. Research Grant GM21510 and N.I.H. Genetics Training Grant GM07091

INTRODUCTION

Eukaryotic chromosomes are composed of single linear fibers of DNA complexed with histones and other chromosomal proteins. Chromosomes replicate once per cell cycle, initiating DNA replication at multiple sites along the molecule, and daughter chromosomes are faithfully segregated to daughter cells at mitosis. Three types of cis-acting DNA elements are known to function in chromosome replication and/or segregation.

Centromeres provide the chromosomal attachment site for the mitotic and meiotic spindle, and thus function in segregation. Yeast centromeres were cloned and identified on the basis of their ability to stabilize autonomously replicating plasmids in yeast (reviewed in 2,3). Functional yeast centromeres (CENs) are on the order of 150 bp in length and are highly homologous to each other. In fact, CEN3 and CEN11 are functionally interchangable.

Telomeres are the physical ends of chromosomes. Yeast telomeres were cloned on the basis of their ability to allow a plasmid to replicate as a stable, linear molecule (reviewed in 2). Yeast telomeres are highly homologous and terminate with several hundred base pairs of the irregular repeat $C_{1-3}A$. These sequences must function to allow the replication of the 3' terminal portions of the strands of a linear of a DNA duplex, which cannot be accomplished by any known DNA polymerase acting alone. In addition, these sequences must allow for the stable maintenance of telomeres, which unlike other free DNA ends, are stably maintained.

The third class of elements, autonomously replicating sequences (ARS) were identified on the basis of their ability to confer on yeast plasmids the ability to replicate extra chromosomally (10,16). ARS elements have many of the properties expected of replication origins, but definitive proof of their function is lacking. From measures of the fraction of chromosomal DNA restriction fragments which carry ARS elements, it has been estimated that there is one ARS per 32-40 kb (1,4), a number in close agreement with the spacing of replication origins estimated from electron microscopic analysis of replicating chromosomal DNA (14). In addition, ARS1-containing plasmids are found in the nucleus (10) and in several in vitro replication systems, replication appears to initiate preferentially at or near an ARS (reviewed in 20). However, a recent report (9) suggests that at least some in vitro systems may not be initiating DNA replication but may be elongating RNA primers present in plasmids isolated from E. coli.

If ARS's are not replication origins, then what are they? One possibility is that any DNA sequence introduced into the nucleus can replicate there. This is suggested by experiments which demonstrated that any DNA sequence injected into Xenopus oocytes is replicated in a controlled way (7). ARS elements could then be sequences that allow a plasmid to be maintained in the nucleus and segregated sufficiently well at cell division to allow transformants containing unintegrated copies of the plasmid to grow into colonies under selective conditions. This putative segregation function performed by the ARS must be different from the segregation function performed by CEN sequences, and could allow compartmentalization of the plasmid in the nucleus as a nuclear matrix binding site, for example. It is also possible that ARS elements are required in cis for elongation or termination steps in DNA synthesis (e.g. as topoisomerase binding sites).

Cloned elements of these types have been combined in plasmids to produce artificial minichromosomes. One of the most striking features of these minichromosomes is that they are at least two orders of magnitude less stable than natural chromosomes (8,13). This may be an effect of their size: there is evidence that minichromosomes become increasingly stable as they increase in size (8,13), and deletion derivatives of chromosome III have reduced mitotic stability (18). However, there may be other elements, as yet unidentified, which account for the increased stability of natural chromosomes.

As an approach to understanding the functional organization of a eukaryotic chromosome and determining how ARS elements function in stabilizing chromosomes, we have cloned and prepared a restriction map of a 200 kb circular derivative of yeast chromosome III which contains the sequences between HML and MAT (17). Based on the average spacing between replication origins, this chromosome is expected to have 5 or 6 origins. In addition one of the cloned fragments carries CEN3 and others may carry additional, as yet unidentified, elements which are important for chromosome stability.

RESULTS AND DISCUSSION

Cloning and Mapping Ring Chromosome III

Our strategy for cloning and mapping the ring chromosome was to construct a library of fragments enriched for ring chromosome in the URA3 vector YIp5. The integrating vector was selected so that we could directly test fragments for ARS activity by determining the efficiency with which the plasmids transform yeast. For construction of the libraries, covalently closed circular DNA was prepared from strain XG1#24 (17) by a gentle alkaline lysis method (5) and digested to completion with BamHI. BamHI was chosen because it was expected to cut the ring chromosome at intervals of, on average, 9.3 kb, and because it does not cleave the major known contaminants in the cccDNA extracts, 2 µm DNA and ribosomal DNA. To facilitate mapping, we screened a library of a partial EcoRI digest of DNA from the yeast strain A364A in λ-Charon 4A with isolated ring chromosome III as probe to obtain fragments which overlap the BamHI fragments obtained by direct cloning. Initially we were able to determine a restriction map of 161 kb of the chromosome, using cloned fragments containing the HIS4, LEU2, CEN3, PGK1 and MATα loci to screen the λ and YIp5 libraries, and "walking" from those regions by overlap hybridization techniques.

However, two gaps remained in the map, and we were unable to obtain fragments which overlapped the BamHI sites at each end of the gaps by carefully screening two additional libraries in λ vectors and a library constructed in a CEN-containing plasmid. On the assumption that the sequences in the gaps might be deleterious or lethal to E. coli, we constructed plasmids designed to clone the sequences directly in yeast by exploiting the ability of yeast to repair "gapped" plasmids (15). Thus by cloning fragments from the ends of the gaps in their chromosomal orientation we constructed a plasmid which could be linearized at the junction of the gap termini. When this plasmid was used to transform yeast, it could be repaired by gene conversion, using the homologous yeast chromosome to provide information in the gap. Using this strategy, we were able to recover circular plasmids from yeast which carried the sequences in the gaps. The restriction map of the 200 kb ring chromosome, which is summarized in Figure 1 (Newlon et al., in prep.)

One interesting feature of the restriction map is that it

reveals a five-to ten-fold variation in the ratio of physical to genetic distance. The average ratio for the whole ring chromosome is 2.7 kb/cM, with values ranging from 1 kb/cM in the HIS4 to LEU2 interval to 5-10 kb/cM in the CRY1 to MAT interval. The basis for this variation has not been studied, but the data suggest that there may be a hot spot for recombination between HIS4 and LEU2 (Newlon et al., in prep).

In the process of constructing the restriction map, we developed detailed maps for four laboratory strains. Of more than 100 restriction sites sampled, only four are polymorphic in a way that could have resulted from single base changes. The major differences between the strains result from polymorphisms related to the transposable element, Ty1. The ring chromosome from XG1#24 shown in Figure 1 has three Ty's and four delta sequences (presumably left from Ty excisions). Other strains have from one to six Ty's in the ring chromosome. All the Ty's we have discovered are in the intervals between HIS4 and LEU2 and between PGK and CRY1 (Newlon et al., in prep).

Identification of ARS Elements on the Ring Chromosome

ARS-containing plasmids transform yeast with efficiencies 10^2- to 10^3- fold higher than integrating plasmids (16). The eighteen BamHI fragments from the ring chromosome have been tested; six of the fragments have ARS activity. Subcloning these fragments has revealed a total of at least twelve ARS elements: three fragments have one each, two fragments have two each and, surprisingly, the fragment carrying the MAT/HML fusion (E5F) has at least five ARS elements (Figure 2). The location of the ARS elements is summarized in Figure 1, with each ARS named for the fragment in which it resides. Three of the ARS elements are weak: transformants carrying plasmids containing them grow under selective conditions with doubling times from 2- to 10-fold longer than transformants carrying plasmids with strong ARS elements.

Our measurement by electron microscopy of the average spacing between replication origins on chromosomal DNA is 36 kb (14). The number of ARS elements found on the ring chromosome is thus about twice the number expected if ARS elements are replication origins. However, it would be difficult to visualize by electron microscopy replication bubbles spaced as closely as the cluster of ARS elements in E5F. If this cluster is treated as a

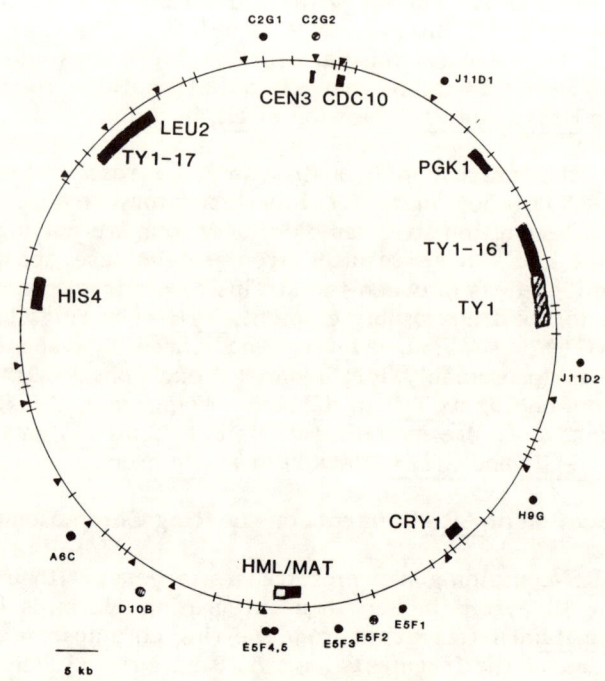

FIGURE 1 Restriction map of ring chromosome III and location of ARS's. The map shown is for the chromosome from strain XG1#24 (17). The ring chromosome contains eighteen BamHI fragments, ranging in size from 150 bp to 40 kb. The 150 bp fragment is not shown but is at the BamHI site to the left of Ty1-17. The locations of ARS's are shown by circles above the restriction map. Filled circles represent "strong" ARS elements which support the growth of transformants under selective conditions at normal rates. Shaded circles represent "weak" ARS elements which support the growth of transformants under selective conditions at two to ten times the normal growth rate. Triangles represent BamHI sites and vertical lines represent EcoRI sites.

single ARS, and weak ARS's are ignored, then the number and spacing of ARS elements on the ring chromosome are broadly consistent with the expectation for the number and spacing of replication origins.

Relative Efficiencies of Chromosome III ARS's

Plasmids carrying only an ARS and a selectable gene are highly unstable in yeast. In cultures growing under selective conditions only 5 to 50% of the cells contain the plasmid, but the plasmid is present in high copy number (20-50 copies per cell) in the cells which contain it (reviewed in 21). The mitotic instability and the high copy number of ARS result from the failure of such plasmids to segregate properly during mitosis. Results from pedigree analysis indicate that in 0.3 to 0.6 of cell divisions one cell fails to get plasmid and the bias in favor of the

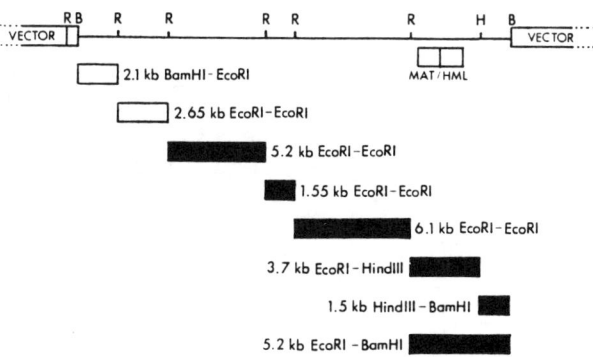

FIGURE 2 Analysis of ARS location in plasmid ESF. The 23 kb BamHI fragment carrying the MAT/HML fusion cassette was subcloned as indicated in YIp5 and the resulting plasmids were tested for ARS activity. Filled rectangles indicate fragments showing ARS activity (high frequency transformation and mitotic instability) while the open rectangles are fragments devoid of ARS activity.

mother cell getting all copies is 19:1 (12). It is difficult to quantitate the relative efficiencies of ARS elements by measuring rates of loss of such plasmids because of the segregation bias and because such measurements are greatly affected by plasmid integration. A substantially improved assay makes use of centromere-containing plasmids to control copy number and provide efficient segregation and makes use of a colony color assay (the red pigment formed by an ade2 mutant) to measure plasmid stability (8,11). In one of these systems (8) the plasmid carries CEN4, URA3, and SUP11, which encodes an ochre suppressor tRNA. When a diploid homozygous for an ade2-ochre mutation is transformed with this plasmid, plasmid copy number is revealed by colony color: transformants with two copies of the plasmid produce white colonies, those with one copy produce pink colonies, and cells which have lost the plasmid produce red colonies. By scoring the frequency of half-sectored colonies arising from cells which were initially part of a pink colony, it is possible to obtain a direct measure of the rate of 2:0 segregations of the plasmid (white/red half sectors) and 1:0 segregations (pink/red half sectors). The 2:0 class includes nondisjunctions and the 1:0 class includes failures of replication and plasmid loss during mitosis.

Four chromosome III ARS elements have been cloned into the BamHI site of this vector by adding BamHI linkers to appropriate restriction fragments. The results are summarized in Table I. This plasmid carries ARS3, so measurements are made in conjunction with that ARS. However, ARS3 is so weak that effects of adding additioal ARS elements to the plasmid are readily seen. Controls are the ARS3 plasmid alone and that plasmid with ARS1 inserted. ARS1 is located on chromosome IV and was the first ARS described (10,16). It appears that the ARS to the left of CEN3 (C2G1) is similar in strength to ARS1, while the ARS in D10B, which appears weak in YIp5 on the basis of increased doubling time, is at least 6-fold weaker as measured by the 1:0 segregation rate. While the ARS elements in H9G and A6C cannot be distinguished from the C2G1 ARS when cloned in YIp5, they are 2- to 3-fold weaker than the C2G1 ARS by this assay.

These data show that plasmids carrying the C2G1 ARS appear to undergo nondisjunction at a rate lower than the other plasmids. This result has been confirmed using a different plasmid carrying SUP11 and CEN3 (data not shown). The 530 bp

C2G1 fragment may therefore carry an element which aids in plasmid segregation. Experiments to test this hypothesis are in progress.

Effects of Deleting ARS's from Chromosome III

Whatever the role of ARS elements, be it function as replication origins, nuclear matrix binding sequences, or something else, it is of interest to determine whether they play an essential role in chromosome function. An approach to this question is to systematically delete ARS elements from a chromosome and measure the effect on chromosome stability. For this purpose we are using three derivatives of chromosome III. One is a 66 kb ring chromosome III which resulted from a recombination between Ty1-17 and Ty1-161 (19) and contains only two strong ARS elements, C2G1 and J11D1 (see Fig. 1). (The C2G2 ARS can be seen only in a CEN-containing plasmid; it is not

TABLE 1
RELATIVE EFFICIENCIES OF CHROMOSOME III ARS'S[a]

ARS	Rate of 1:0 segregation	Rate of 2:0 segregation
ARS3	> 0.5	n.d.
ARS1 + ARS3	0.027	0.02
C2G + ARS3	0.028	0.002
A6C + ARS3	0.050	0.018
H9G + ARS3	0.088	0.028
D10B + ARS3	0.165	0.011

[a]Chromosome III ARS elements were cloned as approximately 1 kb fragments into the BamHI site of YRp14-CEN4 (8). ARS3 designates the plasmid YRp14-CEN4 and ARS1 + ARS3 the plasmid YRp14-CEN4-ARS1 (8). Cells from a single pink colony were grown nonselectively for 5-10 generations, diluted and plated on color assay plates (8). Data were calculated as the fraction of red/pink (1:0) or red/white (2:0) half-sectored colonies per total pink colonies. The numbers are based on scoring at least 2000 colonies derived from cells with plasmid.

sufficiently strong to yield stable transformants when cloned in YIp5). The second is the 200 kb ring which has at least twelve ARS elements. The third is the wild type chromosome which has been estimated to be 350 kb in length (3a). We have individually deleted a 500 bp fragment containing the C2G1 ARS and a 1.7 kb fragment containing the J11D1 ARS from the 66 kb ring and the wild type, linear chromosome. The deletions were constructed by integrating a URA3 plasmid carrying the desired deletion plus its flanking sequences. Rare homologous recombination between the duplicated flanking sequences results in loss of the plasmid, leaving behind either the deletion or the wild type ARS; cells which have lost the plasmid can be selected by plating on medium containing 5-fluoro-orotic acid, which poisons $URA3^+$ cells.

Table 2 summarizes experiments in which the frequency of loss of deletion-bearing derivatives of the 66 kb ring was measured. Loss of the ring, which carries the $LEU2^+$ allele, can be detected because it uncovers the $leu2^-$ allele on the wild type chromosome. While rates of loss still need to be measured, it is apparent from these frequency measurements that deletion of either the J11D1 or the C2G1 ARS elements significantly destabilizes the 66 kb ring chromosome. Experiments are in

TABLE 2
STABILITY OF ARS-DELETION DERIVATIVES OF 66 KB RING[a]

66 kb	Colonies scored	Leu⁻ colonies	% loss
wild type	352	3	0.85
" "	319	0	0
Δ J11D1	1939	103	5.3
Δ J11D1	1203	91	7.6
wild type	747	9	1.20
wild type	1028	6	0.59
Δ C2G1	712	20	2.8
Δ C2G1	1370	40	2.92

[a]Cells from a single colony grown on selective plates were grown nonselectively for five generations and then plated on YEPD plates. Colonies were then replicated to plates without leucine and the number of leu⁻ colonies was scored.

progress to try to delete both ARS's from this chromosome.

The same two deletions have been constructed in the MATα chromosome of a strain disomic for the wild type linear chromosome III. In this case, loss of one copy of chromosome III converts a nonmating cell to a cell capable of mating. Fluctuation analysis of quatitative mating experiments (6) was used to measure the rate of loss of wild type and ARS-deletion chromosomes in this strain. No significant difference was seen in the rates of loss of wild type and ARS-deletion chromosomes (data not shown). Thus deletion of a single ARS from a chromosome which has only two strong ARS elements reduces its mitotic stability, while deletion of the same ARS elements from the wild type linear chromosome, which probably has at least twenty such elements, has no measureable effect. It will be of interest to determine how many ARS elements can be removed from the wild type chromosome before an effect on stability can be seen.

In the case of the 66 kb ring chromosome, the effects of ARS deletion are not large. Perhaps this is not surprising since other studies have shown that a single ARS, ARS1 is sufficient to allow replication of a 137 kb plasmid. In fact the 137 kb plasmid with a single copy of ARS1 is more stable than a similar plasmid of only 45 kb (8).

The failure to detect an effect of deleting the C2G1 ARS on mitotic stability of the wild type chromosome suggests that, if the C2G1 ARS functions in chromosome segregation as it appears to in plasmid segregation, then there must be other such elements in the chromosome which compensate for its deletion. Screening the available cloned fragments may identify such elements.

Prospects

The information we have obtained about the organization of chromosome III and the location of ARS elements on this molecule will be useful in further studies of the role of ARS elements in chromosome stability. If ARS elements function as replication origins, it should be possible to correlate their position with that of replication bubbles. In addition, it should be possible to show that replication of ARS elements precedes that of adjacent non-ARS sequences using density transfer experiments.

ACKNOWLEDGEMENTS

We thank Polly Ferguson and Ann Dershowitz for contributions to the ARS-deletion experiments and Diane Clark and Brenda Lemon for typing the manuscript.

REFERENCES

1. Beach D, Piper M, Shall S (1980). Isolation of chromosomal origins of replication in yeast. Nature 284:185.
2. Blackburn EH, Szostak JW (1984). The molecular structure of centromeres and telomeres. Ann Rev Biochem 53:163.
3. Carbon J (1984). Yeast centromeres: structure and function. Cell 37:351.
4. Carle GF, Olson MV (1985). An electrophoritic karyotype for yeast. Proc Nat Acad Sci USA 82:3756
5. Chan CSM, Tye B-K (1980). Autonomously replicating sequences in Saccharomyces cerevisiae. Proc Nat Acad Sci USA 77:6329.
6. Devenish RJ, Newlon CS (1982). Isolation and characterization of yeast ring chromosome III by a method applicable to other circular DNAs. Gene 8:277.
7. Dutcher SK, Hartwell LH (1982). The role of Saccharomyces cerevisiae cell division cycle genes in nuclear fusion. Genetics 100:175.
8. Harland RM, Laskey RA (1980). Regulated replication of DNA microinjected into eggs of X. laevis. Cell 21:761.
9. Hieter PW, Mann C, Snyder M, Davis RW (1985). Mitotic stability of yeast chromosome: a colony color assay that measures nondisjunction and chromosome loss. Cell 40:381.
10. Jong AYS, Scott JF (1985). DNA synthesis in yeast cell-free extracts dependent on recombinant DNA plasmids purified from Escherichia coli. Nuc Acids Res 13:2943.
11. Kingsman AJ, Clarke, Mortimer RK, Carbon J (1979). Replication in Saccharomyces cerevisiae of plasmid pBR313 carrying DNA from the yeast TRP1 region. Gene 7:141.
12. Koshland D, Kent JC, Hartwell LH (1985). Genetic analysis of the mitotic transmission of minichromosomes. Cell 40:393.
13. Murray AW and Szostak JW (1983). Pedigree analysis of plasmid segregation in yeast. Cell 34:961.

14. Murray AW, Szostak JW (1983). Construction of artificial chromosomes in yeast. Nature 305:189.
15. Newlon CS, Burke WG (1980). Replication of small chromosomal DNAs in yeast. ICN-UCLA Symp Molec Cell Biol 19:399.
16. Orr-Weaver TL, Szostak JW, Rothstein RJ (1983). Genetic applications of yeast transformation with linear and gapped plasmids. Meth Enzym 101:228.
17. Stinchcomb DT, Struhl K, Davis RW (1979). Isolation and characterization of a yeast chromosomal replicator. Nature 282:39.
18. Strathern J, Newlon CS, Herskowitz I, Hicks J (1979). Isolation of a circular derivative of yeast chromosome III: implications for the mechanism of mating type interconversion. Cell 18:309.
19. Surosky RT, Newlon CS, Tye B-K (1986). The mitotic stability of deletion derivatives of chromosome III in yeast. Proc Natl Acad Sci USA, in press.
20. Surosky RT, Tye B-K (1985). Resolution of dicentric chromosomes by Ty-mediated recombination in yeast. Genetics 110:397.
21. Williamson DH (1985). The yeast ARS elements, six years on: a progress report. Yeast 1:1.

STRUCTURE AND FUNCTION OF CENTROMERES[1]

Ray Ng, Susan Cumberledge, and John Carbon

Department of Biological Sciences
Univ. of California, Santa Barbara, CA 93106

The centromere is the region of the chromosome where the mitotic apparatus attaches to ensure proper segregation. Several functional centromere DNAs from S. cerevisiae have been isolated and characterized. Altered centromere DNA sequences have been constructed in vitro and their in vivo effects on mitotic and meiotic chromosome segregation have been determined to provide insight into the molecular mechanisms involved in centromere function.

OVERVIEW

Successful cell division requires the partitioning of the newly replicated genetic material between the progeny. The prokaryotic cell need only ensure that each daughter cell receives a copy of the single chromosome. This is readily accomplished by providing two chromosomal attachment sites on the membrane which are separated by inward growth during cell fission. Eukaryotic cells must solve the more complex problem of sorting and localizing groups of chromosomes. In addition, a diploid life cycle and the meiotic process require that groups of chromosomes be paired and segregated through equational and reductional divisions. Proper chromosome segregation occurs with astounding fidelity: in Saccharomyces cerevisiae, a

---

[1] A portion of the research reported in this paper was supported by a research grant (CA-11034) from the National Cancer Institute (NIH). R. Ng holds a postdoctoral fellowship awarded by the American Cancer Society (#PF-2291).

chromosome is lost only once in every 10^5 cell divisions. Our work attempts to elucidate the role centromeres play in the choreography of chromosome movement during both mitosis and meiosis.

The centromere is the region of the chromosome where the mitotic and meiotic spindle attaches to ensure proper segregation (1). Early cytogenetic work on a variety of cell types has shown that during mitosis each newly replicated chromosome consists of two sister chromatids which remain attached at the centromere through prophase. At that time, a kinetochore-spindle fiber complex assembles on the centromere pulling the separated chromosomes towards opposite poles. Meiotic division is more complex. The sister chromatids remain bound at the centromere throughout meiosis I, while they are paired and sorted. Subsequently, in meiosis II the sister chromatids are separated and segregated in a manner apparently analogous to mitosis. The recent development of sophisticated molecular and genetic techniques in S. cerevisiae has made possible a detailed study of the structure and function of the yeast centromere. These studies have been reviewed elsewhere (2,3,4).

ISOLATION OF FUNCTIONAL CENTROMERES

Clarke and Carbon (5) isolated the first centromere from S. cerevisiae by cloning overlapping DNA fragments located between two tightly linked centromere markers (LEU2 and CDC10), which occur at opposite sides of the centromere on chromosome III. These two genes are within eight map units (~20kb) of each other. The centromere was localized within this region by cloning various restriction fragments into ARS plasmid vectors and subsequently assaying these plasmids for their ability to segregate properly through mitosis and meiosis.

Different assays have been used to measure the effect of a centromere on plasmid stability (2). The assay for plasmid mitotic stability is performed by determining the fraction of cells losing plasmid per generation, i.e. the segregation rate. Autonomously replicating sequence (ARS) plasmids containing genetic markers, such as LEU2, can be easily introduced into S. cerevisiae cells by genetic transformation. These plasmids replicate independently of the chromosome yet lack the information necessary for proper segregation. Cells containing the plasmid are easily determined by scoring for a plasmid marker.

Recently, a color colony sector assay, based on the number of half sectored colonies in a population, has been used to accurately determine the rate of plasmid loss (6,7).

The ARS plasmids are highly unstable and are normally lost at a rate of 30% per generation (8,9). During cell division an ARS plasmid segregates predominantly with the mother cell (10). Since the plasmid replicates only once per cell cycle (11), as the population expands fewer cells contain the plasmid. Those cells that retain the plasmid accumulate as many as 50 copies per cell (10), due to the bias in plasmid partition. The presence of a centromere (CEN) sequence on circular or linear ARS plasmids improves the plasmid loss frequency to 10^{-2}-10^{-3}/cell division (5,12). Most of the cells in the population retain the CEN plasmid, and due to proper segregation, the copy number is maintained at 1-2 copies per cell. Even in the event of segregation failure, the copy number of the CEN plasmid remains low, due to the apparent toxicity of excess functional centromere sequences (2).

A second important property of the centromere, proper meiotic segregation, is easily measured by exploiting the life cycle of S. cerevisiae. Unlike most eukaryotes, S. cerevisiae is stable as a haploid (a or α) or a diploid (a/α). When a/α diploid cells are starved for nitrogen they enter a sporulation pathway and undergo meiosis. The four gametes or spores can be isolated, grown and scored for the presence of genetic markers. To follow plasmid segregation, a haploid strain containing an ARS plasmid is crossed with a haploid (without plasmid) of the opposite mating type. The resulting diploid is sporulated and each of the four progeny are tested for the presence of a genetic marker on the plasmid (see Fig. 1). The sister spores, resulting from the second meiotic division, can be identified by constructing the diploid so that it contains a heterozygous centromere-linked marker. The ARS plasmids segregate randomly through meiosis; most of the tetrads show a non-Mendelian plasmid distribution.

The addition of a centromere radically alters ARS plasmid segregation during meiosis. The CEN plasmids segregate predominantly 2+:2-. The plasmid replicates, but remains paired through meiosis I, then separates and segregates to opposite poles during meiosis II. Therefore, two sister spores retain a plasmid and the other two sister spores are left without a plasmid. This meiotic segregation pattern is typical of tightly centromere-linked genes, where recombinative events have not occurred between

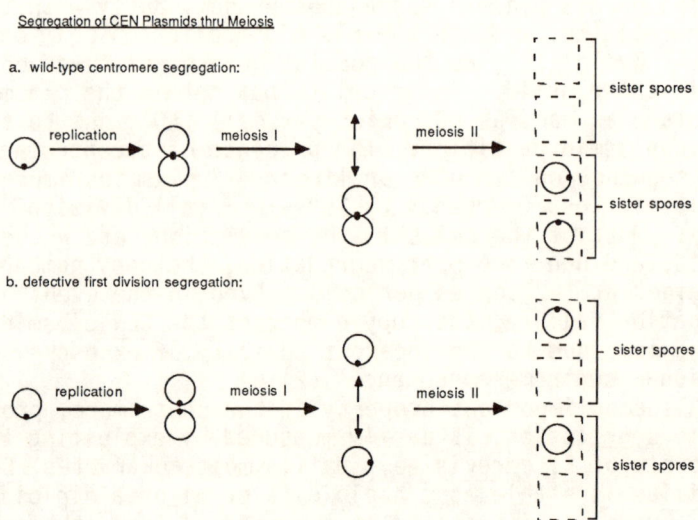

Figure 1. (a) Normal CEN plasmids segregate faithfully through the meiotic cycle of a sporulated yeast cell, appearing always in two sister spores. (b) Defective CEN plasmids exhibit premature separation at the first meiotic division. Thus, a high percentage of the plasmids are found in non-sister spores.

the marker and the centromere. Occasionally, due to segregation or replication errors, tetrads with 4+:0-, 3+:1-, 1+:3- and 0+:4- plasmid distributions are observed. However, the majority of cells show 2+:2- segregation and essentially all of the 2+:2- tetrads are parental and nonparental ditypes.

CENTROMERE STRUCTURE

Several yeast centromeres have been isolated based on the ability of the CEN sequences to stabilize ARS plasmids, even when the host cells are grown under non-selective conditions (2). Recently, Hieter et al. (13) isolated the centromeres from six different yeast chromosomes by using a direct selection procedure. Random genomic DNA fragments from yeast were cloned into an ARS plasmid that is lethal in high copy number due to the presence on the plasmid of an ochre suppressor tRNA. The CEN fragments were isolated by their ability to overcome in cis the toxic effect of the plasmid. Presently, a total of eleven centromeres from yeast have been isolated, and ten have been sequenced (2,3,13). All CEN fragments ensure proper mitotic and meiotic segregation of ARS plasmids and lower their copy number to 1-2 per cell. Although centromeric sequences fail to cross-hybridize with each other, they are interchangeable. The replacement of the 624 bp CEN3 fragment with CEN4, CEN5 or CEN11 does not affect the mitotic or meiotic stability of chromosome III (2). Orientation of the centromere in the yeast chromosome apparently has no affect on chromosome behavior (15). A striking organizational homology is found among the known CEN sequences (see Fig. 2). The common core centromere region is divided into three sequence domains. Element II, a 78-86 bp region of high A+T (>90%) is the largest domain. This sequence is flanked on its left by sequence element I, a conserved 8 bp PuTCACPuTG sequence. Immediately to the right of element II is element III, a 25 bp conserved sequence domain. A palindromic sequence, representing approximately two turns of the helix, is found in element III (13). This highly conserved region is a plausible site for a unique hairpin structure and/or a possible protein(s) binding site.

Similarities between the centromere core sequence and the Drosphila melanogaster DM359 satellite DNA sequence

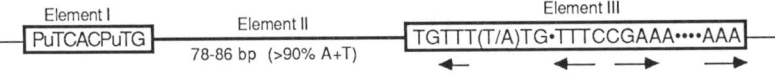

Figure 2. Consensus sequence of S. cerevisiae centromeres (2,13). The arrows indicate the region of dyad symmetry found in sequence element III.

have been noted (14). The DM359 DNA sequence, which occurs in the centromere region of Drosphila sex chromosomes, contains element I and III regions that share strong homology with the yeast centromeres, although the element II region is larger (265 bp vs 89 bp) and much lower in A+T content (70%). When cloned into an ARS plasmid, the DM359 sequence does not function as a centromere in yeast. However, it is speculated that this sequence may form an efficacious kinetochore region in the sex chromosomes in Drosphila, since its deletion leads to loss of faithful meiotic segregation and pairing of the X and Y chromosomes (2).

STRUCTURE-FUNCTION STUDIES:MITOTIC FUNCTION

A detailed in vitro mutational analysis has been carried out to determine the sequences important for centromere function. Structurally altered CEN regions are assayed for function either on ARS plasmids or by replacement of the resident CEN sequence in the genome with the mutant CEN, and then following the segregation behavior of the resulting altered chromosomes (15,17). The genomic centromere replacement is carried out by using the fragment-mediated transformation procedure devised by Rothstein (16). Chromosome loss can be assayed by several methods. Our experiments have been carried out on chromosome III, which contains the mating type locus, MATa or MATα. Loss of chromosome III in an a/α diploid cell results in a 2N-1 aneuploid that is either MATa or MATα. One can determine the rate of chromosome III loss by using mating type testers to measure the number of competent "mater" cells in the population (15). Disappearance of a genetic marker can also be used to determine chromosome loss. By following chromosome loss instead of plasmid stability this genome replacement assay results in a 1000-fold increase in sensitivity, since normal yeast chromosomes are lost at a frequency of only 10^{-5}.

A precise determination of the functional boundaries of the yeast centromere has been carried out in several laboratories (3,8). Subcloning and deletion analysis of CEN3 and CEN11 has demonstrated that sequences to the left of element I are not required for normal centromere function (2). Various deletions and rearrangements of sequences in element I and II have been generated. In general, larger deletions of the (A+T)-rich sequences result in progressively lower chromosome mitotic stability (see Fig. 4)(17).

The rightward boundary has been analyzed in greater detail. A CEN3 fragment, containing elements I, II, and III, plus 28 bp beyond the conserved sequences in element III, has been generated by Bal31 deletion (M. Fitzgerald-Hayes, personal communication). This fragment retains normal centromere activity when substituted into the genome. Experiments on CEN14 show that the right-hand boundary is no further than 25 bp from the conserved element III region (18). Moreover, deletions extending 4 nucleotides into the conserved element III region of CEN3 reduce the mitotic stability of chromosome III by 1000-fold (see Fig. 3).

To determine if the functional centromere sequences extend past the conserved nucleotides of element III, we have generated by oligonucleotide site-directed mutagenesis a unique BglII restriction site four nucleotides downstream of all apparent conserved sequences in the element III region (see Fig. 3). The 130 bp centromere fragment thus created was tested for centromere activity. Surprisingly, the fragment when substituted into chromosome III showed reduced centromere activity; chromosome III loss was five times greater than is observed with the normal chromosome. However, a larger CEN fragment containing the base substitutions had no affect on chromosome loss (Fig. 3). These results suggest that nucleotide sequences immediately adjacent to element III still influence centromere function, perhaps by conferring an optimal structure to the region. In addition, neighboring regions around the centromere core have not been positively excluded from possible centromere function, since they are still present in the genome when small centromere regions are tested.

Because element III is a highly conserved sequence domain, its role in CEN function has been tested by several groups. Panzeri, Stotz and Philippsen (3), working with CEN6, have constructed a 14 bp deletion, 11 bp into element III plus 3 bp of adjacent DNA, replacing that region with an 8 bp linker sequence. This alteration abolishes centromere function when tested on plasmids. Similar deletions in CEN3 in the five conserved A nucleotides at the right-hand end of element III result in a dramatic reduction in centromere activity when tested in the genome (see Fig. 3). Several point mutations have also been generated in vitro, and their effects have been tested on plasmids as well as in chromosome III. The results are

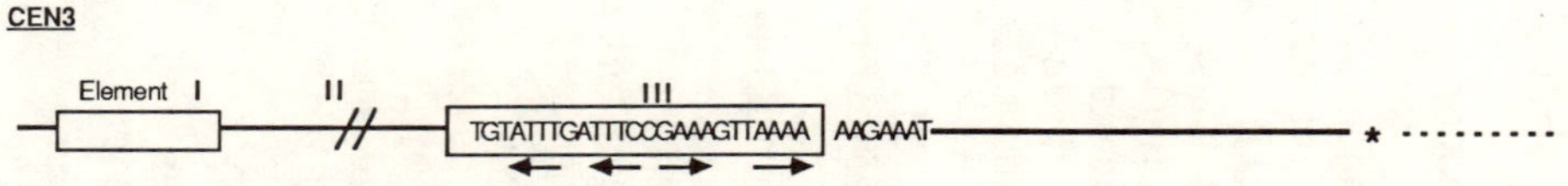

CEN3

Element I — II — III — AGAAAT ·········*

Wild-type (290 bp): TGTATTTGATTTCCGAAAGTTAAAA

Point Mutants:
#1 TGTATTTGATTTCTGAAAGTTAAAA
 2 TGTATTTGATTTTCGAAAGTTAAAA
 3 TGTATTTGATTTCCAAAAGTTAAAA
 4 TGTATTTGATTTCCGAAAGTTAAAA
 5 TGTATTTAATTTCCGAAAGTTAAAA
 6 TGTATTTGATTTCCGAAAGTTAACA
HpaI 7 TGTATTTGATTTCCGAAAGTTAAAA AAGATCT
BglII

Deletions:
BglII #1 TGTATTTGATTTCCGAAAGTT**CATC**--------
HpaI 2 TGTATTTGATTTCCGA**ACACTT**--------
       3 TGTATTTGATTTCCG**CCACTT**--------

	Plasmid Loss/Generation	Chromosome III Non-Disjunction per Cell Division	Plasmid Meiotic Segregation (sister:non-sister)
Wild-type	.06-.09	0.7×10^{-5}	17:0
#1	[90%]	ND	ND
2	[38%]	3.6×10^{-2}	ND
3	.13-.22	3.0×10^{-1}	7:5
4	.12-.20	1.7×10^{-1}	15:2
5	ND	1.0×10^{-5}	18:1
6	ND	0.9×10^{-4}	ND
7	ND	1.9×10^{-5}	15:2
BglII #1	ND	0.7×10^{-4}	10:0
HpaI 2	.20	2.5×10^{-2}	ND
3	.20	ND	ND

Figure 3. Various point mutations and deletions in CEN3 element III have been generated and their effects on CEN activity measured by mitotic and meiotic assays. The dyad symmetry in element III is marked by arrows. Dashed lines (- -) represent sequences beyond the AluI restriction site (*) found approximately 140 bp downstream of element III. Point mutations #1 and 2 were made by J. McGrew. The percentage of cells without the plasmid after overnight growth in selective media are listed in parentheses (J. McGrew, B. Rayala, and M. Fitzgerald-Hayes, manuscript submitted). ND: no data.

listed in Figure 3. Five conserved C:G base pairs are present in element III (see Fig. 2). A C to T transition in the center of element III (fourth C:G bp from the left end) results in the complete inactivation of centromere function when tested on plasmids (J. McGrew, B. Rayala, and M. Fitzgerald-Hayes, manuscript submitted). The importance of this nucleotide may be related to its pivotal central position in the observed dyad symmetry (Fig. 2). Alterations of the two adjacent C:G base-pairs do not have as dramatic an effect on centromere activity, although the stability of the altered chromosomes is reduced by several orders of magnitude (Fig. 3). Changing the second C:G basepair from the left end to a T:A, however, has little effect on CEN activity. To date, six specific point mutations have been examined (summarized in Fig. 3). Of these, the most dramatic effect is seen when the pivotal C:G base-pair in the dyad symmetry region is changed. Modifications outside the palindromic sequence have minor effect. Clearly, sequence element III is an essential component for centromere function; however, results in our laboratory show that it alone is insufficient to confer significant mitotic stability on the yeast chromosome (17).

STRUCTURE-FUNCTION STUDIES:MEIOTIC FUNCTION

In general, mutations that affect mitotic function also cause defects in meiotic segregation. ARS plasmids carrying various deletion mutations in element I, II or III have been assayed for proper meiotic segregation. As shown in Figure 4, in a large percentage of the 2+:2- tetrads, the plasmid segregates to non-sister spores, indicative of precocious first division segregation in meiosis I. In fact, meiotic behavior appears to be much more sensitive to CEN structural changes than is mitotic behavior. For example, a deletion of element I and about 10 bp of element II causes mitotic chromosome non-disjunction to occur only once in every 10^3 cell divisions, or a plasmid segregation rate of 13% loss/gen., yet in 50% of the 2+:2- tetrads scored, the plasmid segregates to non-sister spores (2,17).

The effect of these mutations on meiotic segregation of the CEN-substituted chromosomes is more ambivalent (2,17). Heterocentric diploids, in which one copy of chromosome III contains a wild-type centromere and the other bears an altered CEN3, show relatively normal meiotic behavior. However, homocentric diploids, in which both

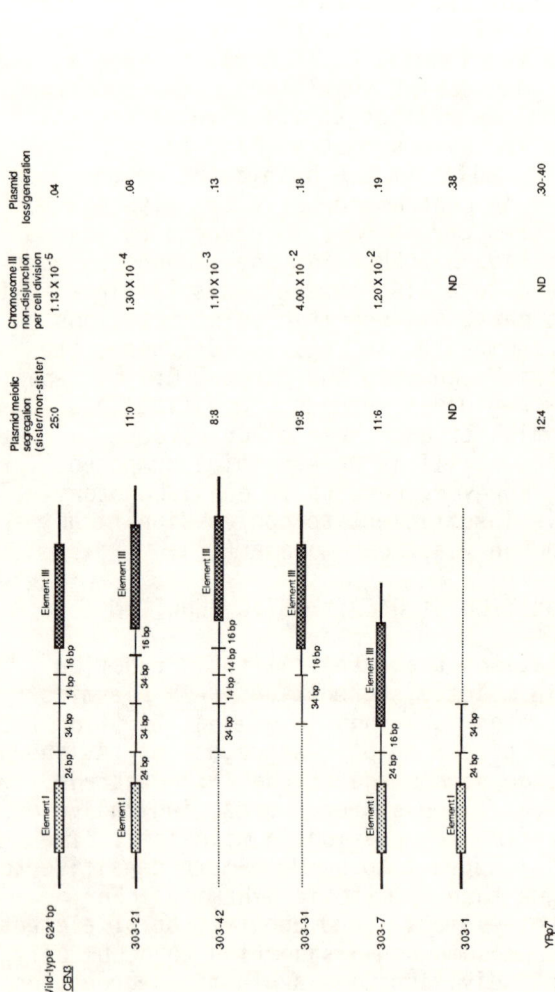

Figure 4. The importance of the element I-III sequence in the 624 bp CEN3 fragment has been tested by using the restriction enzyme AhaIII, which cleaves the element II region into four segments. Various combinations of the AhaIII fragments have been constructed and tested for CEN function when substituted into the genome in place of the normal CEN3 sequences (17). These altered CEN fragments have also been cloned into ARS plasmids to determine their effect on plasmid mitotic and meiotic segregation. Dotted lines (...) are regions that have been deleted in CEN3. See ref. 17 for details. ND: no data.

copies of chromosome III contain the same mutated centromere sequence, give abnormal segregation patterns through meiosis, including separation and random segregation of sister homologs during the first meiotic division (17). How the wild-type centromere sequence is able to compensate for the defective centromere in trans is not understood.

CENTROMERE CONTROL OF 2 MICRON PLAMSID REPLICATION

The centromere sequence when placed on a yeast 2 micron plasmid is dominant over the amplification system that drives that plasmid to high copy number (19). Composite 2 micron-CEN plasmids are maintained at a copy number of 1-2. They faithfully segregate in meiosis and mitosis, but show a high frequency of recombination yielding deletions in the centromere (19). Plasmids with two CEN sequences are subjected to rearrangement presumably due to independent attachment to the mitotic spindle, consequently pulling the plasmid apart (20). A theory to explain how the centromere prevents over-replication of the 2 micron plasmid has been proposed (19). According to this model, the centromere region remains unreplicated or blocked until late in S phase, resulting in a physical block of the replicative fork on the 2 micron plasmid.

CENTROMERE PROTEINS

The chromatin structure of yeast centromere regions has been examined by partial digestion of chromatin with micrococcal nuclease or DNase I (21). A unique 220-250 bp nuclease-resistant centromere core, containing the sequence element I through III region, was shown to exist in chromosomes III, IV, and XI (10,21). Hypersensitive nuclease cleavage sites were found flanking the protected core. This protected region is somewhat larger than the functional size of the centromere (about 150 bp) as defined by the deletion analysis mentioned earlier. The nucleosomes surrounding the core appear to be highly ordered. When plasmids containing large deletions in the centromere are examined, no protected core or nuclease hypersensitive sites are observed. By testing some of the available point mutations in element III, it should be possible to determine if the nuclease protection pattern is only seen in the presence of a functional kinetochore.

Major interest has been focused on the identification and purification of protein(s) that interact specifically with centromere DNA. It is clear that microtubules do not interact with centromeric DNA alone. Since centromere DNA from different chromosomes can be interchanged, the factors responsible for centromere activity are most likely used in common on all chromosomes. Centromere binding proteins (CBP's) must be partially responsible for the nuclease-resistant core in centromere chromatin, since its characteristic structure can be abolished by washing chromatin with high-salt buffers (21). A biochemical analysis of the centromere-protein complex has been severely restricted by the inability to identify and purify specific protein factor(s) responsible for centromere activity. Since yeast cannot tolerate many extra copies of centromere plasmids, the failure to isolate CBP's may be a reflection of the limited amounts of these proteins present in a cell.

The CEN chromatin structure is found throughout the mitotic and meiotic cell cycle (2). Microtubule spindle fibers may be attached throughout most of the cell cycle. This would be in direct contrast to the transient kinetochore complex found in higher mammalian cells (1). Continued connection to the spindle pole bodies throughout the cell cycle may account for the unequal partition of genetic material found in yeast. ARS plasmids partition preferentially with the mother cell (10). Williamson and Fennell (22) have reported that sister chromatids do not segregate randomly. In addition, only cells that have been "mothers" can switch mating types in homothallic strains. The ability to switch is dependent on the expression of the HO gene. This unusual behavior may result from the preferential segregation to the mother cell of the sister chromatid containing an HO gene that is competent for expression (12).

To discover more about the molecular composition of the kinetochore, several groups have developed genetic strategies to complement the molecular analysis. For example, the element III point mutations can now be used to search for second site suppressor mutations in genes specifying proteins that are necessary for centromere activity. In addition, a selection scheme based on a colony color assay has been devised to isolate mutations that affect mitotic functions in trans (6,23). Two classes of mutants have been found: non-disjunction (2:0) and replication mutants (1:0). These mutants are presently being sorted into complementation groups. A study of those gene products should provide further insight into our understanding of chromosome segregation as well as lead to the identification of proteins that interact specifically with the centromere.

FINAL COMMENTS

Yeast has several distinct advantages over other eukaryotic organisms for studying chromosome segregation. The availability of telomere, ARS and CEN sequences enables one to generate altered chromosome structures in vitro and then test their in vivo function by direct substitution into the parent chromosome. The well developed molecular genetic system available in yeast should eventually lead to the identication of proteins involved in cell division and chromosome segregation. Many cell division cycle mutants that affect mitosis and meiosis already exist (3). Experimental work using other organisms has shown that kinetochore microtubules do not generate the force necessary for chromosome movement (24). Therefore, other microtubules and/or cytoskeletal structures associated with the kinetochore microtubules must be involved. In yeast, the genes for tubulin and actin have been cloned and characterized (25,26,27). Numerous mutations in those genes have been generated and used to identify genes that specify proteins that probably interact with kinetochore microtubules (28,29,30). Of recent interest, J. H. Thomas in D. Botstein's laboratory (manuscript submitted) have identified a nuclear division cycle gene (NDC1) in which mutations can result in defective chromosome segregation to the daughter cell. Overall, the experimental advantages of the yeast system are manyfold, and are directly responsible for the rapid expansion of our understanding of the machinery and mechanisms involved in cell division and chromosome movement.

Several different types of kinetochore structures are known (1). For example, most monocotyledon plants and arthropods display a diffuse kinetochore structure, where spindle fiber attachment occurs along the entire length of a chromatid. Thus, Philaenum and Bombyx mori have multiple kinetochore attachment sites. In the lower eukaryotes, Neurospora and Dictyostelium, two or more microtubules are believed to bind to a single kinetochore region. In contrast, yeast has a localized kinetochore region, which binds to a single microtubule fiber (2,3). No repetitive

DNA surrounds the centromere region in yeast, whereas in higher eukaryotes, the centromere region is noted for its heterochromatic character. The CEN sequences from S. cerevisiae fail to act as functional centromeres in other organisms tested; although yeast centromeres do contain strong sequence homologies to various satellite sequences from higher eukaryotes (2). It is uncertain whether the different kinetochore systems seen in nature represent elaborations on a basic kinetochore structure, or are truly different systems of independent design. In that regard, attempts to isolate centromeres from other organisms, such as S. pombe, Neurospora crassa, and Aspergillus nidulans are in progress (2, 3). Some key component(s) of the various systems should be similar, since all involve microtubule attachment to a DNA-chromatin fiber. Thus, it may be possible that a conserved DNA sequence element from yeast could be used to isolate centromeres and CBP's from other organisms.

ACKNOWLEDGMENTS

We thank K. Luehrsen for his help with the figures.

REFERENCES

1. Rieder, CL (1982) Int Rev Cyt 79:1.
2. Clarke, L and Carbon, J (1985) Ann Rev Gen 19:in press.
3. Blackburn, EH, and Szostak, JW (1984) Ann Rev Biochem 53:163.
4. Carbon, J (1984) Cell 37:351.
5. Clarke, L, and Carbon, J (1980) Nature 287:504.
6. Heiter, PA, Mann, C, Synder, MP, and Davis, RW (1984) Cell 40:381.
7. Koshland, D, Kent, JC, and Hartwell, LH (1984) Cell 40:393.
8. Bloom, KS, Fitzgerald-Hayes, M, and Carbon, J (1982) Cold Spring Harbor Symp Quant Biol 47:1175.
9. Stinchcomb, DT, Struhl, K, and Davis, R (1979) Nature 282:39.
10. Murray, AW, and Szostak, JW (1983) Cell 34:961.
11. Fangman, WL, Hice, RH, and Chiebowicz-Stedziewska, E (1983) Cell 32:831.

12. Murray, AW, and Szostak, JW (1983) Nature 305:189.
13. Hieter, P, Pridmore, D, Hegemann, JH, Thomas, M, Davis, RW, and Philippsen, P (1985) Nature in press.
14. Fitzgerald-Hayes, M, Clarke, L, and Carbon, J (1982) Cell 29:235.
15. Clarke, L, and Carbon, J (1983) Nature 305:23.
16. Rothstein, RJ (1983) Meth Enzymol 101:202.
17. Carbon, J, and Clarke, L (1984) J Cell Science, Suppl. 1:43.
18. Neitz, M, and Carbon, J (1985) Mol Cell Biol in press.
19. Tschumper, G, and Carbon, J (1983) Gene 23:221.
20. Mann, C, and Davis, RW (1983) Proc Natl Acad Sci USA 80:228.
21. Bloom, KS, and Carbon, J (1982) Cell 29:305.
22. Williamson, DH, and Fennell, DS (1981) In D. von Wettstein et. al., (eds): "Molecular Genetics in Yeast, Alfred Benzon Symp" 16 (Munksgaard, Copenhagen), p 89.
23. Synder, M, Hieter, P, Mann, C, and Davis, RW (1984) In "Twelfth Int Conf in Yeast Genetics and Mol Biol Abstr" Edinburgh p 53.
24. Pickett-Heaps, J, Spurck, T, and Tippit, D (1984) J Cell Biol 99:137s.
25. Gallwitz, D, and Sures, I (1980) Proc Natl Acad Sci USA 77:2546.
26. Ng, R, and Abelson, J (1980) Proc Natl Acad Sci USA 77:3912.
27. Neff, NF, Thomas, JH, Grisafi, P, and Botstein, D (1983) Cell 33:211.
28. Shortle, D, Novick, P, and Bostein, D (1984) Proc Natl Acad Sci USA 81:4889.
29. Novick, P, and Botstein, D (1985) Cell 40:405.
30. Thomas, JH, and Botstein, D (1983) In "The Molecular Biology of Yeast Abstr." Cold Spring Harbor Laboratory, C.S.H. New York, p 7.

GENES AT TELOMERES: FERMENTATION GENE FAMILIES[1]

Marian Carlson

Department of Human Genetics and Development
and Institute for Cancer Research
Columbia University, College of Physicians and Surgeons
New York, NY 10032

Saccharomyces species are able to ferment a variety of carbon sources. Fermentation ability is controlled by dispersed gene families, such as the SUC, MAL, MEL, MGL and STA families. A striking feature of these families is that different Saccharomyces strains carry different active members of each family. Studies of the molecular basis for this variability indicate that these genes are present at different chromosomal locations in different genomes. Members of the SUC, MAL and MGL families have been shown to reside near chromosome telomeres. Analysis of the SUC family indicates that dispersal of SUC genes to different chromosomes occurred by rearrangements of telomeres.

The yeasts Saccharomyces are able to ferment a wide variety of sugars including disaccharides and oligosaccharides (see refs. 1 and 2 for review) and also starch and dextrins. Fermentation of these carbohydrates is controlled by dispersed families of genes. Genetic analysis of different yeast strains has led to the identification of the SUC (sucrose), MAL (maltose), MEL (melibiose), MGL (α-methylglucoside) and STA (starch) or DEX (dextrin) gene families. Each family includes multiple, functionally equivalent, unlinked loci that control ability to ferment a spectrum of sugars or starch.

For example, the SUC family includes six unlinked SUC genes, each a structural gene for invertase, an enzyme that

---

[1]This work was supported by Public Health Service grant GM32065 from the National Institutes of Health.

hydrolyzes sucrose, raffinose and related sugars. Similarly, the MAL family includes five unlinked loci, and each active MAL locus confers ability to ferment maltose and related sugars. Recent studies have revealed that a MAL locus includes several genes involved in maltose fermentation.

The common property of these families that is most interesting from the viewpoint of genomic organization and chromosome structure is that closely related Saccharomyces strains carry different active members of each family. This observation was made in the early years of yeast genetics when fermentation genes were among the few genetic markers available. Saccharomyces strains were found to differ in their ability to ferment different sugars, and the genetic basis of these differences was investigated by genetic analysis of crosses between strains with different phenotypes (see ref. 3 for review). The segregation of ability to ferment sugars was followed in tetrads, and multiple, or polymeric, fermentation genes were identified. For example, three distinct SUC genes were identified in S. chevalieri, whereas no active SUC allele was present in S. italicus (4, 5). Three MAL genes were found in S. cerevisiae, one in S. italicus and none in S. chevalieri (4, 6). Similar genotypic variation was reported for MEL and MGL genes. Thus, closely related Saccharomyces strains differ widely in genotype for fermentation genes. This article reviews recent findings regarding the molecular structure of fermentation gene families and the basis for the observed genotypic variability.

Before considering their structure in detail, let us first consider these gene families from a biological perspective. The fact that Saccharomyces genomes carry dispersed families of fermentation genes may reflect the biology and ecology of the organism. What advantage might the multiplicity of fermentation genes and the observed genotypic variability offer to the organism? Yeasts are fermentative organisms, highly dependent on sugars for growth. The presence of multiple genes may allow increased production of relevant hydrolytic enzymes and provide a competitive advantage. Multiple genes also provide insurance against loss of ability to utilize a particular sugar due to mutation. Yeasts survive in environments in which the relative abundance of different sugars varies widely, and from an evolutionary standpoint, dispersed gene families would appear to offer the organism great versatility in adaptation to different environments. Variability in fermentation genotype could contribute to expression of optimal amounts of different hydrolytic enzymes. The organization of fermentation gene families may also facilitate relatively rapid genotypic change, for example, in response to environmental change resulting from transport of yeast populations by insects, birds and other animal vectors. As we shall see below, in the case of the SUC gene family, which is the family for which the most detailed structural information is currently available, the evidence suggests that changes in SUC genotype could perhaps occur relatively rapidly within a population under selective pressure.

SUC Gene Family

The molecular basis for the variation in genotype among Saccharomyces strains has been investigated in detail for the SUC gene family (SUC1-SUC5, SUC7). Briefly, the findings can be summarized as follows. Each active SUC locus encodes the enzyme invertase (7) and shows sequence homology to other SUC loci (8). Strains that do not carry an active SUC allele at a known SUC locus usually have no SUC gene sequences at that locus (8); the naturally occurring negative alleles are designated suc^o. The exception is the SUC2 locus; the $suc2^o$ allele is a mutant SUC gene that is transcribed normally, but does not encode invertase activity (8, 9). These findings that the differences in SUC genotype among closely related strains reflect the presence or absence of a SUC gene at a given locus indicated that SUC genes moved from one chromosomal locus to another in recent evolutionary history. Rearrangements of SUC genes appear to have occurred much more frequently than rearrangements of most other genetic markers, except for transposable elements such as Ty.

Physical analysis of the SUC loci has revealed a likely molecular mechanism for the movement of SUC genes among chromosomes. Molecular cloning and blot hybridization analysis of SUC loci has shown that all of the SUC loci except for SUC2 are located very near chromosome telomeres (10). These telomeric SUC genes are embedded within known telomere-adjacent sequences, the X and Y⁻ elements identified by Chan and Tye (13). Figure 1a shows the structure of a typical telomere. The SUC3, SUC4, SUC5 and SUC7 genes are located on approximately 7-kb elements between X and Y⁻ sequences (Figure 1b); the SUC1 gene is similarly situated except that the 5⁻ flanking region includes a sequence substitution (10). Genetic mapping has also located SUC genes near the ends of chromosomes: SUC1, chromosome VII (11); SUC3, chromosome II (12); and SUC5, chromosome IV (12).

a

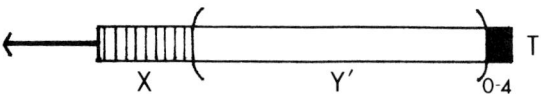

b

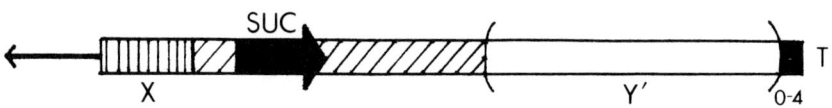

FIGURE 1. Structures of telomeric SUC and suc^o loci. (a) Structure of a typical telomere and suc^o locus (adapted from refs. 10, 13, 14). (b) Structure of a typical telomeric SUC locus. The thin line represents unique sequences; the arrowhead points towards the centromere. The vertically-hatched bar indicates X sequence. The open bar is the Y´ element, which may be present in 0-4 copies. The dark bar represents $C_{1-3}A$ repeats at the telomere (T). The diagonally-hatched bar represents the 7-kb element containing a SUC gene (dark arrow pointing in the direction of transcription).

Analysis of the telomeric $suc1^o$, $suc3^o$, $suc4^o$ and $suc7^o$ loci showed that they resemble normal yeast telomeres and, as expected, lack the SUC gene-containing element (Figure 1a). Comparison of SUC and suc^o loci suggested that the original telomeric SUC locus evolved by insertion of a 7-kb SUC gene-containing element into a suc^o locus, or perhaps by its substitution for a small portion of the X sequence at a suc^o locus (10). Dispersal of the SUC genes to other chromosome telomeres could then have occurred by rearrangements of telomeres. The most likely mechanism accounting for such rearrangements is recombination between sequences centromere-proximal to the SUC gene and homologous sequences at suc^o loci. All of the cloned suc^o loci ($suc1^o$, $suc3^o$, $suc4^o$ and $suc7^o$) display sequence homology to other suc^o and SUC loci in appropriate regions; for example, most SUC and

suc^o loci share homologous X sequences as shown in Figure 1. Such recombination events between SUC and suc^o telomeres of different chromosomes would result in the dispersal of SUC genes to different chromosomes.

The SUC2 gene is an unusual member of the SUC gene family. Although its genetic map position is at the end of chromosome IX (12), analysis of cloned DNA showed that the gene is located at least 14 kb from telomere-adjacent X sequences and does not share extensive flanking homology with the other SUC loci (10). Nucleotide sequence comparison of SUC2 and one of the telomeric genes, SUC7, showed that the homology between the two extends from about position -624 (relative to the translational start) to +1791 (15), which is close to the 3´ end of the mRNA (16). The conserved 5´ noncoding sequence includes the upstream regulatory region (UAS). Thus, it appears that a duplication of the SUC structural gene and flanking regulatory sequences occurred in the yeast genome. The mechanism by which this duplication was effected remains unclear; however, because most, if not all, strains carry SUC gene sequences at the SUC2 locus (either SUC2 or the defective $suc2^o$ allele), while many strains carry no telomeric SUC sequences, we suggest that the SUC2 gene was the progenitor of the SUC gene family.

MAL Gene Family

The MAL gene family includes five loci (MAL1-MAL4, MAL6). A dominant MAL allele at any one locus is sufficient to confer ability to ferment maltose and related glycosides. Recent physical and genetic analysis of MAL loci has revealed that each MAL locus includes at least three genes involved in maltose utilization, which probably encode maltase, a maltose transport protein and a regulatory protein (17, 18). Moreover, many Saccharomyces strains carry cryptic MAL loci, which carry functional copies of some, but not all, of these genes (19, 20, 21) These cryptic alleles were designated MALg and MALp alleles in the absence of precise information regarding their functions; MALg and MALp alleles complement to confer ability to ferment maltose (22). It is likely that MALp encodes a functional regulatory protein, and that MALg encodes both maltase and maltose permease (17, 18). Such cryptic alleles have been identified at the MAL1 and MAL3 loci.

Although the MAL loci are more complex than the SUC loci, the overall organization of the MAL gene family is similar to that of the SUC family. At three of the MAL loci

(MAL2, MAL4, MAL6), most Saccharomyces strains either carry an active MAL allele or no MAL sequences (20, 21). In contrast, most strains carry MAL-related sequences, either active or cryptic, at MAL1 and MAL3 (20, 21); the SUC family is similar in that most strains carry SUC-gene information at the SUC2 locus, either SUC2 or the defective suc2° allele (8). A further similarity between the two families is that MAL genes also map genetically to the ends of chromosomes. The four MAL loci that have been mapped are each at the end of the right arm of a chromosome: MAL1, chromosome VII (11); MAL2, III; MAL3, II; and MAL4, XI (12). Interestingly, two of the MAL loci are tightly linked to SUC loci: MAL1 to SUC1, and MAL3 to SUC3 (12). Molecular cloning has indicated that these pairs of MAL and SUC loci are physically very close. Sequences homologous to cloned MAL6 DNA were located about 3 kb centromere-proximal to the SUC1 gene on cloned DNA from the SUC1 locus (10). Sequences homologous to MAL and conferring MALg function were also found close to the suc1° and suc3° loci cloned from a maltose-nonfermenting strain (10). These findings suggest that the MAL1 and MAL3 loci are close to telomeres.

MEL, MGL and STA Gene Families

The other fermentation gene families have not been extensively studied at the molecular level. The MEL gene family is involved in the fermentation of melibiose and raffinose. Early studies demonstrated that some strains, such as S. cerevisiae, are melibiose nonfermenters, and others, such as S. italicus var. melibiosi, carry multiple (at least six) MEL genes (23). The MEL1 gene has been cloned, and the evidence suggests that it is a structural gene for α-galactosidase, or melibiase (24). A mel° strain was examined and found to have no detectable MEL sequences (24), suggesting that variability in MEL genotype reflects variability in the presence of MEL sequences within the genome. These findings suggest that MEL genes, like SUC and MAL genes, have moved within Saccharomyces genomes in recent evolution.

The family designated MGL probably comprises two subfamilies because pairs of MGL genes are required for α-methylglucoside utilization (see refs. 1 and 3 for review). For example, MGL2 in combination with either MGL1 or MGL3 confers ability to ferment α-methylglucoside. It seems likely that further study of MGL genes will reveal a story as complex as that of the MAL loci. Little is known about the genomic organization of the MGL family, but it is curious that the only MGL gene that has been genetically mapped, MGL2, is tightly linked to SUC3 and MAL3 at the end of the right arm of chromosome III (12).

S. diastaticus genomes have been found to carry a family of three unlinked STA (starch) or DEX (dextrin) genes (STA1-STA3; 25). Each of the STA genes confers ability to produce extracellular glucoamylase and therefore to ferment starch or dextrins. Molecular cloning of the STA1 and STA3 genes has shown that the two genes are highly homologous and encode glucoamylase (26); DEX1, which may be allelic to STA1, has also been cloned (27). Analysis of sta^o genomes revealed the presence of two loci containing sequences homologous to the cloned STA DNA, one possibly encoding sporulation specific glucoamylase (26). Yamashita et al. have proposed that recombination among the sequences present in the sta^o genome of S. cerevisiae could have generated a STA gene.

Chromosomal Locations of Fermentation Genes

It is a striking feature of the Saccharomyces genetic map that all of the fermentation markers are at the ends of chromosomes (11, 12). These mapped genes include four SUC genes, four MAL genes and MGL2. In addition, molecular cloning studies summarized above have provided physical evidence that all of the SUC genes, except for SUC2, and both MAL1 and MAL3 are located very near telomeres. Mapping data are not available for other fermentation gene families, except for the hexokinases, which play a more general role in fermentation. The genes for both hexokinases, HXK1 and HXK2, and also the gene for the related enzyme glucokinase, GLK1, map near the ends of chromosomes, the left arms of VI, VII and III, respectively (12); their physical proximity to telomeres or telomere-adjacent sequences has not been established. These observations raise the possibility that it will prove to be the general case that fermentation genes are located near ends of chromosomes.

What would be the advantage to the organism of such an arrangement? We speculate that the movement of fermentation genes near telomeres proved advantageous because it facilitated the evolution of gene families and provided a mechanism for generating Saccharomyces genomes with variable fermentation gene components. We have already argued that multiplicity of fermentation genes and genotypic variability may have special benefits for yeasts. The location of a gene near a telomere facilitates the evolution of a gene family for two reasons. First, dispersal to different chromosomes can be effected by rearrangements of telomeres. In the absence of any essential gene distal to the fermentation marker, such a rearrangement is less likely than other gross chromosomal rearrangements to have deleterious consequences for the organism. Second, telomeres carry repeated telomere-adjacent sequences (13) that can undergo homologous recombination with sequences at other telomeres to cause rearrangements. Moreover, such a mechanism for rearrangement can perhaps be useful not only for dispersal of fermentation genes on a long-term evolutionary time scale, but also for rapid genomic change in response to environmental stress.

In summary, the fermentation gene families are a unique feature of the Saccharomyces genome. We suggest that these families are particularly advantageous to the organism and that the telomeric location of many fermentation genes is not accidental. The evidence suggests that for the SUC gene family, and probably the MAL gene family, the movement of a progenitor gene near to a telomere was instrumental in the evolution of the family. It will be interesting to learn whether other fermentation genes are near telomeres and, if so, whether their telomeric location played a key role in the evolution of their respective families.

REFERENCES

1. Fraenkel DG (1982). Carbohydrate metabolism. In Strathern JN, Jones EW, Broach JR (eds): "The Molecular Biology of the Yeast Saccharomyces. Metabolism and Gene Expression," Cold Spring Harbor: Cold Spring Harbor Laboratory, p 1.
2. Barnett JA (1976). The utilization of sugars by yeasts. Adv Carbohydr Chem Biochem 32:125.
3. Mortimer RK, Hawthorne DC (1969). Yeast genetics. In Rose AH, Harrison JS (eds): "The Yeasts," New York: Academic Press, p 385.
4. Gilliland RB (1949). A yeast hybrid heterozygotic in four fermentation characters. CR Trav Lab 24:347.
5. Winge O, Roberts C (1952). The relation between the polymeric genes for maltose, raffinose, and sucrose fermentation in yeasts. CR Trav Lab 25:141.
6. Winge O, Roberts C (1948). The inheritance of enzymatic characters in yeasts, and the phenomenon of long term adaptation. CR Trav Lab 24:263.
7. Carlson M, Taussig R, Kustu S, Botstein D (1983). The secreted form of invertase in Saccharomyces cerevisiae is synthesized from mRNA encoding a signal sequence. Mol

Cell Biol 3:439.
8. Carlson M, Botstein D (1983). Organization of the SUC gene family in Saccharomyces. Mol Cell Biol 3:351.
9. Carlson M, Botstein D (1982). Two differentially regulated mRNAs with different 5´ ends encode secreted and intracellular forms of yeast invertase. Cell 28:145.
10. Carlson M, Celenza JL, Eng FJ (1985). Evolution of the dispersed SUC gene family of Saccharomyces by rearrangements of chromosome telomeres. Mol Cell Biol, in press.
11. Celenza JL, Carlson M (1985). Rearrangement of the genetic map of chromosome VII of Saccharomyces cerevisiae. Genetics 109:661.
12. Mortimer RK, Schild D (1982). Appendix I. Genetic map of Saccharomyces cerevisiae. In Strathern JN, Jones EW, Broach JR (eds): "The Molecular Biology of the Yeast Saccharomyces. Metabolism and Gene Expression." Cold Spring Harbor: Cold Spring Harbor Laboratory, p 639.
13. Chan CSM, Tye B-K (1983). Organization of DNA sequences and replication origins at yeast telomeres. Cell 33:563.
14. Walmsley RW, Chan CSM, Tye B-K, Petes TD (1984). Unusual DNA sequences associated with the ends of yeast chromosomes. Nature 310:157.
15. Sarokin L, Carlson M (1985). Comparison of two yeast invertase genes: conservation of the upstream regulatory region. Nucleic Acids Res, in press.
16. Taussig R, Carlson M (1983). Nucleotide sequence of the yeast SUC2 gene for invertase. Nucleic Acids Res 11:1943.
17. Needleman RB, Kaback DB, Dubin RA, Perkins EL, Rosenberg NG, Sutherland KA, Forrest DB, Michels CA (1984). MAL6 of Saccharomyces: A complex genetic locus containing three genes required for maltose fermentation. Proc Natl Acad Sci USA 81:2811.
18. Cohen JD, Goldenthal MJ, Buchferer B, Marmur J (1984). Mutational analysis of the MAL1 locus of Saccharomyces: identification and functional characterization of three genes. Mol Gen Genet 196:208.
19. Chow T, Goldenthal MJ, Cohen JD, Hegde M, Marmur J (1983). Identification and physical characterization of yeast maltase structural genes. Mol Gen Genet 191:366.
20. Needleman RB, Michels C (1983). Repeated family of genes controlling maltose fermentation in Saccharomyces carlsbergensis. Mol Cell Biol 3:796.
21. Michels CA, Needleman RB (1984). The dispersed repeated family of MAL loci in Saccharomyces spp. J Bacteriol 157:949.

22. Naumov GI (1970). Comparative genetics of yeast. IV. Identification of maltose complementing factors in Saccharomyces. Genetika 6:121.
23. Roberts C, Ganesan AT, Haupt W (1959). Genetics of melibiose fermentation in Saccharomyces italicus var. melibiosi. Heredity 13:499.
24. Post-Beittenmiller MA, Hamilton RW, Hopper JE (1984). Regulation of basal and induced levels of the MEL1 transcript in Saccharomyces cerevisiae. Mol Cell Biol 4:1238.
25. Tamaki H (1978). Genetic studies of ability to ferment starch in Saccharomyces: gene polymorphism. Mol Gen Genet 164:205.
26. Yamashita I, Maemura T, Hatano T, Fukui S (1985). Polymorphic extracellular glucoamylase genes and their evolutionary origin in the yeast Saccharomyces diastaticus. J Bacteriol 161:574.
27. Meaden P, Ogden K, Bussey H, Tubb RS (1985). A DEX gene conferring production of extracellular amyloglucosidase on yeast. Gene 34:325.

STRUCTURAL AND GENETIC ANALYSES OF THE YEAST GENOME[1]

A.J. Lustig, M.G. Goebl and T.D. Petes

Department of Molecular Genetics and Cell Biology
The University of Chicago
Chicago, Illinois 60637

ABSTRACT We describe below two types of analyses concerned with the structure, replication and stability of the yeast genome. The first study involves an analysis of the ends (telomeres) of yeast chromosomes. Previous research indicated that chromosomes ended in a tract of simple sequence DNA (poly $C_{1-3}A$) that was added by an untemplated replication mechanism. Below, we describe the isolation of mutants that have shorter-than-normal tract lengths. The second study concerns the fraction of the yeast genome that is required for normal cell growth. This estimate was obtained by determining what fraction of random gene disruptions resulted in dead or slow growing cells. Of twenty disruptions analyzed thus far, three are haploid lethal and three result in slow growth.

---

[1]This work was supported by grants from the National Institutes of Health (GM 24110 and K 04 AG 00077), National Cancer Institute (CA 19265),and a National Research Service Award (CA 09267-0552) to MGG.

INTRODUCTION

Since the two studies described in this paper concern different aspects of the yeast genome, we will have separate Introduction, Results and Discussion sections for each topic.

The terminal sequences of eukaryotic chromosomes (telomeres) must have two fundamental properties. First, they must be replicated by a mechanism that allows complete duplication. Watson (1) pointed out that since known DNA polymerases require a primer and elongate only in the 5' to 3' direction, one would expect a gradual loss of DNA from the ends of chromosomes. A second expected property of the telomeres is that they may protect essential sequences of the chromosomes from cellular exonucleases.

The telomeres from a number of eukaryotes have recently been analyzed at the molecular level. In each case examined, a simple repetitive 'satellite-like' sequence is present at or near the terminus of the chromosome. This simple sequence is $(C_4A_4)_n$ in <u>Oxytricha</u> macro- and micronuclear DNA (2,3), $(C_4A_2)_n$ in <u>Tetrahymena</u> macronuclear DNA (4), $(C_3TA_2)_n$ in <u>Trypanosoma brucei</u> chromosomal DNA (5,6) and $(C_{1-3}A)_n$ in chromosomes of the yeast <u>Saccharomyces cerevisiae</u> (7,8).

The structure and properties of yeast telomeres have been under active investigation by a number of laboratories. The $(C_{1-3}A)_n$ tract, ranging in size from 200-850 bp, punctuates the ends of most, if not all, yeast telomeres (7-9). Transformation of yeast with linear plasmids containing telomeres of <u>Oxytricha</u> or <u>Tetrahymena</u> results in the acquisition of this $(C_{1-3}A)_n$ tract while retaining at least a portion of the <u>Oxytricha</u> or <u>Tetrahymena</u> ends (7,10,11). Sequencing analysis has revealed that the $(C_{1-3}A)_n$ tract is added to the $(C_4A_2)_n$ strand of <u>Tetrahymena</u> (7).

Additional repeated sequences are present centromere-proximal to the $(C_{1-3}A)_n$ tract (Figure 1). All known telomeres contain a single copy of the repeated X sequence. The majority of telomeres are also associated with the 6.7 kb repeated Y' element (12). Autonomously replicating sequences are located in both X and Y' repeats (12) and may reflect origin of replication activity associated with the telomere. A short $(C_{1-3}A)_n$ tract is also found at the junction between X and Y' as well as between reiterated Y' units (8,13). Two structural classes of yeast telomeres are thus present; those containing the X repeated element alone (X class) and those containing both X and Y' (XY' class).

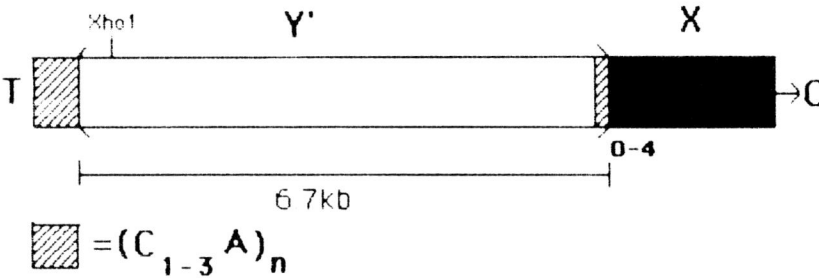

Figure 1. Structure of a yeast telomere. Shown above is the generalized yeast telomere structure. The terminal $(C_{1-3}A)_n$ tract ▨ is preceded by 0 to 4 copies of Y' ☐ and a single copy of X ■ . The conserved XhoI site is shown within the Y' repeat. T and C refer to the telomere and centromere, respectively.

A number of lines of investigation have recently pointed to the dynamic state of the yeast telomere. First of all, both single copy chromosomal telomeres and telomeres of cloned linear plasmids behave as heterodisperse bands when Southern analysis using a telomere specific probe is carried out (10,14). This suggests that even a single telomere is capable of size variance within different cells of the population. Secondly, as pointed out above, yeast cells are capable of altering exogenous telomeric templates in vivo (7,10,11). These data suggest the presence of a specific mechanism for processing telomeres. In addition, Dunn et al (15) have demonstrated that the Y' element and $(C_{1-3}A)_n$ tracts of telomeres of linear plasmids can recombine with chromosomal telomeres. These results indicate that telomeres can interact so as to create structurally altered termini. Finally, our laboratory (14) and the laboratory of Haber (16) have identified variant strains of yeast differing in $(C_{1-3}A)_n$ tract length. As much as a two fold difference in tract length has been observed, and this difference in length is under genetic control (14,16).

To begin to analyze the genes controlling $(C_{1-3}A)_n$ tract length and to understand the functions of telomeric sequences, we have identified mutations altering telomere length. We describe here the characterization of three mutants that demonstrate a dramatic decrease in size of the $(C_{1-3}A)_n$ tract. Based upon this data, we propose a new model for telomere replication.

RESULTS

Previous studies from our laboratory have shown that the tract length of a yeast telomere can be easily assayed by Southern blotting (9,17). The primary XY' class has a conserved XhoI site about 1.3 kb from the chromosomal termini of the strain A364A (9,10,14). Thus, DNA digested with this enzyme and subjected to Southern analysis using a telomere specific probe produces a major diffuse band centered at 1.3 kb as well as a number of higher molecular weight heterodisperse bands. Many of these latter species have been shown to represent single copy X class telomeres (14). PolydGT·polydCA was used as the probe in these studies as this copolymer hybridizes to $(C_{1-3}A)_n$ sequences. Bal31 digestions of yeast DNA, followed by the analysis described above, have indicated that the $(C_{1-3}A)_n$ tract length of most chromosomal termini in A364A is 360 ± 180 bp (14). Thus, variations in size of the major telomere class should represent a general alteration in $(C_{1-3}A)_n$ length.

To identify mutations affecting the length of the telomeric $(C_{1-3}A)_n$ tract we utilized the highly mutagenized yeast strain bank constructed by Cal McLaughlin (18). This bank was derived from A364A by nitrosoguanidine mutagenesis and is temperature sensitive (ts) for growth at 37°C. Cells from approximately 200 mutagenized derivatives of A364A were grown overnight at 23°C followed by a 5-6 hr. shift to 37°C. The DNA was isolated, digested with XhoI, and Southern analysis performed using nick-translated polydGT·polydCA. The mutants were subsequently screened for deviation in $(C_{1-3}A)_n$ tract size from the wild type A364A parent.

Most strains tested maintained the parental tract length. However, a few mutants showed a substantial alteration in the size of the XhoI XY' class fragment. In particular, three mutants, AL1, AL4, and AL5 demonstrated a reduction in fragment size (Figure 2). In both AL1 and AL4 the fragment distribution decreases in size by approximately 300 bp while in AL5 the decrease is about 170 bp. The terminal tract lengths reflected by these decreases are 80 ± 100 bp in AL1, 60 ± 80 bp in AL4 and 190 ± 155 bp in AL5. These ranges in size are likely to reflect a large number of overlying hetero-disperse telomeric bands. A qualitatively similar reduction in size is observed in X class telomeres. These phenotypes are present in all strains at 23°C as well as 37°C. Most of the experiments described below, therefore, were conducted at the permissive temperature.

Each of the mutants was crossed to an isogenic strain of opposite mating type, B364B, and the DNA from the diploids was isolated and digested with XhoI. Southern analysis revealed a distribution of XY' class fragments very close to the size of the original wild type. The X class telomeres behaved in a qualitatively similar fashion. These mutations are thus

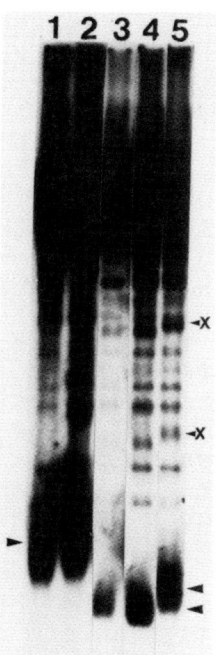

Figure 2. Identification of telomere mutants. Shown here is an autoradiogram of XhoI digested DNA from strain A364A (MAT a, lys2 tyr1 his7 ura1 ade1 ade2 gal2) (lane 1), a congenic strain 5153-11-1 (MAT alpha, met 2) (lane 2) and three mutants designated AL1 (lane 3), AL4 (lane 4), and AL5 (lane 5), probed with the co-polymer polydGT·polydCA at 50°C by the method of Southern (17). The X class telomeres (X) and XY' telomere fragment of the mutants (bold arrow) are denoted by arrows on the right. The wild type XY' class fragment is denoted by a bold arrow on the left. Many of the additional bands are the result of hybridization to internal alternating CA tracts in the yeast genome.

recessive.

Surprisingly, when these diploids were sporulated and the DNA from tetrads isolated and characterized as above, we found that two of the spores contained telomeres that were wild type in length and two spores contained telomeres that were only slightly shorter than wild type. These results suggested that there may be a lag in the expression of the mutant phenotype. To test this possibility, we allowed spore cultures derived from diploids containing the mutations found in AL1 and AL4 to divide vegetatively and analyzed the $(C_{1-3}A)_n$ tract length during this process. As shown in Figure 3, telomeres of mutant length are obtained only after substantial subculturing. We estimate that about 120 generations are required for the mutant length to be reached. A less extensive phenotypic lag appears to take place in AL5. Single-copy X class telomeres show a similar lag in expression. No significant changes in the wild type spores were observed during subcloning. Similarly, extensive subcloning of diploids heterozygous for any of the mutations resulted in no alteration in telomere size.

Each mutant was backcrossed several times to isogenic strains of opposite mating type to eliminate any unrelated mutations present in the original strains. The backcrossed strains were sporulated and each member of at least five tetrads was subcloned extensively. (For simplicity, we will refer to the backcrossed derivatives of the original mutants by identical nomenclature.) This analysis revealed that in two mutants, AL1 and AL5, the telomere mutations segregated 2:2 but were unlinked to a *ts* mutation. In contrast, in each of 10 tetrads, the mutation present in AL4 and a *ts* mutation co-segregated. Thus, although all three mutations behave as alterations in single nuclear genes, only one is tightly linked to a *ts* mutation present in the original strain collection.

Strains carrying the mutations were crossed to each other to determine the number of complementation groups present. Crosses of AL1

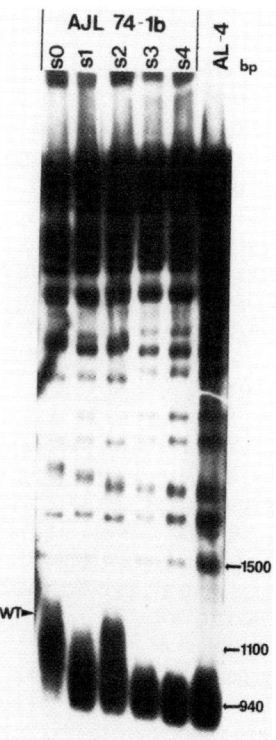

Figure 3. Phenotypic lag in expression of mutations affecting telomere length. Shown above is a Southern analysis of XhoI digested DNA isolated from a mutant spore (AJL 74 1b) at various stages of subcloning after sporulation of a heterozygous diploid. PolydGT·polydCA was used as the probe. AL4 was crossed to an isogenic wild type strain, B363B, and the resultant diploid (AJL 74) was sporulated. The spores were subcloned at the permissive temperature (23°C) four times and the DNA isolated from different stages of the subcloning (s0-s4). s0 represents approximately 35 generations and each subsequent subcloning repesents increments of about 20 generations. The original AL4 mutant and the position of the wild type (wt) fragment are also shown.

to AL4 produced diploids having telomeres identical to the two parent strains. On the other hand, crosses of AL1 or AL4 to AL5 produced a telomere distribution intermediate between the AL5 and wild type phenotypes. This suggests that the mutations present in AL1 and AL4 belong to the same complementation group (<u>tel1</u>) while the mutation carried by AL5 defines a second gene, <u>tel2</u>. For further discussion, we will refer to mutations present in AL4, AL1, and AL5 as <u>tel1</u>-1 <u>tel1</u>-2 and <u>tel</u>2, respectively.

The terminal phenotype of cells containing the <u>ts</u>-linked telomere mutation (<u>tel1</u>-1) was examined by shifting a culture growing at 23°C to 37°C for fourteen hours. Under these conditions, we observed that most of the cells had buds that were approximately the same size as the mother cell, suggesting a block in medial nuclear division. The terminal phenotype was achieved only after several cell divisions. Interestingly, a cell cycle mutant (<u>cdc</u>17) shown by Carson and Hartwell (19) to have longer than normal $(C_{1-3}A)_n$ tracts, has an identical phenotype. By complementation tests, however, we found that <u>cdc</u>17 was not allelic with <u>tel</u>1.

DISCUSSION

We have identified three mutations which significantly decrease the size of the telomeric $(C_{1-3}A)_n$ tract in yeast. The size of this tract in the A364A parent is 360 ± 180 bp. The mutations present in two of the mutants, AL4 and AL1, belong to the same complementation group (<u>tel</u>1) and reduce telomere size by about 300 bp. The third mutation, <u>tel</u>2, present in AL5 is likely to represent a second gene and reduces tract length by about 170 bp. Each mutation is recessive and segregates 2:2 in crosses with the wild type strain. One of the mutations, <u>tel1</u>-1, is tightly linked to a <u>ts</u> mutation. It is still unknown whether this <u>ts</u> mutation represents a defect within the gene affecting telomere length.

The identification of genes involved in the

control of telomere length provides additional evidence that the tract length is genetically regulated. We suggest that the terminal $(C_{1-3}A)_n$ tract is in a state of dynamic equilibrium, an equilibrium maintained through the action of specific gene products.

The mechanism by which eukaryotic cells replicate telomeres has been the subject of extended debate. The telomere must provide at least two functions: a) protection from nucleases, and b) the ability to compensate for the expected loss of the terminal RNA primer following replication. Cavalier-Smith (20) and Bateman (21) suggested that the telomere could be replicated with known activities if the two strands were covalently linked at their termini, forming a "hairpin." After replication was completed, the two daughter DNA molecules would be separated by enzymatic cleavage with sequence- specific endonucleases. Although several viruses have telomeres that are replicated by this mechanism (22-24), we believe that most eukaryotes utilize a different mechanism for telomere replication. In particular, the fact that many known eukaryotic chromosomes end in a simple sequence which is capable of expansion and contraction is not consistent with a simple hairpin model.

Any new model for telomere replication must account for the following observations: 1) yeast cells have a mechanism that allows untemplated addition of $(C_{1-3}A)_n$ residues to certain linear substrates (7,10,11), 2) single chromosomal telomeres are heterogeneous in length (10,14) and 3) yeast strains have been detected that have substantially different terminal tract lengths (14,16). In addition, the model must account for the following properties of mutants that affect tract length: 1) two classes of mutants (those that have shorter-than-normal telomeric tracts and those that have longer-than-normal telomeric tracts) have been isolated and 2) both classes of mutant show long phenotypic lags in expression. We will first describe the general features of a model consistent with these data and then discuss three specific models incorporating these general features.

We suggest that at least a portion of the terminal $(C_{1-3}A)_n$ tract is replicated by normal semi-conservative replication. The alternative possibility that the tract is synthesized <u>de novo</u> each cell cycle is ruled out by the phenotypic lag in expression of the mutations. As discussed previously, we expect that normal semi-conservative synthesis results in a shortening of the linear DNA molecule each cell cycle. We believe that this shortening is normally balanced by an untemplated addition of $(C_{1-3}A)_n$ residues; the <u>tel</u>1 and <u>tel</u>2 mutants may be partially deficient in this process. If the strains containing these mutations have a partially deficient untemplated addition activity, one would expect a long phenotypic lag in the full expression of the mutant phenotype. Thus, the terminal sequences of the chromosome are maintained by a balance between activities that elongate the telomere (untemplated addition of $(C_{1-3}A)_n$ residues) and activities that shorten the terminal tract. Such an equilibrium between average loss and gain of telomeric sequences may also explain the heterogeneity of telomere tract size. The activity that shortens the tract may simply be the loss of the primer during normal replication. Alternatively, there may also be an active process that removes sequences from the end of the chromosome (a tract- specific exo- or endonuclease, for example). Whatever the molecular details of this process, in order to explain the observation that the different mutant strains can have telomeres with different lengths at equilibrium, we also suggest that either the elongation activity or the shortening activity (or both processes) must have some type of feedback regulation. Three more detailed models for the regulation of telomere replication are described below.

In the first model, we propose that the feedback regulation of telomere length occurs at the level of the elongation process. In this model, the rate of elongation is slower when the tract is long. One plausible mechanism to achieve this type of regulation is that the enzyme system that is responsible for adding the $(C_{1-3}A)_n$ residues binds to the tract but is capable of the elongation reaction only when not bound to the tract. Thus, when the terminal tracts are long, more enzyme is bound and subsequent elongation occurs less efficiently. By this model, the tel1 and tel2 mutants represent partially deficient elongation activities. One expected phenotype of these mutants is that the tracts would decrease in length until the increased amount of free enzyme compensated for the reduced elongation activity.

In the second model, the feedback regulation operates at the level of degrading the tract rather than affecting elongation. We propose that loss of terminal sequences occur through the action of a tract-specific endonuclease, in addition to losses resulting from removal of the primer, and that these losses are balanced by the elongation activity. We predict that such a mechanism would be self-regulating since the frequency with which such an endonuclease would act should be dependent on the size of the tract. By this model, a mutation in the elongation activity would result in shortening of the terminal tract until the tract length is reduced to a size at which the endonuclease activity is balanced by the elongation activity.

In the third model, the tract length is controlled by structural proteins at the telomere. We suggest that these proteins protect the end from nuclease action. A mutation within this protein may result in a smaller region of protection and, therefore, a smaller terminal tract length. The finding of Gottschling and Cech (26) that non-nucleosomal nuclease protected region are present at the termini of Oxytricha DNA is consistent with this model.

We stress that the models described above are not mutually exclusive nor are they the only reasonable alternatives. In addition, it should be pointed out that the molecular details of these processes are very unclear. For example, it is not known whether the elongation process occurs by a recombinational mechanism (8) or a terminal transferase-like enzymatic reaction (7,9). The recent observation of Szostak (27) that cloned telomeres containing sequences other than CA-rich sequences can act as substrates for terminal addition of $(C_{1-3}A)_n$ is most easily explained by a terminal transferase reaction. Whatever the details of these mechanisms, our data, as well as those of Carson and Hartwell (19), suggest that the terminal sequences of the chromosome are maintained by a balance between mechanisms that extend and degrade the terminal sequences. This mechanism of "replicating" DNA is clearly quite different from normal semi-conservative replication.

In the following section of the paper, we will discuss a different topic relating to the structure and replication of the yeast genome, the fraction of the DNA in the yeast genome that is required for normal cell growth.

INTRODUCTION

For many years, it has been clear that there is not a direct relationship between the amount of DNA per cell and the genetic complexity of the organism as measured by other criteria. For example, the African lungfish has about 40 times more DNA per cell than the human (28). Part of the discrepancy between the amount of DNA per cell and genetic complexity is explained by the presence of repeated sequences in the eukaryotic genome (29) since these sequences increase the amount of DNA per cell without increasing the genetic complexity. However, the existence of repeated sequences does not completely explain the paradox. Genetic studies in Drosophila indicate that there are approximately 5000 essential genes, approximately one gene per band on a polytene chromosome (30). Studies of mRNA complexity, however, indicate that there are more than 17,000 distinct transcripts (31). This result indicates that either there are a large number of genes that are transcribed but are not essential (at least under laboratory conditions) or that the methods for making mutational changes avoids certain regions of the genome.

We decided to investigate these issues in yeast. Previously, other workers had pointed out that a "gene number" paradox existed for yeast. For example, although only 50 mutants that affect the cell cycle have been isolated (32), there are at least 4000 to 5000 discrete transcripts in vegetatively growing yeast cells (33). In a study designed to investigate this issue, Kaback et al (34) attempted to determine the number of genes on yeast chromosome I that were mutable to temperature sensitivity. These workers identified only 3 genes; one interpretation of this study is that there are only 250 genes essential for vegetative growth in yeast. Since this number seemed very low, we decided to re-investigate this problem by making random gene disruptions in vitro in cloned DNA segments and determining what fraction of these disruptions are haploid-lethal when transformed back into yeast. As described below, this approach indicates that about 30% of the yeast genome is important for vegetative growth.

RESULTS

Our genetic analysis involved several steps: 1) BamHI restriction fragments of yeast DNA were inserted into the plasmid vector pBR322, 2) 100 recombinant plasmids were screened for those that contained a single restriction site for the restriction enzymes BclI, BglII, XbaI or XhoI, 3) the selectable yeast genes ura3 (for BclI, BglII or XbaI) or leu2 (XhoI) were inserted into these plasmids, 4) the yeast DNA inserts were cleaved out of the recombinant plasmids and, in individual experiments, used to transform the diploid strain MGG3 (which was constructed by a mating type switch and was homozygous for mutations at the ura3 and leu2 loci); this protocol (35) results in the replacement of uninterrupted genomic sequences with the disrupted sequences derived from the plasmid on

one of the two homologous chromosomes in the
diploid and 5) the diploid was sporulated and
the effect of the disruption monitored in the
haploid spores. The phenotype of the disruption
was monitored on rich growth media and minimal
media. The presence of the disruption for all
transformants was confirmed by Southern
analysis.

We found three different phenotypes among the
20 transformants analyzed. For 14 of the 20
transformants, no obvious growth defects on
either type of solid media were observed. For 3
transformants, only two spores from each tetrad
grew and these spores were those that lacked the
disruption; thus, 3 of 20 disruptions were in
genes essential for vegetative growth or spore
germination. For an additional 3 transformants,
the two spores of the tetrad that contained the
disruptions grew more slowly than the two spores
without the disruptions. Thus, these transformants represent disruptions in genes that are
useful but not essential under lab conditions
for vegetative growth. Presumably, outside of
the lab, strains containing these disruptions
would be strongly selected against. We conclude
from these preliminary results that about 30% of
the yeast genome is used during vegetative
growth (the 95% confidence limits of this
estimate are 12% to 54%).

One possible explanation of the lack of
effect of the disruptions in the majority of the
transformants is that many of the disruptions
arc in repeated yeast genes. This possibility
was ruled out by our Southern analysis of DNA
from strains containing the disruptions. Only
four of the twenty insertions examined in our
study contained repeated genes.

DISCUSSION

Our estimate of the fraction of the yeast
genome that is important for vegetative growth
(30%) is similar to estimates of the fraction of
the yeast genome that is transcribed (40%; ref.
33). Thus, although our studies indicate that a

substantial fraction of the yeast genome is not required for growth under lab conditions, we do not find the small number of essential genes detected in a previous study (34). As suggested by those researchers (34), the small number of essential genes detected may be the result of the mutagen used or the extent to which yeast genes can be mutated to a temperature sensitive phenotype.

Our estimate of the fraction of "essential" yeast DNA is also subject to several reservations. First, our study investigated only effects that significantly altered vegetative growth. A subtle effect on growth or effects on other phases of the life cycle would not have been detected. Second, we assume that the restriction sites that were used to create the disruptions are randomly placed with respect to essential and inessential regions of the genome. Although the density of restriction sites within the yeast genome offer some support for this assumption, we are also investigating the properties of disruptions constructed using transposable elements. Third, the number of individual disruptions investigated thus far is small.

SUMMARY

In conclusion, we have reported two types of studies concerning the structure of the yeast genome. In the first part of this paper, we described the isolation of mutants that affect the telomeres of the yeast chromosome. The analysis of these mutants indicates that the telomeres are replicated by a novel duplication mechanism involving a dynamic equilibrium between mechanisms that extend and degrade terminal sequences. In the second section, we presented evidence that suggests that about 30% of the yeast genome is required for normal vegetative growth.

ACKNOWLEDGMENTS

We would like to thank Dr. Sue Jinks Robertson for critical reading of this manuscript. We would also like to thank Drs. D. Koshland, M. Çarson and L. Hartwell for providing us with strains containing the wild type cdc17-1 mutations as well as the wild type strains 5153-11-1 and B364B.

REFERENCES

1. Watson JD (1972). Origin of concatameric T7 DNA. Nature New Biol 239:197.
2. Dawson D, Herrick G (1984). Telomeric properties of C_4A_4-homologous sequences in micronuclear DNA of Oxytricha fallax. Cell 36:171.
3. Klobutcher LA, Swanton MT, Donini P, Prescott DM (1981). All gene-sized DNA molecules in four species of hypotrichs have the same terminal sequence and unusual 3' terminus. Proc Natl Acad Sci U.S.A. 78:3015
4. Blackburn EH, Gall JG (1978). A tandemly repeated sequence at the termini of the extrachromosomal RNA genes in Tetrahymena. J Mol Biol 120:335.
5. Blackburn EH, Challoner PB (1984). Identification of a telomeric DNA sequence in Trypanosoma brucei. Cell 36:447.
6. Van der Ploeg LHT, Liu AYC, Borst P (1984). Structure of the growing telomeres of trypanosomes. Cell 36:459.
7. Shampay J, Szostak JW, Blackburn EH (1984). DNA sequences of telomeres maintained in yeast. Nature(London) 310:154.
8. Walmsley RM, Chan CSM, Tye B-K, Petes TD (1984). Unusual DNA sequences associated with the ends of yeast chromosomes. Nature(London) 310:157.
9. Walmsley RM, Szostak JW, Petes TD (1983). Is there left-handed DNA at the ends of yeast chromosomes? Nature(London) 302:84.

10. Szostak JW, Blackburn EH (1982). Cloning yeast telomeres on linear plasmid vectors. Cell 29:245.
11. Pluta AF, Dani GM, Spear BB, Zakian VA (1984). Elaboration of telomeres in yeast: recognition and modification of termini from Oxytricha macronuclear DNA. Proc Natl Acad Sci U.S.A. 82:1475.
12. Chan CSM, Tye B-K (1983). Organization of DNA sequences and replication origins at yeast telomeres. Cell 33:563.
13. Chan CSM, Tye B-K, personal communication.
14. Walmsley RM, Petes TD (1985). Genetic control of chromosome length in yeast. Proc Natl Acad Sci U.S.A. 82:506.
15. Dunn B, Szauter P, Pardue ML, Szostak JW (1984). Transfer of yeast telomeres to linear plasmids by recombination. Cell 39:191.
16. Horowitz H, Thornburn P, Haber JE (1984). Rearrangements of highly polymorphic regions near telomeres of Saccharomyces cerevisiae. Mol Cell Biol 4:2509.
17. Southern EM (1975). Detection of specific sequences among DNA fragments separated by gel electrophoresis. J Mol Biol 98:503.
18. Klyce HR, McLaughlin CS (1973). Characterization of temperature-sensitive mutants of yeast by a photomicrographic procedure. Exp Cell Res. 82:47.
19. Carson M, Hartwell L, personal communication.
20. Cavalier-Smith T (1974). Palindromic base sequences and replication of eukaryotic chromosome ends. Nature (London) 250:467.
21. Bateman AJ (1975). Simplification of palindromic telomere theory. Nature (London) 253:379.
22. Astell CR, Smith M, Chow MB, Ward DC (1979). Structure of the 3' termini of four rodent parovirus genomes: nucleotide sequence at origins of DNA replication. Cell 17:691.
23. Baroudy BM, Venkatesan S, Moss B (1982). Incompletely base-paired flip-flop terminal lops link the two DNA strands of the vaccinia virus genome into one uninterrupted polynucleotide chain. Cell 28:315.

24. Tattershall P, Ward DC (1976). Rolling hairpin model for replication of parovirus and linear chromosomal DNA. Nature (London) 263:106.
25. Singh H, Dumas LB (1984). A DNA primase that copurifies with the major DNA polymerase from the yeast Saccharomyces cerevisiae. J Biol Chem 259:7936.
26. Gottschling DE, Cech TR (1984). Chromatin structure of the molecular ends of Oxytricha macronuclear DNA: phased nucleosomes and a telomeric complex. Cell 38:501.
27. Szostak JW, personal communication.
28. Pedersen RA (1971). DNA content, ribosomal gene multiplicity and cell size in fish. J Exp Zool 177:65.
29. Britten RJ, Kohne DE (1968). Repeated sequences in DNA. Science 161:529.
30. Lefevre G (1974). The relationship between genes and polytene chromosome bands. Ann Rev Genet 8:51.
31. Levy LS, Manning JE (1981). Messenger RNA sequence complexity and homology in developmental stages of Drosophila. Dev Biol 85:141.
32. Pringle JR, Hartwell LH (1981). The Saccharomyces cerevisiae cell cycle. In Strathern JN, Jones EW, Broach, JR (eds): "The Molecular Biology of the Yeast Saccharomyces: Life Cycle and Inheritance", Cold Spring Harbor: Cold Spring Harbor Laboratory, p 97.
33. Hereford LM, Rosbash M (1977). Number and distribution of polyadenylated RNA sequences in yeast. Cell 10:453.
34. Kaback DB, Oeller PW, Steensma HY, Hirschman J, Ruczinsky D, Coleman, KG, Pringle J (1984). Temperature-sensitive lethal mutations on yeast chromosome I appear to define only a small number of genes. Genetics 108:67.
35. Rothstein RJ (1983). One-step gene disruptions in yeast. Methods in Enzymology 101:202.

ELECTROPHORETIC KARYOTYPING OF SACCHAROMYCES CEREVISIAE[1]

Georges F. Carle

Department of Genetics
Washington University School of Medicine
Saint Louis, Missouri 63110

ABSTRACT A new method of separation for high molecular DNA molecules has been used to establish an "electrophoretic karyotype" of Saccharomyces cerevisiae. The separation and characterization of individual yeast chromosomes was performed by Orthogonal-Field-Alternation Gel Electrophoresis followed by transfer to nitrocellulose membrane and hybridization to single-copy yeast DNA probes. In our standard laboratory strain, twelve individual bands can be detected by ethidium-bromide staining of the gels; nine of these bands arise from a single chromosome and the other three appear to be doublets. Interstrain chromosome-length polymorphisms (CLP's) have been used to resolve the three doublets into singlets. An example of an electrophoretic karyotype of the progeny from a cross between two strains presenting a CLP is shown. Further fractionation of the yeast genome into chromosome-size fragments was performed using 8 base-pair recognition sites restriction enzymes, followed by OFAGE separation.

INTRODUCTION

The development of a new electrophoretic technique has lead to the separation and visualization of intact chromosomal DNA molecules from several organisms (1-8). For the systems where little genetics and cytogenetics is known,

[1]This work was supported in part by a grant from the National Institutes of Health (GM28232).

this approach should provide new information about the organization of their genomes. In yeast, it should complement genetic data by giving a physical insight into events such as translocations, gene amplification, and recombination. Genetic mapping has suggested that the yeast genome contains approximately 17 chromosomes (9) with an average size of 500-1000 kilo-base pairs (kb). Using high resolution epifluorescent microscopy and synchronous sporulation of spheroplasts, Kuroiwa et al. recently presented cytological evidence that the haploid chromosome number in S. cerevisiae is about 16 (10).

For years conventional electrophoresis has been unsuccessful in the separation of very large DNA molecules. The first successful attempt was reported in 1982 by Schwartz et al. when they were able to resolve molecules of several hundred kilo-base pairs on agarose gels in the presence of two alternately applied, approximately orthogonal electric fields.

Since then, we have independently developed our own implementation of the technique, and by using an agarose imbedded cell lysis protocol to preserve the intactness of the largest yeast chromosomes (2,5) we were able to show that several single-copy DNA probes would hybridize to specific bands separated by the orthogonal-field-alternation technique. Fifteen of the seventeen genetically defined chromosomes were assigned in that way (5). What appears to be the largest yeast chromosome (XII) has been resolved only recently by using low agarose percent gels combined with linear switching interval gradients (unpublished data). Therefore only chromosome XVII, for which no hybridization probe is presently available, is yet to be assigned.

RESULTS

Typical OFAGE pattern of S. cerevisiae:

Fig.1 shows the separation of chromosomal DNA molecules in our standard laboratory strain AB972 by OFAGE and summarizes the assignment of chromosomes I-XI and XIII-XVI. Nine of the bands are singlets as shown by hybridization with single copy gene DNA probes (4,5), and three are doublets. More recently a thirteenth band with a lower mobility than band 12 (chromosome IV) has been reproducibly observed in 0.4% agarose gels; by hybridization band 13 was shown to correspond to chromosome XII.

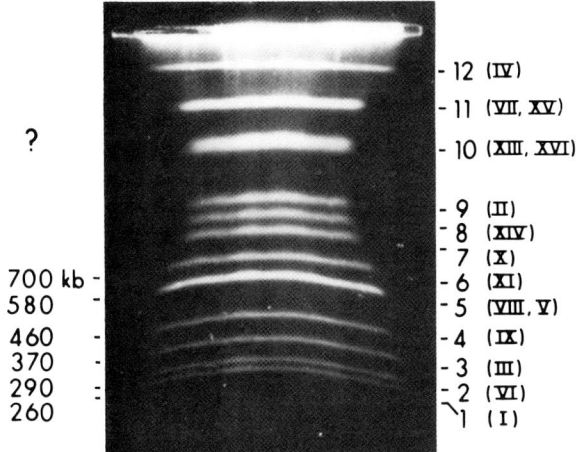

FIGURE 1. An ethidium-bromide-stained agarose gel on which the chromosomal DNA molecules of yeast have been resolved (strain AB972). The size estimates on the left and the band numbering on the right are from ref.4. This particular gel was run at 300v for 18hr at 13° C with a switching interval of 50 sec (from ref.5).

This result completes the separation and assignment of the 16 chromosomes for which probes are available.

In the process of analyzing different laboratory strains, we were able to identify chromosome-length polymorphisms (CLP's) affecting all three doublets. Fig.2 shows the result of one of those CLP's for band 5: in our standard laboratory strain AB972, band 5 appears as a doublet on the ethidium-bromide stained gel and hybridizes with specific probes for chromosomes V and VIII. In the strain A364a, band 5 can be resolved into two singlets (bands 5A and 5B), which correspond to chromosomes VIII and V, respectively.

Two other strains (YNN281 and DCO4α) contain CLP's for bands 10 and 11, respectively, and these CLP's have also been analyzed by DNA-DNA hybridization. Fig.3 shows a comparison of the banding pattern for AB972, A364a, YNN281, and DCO4α .

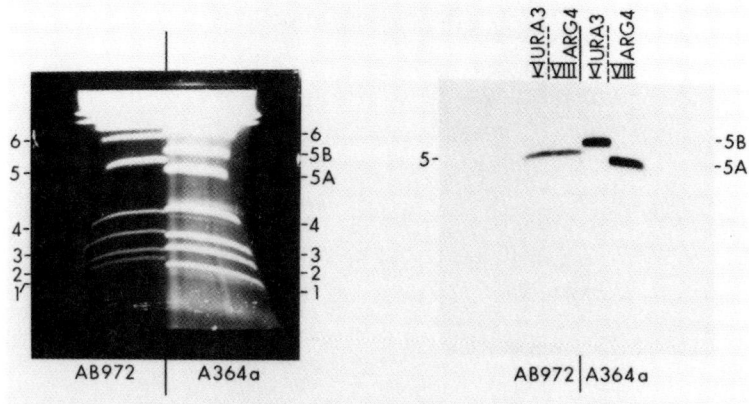

FIGURE 2. Resolution of band 5 into two components in strain A364a. The experimental conditions were identical to the ones described in the legend to Fig.1 besides for a 30 sec switching interval to optimize resolution in the band 5 region. After transfer to nitrocellulose, the membrane was then cut into four strips that were separately hybridized with URA3 (V) and ARG4 (VIII). The filterstrips were then positioned in their original alignment before the autoradiogram was exposed (from ref.5).

Application of OFAGE to tetrad analysis:

Fig.4 shows the result of a cross between two strains containing a chromosome-length polymorphism (AB972 and A364a). The four members of a dissected tetrad were grown in liquid culture, embedded in agarose as previously described (5), and analyzed by OFAGE. This example, among others, shows that detectable recombination events are taking place for several chromosomes in the different progeny. The appearance of the chromosomes in the progeny that differ in length from the parental chromosomes suggests that the parental chromosomes differ in structure at more than one site.

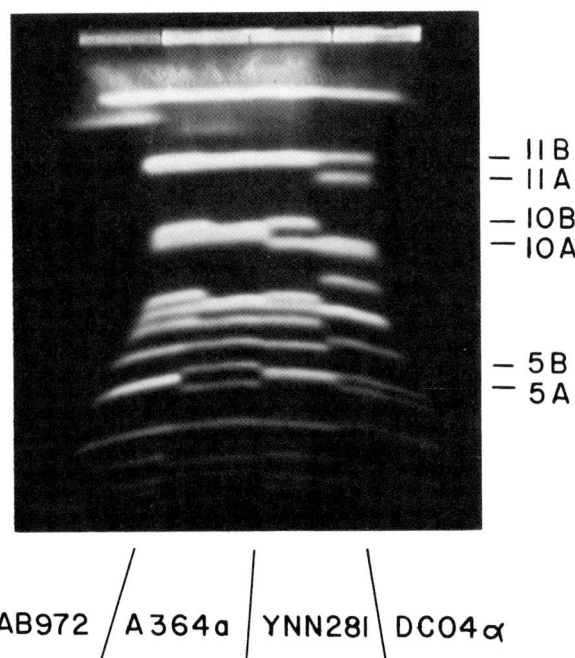

FIGURE 3. Chromosome-length-polimorphisms resolving the three doublets, band 5, 10, and 11. This gel had a 1.2% agarose concentration and was run with a linear switching interval gradient ranging from 40 sec to 60 sec. AB972 is our standard strain, A364a has a CLP for band 5, YNN281 for band 10, and DC04α for band 11.

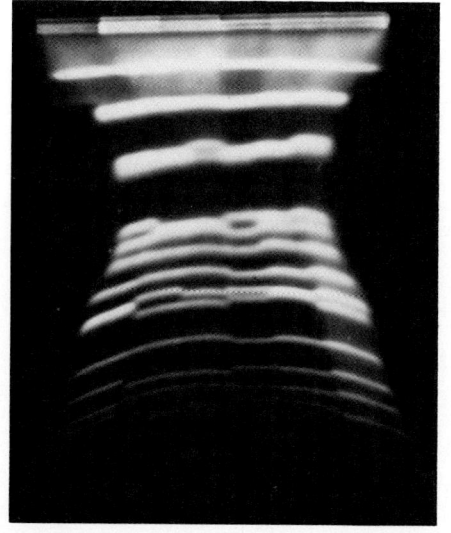

FIGURE 4. OFAGE pattern of a tetrad resulting from a cross between two strains presenting CLP's (AB972 and A364a).

Restriction digest of yeast chromosomes with Sfi I and Not I:

Fig.5 shows a restriction digest of total DNA from the strain AB972 by two restriction enzymes with 8 base-pairs recognition sites. The fragments, which range in size up to 600 kb, are readily separable by OFAGE. The switching interval was 30 seconds in this case in order to expand the resolution in the 200-700 kb region. The Sfi I digest shows a very intense band at about 9 kb corresponding to the ribosomal DNA repeat, which is cleaved once by this enzyme (unpublished data).

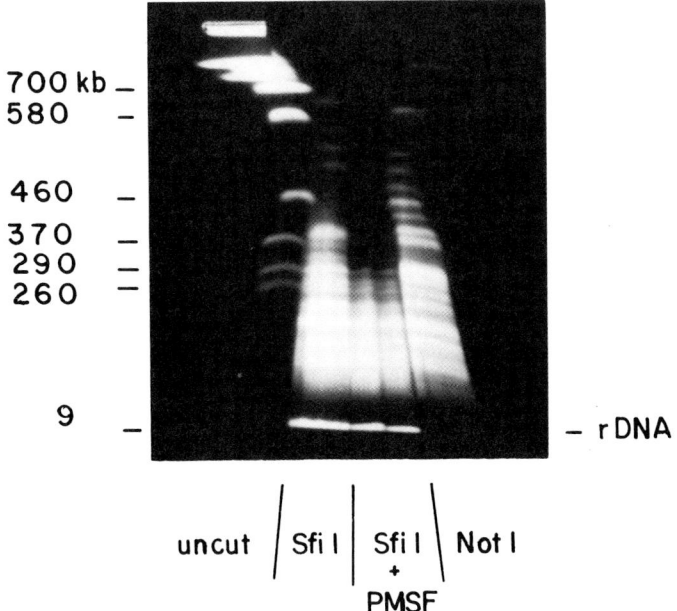

FIGURE 5. Restriction digest of total DNA from the strain AB972 with Sfi I and Not I. The switching interval was 30 sec in order to expand resolution in the lower size range.

DISCUSSION

We have completed the development of an electrophoretic karyotype for Saccharomyces cerevisiae by assigning each chromosome to a specific gel band (5). This technique should provide a convenient method of mapping any cloned genes to a particular chromosome by hybridization to southern transfers of OFAGE gels. As an example of this procedure, we were able to assign SUP5 and GAL80 to chromosome XIII (5); these genes had previously been shown to be linked to a fragment designated as F8, which had been difficult to map by standard methods.

The application of OFAGE to tetrad analysis opens the possibility of analyzing major recombination events and karyotype abnormalities at a molecular level.

Finally, the use of restriction enzymes with few sites per chromosome should provide a method of physical mapping at higher resolution than is possible with intact chromosomes. These enzymes also provide a method of estimating the size of those chromosomes for which we do not yet have reliable estimates. More generally, when used to analyze genomes from other organisms, OFAGE will provide a new approach to establishing physical linkage between genes (11).

ACKNOWLEDGMENTS

We are grateful to Maynard V. Olson, and Jennifer K. Lodge for critical reading of the manuscript.

REFERENCES

1. Schwartz DC, Saffran W, Welsh J, Haas R, Goldenberg M, Cantor CR (1982). New techniques for purifying large DNAs and studying their properties and packaging. Cold Spring Harbor Symp. Quant. Biol. 47:189.
2. Schwartz DC, Cantor CR (1984). Separation of yeast chromosome-sized DNAs by pulse field gradient gel electrophoresis. Cell 37:67.
3. Van der Ploeg LHT, Schwartz DC, Cantor CR, Borst P (1984). Antigenic Variation in Trypanosoma brucei analyzed by electrophoretic separation of chromosome-sized DNA molecules. Cell 37:77.
4. Carle GF, Olson MV (1984). Separation of chromosomal DNA molecules from yeast by orthogonal-field-alternation gel electrophoresis. Nucleic Acids Res 12:5647.
5. Carle GF, Olson MV (1985). An electrophoretic karyotype of yeast. Proc Natl Acad Sci USA 82:3756.
6. Van der Ploeg LHT, Cornelissen AWCA, Michels PAM, Borst P (1984). Chromosome rearrangements in Trypanosoma brucei. Cell 39:213.
7. Altschuler MI, Yao Meng-Chao (In preparation). Macronuclear DNA of Tetrahymena thermophila exists as defined subchromosomal-sized molecules.
8. Kemp DJ, Corcoran LM, Coppel RL, Stahl HD, Bianco AE, Brown GV, Anders RF (1985). Size variation in chromosomes from independent cultured isolates of Plasmodium falciparum. Nature 315:347.
9. Mortimer RK, Schild D (1980). Genetic map of Saccharomyces cerevisiae. Microbiol Rev 44:519.
10. Kuroiwa T, Kojima H, Miyakawa I, Sando N (1984). Meiotic karyotype of the yeast Saccharomyces cerevisiae. Exp Cell Res 153:259.
11. Policastro PF, Daniels-McQueen S, Carle GF, Boime I, (In preparation) A map of the hCGβ-LHβ cluster.

CONTROL MECHANISMS OF CHROMOSOME MOVEMENT IN MITOSIS OF FISSION YEAST[1]

Mitsuhiro Yanagida, Yasushi Hiraoka[2], Tadashi Uemura, Sanae Miyake, and Tatsuya Hirano

Department of Biophysics, Faculty of Science, Kyoto University, Sakyo-ku, Kyoto 606, Japan

ABSTRACT   A number of conditional-lethal mutants that are blocked in the nuclear division of fission yeast were isolated, and some of the genes required for chromosome condensation, segregation and separation were identified, including α- and β-tubulin, and type II DNA topoisomerase genes. Individual chromosomes are visualized in a tubulin mutant, and a pathway for mitotic chromosome movement is deduced from temperature-shift experiments using DAPI staining and indirect immunofluorescence microscopy. Distinct in vivo roles for type I and type II topoisomerases are proposed from analyses of defective phenotypes of the mutants. Our study showed that processes of spindle dynamics and chromosome condensation/segregation are independent at certain steps of the nuclear division.

INTRODUCTION

Chromosome movement during mitosis is controlled by many genes. The identification of the gene products and functions is a prerequisite for understanding the molecular mechanisms of chromosome condensation, segregation, and

---

[1]This work was supported by grants from the Ministry of Education, Science and Culture of Japan, Yamada Science Foundation, Kurata Foundation and Takeda Science Foundation.
[2]Present address: Department of Biochemistry and Biophysics, University of California, San Francisco.

separation. The fission yeast Schizosaccharomyces pombe shows a series of stages during nuclear division (1-5). The interphase hemisphere of the nuclear chromatin region (defined as the nuclear area stained with a DNA-specific fluorescent probe, DAPI) changes into a condensed ellipsoid, then segregates into a U-shaped intermediate structure and separates into two smaller hemispheres, which move to opposite ends of the cell (3). The nuclear division takes place in about 20 min at 30°C in a rich medium, taking about 15% of the time required for a single cell cycle. Soon after the nuclear division, the septum is formed and the cell enters the S-phase (DNA synthesis); cell separation occurs in mid S-phase. The G1 phase is short in S. pombe (6).

To identify the genes required for chromosome movement in mitosis, we employed three different approaches. [1] Isolation of conditional-lethal mutants that are blocked in nuclear division. Their gene functions and products may be identified by studying the mutant phenotypes and the cloned genes. [2] Isolation of mutants defective in a known gene product (such as DNA topoisomerase or protein kinase) by assaying a large number of extracts of the mutagenized cells. The mutants may be arrested at a specific stage of the nuclear division. [3] Cloning of the genes (such as histone H1) with products possibly involved in the nuclear division but which are difficult to assay. Integration of the in vitro mutagenized gene onto a chromosome to replace the normal one may produce defective phenotypes in the nuclear division.

In this paper we report our recent results obtained from [1] and [2] and describe some new insights on the control mechanisms of chromosome movement in mitosis.

GENES REQUIRED FOR NUCLEAR DIVISION OF FISSION YEAST

In Table 1, 24 genes known to be essential for the nuclear division of S. pombe are listed. The cdc (cell division cycle) mutants are temperature-sensitive (ts) and those defective in nuclear division have phenotypes wherein DNA synthesis proceeds and the cell is elongated, but the nucleus remains undivided (7) (Fig. 1). The nda (nuclear division arrest) mutants are cold-sensitive (cs) and were isolated by the same criteria as the cdc mutants defective in the nuclear division (4). Interestingly, none of the ts cdc mutations fell in any one of the genes defined by the cs nda mutations (4, unpublished result).

Table 1

GENES REQUIRED FOR THE NUCLEAR DIVISION OF S. pombe[a]

Genes	Mutation	Locus	Product	Reference
cdc1	ts	I		(7)
cdc2	ts wee	II	PKase	(7,8,9)
cdc5	ts			(7)
cdc6	ts	(II)		(7)
cdc13	ts	(II)		(7,13)
cdc25	ts	I		(14)
cdc27	ts			(7)
cdc28	ts			(7)
nda1	cs	II		(4)
nda2	cs ss	II	α-tubulin	(4,11)
nda3	cs ss ben	II	β-tubulin	(4,5)
nda4	cs	I		(4)
nda5	cs	(I)		(4)
nda6	cs	(I)		(4)
nda7	cs	(I)		(4)
nda8	cs	(I)		(4)
nda9	cs	(I)		(4)
nda10	cs	(III)		(4)
nda11	cs	III		(4)
nda12	cs			(4)
top2	ts	II	top II	(12)
cut1	ts	III		
cut2	ts	II		
nuc2	ts			

[a] I, II, III, chromosomal location; with parenthesis, precise location unknown; alleles in nda4-nda14: nda4-108, nda5-4, nda6-48, nda7-465, nda8-476, nda9-525, nda10-3, nda11-138, nda12-170 (4).

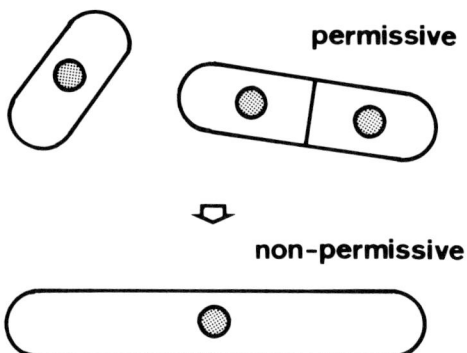

FIGURE 1. ts cdc and cs nda mutants defective in nuclear division accumulate singly nucleated, elongated cells in cultures under the restrictive condition.

The cdc2 gene not only determines the timing of mitosis but also controls the start of the cell cycle (8), and appears to encode a protein kinase (9). The gene products of nda2 and nda3 that act at later stages of the nuclear division (10) were identified as α- and β-tubulin, respectively (5,11). No other cdc or nda gene functions required for the nuclear division are understood at the molecular level. nda4-108 and nda11-138 were recently mapped in chromosome I (3.4 cM from his6) and in III (linked to arg1; Y. Nakaseko, unpublished), respectively.

The ts top2 mutants were originally isolated by mass screening; the mutant extracts contain heat-sensitive type II DNA topoisomerase (12). Under the restrictive condition, the nucleus of top2 was not divided. Detailed phenotypes of top mutants as well as newly isolated cut and nuc will be described in later sections.

PHENOTYPES OF TUBULIN MUTANTS

Cells of a cs nda2-52 mutant showed multiple defects in microtubular functions at non-permissive temperature (4,10). The nucleus was displaced from the center of the cell, no spindle was observed, and the duplicated spindle pole bodies only migrated partially on the constricted nuclear membranes. Cell branching occasionally occcurred. The nuclei of the arrested cells stained with DAPI showed two divided chromatin regions in a short distance or three distinct chromosomal domains (4). At the permissive temperature, the mutant was found to be supersensitive (ss) to TBZ (thiabendazole) and related benzimidazole compounds (10), known as fungal microtubular inhibitors (16). A single mutation caused cs as well as ss phenotypes. Further genetic analyses showed that nda2 is one of the two major loci that determine supersensitivity to the drugs (10). The other one was ben1 (identical to nda3), which also controls drug resistance (15). In contrast, none of the resistant mutants were mapped at nda2.

Cells of a cs nda3-311 mutant showed phenotypes similar but not identical to nda2-52 (4, 10). The nuclear division was blocked and three condensed chromosomes were observed (17). At 20°C, 80-90% of the arrested cells showed such chromosomes (5). Because of morphological phenotypes and of genetic control on the drugs, we concluded that the gene products of NDA2 and NDA3 are closely related to nuclear and cytoplasmic microtubular functions (4,10).

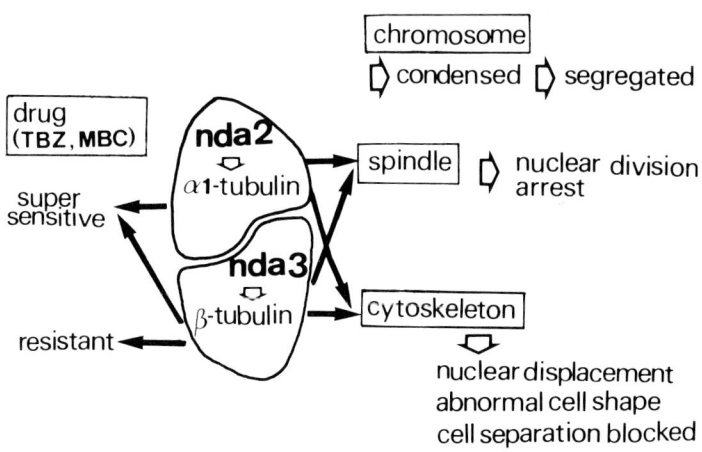

FIGURE 2. Pleiotropic control of NDA2 (coding α-tubulin) and NDA3 (coding β-tubulin) on spindle and cytoskeletal microtubular functions. Sensitivity to benzimidazole fungicides is also altered in the mutants. Chromosome condensation and segregation into sister chromatids proceed in the absence of tubulin gene functions.

Pleiotropic control of nda2 and nda3 on cell cycle, cell structure and sensitivity to benzimidazole compounds are illustrated in Fig. 2. The immediate cause for the cell cycle arrest seems to be the failure in spindle formation. The nuclear displacement, abnormal cell shape and the block of cell separation are probably due to the defect in cytoplasmic microtubules.

The genes of NDA2 and NDA3 were molecularly cloned by transformation and were identified to encode α- and β-tubulins, respectively (5,11). For the NDA2 gene cloning, two different genomic sequences that complemented both cs and ss phenotypes were isolated (11). Nucleotide sequence determinations showed that both the clones encode α-tubulins differing in molecular weights and amino acid sequences (homology to porcine α-tubulin was 76% in both cases). By chromosomal integration of the cloned genes, one (designated as α1-tubulin gene) was determined to be derived from the NDA2 gene itself and contained a 90 bp intron (Fig. 3). The other (α2-tubulin gene) was not linked to nda2 and did not contain intron. These two α-tubulin genes were transcribed, as judged from their Northern blots (11) and were also translated, from the expression of β-galactosidase in the fused genes constructed with the 5' sequences of α1- and α2-tubulin genes (Y. Adachi et al., in prep.).

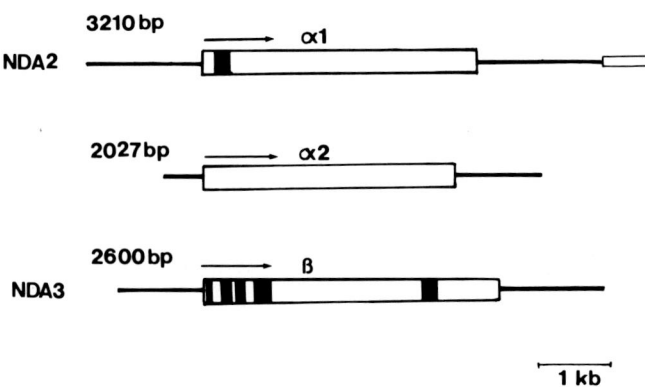

FIGURE 3. The cloned genomic sequences that complemented nda2 and nda3 (5,11). Coding regions are indicated by the boxes. Filled parts are introns. Arrows indicate the direction of transcription.

For the NDA3 gene cloning, the 2.6 kb Hind III genomic DNA fragment that complemented the cs and ss phenotypes of nda3 mutants was isolated (5). By chromosomal integration, the cloned sequence was found to be derived from the NDA3 gene. It has a coding frame split with five short introns. The predicted amino acid sequence contained 448 residues, and was 75% homologous to that of chicken β-tubulin. Genomic Southern hybridization indicated that the sequence is present as a single copy. Thus, the genome of S. pombe has two α-tubulin and one β-tubulin genes (Fig. 3). Our recent experiments of gene disruption showed that the α1-tubulin (nda2) gene is essential whereas the α2-gene is dispensable. Such difference in the expression of the two genes appears to lie in the regulatory sequences and not in the coding sequences (Y. Adachi et al., in prep.).

CHROMOSOME MOVEMENT IN nda3-311 BY TEMPERATURE SHIFT-UP

Cells of the cs β-tubulin mutant were uniformly arrested in mitosis at the non-permissive temperature (5). DAPI staining and indirect immunofluorescence microscopy showed three condensed chromosomes but no spindle. We found a rapid, reversible reactivation of the cs gene product by the shift to a permissive temperature (36°C). After temperature shift-up, the chromosomes moved rapidly to opposite ends of the cell (Fig. 4). Six minutes after the shift-up, the spindle appeared and elongated. The chromosomes were separated at a constant speed (relative velocity 1 μm/min), and the spindle disappeared after the chromosomes reached opposite ends of the cell. The separation took 12 min, and the completely separated chromosomes were an average of 12 μm apart.

A schematic illustration of the spindle formation is shown in Fig. 5a. The short spindle first seen at 6 min was oblique to the cell axis. This was in accordance with the initial chromosome movement after the shift-up. The fully extended spindle elongated and became parallel to the cell axis. The nuclear membranes are known to be constricted and make a dumbbell shape at the late stage of mitosis (1).

The spindle movement in the wild type cells was similar but not identical to that observed in nda3-311 after the temperature shift-up (5)(Fig. 5). A tiny spindle was formed before the nuclear chromatin region began to be condensed, and then elongated to a short, oblique spindle of increasing size at the stage of the U-shaped chromatin region (Fig. 5b).

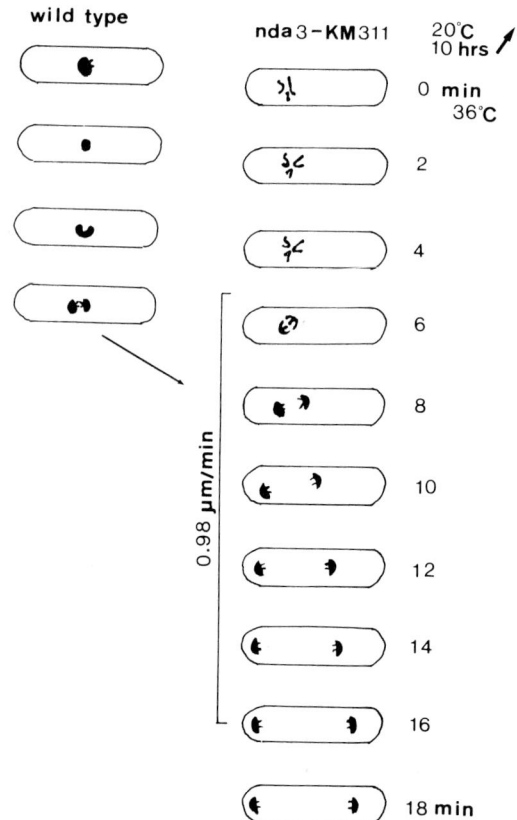

FIGURE 4. Mitotic chromosome movements in the wild type and the cells of nda3-311 after the temperature shift-up (5). Morphological changes in the nuclear chromatin region stained with DAPI are shown. The cells of nda3-311 were first grown at 36°C, then incubated at 20°C for 10 hr and transferred to 36°C. Results of the wild type were obtained by selection synchrony (3). The relative velocity of the chromosome movement is indicated.

Immunofluorescence was brighter and thicker at the two ends than in the middle of the short spindle. The brighter parts may be related to the chromosomal microtubules (Y. Hiraoka et al., in prep.). The tiny spindle was never observed in the nda3 mutant; the short spindle (about 2 μm, equivalent

Genes Required for Mitosis 287

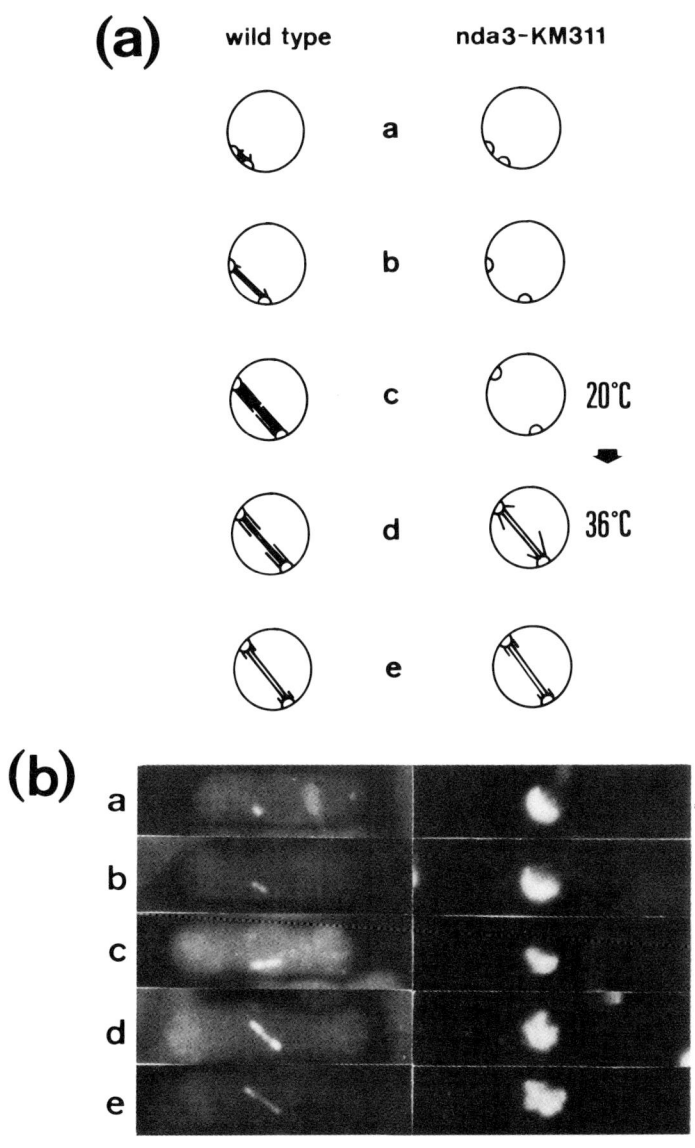

FIGURE 5. The spindle formation in the wild type and in the cells of nda3-311 after the temperature shift-up. (a) Proposed pathways. See text. (b) Immunofluorescence (left) and DAPI-stained (right) micrographs of the wild type cells. Bar indicates 10 μm.

to the diameter of the nucleus) was the shortest found, and the ends were not brighter than the middle, possibly because the chromosomal microtubules were radiated due to the dispersed chromosomes (see discussion in later section). An important conclusion was that the S. pombe spindle was formed only during the nuclear division and, in this aspect, is similar to that of higher eucaryotes. Spindle dynamics (formation, elongation and disappearance) is much faster than that of S. cerevisiae (18).

GENES REQUIRED FOR THE EARLY STEPS OF NUCLEAR DIVISION.

Mutants of nda1-376 and nda4-108 are blocked at early stages of the nuclear division (4). Many cells of nda1 at the restrictive temperature showed a condensed ellipsoidal chromatin region, indicating that the segregation into the U-shaped structure was blocked. In contrast, the cells of nda4 contained the U-shaped nuclear chromatin region that showed granules or dot-like structures (Fig. 6). A step in the transition from the U-shaped intermediate to the two daughter chromatin regions might be blocked in nda4. Alternatively, chromosome condensation may be abnormal.

Double mutants of nda1-nda3 and nda1-nda2 showed the phenotypes of nda1. Similarly, phenotypes of nda4-nda3 and nda4-nda2 showed the phenotypes of nda4. Therefore, the products of nda1 and nda4 seem to act before those of nda2 (α1-tubulin) and nda3 (β-tubulin).

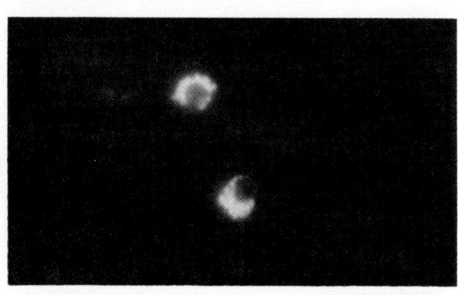

FIGURE 6. A fluorescence micrograph of nda4-108 incubated at the restrictive temperature for 10 hr. The cells were stained with DAPI. Bar indicates 10 μm.

Genomic sequences which complemented nda1 and nda4 were cloned by transformation. 5-kb and 3.6-kb genomic DNA sequences complemented cold sensitivity of nda1 and nda4 mutants, respectively. The cloned DNAs were integrated onto a chromosome and were found to be tightly linked to the respective mutant genes. Nucleotide sequences are being determined.

THE NUCLEAR DIVISION BLOCKED IN TYPE II DNA TOPOISOMERASE MUTANTS

Mutants defective in DNA topoisomerases were isolated by screening individual extracts of mutagenized cells (12). Type I topoisomerase mutants (top1) were viable and did not show any defect in mitotic growth, sporulation or UV sensitivity. In the extract of top1, Mg^{++}- and ATP-dependent type II activity could be detected. Temperature-sensitive mutants having heat-sensitive (hs) type II enzymes were isolated and the ts marker cosegregated with the hs activity (12). All the ts mutants fell in one locus (top2), tightly linked to leu1 in chromosome II. Among 600 ts mutants tested by in vitro assays, five were top2. A genomic sequence that complements top2 and is homologous to the S. cerevisiae TOP2 has recently been cloned.

The nuclear division was arrested in the cells of top2 mutants at the restrictive temperature (12). No indication of chromosome condensation-segregation was obtained. The septum, however, was formed and the cell separated into two halves so that the undivided nucleus was cut, leading to the cell death.

This surprising suicidal phenotype was further investigated and details will be published elsewhere (T. Uemura et al., in prep.). First, indirect immunofluorescence microscopy showed that the spindle was formed at the restrictive temperature, apparently at normal timing. Small parts of the nuclear chromatin were pulled along with the spindle (Fig. 7). Secondly, the nuclear division was also blocked in the double mutant top2-cdc11. The cells of a single cdc11 mutant, which was defective in septum formation but not in nuclear division (7), became multinucleated at the restrictive temperature. In the double mutant, however, the spindle was formed but the nucleus was not divided. The size of the nuclear chromatin region increased strikingly. Thirdly, DNA synthesis did not seem to be blocked in top2.

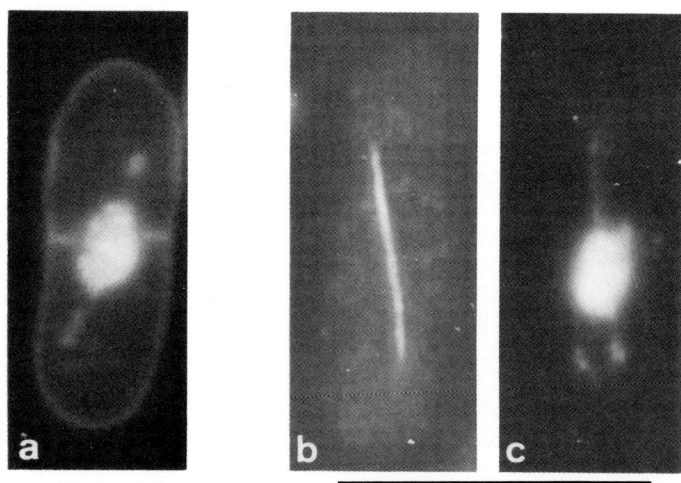

FIGURE 7. DAPI-stained and immunofluorescence micrographs of the cells of top2. (a) DAPI stained. (b) observed by indirect immunofluorescence microscopy using monoclonal antibody against yeast tubulin (19). (c) Ther same cell observed by DAPI staining. Bar indicates 10 μm. mutants at the restrictive temperature.

These results clearly demonstrate that top2 mutations block chromosomal condensation-segregation but neither DNA synthesis, spindle formation, septum formation nor cell separation were blocked. Basically the same results were obtained from independently isolated mutants.

DISTINCT IN VIVO ROLES OF TYPE I AND II DNA TOPOISOMERASES

The double mutants top1 top2 showed phenotypes strikingly different from those of the single top2 mutants (2). Most cells of the double mutant incubated at the restrictive temperature were rapidly arrested at various stages of the cell cycle, showing an altered nuclear chromatin region which looked like a ring or a hollow structure. Transformation from the regular hemisphere occurred at any stage of the cell cycle so that the phenotype is not cdc. Only 5% of the cells showed the morphological phenotype of top2 single mutant.

We have postulated the differential roles of type I and type II topoisomerases (12) as schematized in Fig. 8. Type II enzyme has dual functions: One is to control, throughout the cell cycle, the superhelical density in chromatin together with the type I enzyme by relaxing negative or positive supercoils. The other is to control the chromosome condensation or segregation in mitosis by unknotting and decatenating, or knotting and catenating the chromosomal DNA entangled in the nuclear domains. Therefore, the defect in type I enzyme can be complemented by type II enzyme; top1 is viable. The defect in type II (top2) enzyme causes the arrest in nuclear division, specifically at the chromosome condensation or segregation stage. The defect in controlling superhelical density can only be realized in the double mutant, and produces a large structural alteration in the nuclear chromatin region. The defect in maintaining superhelical density appeared so rapidly that the phenotype of the double mutant overdominated that of the single top2 mutant. In short, relaxing activity of type I and II enzymes is required throughout the cell cycle, whereas activity for the resolution of catenated or knotted DNA is required only during the nuclear division. This hypothesis is examined by investigating the topological forms of plasmid in the mutant cells, and compatible results are obtained (T. Uemura et al., in prep.).

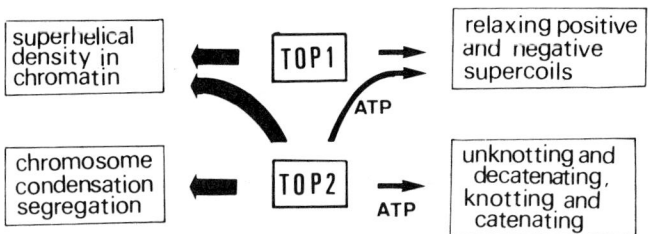

FIGURE 8. Proposed in vivo roles for type I (TOP1) and type II (TOP2) DNA topoisomerases. Proper superhelical density in chromatin is maintained by TOP1 and also by TOP2 throughout the cell cycle. Mitotic chromosome condensation/segregation is controlled by TOP2 which produces transient double strand breakages.

ISOLATION OF cut AND nuc MUTANTS BY CYTOLOGICAL SCREENING

Our previous criteria for isolation of the mutants defective in nuclear division did not include those similar to top2 (Fig. 1). Therefore, we rescreened mutants, showing a morphological phenotype similar to top2, that is, the undivided nucleus cut across by the septum. 600 ts mutants were individually grown, incubated at the restrictive temperature and observed by DAPI staining method (T. Hirano et al., details published elsewhere). By this cytological screening method, we obtained a number of mutants showing such phenotype (Fig. 9). Five were identified as top2 by genetical crosses and the remaining were classified into ten new complementation groups. These were called cut mutants (abbreviation of "cell untimely teared"). cut1 and cut2 were mapped in chromosomes III and II, respectively (Table 1). In the double mutant with cdc11, the nuclear division was blocked. The phenotypes of cut1 and cut2 are not identical to top2, because they did not show the altered phenotype of the double mutant top1-top2 when combined with top1. Further investigations of these and other cut mutants are being carried out.

By the cytological screening method, we obtained a mutant nuc1 (alteration in nuclear structure) with a phenotype similar to that of the double mutant top1-top2, namely, the alteration to a hollowed chromatin region (Fig. 10). The phenotype is not cdc. The mutant extracts contained a normal level of type I and II topoisomerase activities. nuc1 was unlinked to top2.

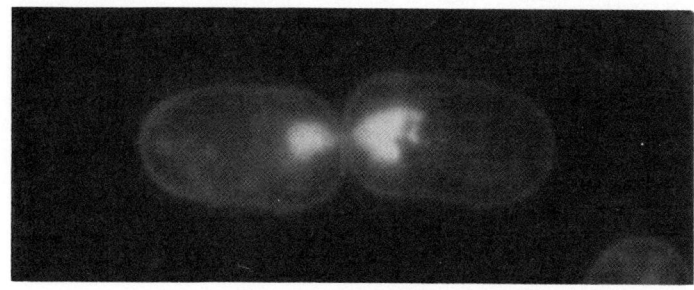

FIGURE 9. A fluorescence micrograph of DAPI-stained cut1-206 incubated at the restrictive temperature for 4 hr. Bar indicates 10 μm.

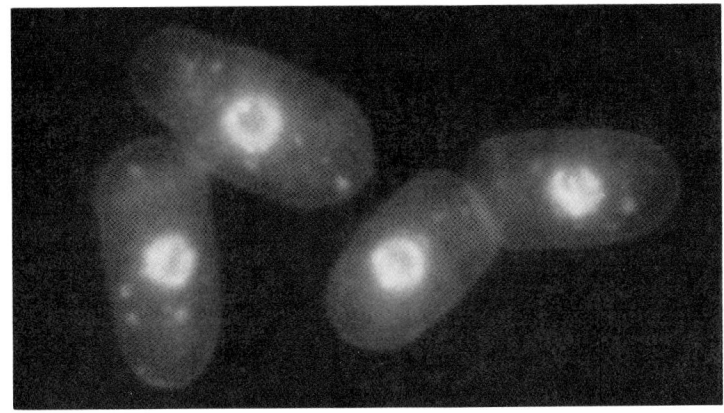

FIGURE 10. A fluorescence micrograph of DAPI-stained nuc1-632 incubated at the restrictive temperature for 2 hr. The nuclear chromatin region is altered to a ring-like, hollow structure and the altered region is indistinguishable from that of top1-top2 double mutant. Bar indicates 10 μm.

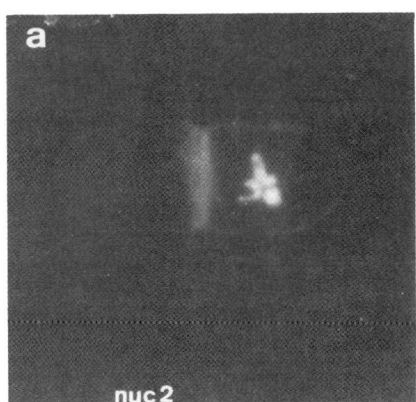
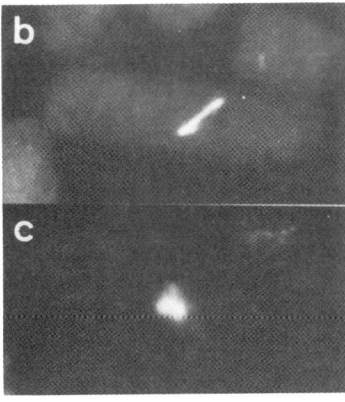

FIGURE 11. DAPI-stained and immunofluorescence micrographs of nuc2-663. (a) DAPI-stained. Three condensed chromosomes can be seen. (b) observed by indirect immunofluorescence microscopy, using monoclonal antibody against yeast tubulin. (c) The same cell observed by DAPI staining. The short spindle oblique to the cell axis is present. Bar indicates 10 μm.

The other, novel mutant nuc2 showed phenotypes similar (but not identical) to those of nda3; three condensed chromosomes were seen and the nucleus was displaced from the center of the cell at the restrictive temperature. However, the short spindle was formed in the arrested cells (Fig. 11); its size was highly uniform, indicating that the spindle dynamics was blocked at a specific stage. The NUC2 product seems to be directly involved in the spindle elongation. The genomic sequence that complemented nuc2 was cloned, and the nucleotide sequence is being determined.

CONTROL MECHANISMS IN NUCLEAR DIVISION

Hartwell and his colleagues (20) showed that mutants of S. cerevisiae blocked in nuclear division were also blocked in further cytokinesis and DNA replication, and mutants blocked in DNA replication became blocked in further nuclear division and cytokinesis. In contrast, mutants blocked in cytokinesis continued to undergo DNA replication and nuclear division. A similar set of dependent sequences has been found for S. pombe (2,7).

The main conclusion obtained from our study is that spindle dynamics and chromosome condensation-segregation of S. pombe are mutually independent in certain steps of the nuclear division. In the cells of top2 at the restrictive temperature, neither chromosome condensation nor segregation took place, although the spindle was formed and further cytokinesis took place, and DNA synthesis also occurred. The formed spindle failed to separate chromosomes but pulled parts of the chromosomes and disappeared at normal timing, indicating that the spindle dynamics proceeded normally. Consistently, in the top2-cdc11 double mutant, nuclear division and cytokinesis were blocked but spindle was formed and DNA synthesis occurred. The structural changes of the nuclear membrane in top2 are unknown, and ultrathin-sectioning for electron microscopy is required to understand the processes in more detail.

The phenotypes of top2 are not exceptional among the mutants defective in nuclear division, because the cut mutants which are classified in ten complementation groups showed phenotypes similar to those of top2. Thus, many gene products may be involved in making the chromosomal structures ready for segregation and separation into sister chromatids, and the defects in these chromosomal functions might not block the spindle dynamics (Fig. 12).

In the cells of the β-tubulin mutant at the restrictive temperature, the nuclear division was arrested at a stage similar to prophase; individually condensed chromosomes could be seen (5,17). In the α-tubulin mutant, the nuclear chromatin region was divided into two parts within a short distance (4). Therefore, we concluded that condensation and individualization (and possibly segregation) of chromosomes do not require nuclear and cytoplasmic microtubular functions. The defects in cdc and nda genes generally blocked further cytokinesis.

The mitotic pathways of wild type and nda3-311 (β-tubulin mutant) after the temperature shift-up were not the same. Hypothetical pathways to explain the spindle and chromosome movements in the wild type and β-tubulin mutants are shown in Figs. 4 and 5a. In the wild type, a tiny spindle was already formed between the duplicated SPBs; it elongated according to the migration of SPBs on the nuclear membranes (see also Fig. 5b). In the nda3 mutant at the restrictive temperature, the spindle was not formed, but the SPB might duplicate and migrate to the opposite ends of the nucleus. After the temperature shift-up, microtubules assembled onto the SPBs, forming the spindle which quickly elongated. This pathway predicts an distinct intermediate step where the spindle can be reversibly assembled on the migrated SPBs and elongated further. The nuc2 mutant is indeed blocked at such a transitional step due to the failure of spindle elongation (Fig. 12).

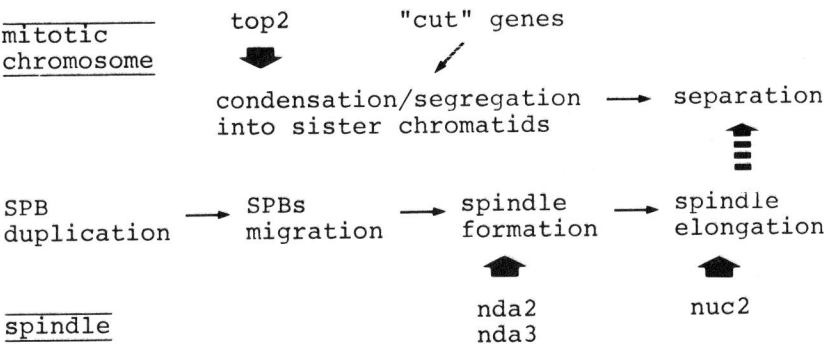

FIGURE 12. Pathways for chromosome movement and spindle dynamics that are independent in certain steps of mitosis. See text.

ACKNOWLEDGEMENTS

We thank Dr. J. Kilmartin for antibody and Dr. P. Nurse for strains.

REFERENCES

1. McCully EK, Robinow DF (1971). Mitosis in the fission yeast Schizosaccharomyces pombe: a comparative study with light and electron microscopy. J Cell Sci 9:475
2. Nurse P (1981). Genetic analyses of the cell cycle. In Glover SW, Hopwood DA (eds): "Genetics as a Tool in Microbiology" Cambridge: Cambridge University Press, p.291.
3. Toda T, Yamamoto M, Yanagida M (1981). Sequential alterations in the nuclear chromatin region during mitosis of the fission yeast Schizosaccharomyces pombe. J Cell Sci 52:271.
4. Toda T, Umesono K, Hirata A, Yanagida M (1983). Cold-sensitive nuclear division arrest mutants of the fission yeast Schizosaccharomyces pombe. J Mol Biol 168:251
5. Hiraoka Y, Toda T, Yanagida M (1984). The NDA3 gene of fission yeast encodes β-tubulin: a cold-sensitive nda3 mutation reversibly blocks spindle formation and chromosome movement in mitosis. Cell 39:349
6. Mitchison JM (1970). Physiological and cytological methods for Schizosaccharomyces pombe. In Prescott DM (ed): "Methods in Cell Physiology" vol 4. p 131.
7. Nurse P, Thuriaux P, Nasmyth K (1976). Genetic control of the cell division cycle in the fission yeast Schizosaccharomyces pombe. Molec Gen Genet 146:167.
8. Nurse P, Bissett Y (1981). Gene required in G1 for commitment to cell cycle and in G2 for control of mitosis in fission yeast. Nature 292:558
9. Beach D, Durkacz B, Nurse P (1982). Functionally homologous cell cycle control genes in budding and fision yeast. Nature 300:706
10. Umesono K, Toda T, Hayashi S, Yanagida M (1983). Two cell division cycle genes NDA2 and NDA3 of the fission yeast Schizosaccharomyces pombe control microtubular organization and sensitivity to anti-mitotic benzimidazole compounds. J Mol Biol 168:271.

11. Toda T, Adachi Y, Hiraoka Y, Yanagida M (1984). Identification of the pleiotropic cell division cycle gene NDA2 as one of two different α-tubulin genes in Schizosaccharomyces pombe. Cell 37:233.
12. Uemura T, Yanagida M (1984). Isolation of type I and II DNA tipoisomerase mutants from fission yeast: single and double mutants show different phenotypes in cell growth and chromatin organization. EMBO Journal 3:1737
13. Nasmyth K, Nurse P (1981). Cell division cycle mutants altered in DNA replication and mitosis in the fission yeast Schizosaccharomyces pombe. Mol Gen Gent 182:119
14. Fantes P (1979). Epistatic gene interactions in the control of division in fission yeast. Nature 279:428.
15. Yamamoto M (1980) Genetic analysis of resistant mutants to antimitotic benzimidazole compounds in Schizosaccharomyces pombe. Mol Gen Genet 180:231.
16. Kilmartin JV (1981). Purification of yeast tubulin by self-assembly in vitro. Biochemstry 20:3629.
17. Umesono K, Hiraoka Y, Toda T, Yanagida M (1983). Visualization of chromosomes in mitotically arrested cells of the fission yeast Schizosaccharomyces pombe. Curr Genet 7:123.
18. Byers B, Goetsch L (1975). Behavior of spindles and spindle plaques in the cell cycle and conjugation of Saccharomyces cerevisiae. J Bacteriol 125:511.
19. Kilmartin J, Adams, AEM (1984). Structural rearrangements of tubulin and actin during the cell cycle of the yeast Saccharomyces. J Cell Biol 98:922-933.
20. Hartwell LH, Cullotti J, Pringle JR, Reid BJ (1974). Genetic control of the cell division cycle in yeast: a model. Science 183:46-51.

IV. NUCLEAR ORGANIZATION

GENETIC ANALYSES OF snRNAs and RNA PROCESSING IN YEAST[1]

Christine Guthrie, Nora Riedel, Roy Parker
Harold Swerdlow and Bruce Patterson

Department of Biochemistry and Biophysics
University of California, San Francisco
San Francisco, CA 94143

ABSTRACT

We are exploiting the unique genetic capabilities of Saccharomyces cerevisiae to determine the functions of small nuclear RNAs (snRNAs). SnRNAs are generally thought to mediate a spectrum of RNA processing events in eukaryotic cells. The first step in our genetic approach is to clone the genes, disrupt them by insertion of a yeast selectable marker, and replace each resident chromosomal gene with its disrupted counterpart. Surprisingly, at least two SNR genes are completely dispensable; moreover, the double mutant is not even growth-impaired. In a third case, the snRNA gene product is essential for viability. In a complementary approach, we are testing directly for the involvement of snRNAs in mRNA splicing.

INTRODUCTION

The "U" class of small nuclear RNAs (snRNAs) is a group of abundant, metabolically stable species housed in ribonucleoprotein particles (snRNPs) in the nuclei of eukaryotes (for review, see (1)). Despite abundant structural information, the function of these ubiquitous RNA molecules has, until quite recently, remained almost

---

[1] This work was supported by NIH grant GM21119 and NSF grant PCM8303667.

purely speculative. Based on a variety of circumstantial evidence, snRNAs were proposed to be involved in RNA processing reactions. For example, the association of U3 with nucleolar pre-rRNAs (2) prompted the notion that this snRNA plays a role in ribosomal RNA processing and/or transport. The most detailed molecular model (3,4) describes the participation of U1, the most abundant of the snRNAs, in the splicing of messenger RNA precursors via complementarity between highly conserved nucleotides at the 5' end of U1 and the consensus sequence at the 5' intron/exon boundary of pre-mRNAs. The recent development of in vitro splicing systems has now provided a means to direct experimental verification, and all available data (5,6,7) strongly support this hypothesis. The development of other in vitro (8) and heterologous in vivo (9) systems for different RNA processing reactions has recently led to evidence of snRNA involvement in polyadenylation and in the generation of the 3' termini of histone mRNAs, respectively.

Despite this steadily building case for the extensive participation of snRNAs in the full spectrum of RNA processing reactions, genetic data remain sorely lacking. Information of this type is essential for cases in which in vitro systems are not readily forthcoming (such as for rRNA processing, in particular, or RNA transport, in general). Moreover, specific molecular models of base-pairing demand the rigor of a genetic test, via the creation of compensatory nucleotide alterations. Finally, the availability of specific mutations in snRNAs should provide a powerful route to the identification of other cellular components which interact with these RNAs, by the isolation of unlinked suppressor mutations. With these considerations in mind, we turned to Saccharomyces cerevisiae, where we could take advantage of unique genetic and transformation techniques available in this organism. Our first goal was to identify the yeast snRNA analogues, clone the genes, and perform gene replacements to determine the phenotype of cells lacking a particular snRNA.

YEAST HAS snRNAs

Studies of snRNAs in metazoans have proceeded rapidly since the discovery that antibodies from human patients with the autoimmune disease systemic lupus erythematosus (SLE) could precipitate snRNP particles from species as distantly related as insects (3). Unfortunately, sera from SLE patients fail to precipitate snRNPs from yeast. Moreover, our systematic hybridization surveys of yeast DNA and RNA using available heterologous snRNA clones were negative, presumably due to the absence of long stretches of appreciable homology.

We were thus forced to identify yeast snRNA candidates by the direct analysis of in vivo ^{32}P-labelled RNAs. As described previously (10), this led to the identification of species which shared appropriate properties with snRNAs from higher organisms. In particular, we selected for study those small RNAs possessing the characteristic trimethylated cap (m2,2,7G) at their 5' termini. In so doing, we noted several interesting and intriguing differences between the yeast snRNAs and their metazoan counterparts. First, and of utmost importance for our genetic strategy, the yeast snRNA genes are (in the six cases so far examined) encoded by single copy genes; this contrasts to the situation in higher cells, where there are on the order of 10-100 U1 genes, for example (11). In keeping with the lack of genetic redundancy, the yeast snRNAs are found in very low abundance. We estimate there are on the order of ≤200-500 copies of any particular snRNA in a haploid cell; again, this is in striking contrast to the million molecules of U1 in mammalian cells. In fact, we estimate (10) that a yeast cell contains far less than 1% of the total snRNA found in a mammalian cell -- and less than 10% of that in a slime mold.

Finally, an intriguing potential difference between yeast and metazoan snRNAs is the apparently larger number of unique species in Saccharomyces cerevisiae. Whereas metazoan cells are typically reported to have six U-RNAs, we initially reported (10) a set size of ∿11 in yeast. This number was based on two types of experiments. We first performed cap analyses on in vivo-labelled RNA species in the size range of ∿5.8S to ∿7S which were separated by two-dimensional gel electrophoresis. To confirm that we were indeed looking at the major capped

RNAs in this size range, we also performed in vitro capping reactions on cold RNA (after prior treatment with phosphatase, followed by chemical decapping of the previously capped RNAs). While these profiles were in good agreement with one another, we were nonetheless concerned that RNAs outside this restricted size range had escaped our notice. This likelihood was recently strengthened by Birnstiel's unexpected discovery of U7, an snRNA nominally 60 nucleotides in length (9). This small size, together with the relative low abundance of U7 (ca. 30-fold less than U1 in sea urchin), has raised the possibility that additional snRNAs are still to be found in higher cells. Indeed, recent work by R. Reddy and H. Busch (pers. comm.) suggests that this is the case.

To obtain definitive information on the total number of snRNAs in Saccharomyces cerevisiae, we have now taken advantage of a polyclonal antibody directed against the snRNA trimethylated cap (a generous gift of Doug Black and Joan Steitz). Total low molecular weight RNA was purified from cells labelled in vivo with ^{32}P. Following immunoprecipitation, the eluted RNAs were electrophoresed on a 6% acrylamide/7M urea sequencing gel. As shown in Figure 1 (lane 1), snRNAs cannot be detected in the total RNA population at this exposure; by contrast, the immunoprecipitated sample (lane 2) is dramatically enriched for RNAs in the size range of ∼5.8S to ∼10S. In subsequent experiments (data not shown), immunoprecipitated RNAs were analyzed on the partially denaturing (4M urea) two-dimensional polyacrylamide gel system we have typically employed to resolve the yeast snRNAs (cf. Fig. 2, ref. 10). By superimposing these profiles on those from total RNA (i.e., prior to immunoprecipitation), we find that the patterns are generally similar within the size range previously investigated. As we published (op. cit.), several spots in this region do not contain caps; these species do not appear in the immune precipitate.

Interestingly, we can now also identify a number of species among the immunoprecipitated RNAs which were not present in the size-fractionated RNA population previously examined. One of these (designated snR13) has a mobility similar to 5S, and a second (snR14) a mobility similar to 5.8S; two others (snR17 and snR18) migrate more slowly than the largest snRNA (snR10) we had previously characterized. To confirm that these novel species were indeed discrete snRNAs (and not, e.g., breakdown products of mRNA or rRNA), cap analyses were performed on the isolated species. As described previously (10), the

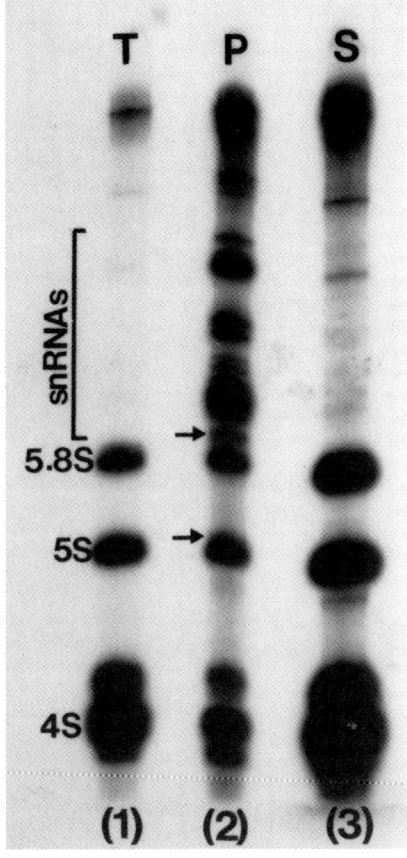

Figure 1. Autoradiograph of 5% acrylamide/7M urea gel. RNA from cells labelled in vitro as described in ref. 10 was electrophoresed directly (lane 1) or following immunoprecipitation with antibody directed against the trimethyl G cap (lane 2); the supernatant of the immune precipitate is run for comparison in lane 3. The arrows indicate the position of snRNA species which closely co-migrate with 5S (snR13) and 5.8S (snR14) rRNAs.

eluted RNAs were digested to completion with a combination of nuclease P1 and nucleotide pyrophosphatase and subjected to two-dimensional thin-layer chromatography. As shown in Figure 2 for snR14 and snR17, this treatment

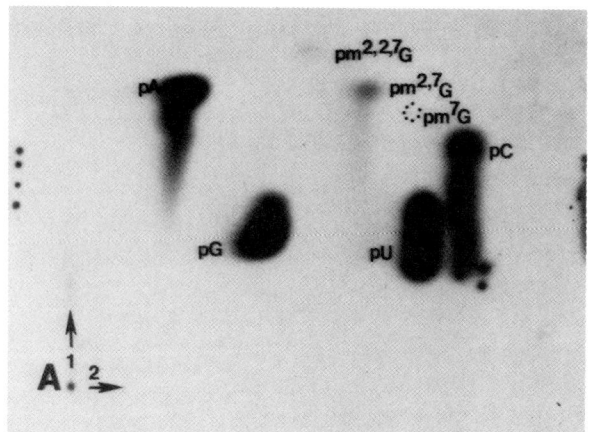

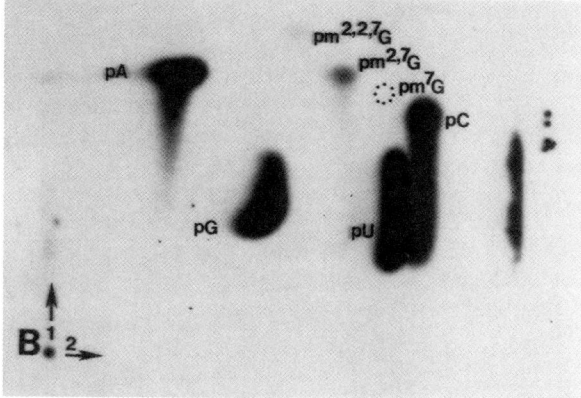

Figure 2. Identification of 5' trimethyl G cap structures by two-dimensional thin-layer chromatography. In vivo-labelled snR14 (a) and snR17 (b) were digested with nuclease P1 and nucleotide pyrophosphatase and chromatographed as described in ref. 10. Marker nucleotides included 5' monomethyl 7-G and were visualized by UV.

generates a spot with the characteristic mobility of m2,2,7G (as well as the related dimethylated species, m2,7G; see ref. 10 for discussion).

From these results we can draw the tentative conclusion that Saccharomyces cerevisiae contains on the order of 16 small capped RNA species. There remain several important caveats, however. One is the lack of information about the possible interrelationships among these species. For example, we have shown (see below) that snR7 is found in two equimolar forms which differ by a 35-nucleotide extension at the 3' terminus. On the other hand, a number of minor species are also visible in the immunoprecipitate, suggesting the possibility that the total number of capped species is larger than 16. Definitive information will require cloning and sequencing of the entire snRNA repertoire, work which is currently in progress. In the meantime, we estimate that the final number of unique snRNAs will be minimally a dozen.

SOME -- BUT NOT ALL -- YEAST snRNAs ARE IMPORTANT FOR GROWTH

Our genetic analysis of snRNA function begins with the introduction of the equivalent of a null allele of a particular snRNA gene into a diploid cell; precise replacement of one of the two chromosomal copies creates a heterozygote for that SNR locus. The SNR/snr transformants are then sporulated. If an essential gene has been successfully inactivated, each tetrad should give rise to only two viable spores, both of which should contain the non-disrupted gene (and can easily be recognized as such because they lack the inserted selectable marker).

As we have reported previously (12), complete elimination of snR3 production (by replacement of sequences encoding it with the yeast LEU2 gene) had no effect on viability. In fact, haploid cells lacking an intact SNR3 locus suffered no selective disadvantage when co-cultured with otherwise isogenic sister spores through some 60 generations of growth in a chemostat! To reconcile this result with the broad evolutionary conservation of snRNAs, we postulated that SNR3 might encode a functionally redundant product. We reasoned that the most likely candidate would be another snRNA with

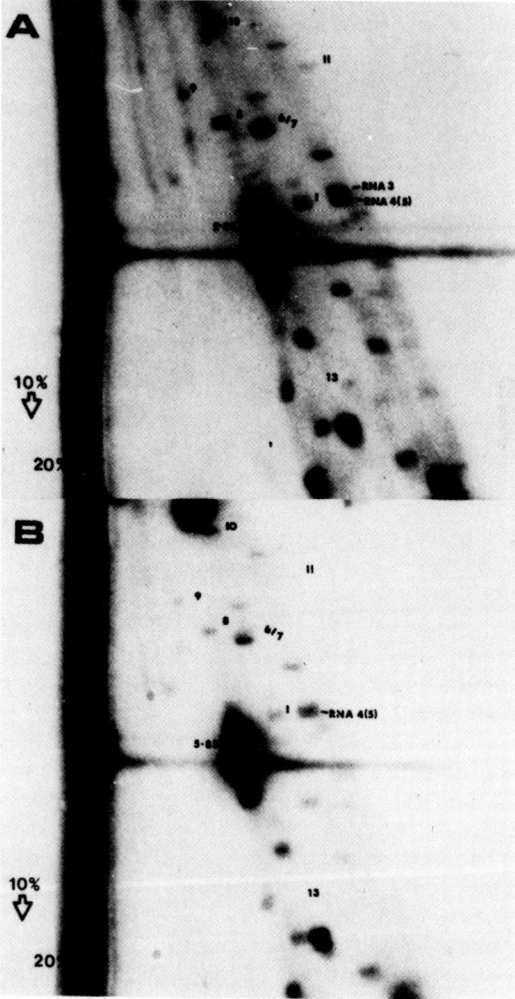

Figure 3. Comparison of in vivo-labelled profiles by two-dimensional electrophoresis in polyacrylamide gels (10%/20%) containing 4M urea. Low molecular weight RNA was extracted from a wild-type haploid (A) and compared with RNA from an otherwise isogenic SNR3::Leu2 strain (B). This reveals that several RNA species (snR4, snR5) with similar 2D mobiles to snR3 are distinguishable when synthesis of snR3 is prevented by gene disruption.

structural similarity to snR3. As shown in Figure 3, haploid cells of the genotype SNR3::LEU2 contain a species (designated snR4) which co-migrates with snR3 on two-dimensional gels. Because of the sieving characteristics of 20% acrylamide and partial denaturation due to the presence of 4M urea, this similarity in mobilities is consistent with the presence of secondary structures of roughly equal stability.

To test the hypothesis that the viability of cells lacking snR3 is due to the presence of snR4, we first constructed a partial deletion of SNR4; as diagrammed in Figure 4A, 125 nucleotides of coding sequence and 119 of 5' flanking sequence were deleted and replaced by a fragment of yIP30 carrying the yeast URA3 gene. Northern analysis of haploid cells carrying SNR4::URA3 (Figure 4B) demonstrates that the gene has indeed been successfully inactivated; that is, no snR4-hybridizing signal can be detected in the two Ura+ (gene-disrupted) spores. These cells grow normally under all conditions tested to date. When SNR4::URA3 haploids were crossed to SNR3::LEU2 haploids and the resulting heterozygotes were sporulated and dissected, all Leu+ Ura+ spores were found to be viable. That is, the double mutant is not lethal, demonstrating that both of these snRNAs are dispensable for growth. In the meantime, we have found yet a third co-migrating snRNA species, which we have designated snR5! Gene disruption experiments to make the SNR5 deletion and, if necessary, the triple mutant, are underway.

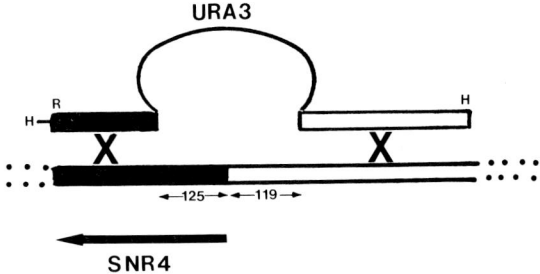

Fig. 4A

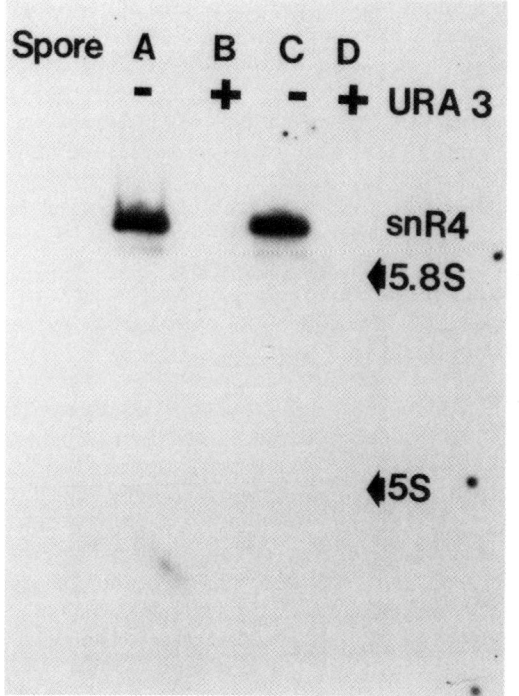

Fig. 4B

Figure 4. Construction and analysis of a null allele of SNR4. (A) A construction was made which resulted in the replacement of 125 nucleotides of coding sequence and 119 nucleotides of 5' flanking sequence of the single copy SRN4 gene with the yeast URA3 gene. After introduction into yeast as described in ref. 12, Northern analysis (B) was used to demonstrate the absence of any hybridization signal with an snRN4 probe in the two Ura+ spores (spores B and D) of a four-spored tetrad.

Analyses of other snRNAs have provided more dramatic results. Cells deleted for SNR10 exhibit cold- and osmotic-sensitive growth (13). The identification of a leaky, conditional biological phenotype associated with the deletion of an snRNA gene was entirely unexpected, since the defect in snR10 synthesis is complete and non-conditional. Experiments to determine the biochemical basis of the snR10- phenotype are underway. Preliminary results suggest that at least one biochemical defect in this strain involves rRNA synthesis and/or processing.

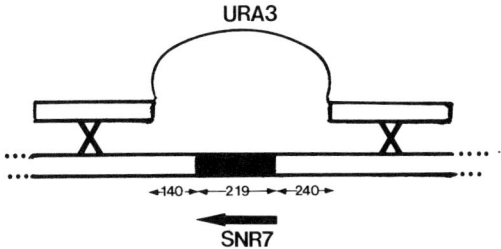

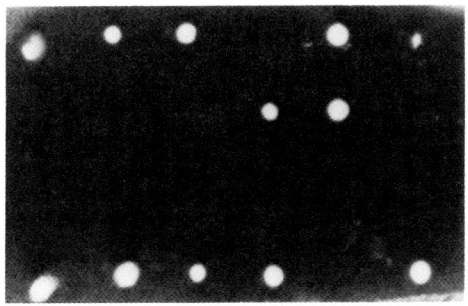

DISSECTED TETRADS

Figure 5. Construction and analysis of a null allele of SNR7. (A) A construction was made which resulted in the substitution of the yeast URA3 gene for the entire SNR7 gene, including 5' (240 nte) and 3' (140 nte.) flanking sequences. The construct was introduced into diploid cells as described in ref. 12. Sporulation and dissection (B) revealed the production of only two viable spores per tetrad. Southern analysis (not shown) confirmed the co-segregation of viability with the non-disrupted SNR7 gene.

In view of the preceding, our most exciting finding to date is that the single-copy gene encoding snR7 is in fact essential for viability. As shown in Figure 5A, we replaced ~600 nucleotides comprising the coding and flanking sequences of SNR7 with the URA3 gene and introduced this construction into diploid cells. Following sporulation and dissection, only two viable spores per tetrad are generated (Fig. 5B); in all cases, the live cells are Ura-. To prove that the lethal defect is directly due to the loss of the SNR7 gene product, we have gone on to show that lethality can be complemented in trans by a small subclone encoding only the snRNA (data not shown). Our highest priority is now to generate conditional alleles in order to probe the biochemical basis of the lethal phenotype. Of particular interest would be the possibility that snR7 participates in pre-mRNA splicing.

ARE snRNAs INVOLVED IN mRNA SPLICING IN YEAST?

In the last year, it has become virtually certain that the mechanism of splicing in yeast and in metazoans is fundamentally the same. The most compelling basis for this important conclusion is the discovery, derived from a detailed analysis of transcripts in mammalian and yeast systems (14,15), that splicing in both cases occurs via a highly unusual intermediate. The so-called "lariat" derives its name from its novel structure, in which the 5' end of the intron is joined by a 2'-5' phosphodiester bond to a site within the intron; the branchpoint this creates is located at a somewhat variable position upstream of the 3' splice junction.

As alluded to earlier, there is now strong experimental support for the essential role of the U1-containing snRNP in metazoan pre-mRNA splicing via recognition of the 5' intron/exon boundary. More recently, evidence has been presented (16,17) for the involvement of the U2-containing snRNP, which is thought to interact with the sequences at or near the branchpoint. By analogy, then, it seems highly likely that particular snRNPs in yeast will play a similar role in pre-mRNA splicing by recognizing and binding to splicing signals within the intron.

The question of the hour concerns the nature of the interactions between the snRNPs and the pre-mRNA "target". For example, while the central appeal of the U1-hypothesis (op.cit.) stems from the direct complementarity between snRNA and pre-mRNA, data from Steitz and her colleagues argue that the proteins in the mammalian U1 snRNP also play an essential role in the process of splicing (18). In considering the relative contributions of RNA and protein to the specificity of the splicing mechanism, it may be instructive to note the most striking distinction between yeast and higher eukaryotes. With the exception of the essentially invariant dinucleotides /GT and AG/ found at the 5' and 3' junctions, respectively, metazoan consensus sequences are characterized by substantial variability. As we shall now describe, lower eukaryotes appear to adhere to much stricter rules at the primary sequence level.

ARE CONSERVED INTRON SEQUENCES THE BINDING SITES FOR YEAST snRNAs?

Sequence analysis of the set of known spliced nuclear genes in Saccharomyces cerevisiae has revealed two elements which exhibit striking conservation at the primary sequence level. These are the hexanucleotide GTAPyGT, located at the 5' intron/exon junction, and the heptanucleotide TACTAAC, located some 4 to 53 nucleotides from the 3' end of the intron (19). Table 1 demonstrates the rather remarkable fact that virtually every position of these signals is absolutely conserved. [One can also observe the presence of additional weakly conserved extensions of the splicing signals (see Table 1). These include a preference for the two nucleotides immediately 5' of the intron (TG) and the conservation of particular residues near the TACTAAC box. For example, the nucleotide two bases 5' of the TACTAAC box is a T residue in 78% of the cases examined. While these weakly conserved elements do not appear to be absolutely required for splicing, and are not found in other fungal introns (see below), nucleotides at these positions may play some subtle role in influencing the efficiency of the splicing process in Saccharomyces cerevisiae.]

What might this high degree of sequence conservation be telling us about the splicing process? One attractive

TABLE 1 Compilation of Conserved Intron Sequences in Saccharomyces cerevisiae

Organism	Gene and intron #	5' junction sequence	TACTAAC box sequence	Distance to AG (a)	Reference
Saccharomyces cerevisiae	Actin	TTCTG/GTATGTTCTAG	TCATGTACTAACATCGA	40	(21)
	L17a	TCTCA/GTATGTTAAAA	GCTTTTACTAACAAAAT	41	(22)
	L25	ATCTG/GTATGTGAACT	TAGTTTACTAACATTAA	38	(22)
	L29	CTCAG/GTATGTAGTTC	TGTTTTACTAACGATTT	40	(23)
	L34A	AAAGA/GTATGTGAAAA	ATCGTTACTAACAAACA	26	(24)
	MATa1-1	ATAAT/GTATGTTTTCA	CAATTTACTAACAATAC	8	(25)
	MATa1-2	TTTGG/GTATGTAATAT	TCCTATACTAACAATTT	7	(25)
	RP16A	GTTAG/GTACGTAAAAA	TCATTTACTAACAATAA	32	(26)
	RP16B	GTTAG/GTACGTATAAC	CTTTTTACTAACAAAAG	31	(26)
	RP28A	AGCTC/GTATGTTCAAR	TTAGTTACTAACATTTT	21	(26)
	RP28B	AGCTC/GTATGTATTGA	CCCTTTACTAACATAAC	37	(26)
	RP29	GAGAG/GTATGTCCCTA	TAGACTACTAACATTTT	42	(27)
	RP51A	AAATG/GTATGTTAATA	CAGTATACTAACAAGTT	55	(28)
	RP51B	AAATG/GTACGTACCAC	GTATTTACTAACTTAAG	34	(29)
	RP59	GTCTA/GTATGTTTAAT	TTGGATACTAACAACAT	46	(19)
	RP73	AGATG/GTCAGTATAAC	TAATCTACTAACAAGTT	37	(b)
	S10A	TGAAG/GTATGTAATAT	AACTTTACTAACAAATG	19	(31)
	TUB1	TAATG/GTATGTTCGAT	TTCATTACTAACTTCAA	38	(c)
	TUB3	TAATG/GTATGTATGCG	CAGTTTACTAACAGTTT	136	(c)

Tabulation of nucleotide frequencies by position.

A. 5' Consensus Sequence

position	-5	-4	-3	-2	-1	G	T	A	T	G	T	+1	+2	+3	+4	+5
total	19	19	19	19	19	19	19	19	19	19	19	19	19	19	19	18
G	4	4	1	2	13	19	0	0	0	19	0	2	1	2	1	2
A	8	6	8	6	3	0	0	18	1	0	0	9	7	9	12	7
T	6	8	4	10	1	0	19	0	15	0	19	7	6	6	3	5
C	1	1	6	1	2	0	0	1	3	0	0	1	5	2	3	4

B. TACTAAC Box

position	-5	-4	-3	-2	-1	T	A	C	T	A	A	C	+1	+2	+3	+4	+5
total	19	19	19	19	19	19	19	19	19	19	19	19	19	19	19	19	19
G	2	1	5	3	1	0	0	0	0	0	0	0	1	1	2	1	3
A	2	7	6	2	3	0	19	0	0	19	19	0	16	11	6	9	5
T	10	6	3	14	13	19	0	0	19	0	0	0	2	7	8	8	9
C	5	5	5	0	2	0	0	19	0	0	0	0	0	0	3	1	2

a. The distance to the AG is measured by counting from the first base 3' of the C at the branch point to the last nucleotide 5' of the AG.
b. J. Warner, pers. comm.
c. D. Botstein, pers. comm.

TABLE 2 Compilation of Conserved Intron Sequences in Fungi

Organism	Gene and intron #	5' junction sequence	TACTAAC box sequence	Distance to AG (a)	Reference
Schizosac-	NDA2	TAACG/GTATGTTCAAT	AATTTAGCTAACATTTC	6	(32)
charomyces	NDA3-1	AGATT/GTACGAGCTAT	TTCTAAGCTAATCAATA	4	(33)
pombe	NDA3-2	TTTTG/GTAAGCAAAAT	CTCTCAACTAACGCTTG	7	(33)
	NDA3-3	GGAAT/GTAGGTCTATT	TTTTTTACTGACTTTTA	8	(33)
	NDA3-4	ACGAG/GTAGGTTTTTT	TTTAGTACTGACGACTG	6	(33)
	NDA3-5	TAATG/GTATGTTTACT	TCATCTGCTAACCTCGA	8	(33)
Schizophyllum	1G2-1	TCGAC/GTGAGTGGTCA	TTGCATGCTGACGCACG	10	(34)
commune	1G2-2	CTTGC/GTGAGTAGCCT	CGACTTGCTGATGTGTC	9	(34)
	1G2-3	AGTTT/GTGAGTGCCAT	TCTCTCACTAACCATCC	8	(34)
Neurospora	histone H3	GTCTC/GTAAGTTTATG	CACGTTGCTAACGCGTC	12	(35)
crassa	histone H4-1	TGGAC/GTAAGTTGTCT	ATCATGACTGACTCGTA	15	(35)
	histone H4-2	TGCCA/GTACGTTATGA	TGGTCAACTAACATGTC	15	(35)
	am-1	CAAGG/GTACGTCTGAG	CATAGAGCTGACTTGAT	15	(36)
	am-2	TGGCC/GTAAGTGACCG	AGATTTGCTGACTCGGC	11	(36)
Aspergillus	glucoamylase-1	GGACT/GTATGTTTCGA	ATTGATGCTGACTGGCG	18	(37)
niger	glucoamylase-2	TGCTT/GTATGTTCTCC	TATGTAGCTGACTGGTC	5	(37)
	glucoamylase-3	ATATG/GTGTGTTTGTT	AGCAGAGCTAACCCGCG	10	(37)
	glucoamylase-4	TTGTG/GTAAGTCTACG	GTGCGTACTAACAGAAG	6	(37)
	glucoamylase-5	CACCA/GTACGTCATCA	CTGAGTGCTGACAAGTA	12	(37)
Trichoderma	CBHI-1	CTGCC/GTAAGTGACTT	CTCCCAGCTGACTGGCC	11	(38)
reesei L27	CBHI-2	ATGAT/GTGAGTTTGAT	CAAGCAGCTGACTGAGA	11	(38)
	EG1-1	AAAAG/GTGAGCCTGAT	TTCCATGCTGACATGGT	10	(b)
	EG1-2	CGACT/GTTCGTATCCC	TGAGATGCTAACCGCTA	11	(b)

Tabulation of nucleotide frequencies by position.

A. 5' Consensus Sequences

position	-5	-4	-3	-2	-1	G	T	A	A	G	T	+1	+2	+3	+4	+5
total	23	23	23	23	23	23	23	23	23	23	23	23	23	23	23	23
G	3	9	7	2	8	23	0	6	2	23	0	5	3	4	2	4
A	6	5	9	6	2	0	0	16	11	0	1	3	5	6	7	4
T	9	7	3	8	7	0	23	1	5	0	20	10	11	7	5	13
C	5	2	4	7	6	0	0	0	5	0	2	5	4	6	9	2

B. TACTAAC Box

position	-5	-4	-3	-2	-1	A/T	Pu	C	T	Pu	A	C	+1	+2	+3	+4	+5
total	23	23	23	23	23	23	23	23	23	23	23	23	23	23	23	23	23
G	1	5	4	5	5	1	16	0	0	13	0	0	5	6	12	4	6
A	5	5	5	5	5	9	7	0	0	10	23	0	5	4	4	2	7
T	10	11	7	8	7	12	0	0	23	0	0	2	8	7	4	12	2
C	7	2	7	6	5	1	0	23	0	0	0	21	5	6	3	5	8

a. The distance to the AG is measured by counting from the first base 3' of the C at the branch point to the last nucleotide 5' of the AG.
b. M. Innis, pers. comm.

hypothesis is that the specificity of splice site selection in a comparatively simple unicellular eukaryote is more "hard-wired" than in metazoans, where the use of alternative splicing pathways can be an effective device for achieving developmental diversity. Mechanistically, this might be achieved by a greater reliance on the RNA-mediated aspects of splicing in the lower eukaryotes. That is, complementarity between snRNAs and pre-mRNA signals might predominate in yeast splice site selection, whereas the snRNP proteins (for example) may play a more determinative role in higher cells.

As a first step in addressing this intriguing question, we would like to identify 1) which of these nucleotides are indeed critical for the splicing process in vivo, and 2) their specific molecular roles. To evaluate the potential significance of a particular residue in a consensus sequence, two approaches have proven instructive: comparative sequence analysis and genetics. As just described, Saccharomyces cerevisiae data are not illuminating for the former approach in that the degree of conservation is virtually absolute. Comparison of intron sequences from other fungi, however, turns out to be more informative.

Compiled in Table 2 are data from the budding yeast Schizosaccharomyces pombe and several filamentous fungi. Examination of the derived 5' consensus sequence illustrates three degrees of sequence conservation. Three residues, the GT at the 5' junction and the G in the fifth position of the consensus, are invariant in fungal introns. Highly conserved, but subject to some variation, are the third and sixth positions of the introns, normally an A and a T respectively. Finally, the least conserved base within this splicing signal is the T found in position 4 in Saccharomyces cerevisiae (Table 1), which in other fungi varies even more considerably (Table 2).

These results are summarized by the consensus shown in Figure 6. It is particularly interesting to note that the invariant nucleotides in the fungal splicing signals correspond to the most highly conserved residues in metazoan introns. This parallel importance of specific residues in both fungi and metazoans reinforces the argument for a fundamentally similar splicing process in the widely separated organisms.

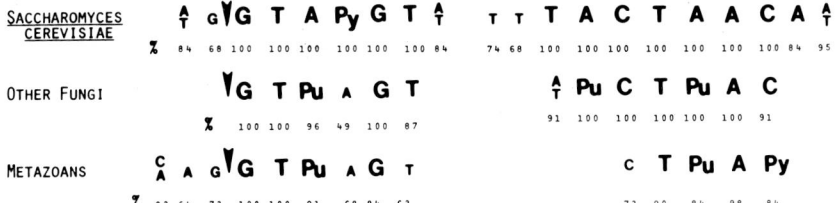

Figure 6. Comparison of splicing signals. The figure compares the splicing signals for different groups of introns. It is important to note that only in a few cases has the location of the branch point been demonstrated experimentally. Large letters indicate nucleotide present in at least 84% of the intron sequences compared. Smaller letters indicate a lesser degree of conservation. The percent of intron sequences conforming to the consensus in a particular position is shown beneath each nucleotide. Consensus sequences for metazoans are taken from Mount (39) and Keller and Noon (40).

GETTING AT THE SPLICING MACHINERY: EXTRAGENIC SUPPRESSORS OF SUBSTRATE MUTATIONS

In order to directly evaluate the contributions of the evolutionarily conserved nucleotides to the splicing process, we have undertaken a systematic genetic analysis. Our strategy relies on our construction of a gene fusion between the yeast actin gene, containing a 309 bp IVS, and the yeast HIS4 gene. As we have recently described (20), this allows us to score biologically for splicing-defective mutants among a randomly mutagenized population. In addition, we are using site-specific mutagenesis to generate point mutations in the consensus sequences. Our results to date are extremely interesting and are consistent with the working hypothesis that the most highly conserved nucleotides in fact play multiple roles in the partial reactions of the splicing pathway.

The information we are gathering about the nature and location of these cis-acting mutations suggests plausible recognition sites for the yeast snRNPs which mediate splicing. With these mutants in hand, we can then look for direct interactions between the substrate and the splicing machinery by the characterization of extragenic revertants. These suppressor mutations should reveal the genes for the snRNAs, the snRNP proteins, and other factors which form (or regulate) the splicing apparatus. It is with this strategy that the greatest virtues of the yeast system promise to be realized.

ACKNOWLEDGMENTS

We thank the other members of the laboratory for their support and encouragement throughout the course of this work. David Tollervey made significant contributions to the early phases of the snRN4 study. We are indebted to Doulas Black and Joan Steitz for their generous gift of antibody and advice on its use. We are grateful to Judy Piccini for her skill, patience, and good humor in the preparation of this manuscript. B.P. is an NSF predoctoral trainee; N.R. is supported in part by a postdoctoral fellowship from the DAAD.

REFERENCES

1. Busch H, Reddy R, Rothblum L, Choi YC (1982). snRNAs, snRNPs, and RNA processing. Ann Rev Biochem 51:617.
2. Prestayko AW, Tonato M, Busch H Low molecular weight RNA associated with 28 s nucleolar RNA. J Molec Biol 47:505 (1970).
3. Lerner MR, Boyle JA, Mount SM, Wolin SL, Steitz JA (1980). Are snRNPs involved in splicing? Nature 283:220.
4. Rogers J., Wall R (1980). A mechanism for RNA splicing. Proc Natl Acad Sci USA 77:1877.
5. Yang VW, Lerner MR, Steitz JA, Flint SJ (1981). A small nuclear ribonucleoprotein is required for splicing of adenoviral early RNA sequences. Proc Natl Acad Sci USA 78:1371.
6. Padgett RA, Mount SM, Steitz JA, Sharp PA (1983). Splicing of messenger RNA precursors is inhibited by antisera to small nuclear ribonucleoptrotein. Cell 35:101.

7. Kraemer A, Keller W, Appel B, Luhrmann R (1984). The 5' terminus of the RNA moiety of U1 small nuclear ribonucleoprotein particles is required for the splicing of messenger RNA precursors. Cell 38:299.
8. Moore CL, Sharp PA (1985). Accurate cleavage and polyadenylation of exogenous RNA substrate. Cell 41:845.
9. Strub K, Gall G, Busslinger M, Birnstiel ML (1984). The cDNA sequences of the sea urchen U7 small nuclear RNA suggest specific contacts between histone mRNA precursor and U7 RNA during RNA processing. EMBO J 3:2801.
10. Wise JA, Tollervey D, Maloney D, Swerdlow H, Dunn EJ, Guthrie C (1983). Yeast contains small nuclear RNAs encoded by single copy genes. Cell 35:743.
11. Lund E, Dahlberg JE (1984). True genes for human U1 small nuclear RNA. JBC 259:2013.
12. Tollervey D, Wise JA, Guthrie C (1983). A U4-like small nuclear RNA is dispensable in yeast. Cell 35:753.
13. Tollervey D, Guthrie C (1985). Deletion of a yeast small nuclear RNA gene impairs growth. EMBO J, in press.
14. Ruskin B, Krainer AR, Maniatis T, Green MR (1984). Excision of an intact intron as a novel lariat structure during pre-mRNA splicing in vitro. Cell 38:317.
15. Domdey H, Apostol B, Lin R.-J, Newman A, Brody E, Abelson J (1984). Lariat structures are in vivo intermediates in yeast pre-mRNA splicing. Cell 39:611.
16. Black DL, Chabot B, Steitz JA (1985). U2 as well as U1 small nuclear ribonucleoproteins are involved in pre-messenger RNA splicing. Cell, in press.
17. Krainer AR, Maniatis T (1985). Multiple factors including the small nuclear ribonucleoproteins U1 and U2 are necessary for pre-mRNA splicing in vitro. Cell, in press.
18. Mount SM, Pettersson I, Hinterberger M, Karmas A, Steitz JA (1983). The U1 small nuclear RNA-protein complex selectively binds a 5' splice site in vitro. Cell 33:509.

19. Teem JL, Abovich N, Kaufer NF, Schwindinger W.F, Warner JR, Levy A, Woolford J, Leer RJ, van Raamsdonk-Duin MM, Mager WH, Planta RJ, Schultz L, Friesen, JD, Fried H, Rosbash M (1984). A comparison of yeast ribosomal protein gene DNA sequences. Nucl Acids Res 12:8295.
20. Parker R, Guthrie C (1985). A point mutation in the conserved hexanucleotide at a yeast 5' splice junction uncouples recognition, cleavage and ligation. Cell 41:107.
21. Ng R., Abelson J (1980). Structure of a split yeast gene: complete nucleotide sequence of the actin gene in Saccharomyces cerevisiae. Proc Natl Acad Si USA 77:2546.
22. Leer RJ, van Raamsdonk-Duin MMC, Hagendoorn MJM, Mager WH, Planta RJ (1984). Structural comparison of yeast ribosomal protein genes. Nucl Acids Res 12:17 6686.
23. Kaufer NF, Fried HM, Schwindlinger WF, Jasin M, Warner JR (1983). Cyclohexamide resistance in yeast: the gene and its protein. Nucl Acids Res 11:10:3123.
24. Schaap PJ, Molenaar MT, Mager WH and Planta RJ (1984). The primary structure of a gene encoding yeast ribosomal protein L34. Current Genetics 9:47.
25. Tatchell K, Nasmyth KA, Hall BD, Astal C, Smith M (1981). In vitro mutation of the mating type locus in yeast. Cell 27:25.
26. Molenaar CMT, Woudt LP, Jansen AEM, Mager WH, Planta RJ (1984). Structure and organization of two linked ribosomal protein genes in yeast. Nucl Acids Res 12: 7345.
27. Mitra G, Warner JA (1984). A yeast ribosomal protein whose intron is in the 5' leader. J Biol Chem 259:9218.
28. Teem JL, Rosbash M (1983). Expression of a beta-galactosidase gene containing the ribosomal protein 51 intron is sensitive to the rna2 mutation of yeast. Proc Natl Acad Sci USA 80:4403.
29. Abovich N, Rosbash M. (1984). Two genes for ribosomal protein S1 of Saccharomyces cerevisiae complement and contribute to the ribosomes. Mol Cell Biol 4:1871.
30. Teem JL, Abovich N, Kaufer NF, Schwindinger WF, Warner JR, Levy A, Woolford J, Leer RJ, van Raamsdonk-Duin MM, Mager WH, Planta RJ, Schultz L, Friessen JD Rosbash M (1984). A comparison of yeast ribosomal protein gene DNA sequences. Nucl Acids Res 12:8295.

31. Leer RJ, van Raamsdonk-Duin MMC, Molenaar CMT, Cohen LH, Mager WH Planta RJ (1982). The structure of the gene coding for the phosphorylated ribosomal protein in S10 in yeast. Nucl Acids Res 10:5869.
32. Toda T, Adachi Y, Hiraoki Y, Yanagida M (1984). Identification of the pleiotropic cell division cycle gene NDA2 as one of two different alpha-tubulin genes in Schizosaccaromyces pombe. Cell 37:233.
33. Hiraoka Y, Toda T Yanagida M (1984). The NDA3 gene of fission yeast encodes beta-tubulin: a cold-sensitive nda3 mutation reversibly blocks spindle formation and chromosome movement in mitosis. Cell 39:349.
34. Dona JJM, Mulder GH, Rouwendal GJA, Springer J, Bremer W, Wessels JGH (1984). Sequence analysis of a split gene involved in fruiting from the fungus Schizophyllum commune. EMBO 3:2201.
35. Woudt L, Pastink A, Kempers-Veebstra AE, Jansen AEM, Mager W, Planta RJ (1983). The genes coding for histone H3 and H4 in Neurospora crassa are unique and contain intervening sequences. Nucl Acids Res 11:5347.
36. Kinnaird JH, Fincham JRS (1983). The complete nucleotide sequence of the Neurospora crassa am (NADP-specific glunate dehydrogenase) gene. Gene 26:253.
37. Boel E, Hansen MT, Hjort I, Hoegh I, Fiil NP (1984) Two different types of intervening sequences in the glucoamylase gene from Aspergillus niger. EMBO J 3:1581.
38. Shoemaker S, Schweickart V, Ladner M, Gelfand D, Kwok S, Myambo K, Innis M (1983). Molecular cloning of exo-cellobiohydrolase I derived from Trichoderma Reesei strain L27. Bio/Technology 1:691.
39. Mount S (1982). A catalogue of splice junction sequences. Nucl Acids Res 10:459.
40. Keller EB, Noon WA (1984). Intron splicing: a conserved internal signal in introns of animal pre-mRNAs. Proc Natl Acad Sci USA 81:7417.

THE REP1 PROTEIN OF 2 MICRON CIRCLE IS ASSOCIATED WITH THE NUCLEAR MATRIX[1]

Ling-Chuan Chen Wu[2], Paul Fisher*, and James R. Broach

Department of Molecular Biology, Princeton University, Princeton, New Jersey 08544, and *Department of Pharmacology, State University of New York, Stony Brook, NY 11794

ABSTRACT Stable propagation of the yeast plasmid 2 micron circle requires an origin of replication, a cis-active locus designated REP3, and two plasmid encoded proteins – the products of the REP1 and REP2 genes. The three REP loci appear to constitute a partitioning system, insuring equidistribution of plasmid molecules to mother and daughter cells following mitosis. We have raised antibodies against the REP1 protein and used them to identify authentic REP1 protein in plasmid bearing yeast cells. We find that REP1 protein is localized in the nucleus and cofractionates with the yeast nuclear matrix. This association suggests several models for the mechanism of REP1 promoted partitioning of 2 micron circle.

INTRODUCTION

Extraction of a eukaryotic nucleus with reagents that remove chromatin, lipids, and RNA leaves a spherical, nucleus-sized structure, termed the nuclear matrix, composed of a proteinaceous network of fibers and granules (1-3). Depending on the extraction procedure, these structures may

---

[1] This work was supported by grants from the NIH to JRB and PF.
[2] Present address: Department of Pharmacology, State University of New York, Stony Brook, NY 11794.

also be obtained surrounded by a fibrous lamin connected to nuclear pore complexes (4,5). Nuclear matrix preparations have been obtained from a wide variety of eukaryotic cells and appear to be ubiqitious (5-9). In addition, results from a number of studies have implicated the nuclear matrix as a key element in a wide variety of cellular processes, including DNA replication, RNA splicing and maturation, RNA transport, and maintenance of nuclear integrity (1,10-14). However, to date these allegations have been based almost exclusively on the rather circumstantial evidence of cofractionation experiments.

As described in this report and in that of a previous study (9), application of appropriate extraction procedures can yield a nuclear matrix preparation from the yeast Saccharomyces cerevisiae with morphological and compositional properties similar to that of matrix preparations from other eukaryotes. We also show in this report that a protein required for segregation of the yeast plasmid 2 micron circle cofractionates with the nuclear matrix. These observations not only suggest a role for the matrix in promoting genetic segregation but also provide an identification of a matrix component whose activity and function have been genetically defined. Thus, these results offer the possibility of applying genetic techniques available in yeast to address the essentiality of the nuclear matrix in various cellular processes.

MATERIALS AND METHODS

Strains, Plasmids, and Media.

DC04 [cir°], DC04 [cir$^+$], and DC04 [pSS4] are isogenic MATa ade1 leu2-04 Gal$^+$ strains that carry respectively either no plasmid, 2 micron circle plasmid, or plasmid pSS4. The structure of plasmid pSS4 is diagrammed in Figure 1. Its construction has been described previously (15). SD and SG media consist of 0.67% yeast nitrogen base supplemented with amino acids, adenine and uracil as described (16) and either 2% glucose or 2% galactose, respectively.

Immunological Techniques.

Rabbit polyclonal anti-REP1 antisera were obtained as previously described (15). Immunological analysis of proteins immobilized on nitrocellulose filters was accomplished by the procedures of Towbin et al (17) and Renart et al (18) as modified Fisher et al (5). Protein samples were fractionate on SDS polyacrylamide gels (19) and then transfered passively to nitrocellulose in 1.14 M glycine, 0.15 M tris-HCl over a period of 24-72 hrs (5). After transfer, the nitrocellulose filters were incubated for 30 min in 140 mM NaCl, 10 mM KPO_4, pH 7.5, 10 mg/ml BSA (Fraction V). The filters were then washed extensively with Tween-PBS (140 mM NaCl, 10 mM KPO_4, pH 7.5, 0.5% Tween 20). Anti-REP1 antiserum was diluted 1/100 in Tween-PBS and incubated with the nitrocellulose filter for 12 hrs in a Seal-a-Meal (Sears) bag. The filter was then washed three times with Tween-PBS, followed by incubation for 2 hrs at 37° with calf alkaline-phosphatase-conjugated goat anti-rabbit IgG diluted in Tween-PBS. Goat IgG fractions were glutaraldehyde conjugated with calf alkaline phosphatase as described (20) and were used at approximately 1:40,000 dilution. After washing the filters, immobilized alkaline phosphatase was visualized using a histochemical stain (21). The filters were first rinsed briefly in 50 mM Na-glycinate pH 9.6 and then incubated at room temperature in 50 mM sodium glycinate, pH 9.6, 0.1 mg/ml nitro blue tetrazolium, 0.05 mg/ml indoxylphosphate, 4 mM $MgCl_2$. Color reaction was allowed to develop at room temperature to the desired intensity and the reagents then removed by washing with Tween-PBS.

N-Terminal Analysis of REP1 Protein.

$[^{35}S]$-methionine-labeled REP1 protein was isolated from a 2 ml culture of strain DC04 [pSS4] grown at 30° to mid exponential phase (10^7 cells/ml) in SG minus leucine minus methionine and labeled for 50 min with 200 μCi $[^{35}S]$ methionine in the same medium. Cells were harvested, washed with 1 M sorbitol, resuspended in 1 ml 1 M sorbitol and converted to spheroplasts by incubation with 1% glusulase. Spheroplasts were harvested, washed, and lysed in 50 μl 1% SDS, 10 mM Tris-Cl, pH 8, 1 mM EDTA, heated for 2 min at 37°, and diluted with 0.5 ml 2% Triton X-100, 50 mM Tris-Cl, pH 8, 150 mM NaCl, 1 mM EDTA. Cell debris was removed by centrifugation. Preimmune or anti-REP1 antiserum (30 μl) was added to 200 μl supernatant and incubated overnight at 0°. S. aureus cells (100 µl of a 10% suspension) were added and incubation continued for 15 min. The immunocolplex was pelleted and washed three times with 1% Triton X-100, 50 mM Tris-Cl, pH 7.5, and 1 M NaCl. The pellet was resuspended in 50 µl 2% SDS, 80 mM Tris-Cl, pH 8, 9 M urea, 5% β-mercaptoethanol and boiled for 2 min. The supernatant was clarified by centrifugation and fractionated by electrophoresis on an 12.5% polyacrylamide gel. The gel was then dried and autoradiographed. That portion of the gel containing labeled REP1 protein was excised and the protein electroeluted with an ISCO sample concentrator (Model 1750) as described (22). A total of 18,000 cpm REP1 protein and 100 µg myoglobin in 240 µl 10 mM ammonium bicarbonate, 0.2% SDS was applied to a Beckman 890C protein sequencer. Sequence analysis was performed as described (23).

Preparation of the Yeast Nuclear Protein Matrix.

The protocol for fractionation of yeast nuclei and preparation of a subnuclear protein matrix is diagrammed in Figure 4. All of the procedures were conducted at 0°, except as indicated, and all volumes refer to the original wet volume of yeast cells used in the preparation (5 X 10^9 cells equals approximately 1 ml). Centrifugations were for 10 min at 12,000 g unless otherwise noted. Cells were resuspended in 2 volumes buffer consisting of 1 M sorbitol, 50 mM potassium phosphate, pH 7.8, 10 mM magnesium chloride, 1 mM PMSF, 2 µg/ml pepstatin A. Cells were then pelleted by centrifugation and resuspended in 2 volumes buffer consisting of 1 M sorbitol, 5 mM magnesium chloride, 2 mM DTT, 1 mM PMSF, 25 mM potassium phosphate, pH 7.8, 25 mM sodium succinate, pH 5.5, and 2 µg/ml pepstatin A. Zymolyase 60,000 was added at 1.5 mg/ml and the cells were digested at 37° for 10 to 15 min, until greater than 95% of the cells were converted to spheroplasts. Spheroplasts were collected by centrifugation at 1000 g for 5 min. The supernatant was clarified by centrifugation and retained as the Spheroplast Supernatant (SpS). The spheroplasts were resuspended in 4.75 volumes extraction buffer consisting of 50 mM Tris-Cl, pH 7.5, 50 mM NaCl, 5 mM magnesium chloride, 1 mM PMSF, 2 µg/ml pepstatin A. Spheroplasts were lysed by addition of 0.25 volumes 20% Triton X-100 followed by ten strokes with a tight pestle Dounce homogenizer to yield a Crude Homogenate (CH). The

CH was centrifuged to yield a Post Nuclear Supernatant (PNS) and a pellet, which was washed twice with extraction buffer to generate two Wash Supernatants (WS1 and WS2) and a Nuclear Pellet (NP).

The Nuclear Pellet was resuspended in 1 volume 20 mM Tris-Cl, pH 7.5, 5 mM magnesium chloride, containing 10 µg/ml DNAse I and 8 µg/ml RNAse and incubated at 37° for 15 min. The Digested Nuclei (DN) were collected by centrifugation and the Nuclease Supernatant (NS) retained. The pellet was resuspended in 0.9 volumes 290 mM sucrose, 10 mM Tris-Cl, pH 7.5, and 0.1 mM magnesium chloride, to which was added 0.1 volume 20% Triton X-100. After incubation at 0° for 10 min, the suspension was centrifuged and the supernatant (TXS) was removed. The pellet was resuspended in 0.5 volumes 100 mM Tris-Cl, pH 7.5, 290 mM sucrose, 0.1 mM magnesium chloride, and 0.5 volumes 2 M NaCl were added. After 10 min at 0°, the suspension was centrifuged and the supernatant (SS1) removed. The salt extraction was repeated to yield a second salt wash supernatant (SS2) and a final pellet, designated the subnuclear fraction (SNF).

Biochemical compositions of subnuclear fractions were determined as follows. Protein analysis was performed as described (24). Nucleic acid compositions were determined after solubilizing samples in 1% SDS and precipitating nucleic acids with cold 5% TCA. After washing the precipitates several times with 5% TCA, they were resolubilized in 0.1 M dibasic sodium phosphate, 0.1% SDS, and incubated in 0.3 M NaOH at 60° for 60 min. The samples were then mixed with cold 5% TCA and centrifuged. RNA contents were calculated from the A_{260} values of the supernatants. The pellets were extracted with 5% TCA at 90° for 20 min and centrifuged. DNA contents were calculated from the A_{260} values of the supernatants. Absorbance units were converted to µg nucleic acid assuming an average nucleotide molecular weight of 330 and an average molar extinction coefficient of 10^4 at 260 nm.

Electron Microscopy

Spheroplasts and subnuclear fractions were fixed either in suspension or as pellets in 3.1% glutaraldehyde, 4.1% paraformaldehyde, 60 mM sodium cacodylate, pH 7.5. Samples were postfixed in 2% osmium tetraoxide in buffer as

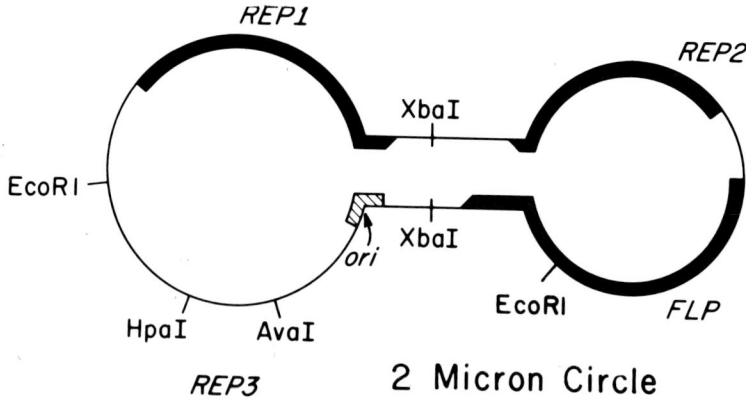

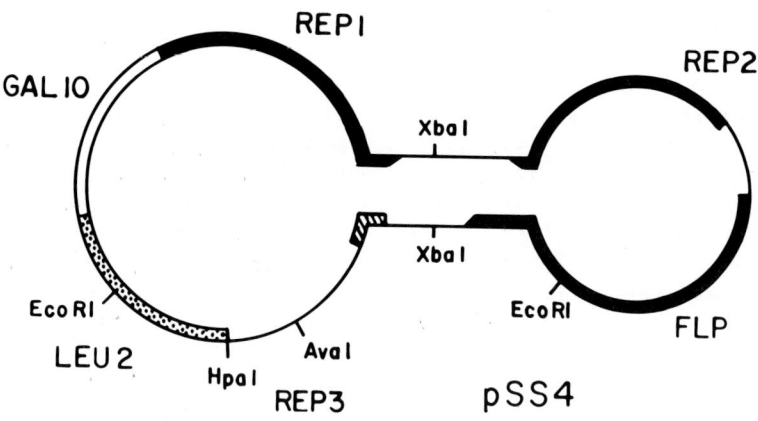

FIGURE 1. Structure of 2 Micron Circle and Plasmid pSS4. On schematic diagrams of the genomes of 2 micron circle and its derivative plasmid, pSS4, are indicated the positions of major plasmid genes (filled lines, tapers at the 3' end), the origin of replication (hatched lines), the GAL10 promoter (open double lines), the chromosomal sequences spanning LEU2 (mottled line), and a number of restriction sites. Plasmid pSS4 is identical to 2 micron circle except that the region between the HpaI site and the beginning of the REP1 gene of 2 micron circle has been replaced in pSS4 by the GAL10 promoter, which is situated to provide sole expression of the REP1 coding region, and the yeast LEU2 gene.

above and then dehydrated in alcohol, embedded in Epon 812, and sectioned onto uncoated copper grids (300 mesh). Grids were strained with 5% aqueous uranyl acetate.

RESULTS

Components of the 2 Micron Circle Stability System

The multicopy yeast plasmid 2 micron circle can be viewed as a chromosomal replicon on which are superimposed plasmid encoded stability functions. Under steady state growth conditions the plasmid behaves essentially as a chromosomal replicon: each plasmid replicates once and only once during each cell cycle, using the same enzymatic machinery used to replicate chromosomal DNA (25-27). In spite of stringent control of its replication, though, the 2 micron circle and hybrid plasmid derived from it are more stable during mitotic growth than hybrid plasmids constructed from ARS fragments. Recent studies have defined the 2 micron circle encoded components that are necessary for stable plasmid propagation (28-29). This stability system, which is diagrammed in Figure 1, consists of two trans-active functions and two cis-acting sites. The trans-acting functions correspond to two proteins encoded by plasmid genes designated REP1 and REP2. One of the cis-acting functions is the replication origin of the plasmid (28). The second cis- acting site, REP3, is located several hundred base pairs away from the origin.

Identification of REP1 Protein.

In order to obtain direct evidence to assess the role of the REP system in plasmid stability, we have undertaken a program to purify and characterize the protein products of the REP loci. As an initial step in this process, we used gene fusion technology to obtain antisera against the REP1 protein. That is, we raised polyclonal anti-REP1 antibodies by injecting rabbits with a hybrid protein derived from a fusion of the 5' two-thirds of the REP1 gene to the E. coli lacZ gene (15). In addition, as a means of regulating expression of REP1 and overproducing REP1 protein, we constructed plasmid pSS4, the structure of which is diagrammed in Figure 1. This plasmid is identical to authentic 2 micron circle except that the REP1 promoter is replaced by the GAL10 promoter. The GAL10 promoter is

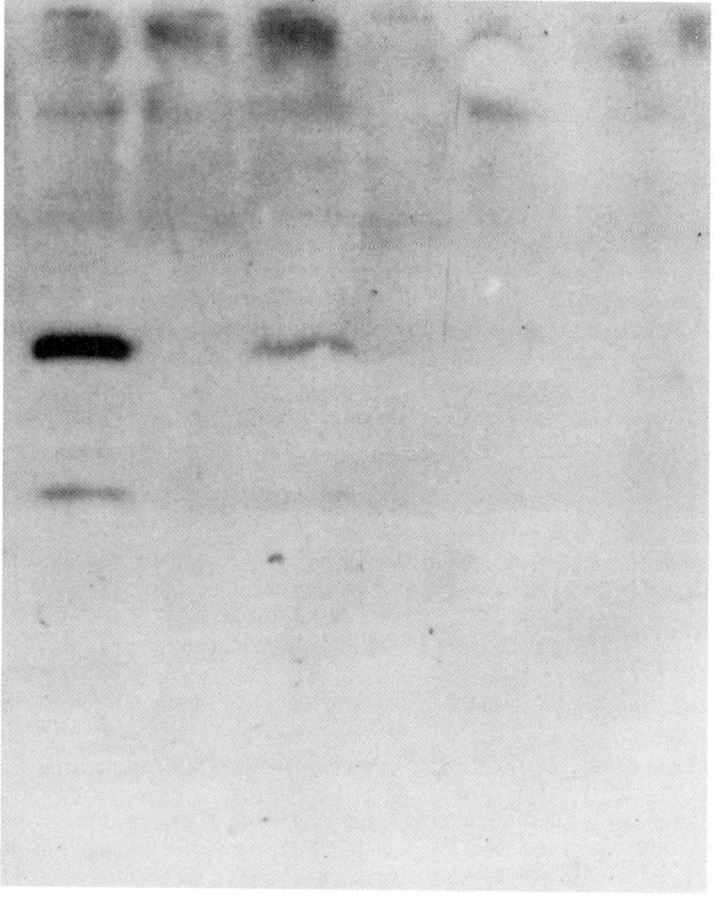

FIGURE 2. Immunoblot Analysis of REP1 Protein. Samples (20 μg) of extracts of strain DC04 [pSS4], DC04 [cir$^+$], or DC04 [cir^o], grown either on glucose or on galactose, were fractionated on a 12.5% polyacrylamide gel and then transfered to nitrocellulose. The filter was then probed with anti-REP1 antibody and the position of the resulting immobilized antibody was visualized as described in Materials and Methods.

quite active in strains grown on galactose but almost completely inactive in strains grown on glucose. The plasmid also carries the yeast LEU2 gene in place of the D coding region of 2 micron circle. Consistent with previous genetic analysis of REP1 function, plasmid pSS4 exhibits high stability when the strain harboring it is grown on galactose but low stability when the strain is grown on glucose (Jayaram and Broach, unpublished observations).

We have used these reagents to identify REP1 protein in extracts of yeast cells. In Figure 2 we present an immunoblot analysis of extracts of strain DC04 containing either no plasmid, authentic 2 micron circle plasmid, or plasmid pSS4. The extracts were fractionated on an SDS polyacrylamide gel, transfered to nitrocellulose and probed with anti-REP1 antibody. As is evident, the antibody binds to a protein that is present in extracts of the [cir$^+$] strain but absent in extracts of the [cir^0] strain. In addition, the antibody binds to a protein of the same molecular weight that is present in the galactose grown, but not glucose grown, strain containing plasmid pSS4. We, therefore, conclude that this protein is the product of the REP1 gene.

The 48,000 dalton molecular weight of this protein, calculated from comparison of its migration to that of cofractionated strandards, agrees closely with the molecular weight of the product of the REP1 gene predicted from the nucleotide sequence of the gene. The smaller molecular weight minor band present in lanes 1 and 3 is immunologically related to the REP1 protein (data not shown) and, thus, most likely represents proteolytic breakdown of the mature protein. The reduced levels of REP1 protein in glucose grown [cir$^+$] cells is puzzling, albeit reproducible. We do not currently know what step in the synthesis or degradation of REP1 protein is sensitive to carbon source variation.

To confirm the identification of the REP1 protein, we performed partial sequence analysis of the product recognized by the anti-REP1 antibodies. Yeast cells containing plasmid pSS4 were grown on galactose and labeled with [^{35}S] methionine. After lysis of the cells, the 48,000 dalton protein was immunoprecipitated with the anti-REP1 antiserum and fractionated on SDS polyacrylamide gel. The purified protein was then subjected to automated sequential Edman degradation and the liberated product of each cycle tested for the presence of label. The results of this analysis are presented in Figure 3. As is evident signifi-

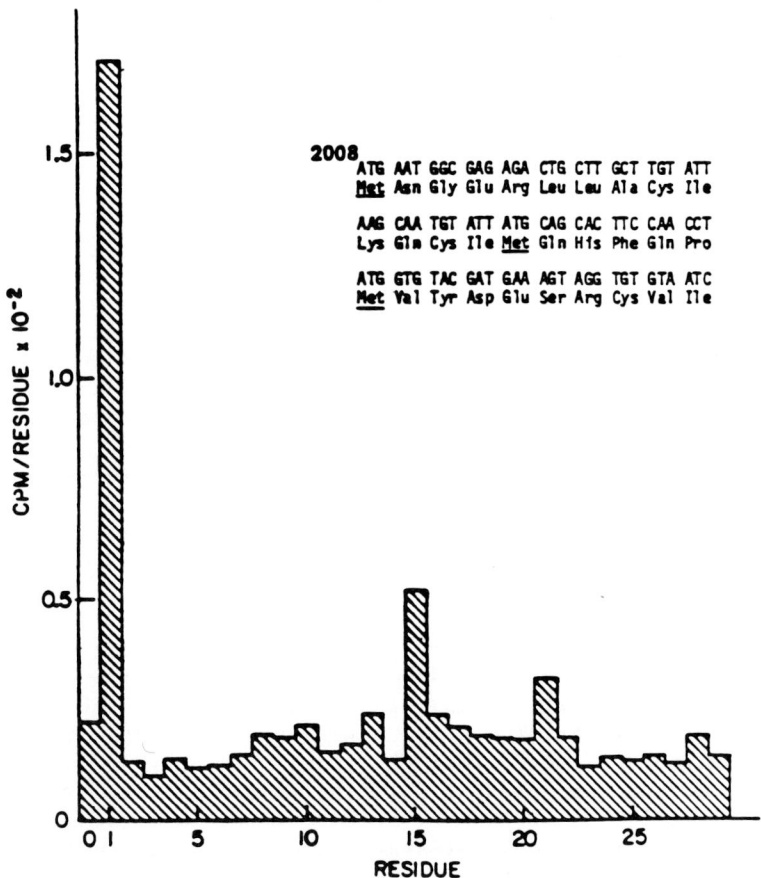

FIGURE 3. Partial Sequence Analysis of REP1 Protein. [^{35}S] methionine labeled REP1 protein was isolated by immunoprecipitation from strain DC04 [pSS4] and subjected to automated sequential Edman degradation as described in Materials and Methods. The amount of label liberated at each cycle is plotted against the cycle number. The nucleotide sequence of the REP1 gene, taken from Hartley and Donelson (33), extending from the initial ATG codon of the extended open reading frame is present in the insert. The predicted amino acid sequence of the first thirty residues is shown, with the methionine residues underlined.

cant amounts of label are liberated in the first, fifteenth, and twenty-first cycles. This agrees precisely with the positions of methionine residues predicted from the nucleotide sequence of the REP1 gene. This results confirms that the 48,000 dalton protein recognized by our anti-REP1 antiserum is the product of the REP1 gene.

The yield of label from sequential degradation of the immunoprecitited protein is substantially less than expected. In the experiment shown, the yield of labeled methionine in the first cycle was only 10% of that predicted form the amount of labeled protein applied. In a separate experiment, the yield was less than 5%. We conclude from this that most of the protein is blocked at the amino end, presumably by acetylation of the N-terminal methionine. This is consistent with the observation by Sherman and Stewart (23) that a modified cytochrome c protein with asparagine as the second residue is acetylated in yeast at the N-terminal methionine. The low repetative yield in the sequential degradation of REP1 protein evident in Figure 3 can be explained by the refractory nature of asn-gly dipeptide bonds to Edman degradation.

REP1 Protein Copurifies with the Yeast Nuclear Matrix.

In preliminary experiments designed to develop a purification protocol for REP1 protein, we have found that REP1 protein is localized in yeast cells to the nucleus (unpublished observations). We demonstrated this both by subcellular fractionation and by indirect immunofluorescence. In addition, by applying to yeast nuclei the fractionation procedure developed by Coffee and his colleagues for isolation of nuclear matrix from mouse L cells (1), we found that a significant proportion of REP1 protein copurified with the yeast nuclear fraction equivalent to the mouse nuclear matrix. However, since the yield of nuclei from yeast is both low and variable, we could not quantitate the fraction of REP1 localized to the matrix or maintain an accounting of the disposition of macromolecules during the fractionation procedure. In addition, residual percoll in gradient-purified nuclei foiled our attempts to obtain reasonable electron mircographs of the products of the fractionation procedure. Therefore, we developed a subnuclear fractionation procedure that circumvented these problems.

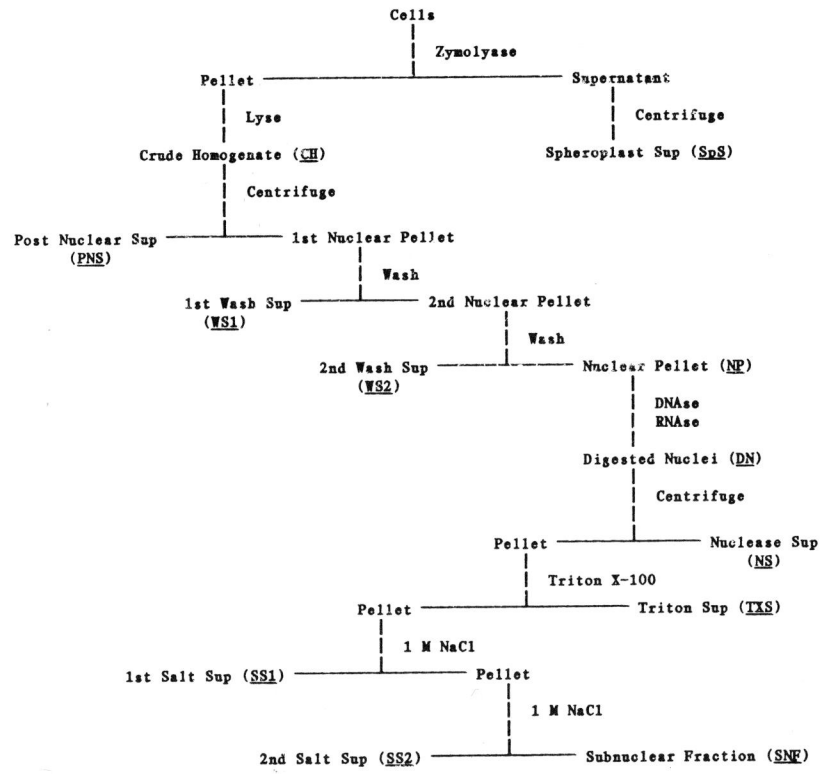

FIGURE 4. Flow Sheet for Subnuclear Fractionation of Yeast. The protocol for subnuclear fractionation of yeast is presented schematically in this diagram. The details of the fractionation are described in Materials and Methods and in the text. The protein patterns of samples from one such fractionation for those fractions designated with underlined abbreviations are shown in Figure 5.

Our protocol for isolation and subfraction of yeast nuclei is diagrammed in Figure 4. The protein, DNA, and RNA content in each fraction of the isolation procedure is presented in Table I. In addition, the profile of proteins in each fraction, following their fractionation by SDS polyacrylamide gel electrophoresis, is shown in Figure 5A. In the example shown we started with cells of strain DC04 carrying plamid pSS4 that were grown in galactose medium.
Cells are converted to spheroplasts and then lysed with Triton X-100 and mechanical shear in hypotonic buffer.

TABLE 1
DISPOSITION OF MACROMOLECULAR CONSTITUENTS
DURING PREPARATION OF THE YEAST NUCLEAR MATRIX

Fraction[1]	Protein (mg)	DNA (µg)	RNA (µg)
SpS	64	56	345
CH	50	347	1730
PNS	46	73	1470
WS1	0.0	48	157
WS2	0.0	0	0
NP	3.0	335	225
DN	2.0	165	134
NS	0.12	0	0
TXS	0.28	0	36
SS1	0.72	150	63
SS2	0.0	16	110
SNF	1.2	60	78

[1]Fraction designations are the same as those used in Figure 4.

This results in lysis of the cell, disruption of the cytoskeleton, and lysis of the most internal organelles, such as mitochondria, vacuoles, and vesicles. The residual pelletable material consists primarily of nuclei, contaminated with some cell wall debris. As shown in Table I, this nuclear pellet fraction represents less than 3% of the total cellular protein but greater than 90% of the cellular DNA. Note that the sample loaded on the SDS-polyacrylamide gel from the nuclear pellet fraction, as well as those from subsequent steps, represents ten times the cellular starting material than those from previous steps.

The nuclear pellet material is subsequently fractionated by sequential extractions. First the nuclear pellet is treated with DNAse and RNAse. Since much of the material remains as nucleoprotein complexes too large to escape from the nuclei, little material is released at this stage into the nuclear supernatant. The sample is then treated with 2% Triton-X100 to remove any residual membranes and membrane associated proteins. Finally, the pellet is ex-

336 Wu, Fisher, and Broach

A.

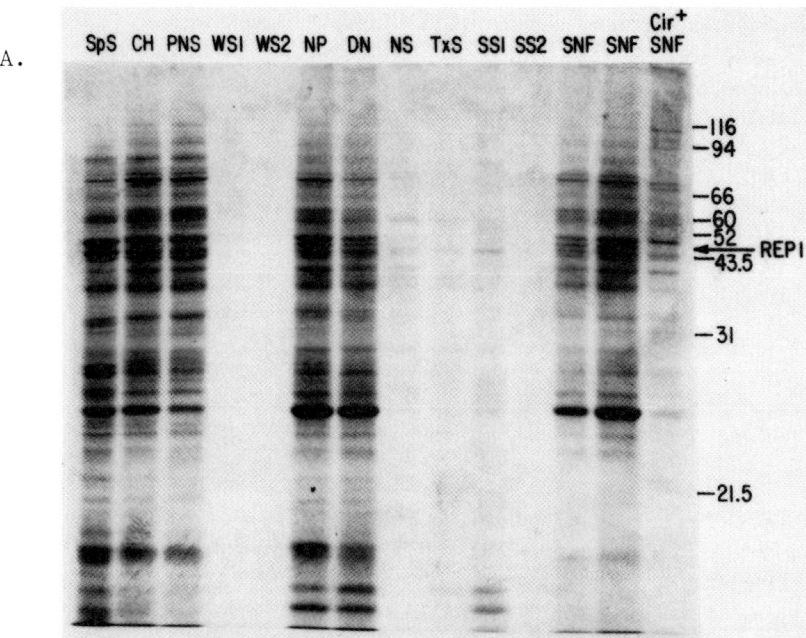

B.

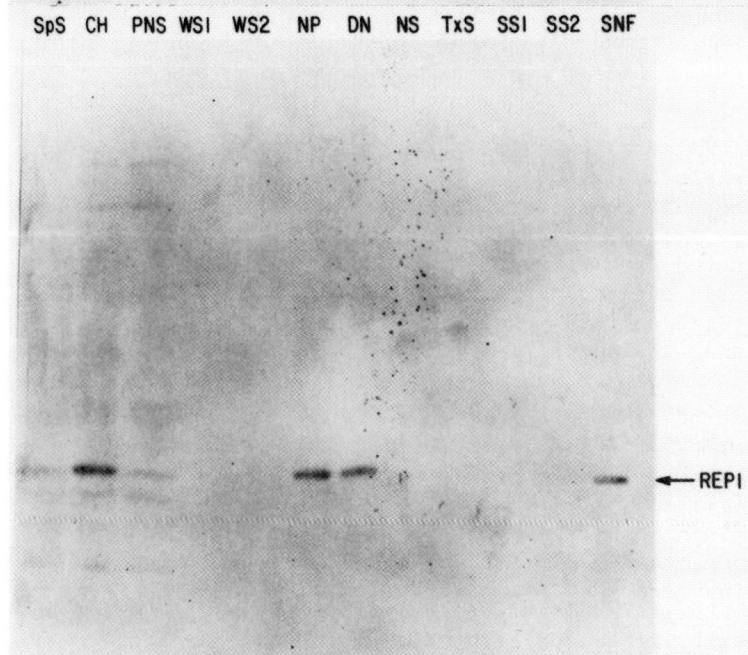

FIGURE 5. Distribution of Proteins during Subcellular Fractionation of Yeast. Subcellular fractionation of strain DC04 [pSS4] grown on SG minus leucine was performed as diagrammed in Figure 4. Samples of each fraction were subjected to SDS-PAGE and either stained with Coomassie Blue (upper panel) or transfered to nitrocellulose and treated with anti-REP1 antibody as described in Materials and Methods (lower panel). Sample designations are the same as those used in Figure 4. For lanes SpS, CH, PNS, WS1, and WS2, 2U of material were loaded on the gel. For lanes NP, DN, TXS, SS1, SS2, Cir [+] SNF, and the first lane of SNF, 10U material were loaded. For the second lane of SNF, 40U were loaded. 1U represents the amount of material derived from 1 µl wet volume yeast cells.

tracted with 1 M NaCl. This releases most of the residual DNA and greater than 90% of the histones as well as a number of chromatin proteins, as evident from the gel fractionation pattern of proteins from the relevant samples. The final subnuclear fraction constitutes 1% of the initial cellular protein and 15% of the initial cellular DNA and consists of 90% protein, 5% DNA, and 5% RNA.

For purposes of comparison, the protein profile of the subnuclear fraction isolated from strain DC04 [cir+] is also shown in Figure 5A. As is evident, the protein profiles of the SNF from the two strains are essentially identical, indicating that the procedure reproducibly yields the same subset of proteins. The only noteworthy distinctions between the two patterns is the increased intensity of two protein bands in the fraction from the pSS4 containing strain: one of unknown identify of approximately 25 Kd molecular weight and one of 48 Kd molecular weight, which we assume to be REP1 protein.

Electron micrographs of sections of whole yeast and of the material present in the subnuclear fraction obtained as described above are presented in Figure 6. The micrographs of the subnuclear fraction are shown at a four fold greater magnification than the whole yeast cells. The structures in B and C are quite prevalent in the subnuclear fraction and are the only discernable structures present. They are generally the same size and shape as yeast nuclei. As is evident, these structures appear amorphously granulated with some aggregation of the granular material. In some of them, a denser body can be seen within the struc-

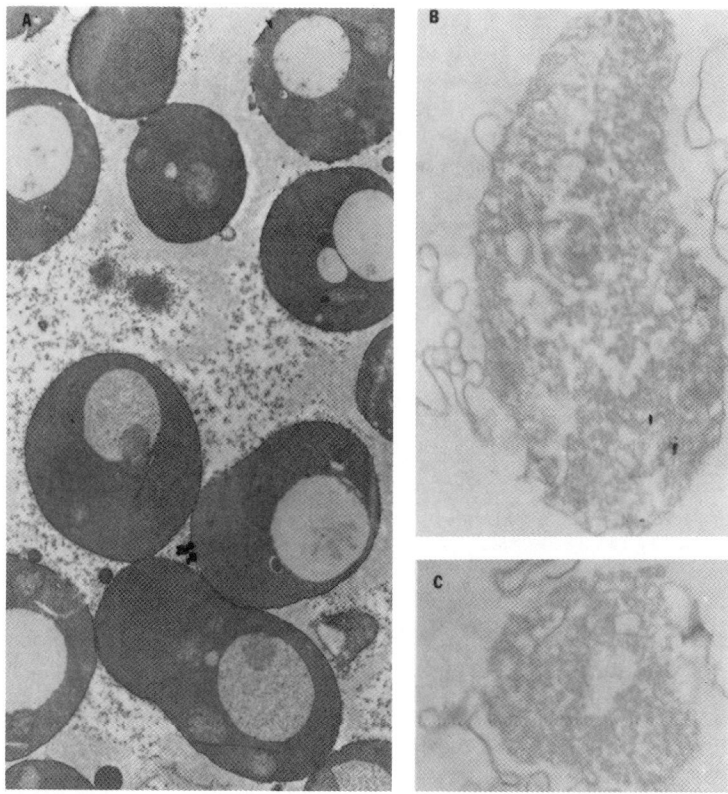

FIGURE 6. Morphological characterization of the yeast subnuclear fraction. Electron micrographs at 10,000 X magnification of thin sections of spheroplasted yeast cells are shown on the left and electron micrographs at 40,000 X magnification of thin sections of two examples of predominant structures present in the subnuclear fraction of yeast are shown on the right.

ture, perhaps representing the residual nucleolus. In addition, like the ones shown in these micrographs, these subnuclear structures are always found surrounded by patches of wavy, laminar-like material. It is not clear whether this material is derived from the original cell wall or represents the nuclear lamin, somewhat peeled away from the matrix.

We examined the partitioning of REP1 protein during subcellular fractionation of yeast by performing immunoblot analysis on SDS-polyacrylamide gel fractionated samples of each of the fractions in the preparation. The results of such an analysis are shown in Figure 5B. As is evident, REP1 protein is present in the crude homogenate and partitions almost exclusively to the nuclear pellet. It is not liberated by treatment with nucleases, non-ionic detergents, or salt, and thus is retained in the subnuclear fraction, or nuclear matrix. As mentioned above, retention of REP1 protein in the nuclear matrix is the same result we observed using Coffee's fractionation procedure starting from purified yeast nuclei. We can further conclude from the analysis present here that under these conditions of expression, REP1 protein is located exclusively in the nuclear matrix fraction. In similar fractionations with [cir$^+$] strains, however, we observe that although all the REP1 protein is nuclear limited, 30-40% of it is salt-extractable from the nuclear pellet. The reason for this difference in behavior of the protein in the two strains is nor clear.

DISCUSSION

Results from several recent studies suggest that 2 micron circle propagates stably because plasmid molecules partition reasonably equally between mother and daughter cells following mitosis, because 2 micron circle has the potential to amplify, and because cellular copy levels of the plasmid are actively maintained. We have shown recently that the proteins encoded by the REP1 and REP2 genes of 2 micron circle, in conjunction with a cis-active site designated REP3, are necessary for stable propagation of 2 micron circle (28,31). Murray and Szostak have demonstrated that hybrid 2 micron circle plasmids lacking a complete REP system exhibit, in a manner similar to that observed with hybrid plasmids constructed from chromosomal ARS elements, a high level of mitotic non-disjunction (32). Specifically, plasmid-bearing cells - even those that apparently contain a large number of plasmids - frequently fail to transmit plasmid molecules to their daughter progeny following mitosis. In other words, such hybrid 2 micron circle plasmids not only fail to partition evenly at mitosis but in fact fail to do so in a highly non-random fashion. In contrast, plasmids in which the 2 micron circle REP system is intact do not exhibit this highly asymmetric inheritance pattern but distribute themselves equitably between mother and daughter cells following mitosis. In addition, Kikuchi observed that the 2 micron circle REP system could enhance the stability of a plasmid without necessarily increasing its total copy level within the population (28). These results taken together suggest that the REP system actively insures equal distribution of plasmid molecules to mother and daughter cells following mitosis.

As an approach to elucidating the mechanism by which the REP proteins promote equipartitioning, we have begun a biochemical analysis of the properties of the REP proteins. In this report we have used anti-REP1 antibodies to identify the in vivo product of the REP1 gene. We find that the REP1 protein has a molecular weight of 48,000 daltons and is translated from the first methionine codon in the REP1 open reading frame. The close agreement between the measured and predicted molecular weights suggests that the protein undergoes little post-translation modification. However, it does appear that the protein has a blocked amino terminus, most likely in the form of an acetylation of the terminal methionine.

We have shown in this study that extraction procedures similar to those used to obtain a nuclear matrix fraction from nuclei of higher cells yield a consistent product when applied to yeast nuclei. This product possesses morphological and compositional features similar to nuclear matrix fractions from other eukaryotic systems. We have also determined that a significant proportion of the REP1 protein cofractionates with the yeast nuclear matrix fraction. In other systems, the nuclear matrix has been implicated as playing a role in DNA replication and RNA synthesis and maturation (1,10-14). The extent to these studies can be related to the yeast cell or even the extent to which the yeast matrix fraction can be considered a monolithic structure awaits further analysis. However, our results clearly indicate that at least one of the components of the partitioning system for 2 micron circle is in a reasonably insoluble complex within the nucleus and is likely associated with the matrix. These observations not only suggest a role for the matrix in promoting genetic segregation but also provide an identification of a matrix component whose activity and function have been genetically defined. Thus, these results offer the possibility of applying genetic techniques available in yeast to address the essentiality of the nuclear matrix in various cellular processes.

Given the genetic analysis of 2 micron circle stability and our results on the fractionation of REP1 protein, we can envision at least two models by which the REP proteins act to insure equi-partitioning of the plasmid at meiosis and mitosis. First, the REP proteins could function as a bivalent connector, promoting attachment of the plasmid, via the REP3 locus, to some component, such as the spindle apparatus, that actively segregates to the two cells at mitosis. On the other hand, the highly nonrandom segregation of rep plasmids suggests that such plasmids are not free to diffuse throughout the nucleus following S phase but rather are stuck or sequestered at some site that remains predominantly with the mother cell at mitosis. Thus, the REP proteins may act merely to pry plasmid molecules loose from this site of sequestration in order that they might be free to distribute themselves randomly throughout the nucleus. Given the relatively high copy number of 2 micron circles in the cell, random distribution of plasmids in the nucleoplasm would almost invariably insure that both mother and daughter cells would acquire plasmids following mitosis.

REFERENCES

1. Berezney R, Coffey DS (1975). Nuclear protein matrix: association with newly synthesized DNA. Science 189:291.
2. Agutter PS, Richardson JCW (1980). Nuclear non-chromatin proteinaceous structures: Their roles in the organization and function of the interphase nucleus. J Cell Sci 44:395.
3. Capco DG, Wan KM, Penman S (1982). The nuclear matrix: Three dimensional architecture and protein composition. Cell 29:847.
4. Dwyer N, Blobel G (1976). A modified procedure for the isolation of a pore complex-lamina fraction from rat liver nuclei. J Cell Biol 70:581.
5. Fisher PA, Berrios M, and Blobed G (1982). Isolation and characterization of a proteinaceous subnuclear fraction composed of nuclear matrix, peripheral lamina, and nuclear pore complexes from embryos of Drosophila melanogaster. J Cell Biol 92:674.

6. Wunderlich F, Herlan G (1977). A reversibly contractile nuclear matrix: its isolation, structure and compostion. J Cell Biol 73:271.
7. Agutter PS, Birchall K (1979). Functional differences between mammalian nuclear matrix and pore lamina preparations. Exp Cell Res 124:453.
8. Long BH, Huang CY, Pogo AO (1979). Isolation and characterization of the nuclear matrix in Friend erythroleukemia cells: Chromatin and hnRNA interactions with the nuclear matrix. Cell 18:1079.
9. Potashkin J, Huberman J (1984). Isolation and initial characterization of residual nuclear structures from yeast. Exp Cell Res
10. Pardoll DM, Vogelstein B, Coffey DS (1980). A fixed site of DNA replication in eukaryotic cells. Cell 19:527.
11. Robinson SI, Nilkin BD, Vogelstein B (1982). The ovalbumin gene is associated with the nuclear matrix of chicken oviduct cells. Cell 28:99.
12. Valenzuela MS, Mueller GC, Dasgupta S (1983) Nuclear matrix-DNA complex resulting from EcoR1 digestion of HeLa nucleoids is enriched for DNA replicating forks Nuc Acids Res.
13. Mariman ECM, van Eekelen CAG, Reinders R, Berns AJM, van Venrooij WJ (1982). Adenoviral heterogeneous nuclear RNA is associated with the host nuclear matrix during splicing. J Mol Biol 154:103
14. Wunderlich F, Giese G, Bucherer C (1978). Expansion and apparent fluidity decrease of nuclear membranes induced by low Ca/Mg. Modulation of nuclear membrane lipid fluidity by the membrane-associated muclear matrix protein? Exp Cell Res 111:479.
15. Broach JR, Li YY, Wu LCC, Jayaram M (1983). Vectors for high level, inducible expression of cloned genes in yeast. In M Inouye (ed): 'Experimental Manipulation of Gene Expression,' New York: Academic Press, p 83.
16. Sherman F, Fink GR, Hicks JB (1983) 'Yeast Genetics Laboratory Manual' Cold Spring Harbor, NY: Cold Spring Harbor Laboratory Press.
17. Towbin H, Staehelin T, Gordon J (1979). Electrophoretic transfer of proteins from polyacrylamide gels to nitrocellulose sheets: procedures and some applications. Proc Nat Acad Sci, USA 76:4350.

18. Renart J, Reisen J, Stark GR (1979). Transfer of proteins from gels to diazobenzyloxymethyl paper and detection with antisera: a method for studying antibody specificity and antigen structure. Proc Nat Acad Sci, USA 76:3116.
19. Laemmli UK (1970). Cleavage of structural protein during the assembly of the head of bacteriophage T4. Nature 227:680.
20. Avrameas S (1969). Coupling of enzymes to proteins with gluteraldehyde. Use of the conjugates of the detection of antigens and antibodies. Immuno Chem 6:43.
21. McGadey J (1970). A tetrazolium method for non-specific alkaline phosphatase. Histochemie 23:180.
22. Bhown AS, Mole JE, Hunter F, Bennett JC (1980). High sensitivity sequence determination of proteins quantitatively recovered from sodium dodecyl sulfate gels using an improved electrodialysis procedure. Anal Biochem 103:184.
23. Anderson CW (1982). Partial sequence determination of metabolically labeled radiaoactive proteins and peptides. In Setlow J, Hillaender M (eds): 'Genetic Engineering, Vol 4' New York: Plenum Publishing Co, p 147.
24. Schaffner W, Weissman C (1973). A rapid, sensitive and specific method for the determination of protein in dilute solution. Anal Biochem 56:502.
25. Livingston DM, Kupfer DM (1977). Control of Saccharomyces cerevisiae. 2 micron DNA replication by cell division cycle genes that control nuclear DNA replication. J Mol Biol 116:249.
26. Petes TD, Williamson DH (1975). Replicating circluar DNA molecules in yeast. Cell 4:249.
27. Zakian VA, Brewer BJ, Fangman WL (1979). Replication of each copy of the yeast 2 micron DNA plasmid occurs during the S phase. Cell 17:923.
28. Jayaram M, Li YY, Broach JR (1983). The yeast plasmid 2 micron circle encodes components required for its high copy propagation. Cell 34:95.
29. Kikuchi Y (1983). Yeast plasmid requires a cis-acting locus and two plasmid proteins for its stable maintenance. Cell 35:487.
30. Sherman F, Stewart JW (1982). Mutations altering initiation of translation of yeast iso-1-cytochrome c: constrasts between eukaryotic and prokaryotic initiation process. In Strathern et al (eds): 'The

Molecular Biology of the Yeast Saccharomyces' Cold Spring Harbor, NY: Cold Spring Harbor Laboratory Press, p.301.
31. Jayaram M, Sutton A, Broach JR (1985). Properties of REP3: a cis-acting locus required for stable propagation of the yeast plasmid 2 micron circle. Mol Cell Biol, in press.
32. Murray AW, Szostak JW (1983). Pedigree analysis of plasmid segregation in yeast. Cell 34:961.
33. Hartley JL, Donelson JE (1980). Nucleotide sequence of the yeast plasmid. Nature 286:860.

YEAST AS A MODEL SYSTEM TO DISSECT THE RELATIONSHIP BETWEEN CHROMATIN STRUCTURE AND GENE EXPRESSION[1]

David S. Gross, Christopher Szent-Gyorgyi, and William T. Garrard

Department of Biochemistry
The University of Texas Health Science Center
Dallas, Texas 75235

ABSTRACT We discuss the utility of yeast for studying the chromatin structure of individual genes and provide as an example selected results from our studies on the heat-shock inducible gene, HSP82. Chromatin footprinting experiments reveal a remarkably organized series of chromatin structures. DNase I hypersensitive sites partition the locus into a variety of chromatin subdomains. Non-transcribed flanking regions appear to possess DNA sequence-positioned whole nucleosomes, while the HSP82 transcription unit itself possesses apparent half-nucleosomal structures. Genomic footprinting experiments suggest that adenine residues within and surrounding the TATA box are highly reactive in vivo to the chemical probe dimethyl sulfate, possibly reflecting an altered helical structure and/or tight binding of a regulatory protein(s).

INTRODUCTION

In this article we describe why yeast is an ideal system for addressing questions on chromatin structure-function relationships. In addition, we briefly review the basic features of yeast chromatin and describe a nuclei isolation procedure that overcomes many of the

---

[1]Supported by grants GM22201, GM29935, GM25829, GM31689 from the NIH and grant I-823 from The Robert A. Welch Foundation.

technical difficulties of earlier methods. Finally, we summarize selected results of our studies on an inducible yeast heat shock gene, HSP82, that indicate the presence of a highly organized chromatin structure to the 13.5 kb locus containing this gene. In this final section we also provide an overview of chromatin and genomic footprinting.

Virtues of Yeast

Saccharomyces cerevisiae is an attractive biological system for investigating the cis-acting DNA sequence and trans-acting factor determinants of chromatin structure, and for determining what role structure plays in dictating gene function. Several reasons exist for this:

Alteration of Cis-Acting DNA Sequences. Unique to eukaryotes, site-directed integration of virtually any DNA sequence is possible with yeast, through the use of positive or negative selection (1-3). Hence, one can alter the DNA sequence within or flanking any non-vital yeast gene in the context of its normal chromosomal environment and assess the impact that a given mutation has on chromatin structure and gene expression (57,58). This approach overcomes potential artifacts caused by foreign DNA sequences, chromosomal position effects, and altered copy number or replication timing.

Alteration of Trans-Acting Factors. Mutations in loci encoding trans-acting regulatory factors can be obtained using classical genetic techniques. Genes encoding such proteins are then amenable to isolation from existing libraries using the mutant host strains for positive selection. This scenario, coupled with in vitro mutagenesis and gene replacement, permits one to engineer alterations in regulatory protein structure, from single amino acid changes to domain shuffling (e.g., see ref. 59), and to assess effects on the chromatin structure and gene expression of the regulated target loci. This approach complements results obtained from altering cis-acting DNA sequences.

Mutant Collections. There exists a large repository of yeast mutants that can be exploited. For example, it would be readily possible to assess the chromatin structure of any specific gene at given stages of the cell cycle by employing CDC mutants (4). Mutations in DNA polymerases (J. Campbell, this volume) and DNA topoisomerases (5,6) have been described and may be useful in chromatin studies as well.

Episomes. The function of DNA sequence elements may also be conveniently examined on yeast plasmids, packaged as minichromosomes and stably maintained through mitotic and meiotic cell divisions at single (or low) copy number. This highly desirable property is due to the availability of vectors containing yeast centromeres (CEN) and replication origins (ARS) (7,8). This allows one to study the chromatin structure and regulation of wild-type or mutant genes in the episomal state, while minimizing the potential for titrating out trans-acting regulatory factors (a possible drawback with multicopy plasmids). Alternatively, if one wishes to isolate biochemical quantities of the chromatin of a particular gene, this is potentially feasible through use of multicopy (non-centromeric) yeast plasmids. Hence, yeast offers a versatility available in no other eukaryotic system.

Haploidy. Yeast can be grown in the haploid state, thus eliminating the uncertainty of potential differences in chromatin structure between alleles (e.g., allelic exclusion, gene dosage effects), and facilitating site-directed integration into a single locus.

Small Genome Size. The genome of yeast is quite small, only threefold that of E. coli, with a haploid size of $\sim 1.5 \times 10^7$ bp (8a). Therefore, it is relatively easy to achieve high resolution mapping of single copy sequences by employing nucleic acid hybridization techniques (see below).

Basics of Yeast Chromatin

Yeast nucleosomes contain a typical complement of core histones (two each of H2A, H2B, H3 and H4), but in marked contrast to other eukaryotes, apparently lack the linker histone H1 (9). Other features of yeast chromatin include: [1] A relatively short nucleosomal repeat length of ~ 165 bp (10,11); [2] Well organized spacing of adjacent nucleosomes with respect to turns and half-turns of the DNA helix (12,13); [3] Highly acetylated core histones (14); [4] The apparent absence of 5-methylcytosine in chromosomal DNA (15); [5] An equivalent sensitivity of bulk chromatin and transcriptionally active chromatin to DNase I (16), in contrast to higher eukaryotes where active chromatin is preferentially DNase I sensitive (17);

[6] Sites hypersensitive to DNase I digestion upstream of transcription units (18-24), or within other critical regulatory regions (18,25); [7] Sequence-specific positioning of nucleosomes (24-26); [8] The presence of certain HMG-like nonhistone chromosomal proteins (27); [9] An apparent lack of heterochromatin (28) and a low proportion of highly repetitive DNA sequences (29); and [10] The ability to form higher order structures of about 250 Å in diameter (30).

A critical difference between the chromatin of yeast and that of higher eukaryotes is the instability of isolated yeast mononucleosomes (31,32). This lability is also evident in reconstitution experiments (31) and may be related to the divergent primary sequence of yeast H3 (33,34), to the hyperacetylated state of yeast core histones (14), and to the apparent conformational flexibility of yeast nucleosomal DNA (60).

Saccharomyces has two different genes for each of the four core histones. Genes encoding H2A and H2B lie head-to-head at one locus, those encoding H3 and H4 lie in the same orientation at another locus, and both loci are duplicated (34). Due to evolutionary divergence, two different primary sequences exist for H2A and H2B; these histone variants differ by either two (H2A) or four (H2B) amino acids (35,36). The subtypes apparently encode no unique function since mutants containing just one copy of either H2A or H2B gene undergo a complete life cycle (37,38). Moreover, genetic experiments indicate that H2A subtypes can associate interchangeably in vivo with H2B subtypes (38). Deletions of up to 19 amino acids at the amino terminus of H2B are nonlethal in yeast lacking wild-type H2B genes, while comparable deletions at the carboxy terminus are lethal (39). An active area of current research is to elucidate the sequences and components necessary to target histones and other chromosomal proteins to the yeast nucleus (see related articles in this volume).

Isolation of Yeast Nuclei

Historically, progress in yeast chromatin research has been impeded by technical difficulties in the isolation of intact, clean nuclei. Problems include contamination with mitochondrial DNA (up to 20% of the

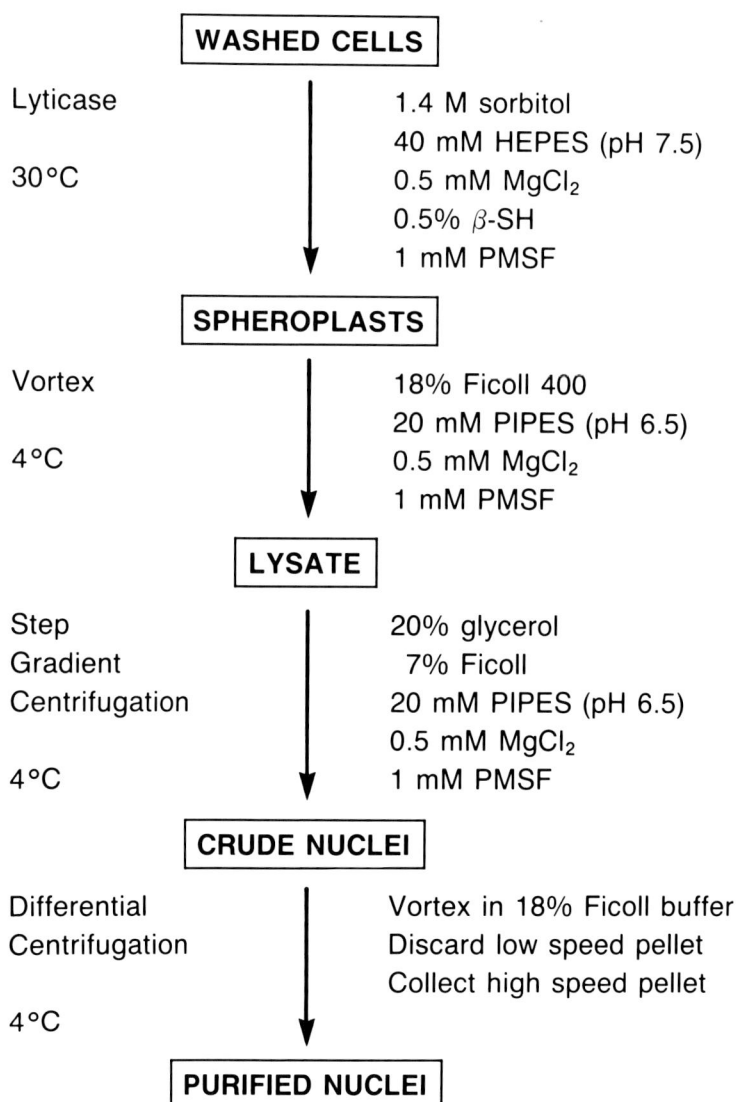

FIGURE 1. Yeast nuclear isolation protocol.

cellular DNA in ρ^+ yeast; ref. 40); co-isolation of attached membranes and ribosomes (RNA can exceed DNA by 100-fold in whole cells; ref. 41); and artifactual proteolysis and nucleolysis. We discuss below an improved method that obviates many of these difficulties.

Figure 1 shows the yeast nuclei isolation procedure that we employ, which is a modification of the method of Szent-Gyorgyi and Isenberg (32). Because of the necessity to remove cell walls enzymatically -- a step requiring prolonged incubation at elevated temperatures -- it is essential that chromatin structure be stabilized throughout this procedure. To eliminate contaminating nucleases and proteases, we purify lyticase, the enzyme used to make spheroplasts, according to the method of Scott and Scheckman (42). Spheroplasts are gently lysed by vortexing, and vacuoles and cytosol are carefully separated from nuclei and cell ghosts by step gradient centrifugation. The final vortexing step breaks nuclei free of cell ghosts and entrapped membranes (especially the abundant rough ER), and is followed by differential centrifugation. We include Mg^{2+} throughout to stabilize chromatin structures and avoid the use of detergents and polyamines that may modify chromatin integrity. By both cytological and biochemical criteria, this protocol yields reasonably clean nuclei that have proven to be an excellent substrate for chromatin footprinting (see below) or for the purification of other nuclear constituents (M. Douglas, personal communication).

CHROMATIN STRUCTURE OF THE HSP82 LOCUS

To understand the cis-acting DNA sequence determinants of transcriptionally poised chromatin structures in S. cerevisiae, we have chosen to study the HSP82 locus, which contains a gene that encodes an 82 kD heat-shock protein (the homologue of Drosophila hsp82). This gene has been cloned, sequenced, and the transcription unit delimited (43). HSP82 is nominally transcribed in the basal state; following heat shock at 39°C, a 14-fold increase in the steady-state level of HSP82 transcripts occurs within 10 min (Szent-Gyorgyi et al., in preparation).

We have employed the indirect end-labeling procedure (44,45) to map those regions of the HSP82 locus protected -- "footprinted" -- from nucleolytic cleavage when yeast nuclei are mildly digested with either DNase I or microccocal nuclease (MNase). This "chromatin footprinting" method is illustrated in Figure 2. Following isolation and mild digestion of nuclei, the DNA is purified, cleaved with a restriction enzyme (R1), electrophoretically resolved on a horizontal agarose gel, transferred to a nylon membrane and hybridized to a short (∼500 bp), ^{32}P-labeled complementary DNA fragment that abuts the R1 site. This effectively labels a nested

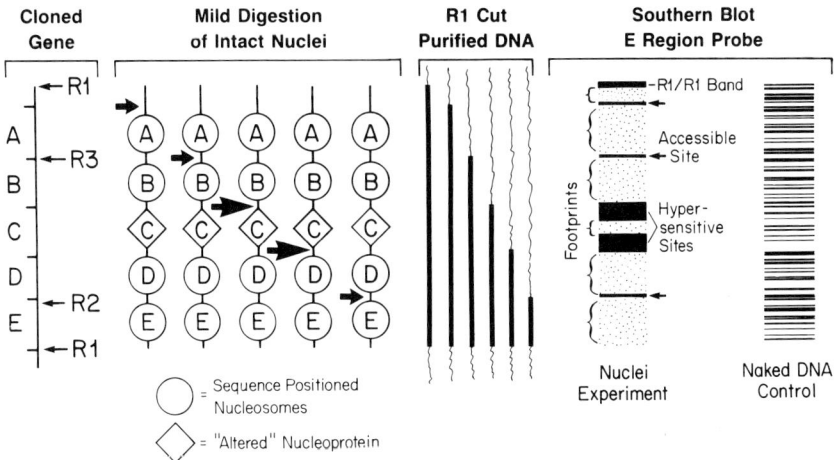

FIGURE 2. Chromatin footprinting technique. R1 restriction endonuclease cleavage of DNA purified from nuclease digested nuclei yields a distribution of fragments which may be analyzed by indirect end-labeling from either R1 site. Here, Southern blotted DNA is visualized by hybridization to the E Region Probe (R2-R1). The same blot can be reprobed (following radioisotopic decay or elution of the first probe) with the A Region Probe (R1-R3). "Altered" nucleoprotein may be a nucleosome with an altered conformation or a nucleosome-free DNA segment bound by a regulatory protein(s). R1/R1 band: the parental, undigested fragment. In the Southern Blot panel, accessible sites and footprints are indicated by arrows and brackets, respectively.

set of daughter fragments defined at one end by the double-stranded DNase I or MNase cut and at the other by the restriction site; nuclease break points ("accessible sites") are then mapped by determining subfragment sizes. Sites highly accessible to nuclease cleavage have been termed hypersensitive sites (44).

Figure 3 summarizes the overall chromatin structure of the HSP82 locus deduced from interpretation of the results of chromatin footprinting experiments to be presented elsewhere (Szent-Gyorgyi et al., in preparation). We find a characteristic set of DNase I hypersensitive sites, which demarcate chromatin subdomains that include DNA sequence-positioned nucleosomes on 5' and 3' flanks, as well as apparent half-nucleosomes within the transcription unit (see below). We emphasize that this chromatin structure exists both before or after heat-shock induction of transcription. Although the HSP82 gene is transcribed

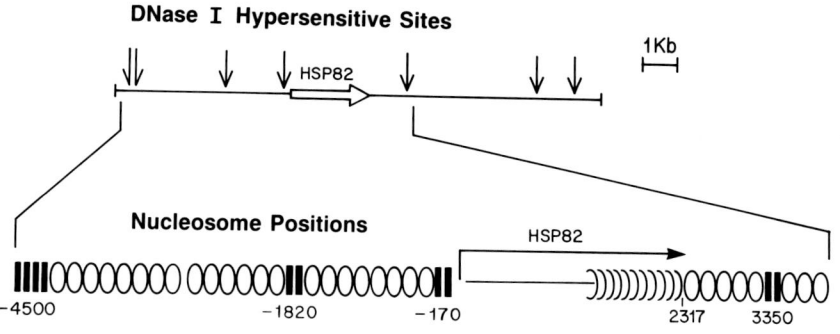

FIGURE 3. Chromatin structure of the yeast HSP82 locus. A schematic representation is illustrated and is based on our interpretation of the results of extensive chromatin footprinting experiments (Szent-Gyorgyi et al., in preparation). Nuclease hypersensitive sites (vertical arrows or double vertical bars), sequence-positioned whole (ovals) and apparent half (semi-ovals) nucleosomes, gaps, and regions of indeterminate structure (thin horizontal line) are indicated.

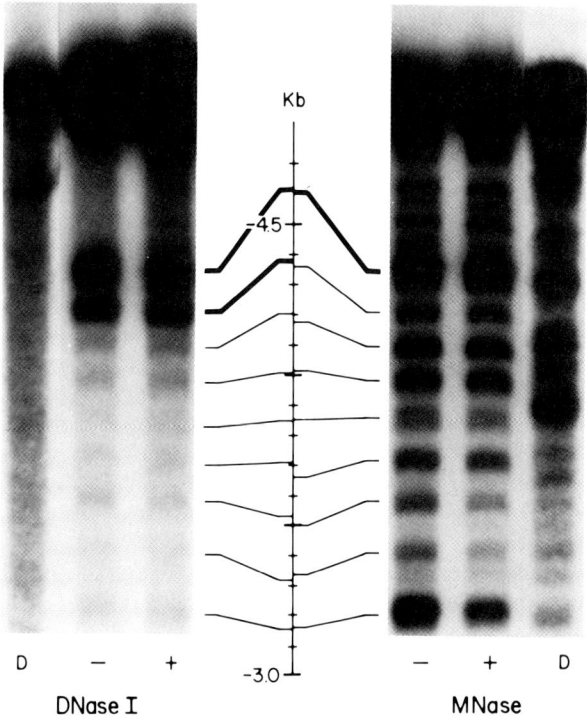

FIGURE 4. Chromatin footprinting a remote region upstream of the HSP82 gene. Nuclei were purified from haploid yeast (strain MY-55) and subjected to mild nuclease digestion. DNA was then purified, restricted, separated on a 2% agarose gel, transferred to Zeta-Probe nylon membrane and subjected to hybridization. The region from 3.1 to 4.6 kb upstream of the HSP82 transcriptional start site was footprinted as indicated by DNase I or MNase digestion of nuclei isolated from non-shocked (-) or heat-shocked (+) cells. Naked DNA controls (D) demonstrate that most sites of nucleolytic cleavage are chromatin-specific. The positions of accessible and hypersensitive sites relative to the transcriptional start site are indicated by thin and thick lines, respectively. (From Szent-Gyorgyi et al., in preparation.)

at a relatively high basal rate before induction, DNase I hypersensitive sites are present prior to transcriptional induction of many other yeast genes which do not exhibit any detectable basal level transcription (18,19,22,23). In addition, heat shock inducible genes of Drosophila possess DNase I hypersensitive sites prior to transcriptional induction (44).

Figure 4 illustrates a typical DNA sequence-positioned nucleosomal array obtained from chromatin footprinting a region far upstream of the HSP82 gene. Seven well-defined footprints between eight accessible sites exist in the MNase digests of nuclei. These nucleosomal footprints are spaced at ~170 bp intervals extending from 3.1 to 4.3 kb upstream of the HSP82 transcription unit. Very similar positions for accessible sites are seen in DNase I digests, but much more faintly. The bands generated by MNase digestion of nuclei are either chromatin specific or greatly enhanced in relative yield as compared to the control digest of naked DNA. The DNase I control also shows little sequence specificity in naked DNA cutting. We interpret these data as indicating that an array of seven DNA-sequence positioned nucleosomes exist 3.1 to 4.3 kb upstream of the HSP82 gene, as has been diagramatically summarized in Figure 3. In addition, two bands hypersensitive to DNase I digestion are centered 4.4 and 4.6 kb upstream of the HSP82 transcription unit (Figure 4). A strong closely spaced doublet band is observed at -4.6 kb for the MNase digest, but the DNase I band at -4.4 kb has only a weak MNase generated counterpart.

We have also examined the chromatin structure within and downstream of the HSP82 transcriptional unit. As shown in Figure 5, a series of five accessible DNase I sites, spaced 80-85 bp apart, define four chromatin footprints at the 3' terminus of the gene. In fact, there appear to be at least ten accessible sites spaced at half-nucleosomal intervals proceeding upstream from position +2.32 kb, the site of the putative transcriptional termination sequence (Szent-Gyorgyi et al., in preparation). The sites of nucleolytic cleavage are largely chromatin specific; nonetheless, we see some preferential cutting of naked DNA at a subset of these sites (see lanes "D"). The structural basis of the short repeat is unknown, but it may reflect altered nucleosomes

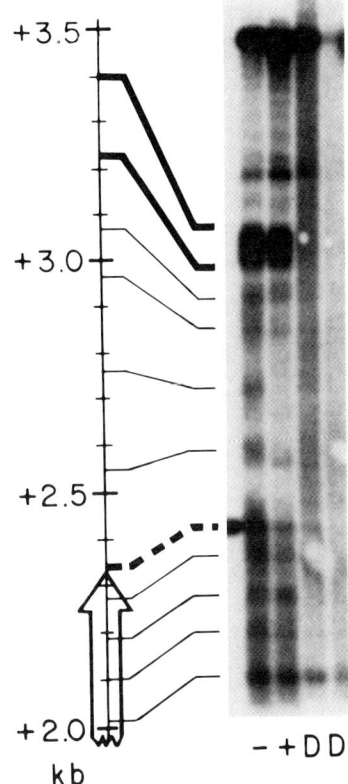

FIGURE 5. Chromatin footprinting of HSP82 transcribed sequences and downstream flank. Nuclear isolation, DNA purification, and blot hybridization were the same as in Figure 4. The region from +2.0 to +5.0 kb downstream of the HSP82 transcription start site was footprinted by DNase I digestion of nuclei obtained from non-shocked or heat-shocked cells. The open arrow indicates the transcriptional unit; other symbols are the same as in Figure 4. The position of an abrupt transition in chromatin structure, corresponding to the putative site of transcriptional termination, is indicated by a heavy broken line. (From Szent-Gyorgyi et al., in preparation.)

susceptible to DNase I cleavage along their dyad axes of symmetry. Whether the presence of apparent half-nucleosomes depends on transcription also remains to be determined.

An abrupt transition between half-nucleosome to whole nucleosome repeat occurs at the transcriptional termination region. As is clear from Figure 5, an array of five sequence-positioned nucleosomes, each ~165 bp in length, extends downstream from the termination sequence. Interestingly, upon heat shock, some chromatin accessible sites are either slightly (+2.55 kb) or strongly (+2.75 kb) protected from nucleolytic cleavage (these sequences are uniformly degraded in the deproteinized DNA controls; lanes "D"). Further downstream, between +3.25 to +3.4 kb, are sequences hypersensitive to DNase I cleavage in chromatin but not naked DNA; beyond this site an array of sequence-positioned nucleosomes, extending 3'-ward, is seen once again.

Although no functional association is known for these hypersensitive sites and DNA sequence-positioned apparent half and whole nucleosomes, we could determine by recombinant DNA technology which sequences within this locus dictate these chromatin structures, taking advantage of the elegant site-directed integration potential of the yeast system. Presently we are addressing related questions but in regions more crucial for HSP82 gene expression. We emphasize that to date only a few studies have taken advantage of the yeast system to address the role of cis-acting DNA sequence determinants on chromatin structure (18,25,26,46-49). Clearly this area remains to be exploited.

Of particular interest is the HSP82 promoter region, which contains the TATA box at position -79 and the heat shock consensus sequence (50) centered at position -173. We have mapped the DNase I double-strand cutting sites within this region to an accuracy of ±20 bp with respect to the underlying DNA sequence. Figure 6 reveals the presence of two broad hypersensitive regions separated by a relatively narrow nuclease insensitive subdomain that is possibly centered at -173. A smaller footprint exists near the TATA box region at position -79. These footprints presumably reflect the binding of regulatory proteins previously demonstrated to bind at analogous sites in the upstream regions of Drosophila heat shock genes (51,52). Sequences between the two footprints

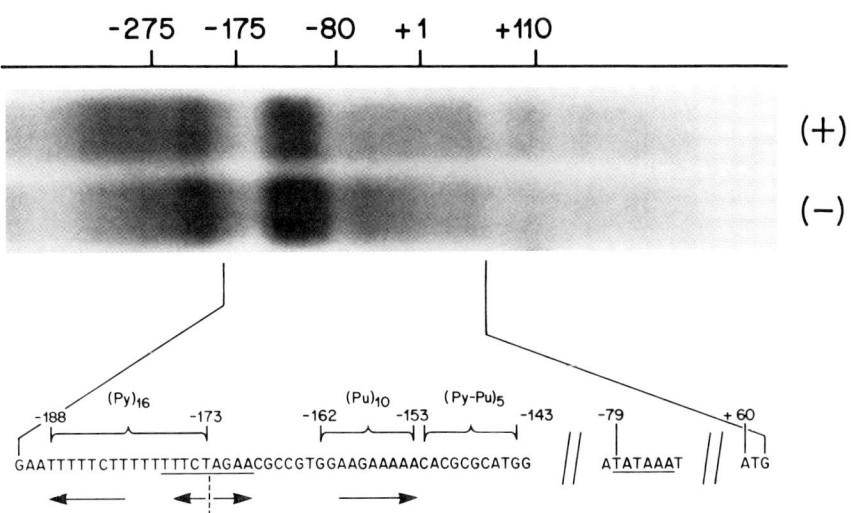

FIGURE 6. DNase I footprint analysis of the HSP82 promoter region. Nuclei from heat-shocked (+) and non-shocked (-) cells were subjected to mild DNase I digestion, and purified DNA was restricted, electrophoretically resolved, and indirectly end-labeled as described in the legend to Figure 4. The hybridization probe was located 1.6 kb downstream of the HSP82 initiation region. Key DNA sequences numbered relative to the transcriptional start site are indicated, including a pallindromic consensus heat shock sequence centered at position -173 that lies between two large inverted repeat elements, and the TATA box at nucleotide -79. (From Szent-Gyorgyi et al., in preparation.)

are especially sensitive to DNase I. The fine structure observed 3' to the transcription start is of unknown significance. The absence of large scale changes in this region upon heat shock parallels observations on hypersensitive sites in the promoter regions of other inducible yeast genes (19-23).

As the resolution of the above analysis is ±20 bp, it is not possible to identify individual nucleotides that are actually in contact with putative regulatory proteins. We therefore have used a modification of the in vivo genomic footprinting technique of Church and Gilbert (53) to achieve single base precision in mapping protein-DNA contact points within the upstream regulatory region of HSP82. Recently, Giniger et al. (47) have also employed this approach to map the in vivo binding sites within an upstream activating sequence of a yeast positive regulatory factor. The experimental strategy is depicted in Figure 7 and involves the use of dimethyl sulfate (DMS) as a probe of accessible purine residues. Because DMS is a small molecule, protection from (or enhancement to) DMS reactivity can be used to infer the location of those nucleotides forming intimate contacts with sequence-specific binding proteins. [It is important to note that DNA packaged into nucleosomes is highly accessible to DMS (54)]. Living cells are briefly exposed to DMS, which rapidly diffuses across the cell wall of haploid yeast and methylates accessible purines; the extent of the methylation is controlled such that, on average, < 1 purine nucleotide per strand is chemically modified along a ~ 500 bp stretch surrounding the HSP82 promoter region. DMS methylates B-form DNA primarily at two sites: the N-7 atom of guanine (in the major groove) and the N-3 atom of adenine (in the minor groove). Subsequent piperidine treatment of deproteinized, restriction endonuclease cleaved DNA results in chain scission at all of the methylated guanine residues; 3-methyladenine is not normally labile to such treatment (55).

Chemically cleaved single-stranded DNA is separated on a sequencing gel and transferred to a positively charged nylon membrane, to which it is UV-crosslinked. It has been our experience that distortion-free, reproducible transfer of DNA fragments <500 b can be achieved by simple vacuum diffusion (such as that generated by a standard gel drier connected to a vacuum pump) coupled with an osmotic gradient, which we term the "hydrokinetic" blot (Gross et al., in preparation). In our hands, horizontal or vertical electrophoretic transfer is more quantitative but often yields distortion.

Figure 8 shows a genomic footprint of the 491 bp segment spanning position -379 to position +112 within

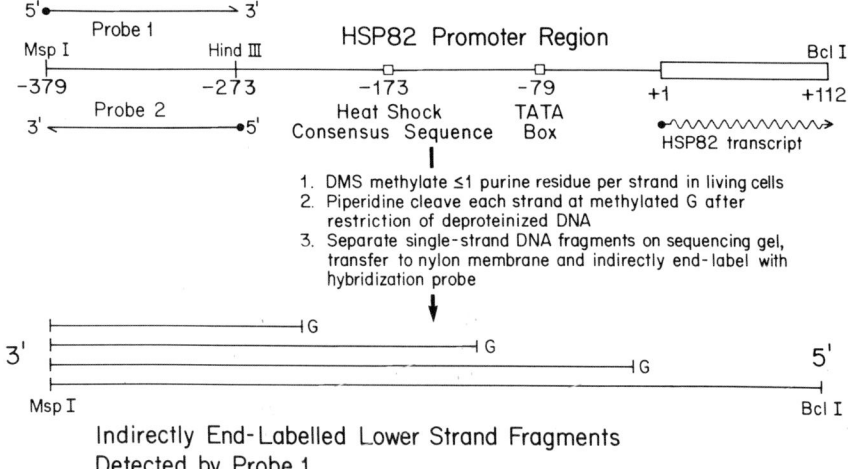

FIGURE 7. Experimental strategy for genomic footprinting of the HSP82 promoter region. Use of strand-specific DNA or RNA hybridization probes 1 and 2 selectively labels either the upper or lower strand. In addition, use of probes abutting the downstream restriction site (Bcl I) permits indirectly end-labeling either strand from the opposite direction (the same blot can be reprobed several times).

the HSP82 promoter region, along with naked DNA controls. These preliminary data suggest that there are few, if any, marked alterations in the reactivity of guanine nucleotides in chromatin versus naked DNA. However, quite unexpectedly, a number of cleavage sites seen only in DNA methylated in vivo map to adenine nucleotides, even though this reaction is normally guanine-specific (55). Strikingly, these unusually reactive adenine residues are clustered in and around the TATA box, and may represent the in vivo points of contact of a transcription factor analogous to that described in studies of Drosophila genes (51,52,56). Additionally, these nucleotides may be in an altered helical conformation, such as a single-stranded loop, which may contribute to their heightened accessibility to DMS. Further studies will be required to establish these points.

CONCLUSION

Yeast is a system with enormous potential for investigating the cis- and trans-acting determinants of chromatin structure and gene function. However, to date this potential has not been adequately exploited. The nucleic acid hybridization techniques illustrated here are sensitive enough to readily map yeast chromatin structures to single nucleotide resolution. Such mapping should delineate nucleotides crucial for regulatory protein binding, for establishing specific chromatin structures, and for concomitant gene expression. The links between these processes can be conclusively elucidated in yeast by in vitro mutagenesis and site-directed integration.

FIGURE 8 (facing page). Genomic footprints of upper and lower strands of the HSP82 promoter region. Also shown are DMS methylated, piperidine cleaved naked DNA controls. The in vivo samples were "hydrokinetically" transferred to Zeta-Probe while the naked DNA samples were transferred to the same membrane in a vertical electroblotting unit. Blots were then indirectly end-labeled with RNA hybridization probes homologous to the upstream region between positions -379 and -273 (see Figure 7). Probes were generated by in vitro transcription of pSP6 constructs containing the Msp I-Hind III fragment subcloned in either the 5' → 3' or 3' → 5' orientation. Haploid yeast concentrated to $\sim 4 \times 10^9$ cells/ml in rich medium, pH 5.8, were treated with 0.5% DMS at 0°C for 0.25 (a), 1.75 (b), or 7.0 (c) minutes. Those sites that display significant cleavage differences from the naked DNA control are labeled; positions are given relative to the HSP82 transcription start. The extremities of the analysis (+112, -230 or -233) are also indicated. The hybridization intensity of the band marked with an asterisk (*) in the lower strand in vivo footprint has not proven reproducible. (From Gross et al., in preparation.)

Chromatin Structure of a Heat Shock Gene

Upper Strand

In Vivo — abc
Naked DNA — 3'

- +112
- (A's)
- > -74 to -76 <
- > -82 to -85 <
- (A's)
- -169 (A)
- -233
- 5'

Lower Strand

Naked DNA — 5'
In Vivo — a b c

- +112
- -79 (A)
- *
- (G) -174
- -176 (A)
- -230
- 3'

ACKNOWLEDGEMENTS

We thank Dr. D.B. Finkelstein for his gifts of recombinant plasmids and stimulating discussions during the early stages of these studies, and Marie Rotondi and Raquel Voss for expert editorial assistance. D.S.G. and C.S.G. were the recipients of National Institutes of Health postdoctoral fellowships.

REFERENCES

1. Rothstein RJ (1983). One-step gene disruption in yeast. Methods Enzymol 101:202.
2. Winston F, Chumley F, Fink GR (1983). Eviction and transplacement of mutant genes in yeast. Methods Enzymol 101:211.
3. Barnes DA, Thorner J (1985). Use of the Lys2 gene for gene disruption, gene replacement, and promoter analysis in Saccharomyces cerevisiae. In Bennett JW, Lasure LL (eds): "Gene Manipulations in Fungi," New York: Academic Press, Inc., in press.
4. Pringle JR, Hartwell LH (1981). The Saccharomyces cerevisiae cell cycle. In Strathern JN, Jones EW, Broach JR (eds): "The Molecular Biology of the Yeast Saccharomyces: Life Cycle and Inheritance," New York: Cold Spring Harbor Laboratory, p 97.
5. DiNardo S, Voelkel K, Sternglanz R (1984). DNA topoismerase II mutant of Saccharomyces cerevisiae: topoisomerase II is required for segregation of daughter molecules at the termination of DNA replication. Proc Natl Acad Sci USA 81:2616.
6. Thrash C, Bankier AT, Barrell BG, Sternglanz R (1985). Cloning, characterization, and sequence of the yeast DNA topoisomerase I gene. Proc Natl Acad Sci USA 82:4374.
7. Clarke L, Carbon J (1980). Isolation of a yeast centromere and construction of functional small circular chromosomes. Nature 287:504.
8. Broach JR, Hicks JB (1980). Replication and recombination functions associated with the yeast plasmid, 2µ circle. Cell 21:501.
8a. Botstein D, Davis RW (1981). Principles and Practice of Recombinant DNA Research with Yeast. In Strathern JN, Jones EW, Broach JR (eds): "The Molecular Biology of the Yeast Saccharomyces: Life

Cycle and Inheritance," New York: Cold Spring Harbor Laboratory, p 607.
9. Certa U, Colavito-Shepanski M, Grunstein M (1984). Yeast may not contain H1: the only known "histone H1-like" protein in Saccharomyces cerevisiae is a mitochondrial protein. Nucleic Acids Res 12:7975.
10. Thomas JO, Furber V (1976). Yeast chromatin structure. FEBS Lett 66:274.
11. Lohr D, Kovacic RT, Van Holde KE (1977). Quantitative analysis of the digestion of yeast chromatin by staphylococcal nuclease. Biochemistry 16:465.
12. Lohr D, Tatchell K, Van Holde KE (1977). On the occurrence of nucleosome phasing in chromatin. Cell 12:829.
13. Lohr D, Van Holde KE (1979). Organization of spacer DNA in chromatin. Proc Natl Acad Sci USA 76:6326.
14. Nelson DA (1982). Histone acetylation in baker's yeast. J Biol Chem 257:1565.
15. Proffitt JH, Davie JR, Swinton D, Hattman S (1981). 5-methylcytosine is not detectable in Saccharomyces cerevisiae DNA. Mol Cell Biol 4:985.
16. Lohr D, Hereford, L (1979). Yeast chromatin is uniformly digested by DNase I. Proc Natl Acad Sci USA 76:4285.
17. Weintraub H, Groudine M (1976). Chromosomal subunits in active genes have an altered conformation. Science 193:848.
18. Nasmyth KA (1982). The regulation of yeast mating-type chromatin structure by SIR: An action at a distance affecting both transcription and transposition. Cell 30:567.
19. Sledziewski A, Young ET (1982). Chromatin conformational changes accompany transcriptional activation of a glucose-repressed gene in Saccharomyces cerevisiae. Proc Natl Acad Sci USA 79:253.
20. Bergman LW, Kramer RA (1983). Modulation of chromatin structure associated with derepression of the acid phosphatase gene of Saccharomyces cerevisiae. J Biol Chem 258:7223.
21. Costlow N, Lis JT (1984). High-resolution mapping of DNase I-hypersensitive sites of Drosophila heat shock genes in Drosophila melanogaster and Saccharomyces cerevisiae. Mol Cell Biol 4:1853.

22. Lohr D (1984). Organization of the GAL1-GAL10 intergenic control region chromatin. Nucleic Acids Res 12:8457.
23. Proffit JH (1985). DNase I-hypersensitive sites in the galactose gene cluster of Saccharomyces cerevisiae. Mol Cell Biol 5:1522.
24. Thoma F, Bergman LW, Simpson RT (1984). Nuclease digestion of circular TRP1ARS1 chromatin reveals positioned nucleosomes separated by nuclease-sensitive regions. J Mol Biol 177:715.
25. Bloom KS, Carbon J (1982). Yeast centromeric DNA is in a unique and highly ordered structure in chromosomes and small minichromosomes. Cell 29:305.
26. Thoma F, Simpson RT (1985). Local protein-DNA interactions may determine nucleosome positions on yeast plasmids. Nature 315:250.
27. Weber S, Isenberg I (1980). High mobility group proteins of Saccharomyces cerevisiae. Biochemistry 19:2236.
28. Gordon CN (1977). Chromatin behaviour during the mitotic cell cycle of Saccharomyces cerevisiae. J Cell Sci 24:81.
29. Fangman WL, Zakian VA (1981). Genome structure and replication. In Strathern JN, Jones EW, Broach JR (eds): "The Molecular Biology of the Yeast Saccharomyces: Life Cycle and Inheritance," New York: Cold Spring Harbor Laboratory, p 27.
30. Rattner JB, Saunders C, Davie JR, Hamkalo BA (1982). Ultrastructural organization of yeast chromatin. J Cell Biol 92:217.
31. Lee KP, Baxter HJ, Guillemette JG, Lawford HG, Lewis PN (1982). Structural studies on yeast nucleosomes. Can J Biochem 60:379.
32. Szent-Gyorgyi C, Isenberg I (1983). The organization of oligonucleosomes in yeast. Nucleic Acids Res 11:3717.
33. Mardian JKW, Isenberg I (1978). Yeast inner histones and the evolutionary conservation of histone-histone interactions. Biochemistry 17:3825.
34. Smith MM, Andresson OS (1983). DNA sequences of yeast H3 and H4 histone genes from two non-allelic gene sets encode identical H3 and H4 proteins. J Mol Biol 169:663.

35. Choe J, Kolodrubetz D, Grunstein M (1982). The two yeast histone H2A genes encode similar protein subtypes. Proc Natl Acad Sci USA 79:1484.
36. Wallis JW, Hereford L, Grunstein M (1980). Histone H2B genes of yeast encode two different proteins. Cell 22:799.
37. Rykowski MC, Wallis JW, Choe J, Grunstein M (1981). Histone H2B subtypes are dispensible during the yeast cell cycle. Cell 25:477.
38. Kolodrubetz D, Rykowski MC, Grunstein M (1982). Histone H2A subtypes associate interchangeably in vivo with histone H2B subtypes. Proc Natl Acad Sci USA 79:7814.
39. Wallis JW, Rykowski M, Grunstein M (1983). Yeast histone H2B containing large amino terminus deletions function in vivo. Cell 35:711.
40. Hall RM, Nagley P, Linnane AW (1976). Biogenesis of mitochondria: genetic analysis of the control of cellular mitochondrial DNA levels in Saccharomyces cerevisiae. Mol Gen Genet 145:169.
41. Duffus JH (1975). The isolation of yeast nuclei and methods to study their properties. In Prescott DM (ed): "Methods in Cell Biology," Vol XII, New York: Academic Press, p 77.
42. Scott JH, Sheckman R (1980). Lyticase: Endoglucanase and protease activities that act together in yeast cell lysis. J Bacteriol 142:414.
43. Farrelly FW, Finkelstein DB (1984). Complete sequence of the heat shock inducible HSP90 gene of Saccharomyces cerevisiae. J Biol Chem 259:5745.
44. Wu C (1980). The 5' ends of Drosophila heat shock genes in chromatin are hypersensitive to DNase I. Nature 286:854.
45. Nedospasov SA, Georgiev GP (1980). Non-random cleavage of SV40 DNA in the compact minichromosome and free in solution by micrococcal nuclease. Biochem Biophys Res Commun 92:532.
46. Costlow NA, Simon JA, Lis JT (1985). A hypersensitive site in hsp70 chromatin requires adjacent not internal DNA sequence. Nature 313:147.
47. Giniger E, Varnum SM, Ptashne M (1985). Specific DNA binding of GAL4, a positive regulatory protein of yeast. Cell 40:767.
48. Struhl K (1982). Promoter elements, regulatory elements, and chromatin structure of the yeast his3

gene. Cold Spring Harbor Symp. Quant. Biol. 47:901.
49. Struhl K (1984). Genetic properties and chromatin structure of the yeast gal regulatory element: an enhancer-like sequence. Proc Natl Acad Sci USA 81:7865.
50. Pelham HRB (1982). A regulatory upstream promoter element in the Drosophila hsp70 heat shock gene. Cell 30:517.
51. Parker CS, Topol J (1984). A Drosophila RNA polymerase II transcription factor binds to the regulatory site of an hsp70 gene. Cell 37:273.
52. Wu C (1984). Two protein-binding sites in chromatin implicated in the activation of heat-shock genes. Nature 309:229.
53. Church G, Gilbert W (1984). Genomic sequencing. Proc Natl Acad Sci USA 81:1991.
54. McGhee JD, Felsenfeld G (1979). Reaction of nucleosome DNA with dimethyl sulfate. Proc Natl Acad Sci USA 76:2133.
55. Maxam AM, Gilbert W (1980). Sequencing end-labeled DNA with base-specific chemical cleavages. Methods Enzymol 65: 499.
56. Parker CS, Topol J (1984). A Drosophila RNA polymerase II transcription factor contains a promoter-region-specific DNA-binding activity. Cell 36:357.
57. Brent R, Ptashne M (1984). A bacterial repressor protein or a yeast transcriptional terminator can block upstream activation of a yeast gene. Nature 312:612.
58. Johnson AD, Herskowitz I (1985). A repressor (Mat α2 product) and its operator control expression of a set of cell type specific genes in yeast. Cell 42:237.
59. Wharton RP, Ptashne M (1985). Changing the binding specificity of a repressor by redesigning an α-helix. Nature 316:601.
60. Saavedra RA, Huberman JA (1986). Flexible chromatin: an intermediate in gene activation? Science (submitted).

ARE SPECIFIC DNA SEQUENCES ASSOCIATED WITH
RESIDUAL NUCLEI?

Judith A. Potashkin[1] and Joel A. Huberman[2]

[1]Cold Spring Harbor Laboratory, P.O. Box 100, Cold Spring Harbor, New York 11724 and [2]Department of Cell and Tumor Biology, Roswell Park Memorial Institute, Buffalo, New York 14263

It is our intention in this article to review what is known and what is speculation in a fascinating area of cell biology: the association of chromosomal DNA molecules with a possible structural framework within eukaryotic nuclei. We shall also point out how the experimental advantages of yeast cells - which are well described throughout this volume - can profitably be employed to help understand this important problem.

DNA LOOPS

It has been known, ever since cytologists discovered lampbrush chromosomes, that the material of which chromosomes are composed (now known to be DNA and associated proteins) can exist, in some cases, as loops of varying size attached to a long, thick backbone. Recent studies (1,2) suggest that organization of DNA into heterogeneously-sized loops is a common feature of eukaryotic chromosomal and nuclear organization. The average sizes of DNA loops in different organisms range from 30 to 100 kb. The DNA loops behave as independent, topologically restrained domains. That is, each loop may be supercoiled or relaxed independently of neighboring loops. In lampbrush chromosomes, each loop behaves (usually) as a single transcription unit: transcription initiates at a site located near the base of a loop and proceeds unidirectionally around the entire loop. The distribution of loop sizes in various organisms correlates with the size of replication units, or replicons, in those organisms. These correlations between DNA loops and units of transcription and replication suggest that these loops may play important organizational roles in both replication and transcription.

Several important and related questions about these DNA loops remain unanswered. What is responsible for the topological restraints on DNA loops? Is the nucleotide sequence of the loops invariant? That is, is the same sequence always found at the base of each loop, at the point of attachment to the "backbone"? And what is the nature of the backbone material? In order to review what is known about the answers to these questions, it will be helpful to first review what is known about another mysterious component of the eukaryotic nucleus: the "nuclear matrix" or "residual nucleus".

RESIDUAL NUCLEI

The terms "nuclear matrix" (3), "nuclear cage" (4), "residual nuclear structure" (5), and "nuclear scaffold" (6) have all been used to describe non-chromatin, possibly structural, internal components of eukaryotic nuclei. Considerable confusion exists as to the exact meaning of these terms. In fact, the term "nuclear matrix" has two meanings. It has been used to refer to that non-chromatin internal component of the nucleus which, formerly, was sometimes referred to as "nucleoplasm" but has more recently been better defined by the use of electron microscopy combined with EDTA regressive staining. This technique minimizes staining of chromatin while enhancing staining of nuclear interchromatin regions. The complex fibrous and particulate structure thus revealed surrounds the chromatin within the nucleus and is therefore properly called a "nuclear matrix" (reviewed in reference 7). The other meaning of the term "nuclear matrix" is essentially the same as the meanings of the terms "nuclear cage", "residual nuclear structure", and "nuclear scaffold". It is an operational meaning. These terms refer to the structure which remains after many of the relatively soluble proteins of the nucleus have been extracted from the nucleus. The extraction procedures employed vary among laboratories, with many of them using high NaCl concentrations (1-2 M). Thus, the precise meaning of these operational terms depends on the laboratory in which they are employed. We shall use the term "residual nucleus" to refer, in general, to structures prepared by extraction of relatively soluble proteins from nuclei. As reviewed below, residual nuclei have a number of features in common, regardless of the method of preparation.

In many laboratories, treatment with nucleases to remove most nuclear DNA and/or RNA is part of the standard procedure used to prepare residual nuclei. If such nuclease treatment is not used, nuclear DNA remains attached to residual nuclei, and, because histones are extracted during preparation of residual nuclei, the DNA is able, to a certain extent, to migrate out of the residual nuclei. The extent of migration is determined by the sizes of the DNA loops mentioned above. Each loop remains attached to the residual nuclear structure at one site or in one region. We shall refer to such DNA-loop-containing residual nuclear structures as "nucleoids" (2). The fact that DNA loops remain attached to residual nuclear structures implies that residual nuclear structures include the "backbone" material discussed above. The major question we are addressing in this article is: do the attachment sites of DNA loops to residual nuclei correspond to specific DNA sequences, or are they random?

Before we deal with this question, however, we wish to describe some additional properties of residual nuclei. Despite the variation noted above in procedures used to prepare residual nuclei and nucleoids, every laboratory that has examined the question has observed that, when nucleoids are treated with DNase, the small amount of DNA remaining associated with residual nuclei is highly enriched for pulse-labeled (replicating) DNA. Quantitative studies suggest that 100% of nuclear replication forks are associated with residual nuclei (reviewed in 7). Although less extensively studied, several other nuclear functions are frequently observed to be associated with residual nuclei, including transcription, RNA processing, and nuclear protein phosphorylation. In addition, a significant fraction of DNA polymerase α, ribosomal precursor RNA, certain snRNAs, heat shock proteins, viral proteins such as T-antigen, hormone binding sites, and calmodulin binding sites are also associated with isolated residual nuclei (reviewed in 7).

These observations <u>suggest</u> that residual nuclei are involved in several very important nuclear processes. However, the observations made so far are simply <u>in vitro</u>

correlations; proof of the in vivo significance of these observations is, as yet, not available. It is now appropriate to discuss how yeast cells can be used to provide definitive information about the in vivo functions of residual nuclei, as well as the specificity or non-specificity of DNA attachments to residual nuclei.

ADVANTAGES OF YEAST

For examination of the association of specific DNA sequences with residual nuclei, yeast cells offer the following advantages. First, they have the smallest genome (of which we are aware) of any eukaryotic organism. This simplifies any experiment in which questions are asked about specific DNA sequences, simply because there are fewer irrelevant DNA sequences present. Second, a large fraction of the yeast genome has been cloned; the cloned DNA molecules are available for use in experiments to test association with residual nuclei. Furthermore, several types of cloned DNA sequences are available in yeast which are not readily available from other organisms. These include possible DNA replication origins (autonomously replicating sequences, or ARS sequences), chromosomal centromere sequences (CEN sequences), and chromosomal telomere sequences. Other papers in this volume provide excellent descriptions of the properties of these sequences. Third, due to the availability of temperature-sensitive mutations, which cause arrest at defined positions in the cell cycle (cell division cycle or CDC mutants; see elsewhere in this volume) and, the fact that the yeast sex pheromones (a and α factor) cause cell cycle arrest in late G1 (see elsewhere in this volume), yeast cells can readily be synchronized in bulk to defined positions within the cell cycle. Fourth, if specific DNA sequences are identified as associated with residual nuclei, then the function(s) of these sequences, and the function(s) of their associations with residual nuclei, can be tested by the techniques of in vitro mutagenesis and gene replacement.

Yeast cells also offer advantages for analysis of the in vivo functions of the proteins of residual nuclei. There are 5 major and numerous minor proteins in yeast residual nuclei (5). The major proteins can be isolated and used for preparation of polyclonal and/or monoclonal antibodies. Monoclonal antibodies can also be prepared against some of the minor proteins of residual nuclei by immunization with unfractionated residual nuclei. The antibodies can then be used to help clone the genes coding for their respective antigens. The use of antibodies to help clone genes has been particularly successful with budding yeast because most of their genes are not interrupted by introns. Thus most genes can be cloned by antibody selection without the need to first clone the corresponding cDNA. Other papers in this volume attest to the relative ease with which yeast genes can be cloned by antibody selection. Once any gene coding for a protein of residual nuclei has been cloned, the in vivo functions of that gene may be tested by the techniques of in vitro mutagenesis and gene replacement discussed elsewhere in this volume.

YEAST RESIDUAL NUCLEI

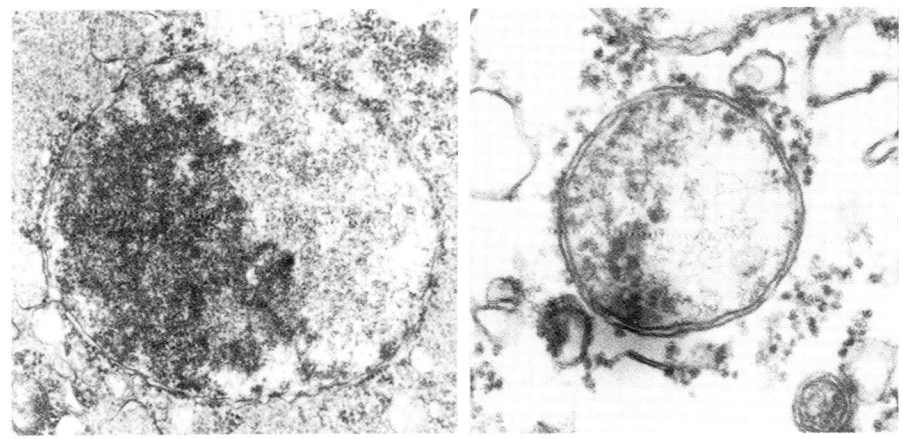

FIGURE 1. Isolated yeast nucleus (left panel) and residual nucleus (right panel). Nuclei and residual nuclei were isolated as previously described (5). The important steps of the isolation procedure involve (a) homogenization of yeast spheroplasts in a low pH, low ionic strength buffer, (b) partial purification of nuclei in a Percoll gradient at neutral pH, (c) preparation of nucleoids by addition of NaCl to 1.95 M, and (d) removal of >90% of the DNA by treatment with DNase 1. The picture of the isolated nucleus (left panel) was previously published (5) and is reprinted here with the permission of Academic Press.

We have recently developed procedures for isolating nucleoids and residual nuclei from yeast cells (5). The first step is isolation of nuclei. The nuclei are then treated with 1.95 M NaCl to produce nucleoids, and the nucleoids are treated with DNase 1 to produce residual nuclei. Figure 1 (left frame) shows an electron micrograph of an isolated yeast nucleus. The light grey material surrounding the nucleus is Percoll from the Percoll gradient used for nuclear enrichment. Most of the Percoll is removed during the extraction and washing steps involved in preparation of residual nuclei (Fig. 1, right frame). Both nuclei and residual nuclei are, however, contaminated by non-nuclear materials, primarily membranes and ribosomes. These can be seen most easily in the right panel. The intact yeast nucleus (Fig. 1, left frame) resembles nuclei from other eukaryotic organisms in most respects: it possesses a double nuclear membrane with embedded nuclear pores, and it also possesses a distinct chromatin region (the light grey region in the right half of the nucleus) and nucleolar region (the darker region in the left half of the nucleus). The major difference visible, at this magnification, between the yeast nucleus and nuclei of other eukaryotic organisms is the relatively large nucleolus in the yeast nucleus. This is a consequence of the fact that the yeast genome contains about as many rDNA genes (over 100) as other eukaryotic nuclei, but yeast nuclei contain a much smaller amount of non-ribosomal DNA than most other eukaryotic organisms. When yeast nuclei are converted to residual nuclei under conditions which leave the nuclear membrane intact (5), more than 50% of the nuclear proteins and more than 90% of the DNA is extracted. Electron microscopy shows that most of the chromatin material is extracted, but a sparse network of thin fibers remains (Fig. 1, right panel). Much material is also extracted from the nucleolar region, but the nucleolar region sometimes appears to remain recognizable as a distinct region (Fig. 1, right panel). Yeast residual nuclei resemble those from other eukaryotic organisms both in general appearance and in the specific association of replicating DNA with the residual nuclei (5).

PREVIOUS STUDIES IN OTHER ORGANISMS

The association of replicating DNA with residual nuclei leads us back to the main question of this article:

are specific DNA sequences associated with residual nuclei? Clearly, the replicating DNA sequences which are associated with residual nuclei are not specific: every DNA sequence in the genome is replicated. In addition, the association of replicating DNA sequences with residual nuclei is temporary; each sequence is associated briefly, once during S phase. We are concerned here with specific associations. Such associations might be "permanent", persisting throughout the cell cycle, or temporary, lasting for only a portion of the cell cycle. Therefore, in order to maximize chances of detecting specific sequence associations, it is important to use cells synchronized to several different portions of the cell cycle.

Most studies previously designed to answer this question have not used synchronized cells. In addition, most previous investigators have used high NaCl (1-2 M) to remove easily soluble nuclear proteins. A wide range of results has been obtained. Cook and Brazell (4) found that the α globin gene was enriched 8-fold in HeLa cell residual nuclei retaining about 13% of original nuclear DNA - the enrichment expected if a tight residual nuclear attachment site is located close to the α globin gene. Other investigators have observed less than theoretical enrichment in residual nuclei of certain actively transcribed sequences (ribosomal DNA (8) and the ovalbumin gene in chick oviduct (9,10)). Still other investigators have found no enrichment at all of transcribed or non-transcribed sequences in residual nuclei (11). Thus, in studies with unsynchronized cells and with NaCl-treated residual nuclei, the results which have been obtained do not provide a clear answer to the question of whether or not specific DNA sequences are attached to residual nuclei.

STUDIES WITH YEAST

We hoped that the use of synchronized yeast cells would provide a clear answer to this question. The details of our experimental procedures and results are being published elsewhere (12). Therefore, we shall provide only a brief summary here. We used two different approaches to determine whether specific DNA sequences are associated with yeast residual nuclei. In the first approach, we used unsynchronized cells. Two different libraries of yeast DNA in phage λ were probed with nick-translated yeast nuclear or residual nuclear DNA. Under the hybridization conditions used, strong signals were obtained only from repeated DNA sequences such as 2 μ plasmid DNA. None of the phage plaques gave a significantly stronger or weaker signal with the residual nuclear probe than with the total nuclear probe. Thus, none of the repeated DNA sequences of unsynchronized yeast is significantly enriched or depleted in residual nuclei prepared with high NaCl.

For our second approach, we purified DNA from whole nuclei or from residual nuclei which had been isolated from cells in G1, G1/S, early S, or nuclear division. This DNA was "dot-blotted" onto nylon membranes and then repeatedly probed with specific cloned, ^{32}P-labeled specific yeast DNA sequences. Because the same DNA samples (as dot blots) were probed with different labeled cloned sequences, even very small differences between the cloned sequences in terms of relative hybridization to the residual nuclear and nuclear DNA could be detected. Despite this potential sensitivity, we could not detect any enrichment or depletion in residual nuclei of 4 cloned sequences corresponding to interesting unique and somewhat repeated genomic DNA sequences. These cloned sequences were: TRP1 (coding for a gene which is constitutively but not abundantly transcribed), ARS1 (coding for a strong ARS element or putative replication origin), CEN6 (coding for the centromere of yeast chromosome VI), and the plasmid pYT14 which contains part of the yeast telomeric DNA repeat (13). However, when we used cloned sequences corresponding to the two most highly repeated DNAs of yeast cells (rDNA and 2 μm plasmid), we did detect cell-cycle-specific enrichment or depletion. We found a slight enrichment (2-3 fold) of rDNA in residual nuclei from late G1, G1/S, and early S phase cells, and we observed a 3-5 fold depletion of 2 μm DNA sequences in residual nuclei from cells undergoing nuclear division or in early G1.

Two conclusions may immediately be drawn from these results. First, the fact that four different genomic DNA sequences were neither enriched nor depleted in residual nuclei suggests that most yeast genomic DNA sequences are probably not specifically associated with residual nuclei prepared by our method with high NaCl. The small enrichment observed with rDNA might be due to its nucleolar location (the other tested sequences are all non-nucleolar) or to its abundant transcription (rDNA is the most heavily transcribed DNA sequence we are aware of in yeast). The depletion observed with 2 μm plasmid DNA might be due to its episomal nature (the other tested sequences are all chromosomal). Second, the cell cycle dependence of the enrichments and depletions detected for rDNA and 2 μm plasmid DNA demonstrates the potential importance of using well synchronized cells for studies of residual nuclear-DNA interactions. Recall that we did not detect significant enrichment or depletion for rDNA or 2 μm DNA when we screened yeast DNA libraries with residual nuclear DNA from unsynchronized cells.

FUTURE WORK

Despite the fact that our observations do not support the concept of specific DNA-residual nuclear interactions, that concept is still an appealing one. It is supported by the reproducible structure of the loops in lampbrush chromosomes and by the correlations noted in the introduction to this paper between DNA loops and units of transcription and replication. Therefore, we are inclined to suspect that our method of preparing residual nuclei may have led to disruption of specific DNA sequence interactions which were present in vivo. This possibility is supported by the recent demonstration by Mirkovitch et al. (6) that a very different method of preparing residual nuclei (by use of the detergent, lithium diiodosalicylate, rather than high NaCl) led to unambiguous demonstration of strong associations between Drosophila melanogaster residual nuclei and specific DNA sequences in the vicinity of histone and heat shock genes. This conclusion is also suggested by the observation that a protein which protects yeast centromere DNA sequences from nuclease digestion can be removed from centromere DNA by NaCl concentrations between 0.7 and 1.25 M (14), well below the 1.95 M NaCl used in our studies. For these reasons, current studies in our laboratory are directed towards exploring the use of alternative agents (such as lithium diiodosalicylate) for preparation of yeast residual nuclei. We anticipate that conditions will be developed which will preserve specific DNA sequence-residual nuclear interactions. It will then be of extreme interest to take advantage of the in vitro mutagenesis and gene replacement techniques described elsewhere in this volume to test the in vivo functions of these specific interactions.

REFERENCES

1. Paulson JR, Laemmli UK (1977). The structure of histone depleted metaphase chromosomes. Cell 12:817.
2. Cook PA, Brazell IA, Jost E (1976). Characterization of nuclear structures containing superhelical DNA. J Cell Sci 22:303.
3. Berezney R, Coffey DS (1975). Nuclear protein matrix: association with newly synthesized DNA. Science 189:291.
4. Cook PR, Brazell IA (1980). Mapping sequences in loops of nuclear DNA by their progressive detachment from the nuclear cage. Nucl Acid Res 8:2895.
5. Potashkin JA, Zeigel RF, Huberman JA (1984). Isolation and initial characterization of residual nuclear structures from yeast. Exp Cell Res 153:374.
6. Mirkovitch J, Mirault ME, Laemmli UK (1984). Organization of the higher-order chromatin loop: specific DNA attachment sites on nuclear scaffold. Cell 39:223.
7. Berezney R (1984). Organization and functions of the nuclear matrix. In Hnilica LS (ed): "Chromosomal Nonhistone Proteins,", Boca Raton: CRC Press, p 119.
8. Pardoll DM, Vogelstein B (1980). Sequence analysis of nuclear matrix associated DNA from rat liver. Exp Cell Res 128:466.
9. Robinson SI, Small D, Idzerda R, McKnight GS, Vogelstein B (1983). The association of transcriptionally active genes with the nuclear matrix of the chicken oviduct. Nucl Acid Res 11:5113.
10. Ciejek EM, Tsai MJ, O'Malley B (1983). Actively transcribed genes are associated with the nuclear matrix. Nature 306:607.
11. Basler J, Hastie ND, Pietras D, Matsui SI, Sandberg AA, Berezney R (1981). Hybridization of nuclear matrix attached deoxyribonucleic acid fragments. Biochem 20:6921.
12. Potashkin JA, Huberman JA (1986). Characterization of DNA sequences associated with residual nuclei of Saccharomyces cerevisiae. Experimental Cell Res, in press.
13. Shampay J, Szostak JW, Blackburn EH (1984). DNA sequences of telomeres maintained in yeast. Nature 310:154.
14. Bloom KS, Amaya E, Carbon J, Clarke L, Hill A, Yeh E (1984). Chromatin conformation of yeast centromeres. J Cell Biol 99:1559.

V. MACROMOLECULAR TRAFFIC I: PROTEIN LOCALIZED IN THE NUCLEUS

IDENTIFICATION OF A NUCLEAR LOCALIZATION SIGNAL OF YEAST RIBOSOMAL PROTEIN L3[1]

Robert B. Moreland*, Hong Gil Nam†,
Lynna Hereford* and Howard M. Fried¶

*Dana Farber Cancer Institute, Boston, MA 02115 and Departments of Chemistry† and Biochemistry¶, University of North Carolina, Chapel Hill, NC 27514

ABSTRACT To begin to understand how ribosomal proteins are transported to the nucleus, we produced in yeast hybrid polypeptides consisting of amino terminal segments of ribosomal protein L3 joined to a carboxy terminal segment of β-galactosidase and located the hybrids by *in situ* immunofluorescence. The first 21 amino acids of L3 were sufficient to localize β-galactosidase to the nucleus. More extensive L3 fusions were also nuclear localized. However, a hybrid protein containing all but 14 amino acids of the carboxy terminus of L3 was not localized; this defect was corrected by inserting a glycine and proline-containing peptide between the L3 and β-galactosidase moieties. The resulting protein was associated with ribosomes, suggesting that it was both transported to the nucleus and assembled into 60S ribosomal subunits which were exported to the cytoplasm.

INTRODUCTION

In eukaryotic cells, ribosomes are constructed in the nucleolus, an unenclosed subdomain of the nucleus which is the site of transcription of ribosomal RNA. In addition to RNA molecules, eukaryotic ribosomes contain about 70 different proteins, which are transported to the nucleus within

---

[1]This work was supported by NSF Grant PCM-8215576, NIH Grant GM33332 and Damon Runyon-Walter Winchell Cancer Fund Fellowship DRW628.

minutes of their synthesis in the cytoplasm to assemble with newly transcribed rRNA. The transport of ribosomal proteins must be rapid and coordinated, since in their absence rRNA is transcribed but rapidly degraded (see Refs. 1-3 for reviews). Furthermore, emerging evidence suggests that the ability of a ribosomal protein to assemble is a major factor in determining its rate of synthesis. This phenomenon is clearly established in *E. coli*, in which unassembled ribosomal proteins repress translation of their mRNAs to prevent unnecessary oversynthesis (4). In yeast, both translation and splicing of some ribosomal protein mRNAs may be repressed by their corresponding proteins, since introduction of additional copies of some ribosomal protein genes leads to inhibition of these two steps in gene expression (5-7). Thus, information about mechanisms which bring about rapid transport and assembly of ribosomal proteins is crucial to an understanding of the regulation of ribosome production.

As with any protein targeted to a specific cellular location, a ribosomal protein must contain information which brings about its transport to the nucleus. For DNA binding proteins or viral proteins several investigators have used gene fusion techniques to identify short stretches of these polypeptides which cause nuclear localization of non-nuclear proteins (8-11). We have applied this technique to yeast ribosomal protein L3 and find that its amino terminal 21 amino acids are sufficient to direct *E. coli* β-galactosidase to the yeast nucleus. Interestingly, from the standpoint of primary sequence, this 21 residue domain shares limited homology to a region of SV40 large T-antigen which is responsible for its nuclear localization.

RESULTS

Yeast ribosomal protein L3 is the product of the gene *TCM1*, which encodes 387 amino acids (12,13). To identify a region of L3 responsible for its nuclear localization, five restriction enzyme fragments of *TCM1*, whose promoter distal ends encompass from 1% to 96% of the coding sequence (see Fig. 1), were joined to the coding sequence of *E. coli* β-galactosidase in plasmid pJT24 (14). The particular fragments were chosen also on the basis of nucleotide sequence which indicated that the ribosomal protein and β-galactosidase coding sequences would be joined in a continuous translational reading frame without additional modification of the

DNA fragments. When introduced into yeast each plasmid led to β-galactosidase production under direction of the *TCM1* transcriptional signals (the parent plasmid pJT24 does not direct β-galactosidase synthesis).

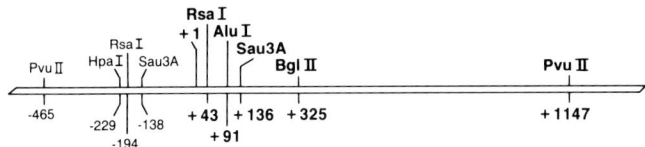

Plasmid Name	Restriction fragment of rpL3 joined to β-GAL	Amino acids of rpL3 joined to β-GAL
pTCM-RR	Rsa I (-194) → Rsa I (+43)	1-5
pTCM-HA	Hpa I (-229) → Alu I (+91)	1-21
pTCM-SS	Sau3A (-138) → Sau3A (+136)	1-36
pTCM-HB	Hpa I (-229) → Bgl II (+325)	1-99
pTCM-PP	Pvu II (-465) → Pvu II (+1147)	1-373
pTCM-PP$_{Nco}$	Pvu II (-465) → Pvu II (+1147)	1-373*

*see legend

FIGURE 1. The gene *TCM1* encodes ribosomal protein L3. Listed are restriction enzyme fragments of *TCM1* which were joined upstream and in the translational reading frame of the *E. coli* β-galactosidase gene. A + refers to distance in nucleotides 3' of the transcriptional start site while - refers to distance 5' of start. Plasmid pTCM-PP$_{Nco}$ is a derivative of pTCM-PP containing 3 tandem NcoI oligonucleotide linkers inserted between the L3 and β-galactosidase DNA segments (see text).

Characterization of L3-β-Galactosidase Genes

Table 1 shows that the amount of β-galactosidase enzyme activity in crude extracts of cells carrying each of the five hybrid genes varies over a 50-fold range. This variation is not due to differences in transcription of the hybrid genes, as each has a promoter of equivalent strength (15). In addition, some differences are too great to be accounted for by plasmid copy number (data not shown). Rather, the variability is likely due to detrimental effects of some of the hybrid proteins upon cell viability since, as Table 1 shows, the presence of most of the plasmids in cells is coincident with significant reductions in growth rate.

TABLE 1
PROPERTIES OF L3-β-GALACTOSIDASE PROTEINS IN YEAST

Plasmid	Amino acids of L3 joined to β-gal	Units of β-gal	Cell doubling time (hrs)	Location of hybrid protein
none	--	--	3.6	--
pTCM-RR	1-5	700	3.8	cytoplasmic
pTCM-HA	1-21	400	6.4	nuclear
pTCM-SS	1-36	15	6.4	nuclear
pTCM-HB	1-99	13	5.4	nuclear
pTCM-PP	1-373	100	4.0	cytoplasmic
pTCM-PP$_{Nco}$	1-373	77	5.4	nuclear (primarily)

To verify that the plasmids direct synthesis of hybrid proteins of expected sizes, cell extracts were incubated with anti-β-galactosidase antibody and the immunoprecipitates were analyzed by SDS-polyacrylamide gel electrophoresis (see Fig. 2). Cells carrying either plasmid pTCM-RR or pTCM-HA yielded a single immunoprecipitable protein with an electrophoretic mobility equal to that of authentic β-galactosidase; these two plasmids join only 5 and 21 amino acids to β-galactosidase, too few to have produced a difference in gel mobility. Cells transformed with pTCM-PP yielded a 157,000 daltons product, close to the predicted size of 155,000 daltons. By contrast, plasmids pTCM-SS and pTCM-HB routinely yielded two immunoprecipitable β-galactosidase proteins, one corresponding in size to the full length hybrid and one the size of β-galactosidase. Figure 2 shows this phenomenon for pTCM-HB. Apparently, the ribosomal protein moieties are cleaved from these two hybrid proteins. In addition, cleavage was not peculiar to L3-β-galactosidase hybrids, since it occurred when the first 110 amino acids of another ribosomal protein, L29, were joined to β-galactosidase (see pCYH-Z in Fig. 2). Because of the difficulty in inhibiting yeast vacuolar proteinases (16), we have not ruled out the possibility that proteolysis occurred *in vitro* during incubation of extracts with antibody. Nonetheless it is clear that the plasmids direct synthesis of hybrid proteins of the correct sizes.

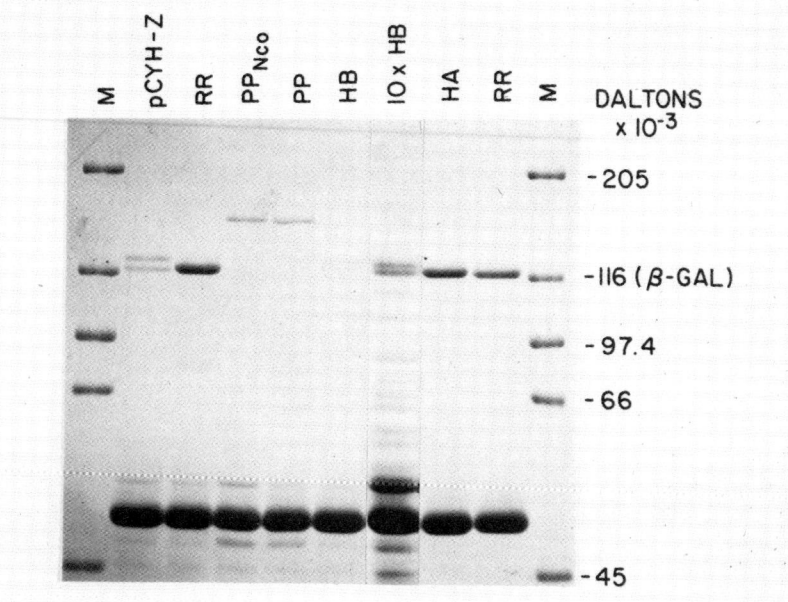

FIGURE 2. Extracts of cells were incubated with anti-β-galactosidase antibody. The immunoprecipitates were electrophoresed in a gel composed of 7% acrylamide-0.35% N, N',-diallyltartardiamide containing 0.1% sodium dodecyl sulfate. RR, HA, HB, PP, PP_{Nco} refer to cells transformed with the plasmids listed in Fig. 1. pCYH-Z refers to cells carrying a 110 amino acid fusion of ribosomal protein L29 joined to β-galactosidase. 10 X HB refers to analysis of 10 fold more cell extract, as these cells produce little hybrid protein.

Subcellular Localization of Hybrid Proteins

To determine the location of the hybrid proteins in yeast, cells were fixed and stained by indirect immunofluorescence with anti-β-galactosidase antibody to detect the hybrid proteins and with DAPI (4,6,diamidino-2-phenylindole) which stains the nucleus by binding to DNA (17,18,8). Figure 3 shows representative fluorescence micrographs for cells transformed with pTCM-RR and pTCM-HA and Figure 4 shows results for pTCM-SS and pTCM-HB. The nuclei of cells transformed with the latter three plasmids showed strong staining with anti-β-galactosidase antibody, while cells transformed with pTCM-RR lacked any localized staining and exhibited fluorescence throughout the cell. Thus, the first 21 amino acids of L3 are capable of localizing β-galactosidase to the nucleus. At this level of resolution, it is not possible to determine whether the proteins are inside the nucleus or sequestered in the nuclear envelope.

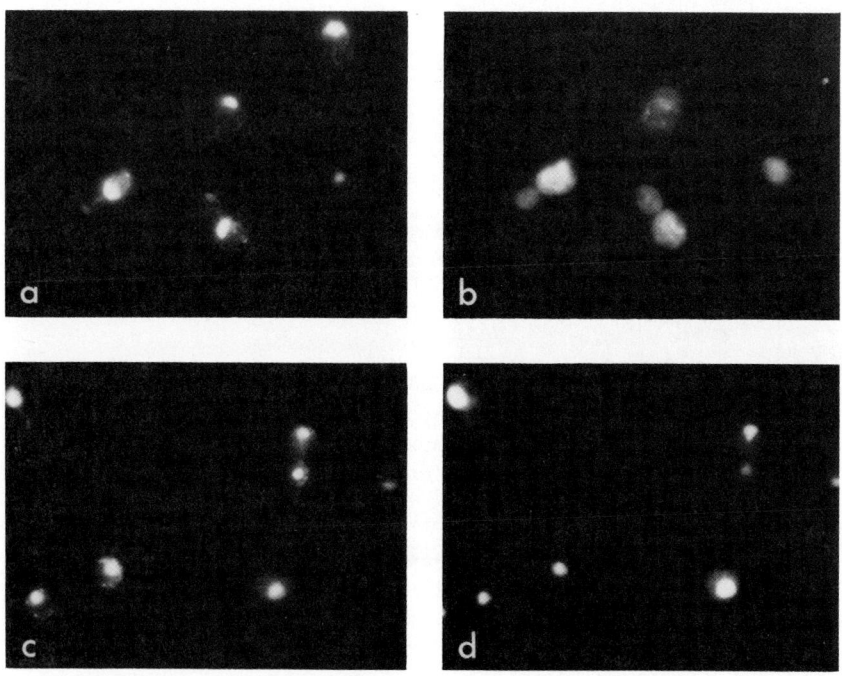

FIGURE 3. Indirect immunofluorescence localization of L3-β-galactosidase proteins. Nuclei were visualized with DAPI and β-galactosidase proteins with rabbit anti-β-galactosidase antibody followed by a fluoresceinated (FITC) goat anti-rabbit IgG antibody. Panel a: pTCM-RR (5 amino acids of L3), DAPI; panel b: pTCM-RR, FITC; panel c: pTCM-HA (21 amino acids of L3), DAPI; panel d: pTCM-HA, FITC.

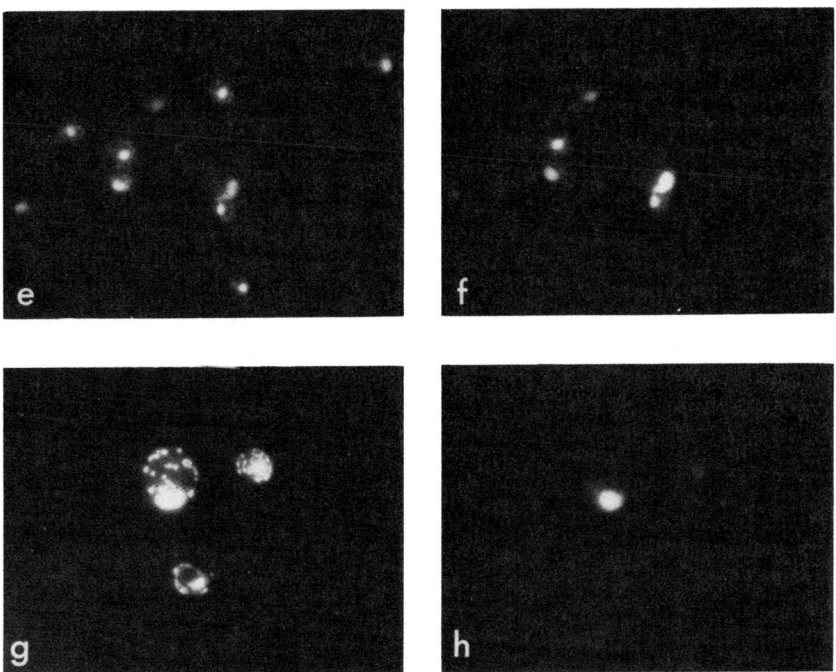

FIGURE 4. Localization of L3-β-galactosidase proteins (as in Fig. 3). Panel e, pTCM-SS (36 amino acids of L3), DAPI; panel f, pTCM-SS, FITC; panel g, pTCM-HB (99 amino acids of L3), DAPI; panel h: pTCM-HB, FITC.

Interestingly, while hybrid proteins containing the first 21, 36, or 99 amino acids of L3 were associated with the nucleus, the fusion containing 373 amino acids of L3 failed to be localized (Figure 5). This result was unexpected, since the 373 amino acid fusion contains all but the carboxy-terminal 14 amino acids of L3 and should most closely resemble L3 itself. However, the protein appeared to be largely insoluble, as most of the β-galactosidase activity was in a low speed pellet after cell disruption. Thus, we speculated that the L3 portion of this protein was prevented from folding properly by its association with β-galactosidase, leading to insolubility and the inability to be transported. This assumption was supported by the finding that insertion of eight additional amino acids between the

L3 and β-galactosidase moieties of this hybrid produced a protein which was targeted to the nucleus (see pTCM-PP$_{Nco}$ in Figure 5). The extra eight amino acids, Ala-His-Gly-Pro-Met-Gly-Pro-Trp, are not derived from L3 but are encoded in three tandem NcoI oligonucleotide linkers inserted at the L3-β-galactosidase junction of pTCM-PP. We arrived at this construction because the NcoI oligomer (CCCATGGG) provided proline and glycine residues with which to bridge the L3 and β-galactosidase moieties. Secondary structure predictions (19) suggested the presence of an α-helix ending at residue 361 followed by a β-pleated sheet structure up to residue 372 and a short (four amino acid) turn at the fusion junction (data not shown). The 8 amino acid insertion was predicted to extend the turn to 10 amino acids, which could act as a hinge between the two proteins and more effectively separate them from one another. This proposition is reasonable

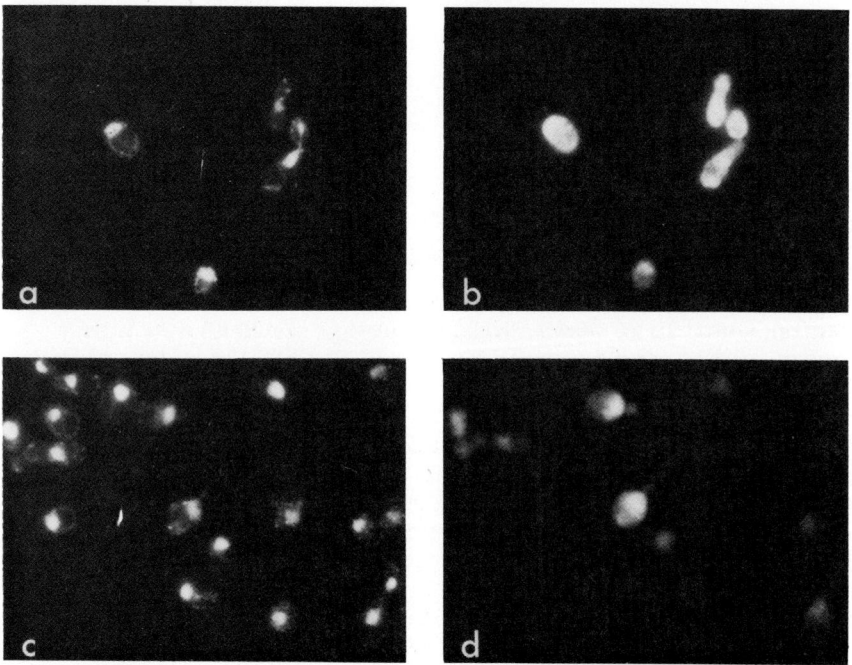

FIGURE 5. Comparison of localization of L3-β-galactosidase protein with and without peptide bridge. Panel a: pTCM-PP (373 amino acids of L3), DAPI; panel b: pTCM-PP, FITC; panel c, pTCM-PP$_{Nco}$ (373 amino acids of L3 plus 8 amino acid peptide bridge), DAPI; panel d: pTCM-PP$_{Nco}$, FITC.

in view of the fact that adjacent secondary structures in proteins have been found to interact (20). Whatever was the defect in the original hybrid protein which prevented its nuclear localization, it is indeed dramatic that an 8 amino acid insertion into a protein of nearly 1400 amino acids would alter its nuclear localization and biochemical properties (see below).

While the peptide bridge between L3 and β-galactosidase permitted the 373 amino acid fusion to localize with the nucleus, in routine examinations of stained cells considerable fluorescence was usually seen in the cytosol, more so than with other nuclear associated proteins. Incomplete localization may simply indicate that the peptide bridge was not an ideal solution to whatever prevented localization. Alternatively, since this fusion contains all but the carboxy-terminal 14 amino acids of L3, we speculated that perhaps acquisition of targeting to the nucleus enabled this protein to assemble into a 60S ribosomal subunit and be exported back to the cytoplasm. To determine if the fusion protein was associated with ribosomes, polyribosomes were sedimented in a sucrose gradient and fractions were assayed for β-galactosidase activity. As a control, ribosomes were also sedimented from cells containing the 5 amino acid fusion. While all of the β-galactosidase activity of the 5 amino acid fusion was found at the top of the gradient, Figure 6 shows that β-galactosidase activity from the large fusion resided not only at the top of the gradient but in a region containing 60S subunits and 80S monosomes and in a region containing polyribosomes. Thus, at least by sedimentation, the hybrid protein behaved as if it were associated with ribosomes. A similar analysis of cells containing the original 373 amino acid fusion (without the NcoI oligomers) did not show β-galactosidase in either the monosome or polysome region (data not shown). A further demonstration of the association of the hybrid protein with ribosomes was found by treating extracts with a small amount of RNase, which converted polyribosomes into 80S monosomes. With such treatment, the polysome peak of β-galactosidase was absent while considerable β-galactosidase activity was still found in the 80S region. The β-galactosidase activity associated with ribosomes is unlikely due to the presence of enzyme molecules in the process of being translated. Absence of as few as the carboxy-terminal 17 amino acids completely inactivate the enzyme (21) and it is unlikely that there were enough chains completed beyond this point, but still unreleased, to account for the fact that nearly half of the β-galactosidase activity was found in the ribosome portions of the gradient. Thus, although we do not know that the hybrid protein was properly assembled, it appeared, nevertheless, to be associated with ribosomes, which probably explains the cytoplasmic fluorescence of antibody stained cells.

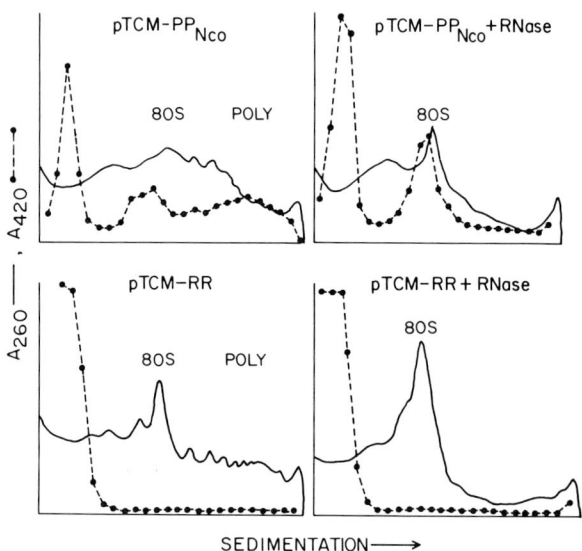

FIGURE 6. Sedimentation analysis of β-galactosidase from cells synthesizing L3-β-galactosidase proteins. Cell extracts were centrifuged in sucrose gradients exactly as described (7). Solid lines (A_{260}) are absorbance of RNA and dashed lines (A_{420}) are the distribution of functional β-galactosidase enzyme as measured by hydrolysis of O-nitrophenyl-β-D-galactoside. +RNAse indicates that extracts were treated with RNAse A before applying to the gradients.

DISCUSSION

In this study we present an initial attempt to determine what information is contained within a ribosomal protein which targets it to the nucleus. Our results show that the amino terminal 21 amino acids of yeast ribosomal protein L3 are sufficient to cause *E. coli* β-galactosidase to localize with the nucleus and therefore this domain must participate in the transport process. The results support the idea that ribosomal proteins should contain signals which direct the proteins efficiently to their site of assembly and they show that this class of proteins, often ignored as far as nuclear transport is concerned, is also amenable to questions concerning protein trafficking.

Clearly many questions do remain to be addressed. We do not know how the amino terminal domain of L3 functions in nuclear targeting. It is generally assumed that proteins enter the nucleus by way of the nuclear pore although the evidence is somewhat limited (22). Perhaps the best evidence comes from a recent study by Feldherr and co-workers in which they injected into *Xenopus* oocytes nucleoplasmin coated gold particles and showed the particles not only accumulating in the nucleus but passing through a nuclear pore complex (23). Possibly the domain we have identified for L3 is involved in nuclear pore recognition and since the first 21 amino acids of L3 carry a net charge of +5, it seems more likely that this region would interact with the pore rather than the hydrophobic barrier presented by the nuclear envelope. However, a large variety of materials also pass out of the nucleus by way of the pore, including ribosomal subunits (24). Whether incoming proteins compete for the same pores is not known and, if they do, it remains to be determined how the direction of traffic is controlled.

It could be argued that the amino terminal region of L3 does not actually target the protein to the nucleus but is only required to keep the protein there once it has entered, perhaps by binding to ribosomal RNA. While this possibility is not excluded at this point, we do not favor it for the following reasons. First, β-galactosidase (MW 116,000) is too large to diffuse freely into the nucleus (25) and the nuclear membrane does not break down in *S. cerevisiae* during mitosis (26). Thus the hybrid proteins should not become associated with the nucleus unless actually targeted there. Indeed, Hall and co-workers showed by cell fractionation methods that β-galactosidase does not associate with the nucleus in yeast (8). Secondly, two studies have shown that a single amino acid substitution in SV40 large T antigen is sufficient to prevent nuclear localization of this protein (27,28). However, the mutant proteins are not altered in either their general affinity for DNA (29) or their ability to recognize the SV40 replication origin (28); thus, failure to be nuclear localized is not due to an inability to interact with an agent likely to cause nuclear retention of the protein. Similarly, despite treatment of HeLa cells with actinomycin D to deplete the pool of newly synthesized ribosomal precursor RNA, ribosomal proteins continue to accumulate in the nucleus (30).

Another question which must be addressed is whether the amino terminal domain of L3 is the only part of the protein required in the overall transport process. We have not presented any data which says that the hybrid proteins are completely inside the nucleus or merely within the nuclear envelope. Cell fractionation and biochemical tests will be necessary to address this issue. Perhaps the amino terminus of L3 is only involved in targeting but additional regions of the protein play a role in other aspects of transport. Pertinent to this question is our recent finding that the first 15 amino acids of ribosomal protein L29 are sufficient to localize β-galactosidase to the nucleus. However, when the first 15 amino acids were deleted (and replaced by [Met]-Asp-Ser) and β-galactosidase was joined near the carboxy terminus of the remaining protein, the hybrid was still nuclear localized (data not shown). Either L29 contains more than a single targeting signal, each one of which can act independently, or perhaps L29 interacts with another (ribosomal ?) protein and is shuttled to the nucleus piggyback. In this vein it is interesting to point out that when four of the five carboxy terminal tails of a nucleoplasmin pentamer are removed, the single tail region is sufficient to transport the complex to the nucleus (31).

Finally, the amino terminus of L3 is of interest from the standpoint of its sequence and predicted secondary structure. As mentioned above the requirements for nuclear localization have been examined in detail with SV40 large T-antigen. A naturally occurring mutant defective for transport of T-antigen has been described (27). A single substitution of asparagine for lysine in position 128 was responsible for the defect. Further, the minimal sequence required for targeting T-antigen to the nucleus has been defined elegantly by Kalderon and co-workers (10,28). The region from residues 126 to 132 has been shown to be sufficient for nuclear localization. In addition the substitution of threonine for lysine 128 destroys nuclear localization while substitutions in positions 129-131 yield partially localized proteins. The primary sequence of SV40 large T-antigen from residues 126-132 (Pro-Lys-Lys$_{128}$-Lys-Arg-Lys-Val) is not strikingly homologous to any part of the amino terminal 21 amino acids of L3. However, the overall motif of a proline residue followed by several basic amino acids (Pro-Arg-Lys$_{20}$-Arg) is noted with strict homology at Lys$_{20}$/Lys$_{128}$. A similar motif is present in yeast histone H2B in a region involved in its nuclear localization (unpublished observation). Further, a common secondary structure can be predicted for the regions of H2B and SV40 large T-antigen that have been determined by mutation to be responsible for their localization (data not shown). Site specific mutation analysis of L3 is in progress to test whether a common nuclear localization signal exists between it and SV40 T-antigen. If the sequences are homologous in function we predict that substitutions at Lys$_{20}$ of L3 (analogous to Lys$_{128}$ of T-antigen) should eliminate its nuclear localization.

ACKNOWLEDGMENTS

The authors are indebted to Drs. Jennifer McKnight and Deepak Bastia for their generous gifts of β-galactosidase antibody, to Dr. Lam Bo Chen for the use of his fluorescence microscope and Dr. John Teem for plasmid pJT24. H.M.F. is especially grateful to Dr. Jonathan Warner for initially suggesting this study.

REFERENCES

1. Siekevitz P, Zamecnik PC (1981). Ribosomes and protein synthesis. J Cell Biol 91:53s.
2. Warner JR, Tushinski RJ, Wejksnora PJ (1980). Coordination of RNA and proteins in eukaryotic ribosome production. In Chambliss G, Craven GR, Davies J, Davis K, Kahan L, Nomura M (eds.): "Ribosomes: Structure, Function, and Genetics," Baltimore, MD: University Park Press, p. 889.
3. Warner JR (1982). The yeast ribosome: structure, function, and synthesis. In Strathern J, Jones E,

Broach JR (eds.): "Molecular Biology of the Yeast *Saccharomyces*," New York: Cold Spring Harbor Press, p. 529.
4. Nomura M, Gourse R, Baughman G (1984). Regulation of the synthesis of ribosomes and ribosomal components. Ann Rev Biochem 53:75.
5. Pearson NJ, Fried HM, Warner JR (1982). Yeast use translational control to compensate for extra copies of a ribosomal protein gene. Cell 29:347.
6. Himmelfarb HJ, Vassarotti A, Friesen JD (1984). Molecular cloning and biosynthetic regulation of the *cry1* gene of *saccharomyces cerevisiae*. Mol Gen Genet 195:500.
7. Warner JR, Mitra G, Schwindinger WF, Studeny M, Fried H (1985). *Saccharomyces cerevisiae* coordinates the accumulation of yeast ribosomal proteins by modulating mRNA splicing, translational initiation, and protein turnover. Molec Cell Biol 5: (in press).
8. Hall MN, Hereford L, Herskowitz I (1984). Targeting of *E. coli* β-galactosidase to the nucleus in yeast. Cell 36:1057.
9. Silver PA, Keegan LP, Ptashne M (1984). Amino terminus of the yeast GAL4 gene product is sufficient for nuclear localization. *Proc Nat Acad Sci USA* 81:5951.
10. Kalderon D, Roberts BL, Richardson, WD, Smith AE (1984). A short amino acid sequence able to specify nuclear location. Cell 39:499.
11. Davey J, Dimmock J, Colman A (1985). Identification of the sequence responsible for the nuclear accumulation of the influenza virus nucleoprotein in Xenopus oocytes. Cell 40:667.
12. Fried HM, Warner JR (1981). Cloning of the yeast gene for trichodermin resistance and ribosomal protein L3. Proc Nat Acad Sci USA 78:238.
13. Schultz LD, Friesen JD (1983). Nucleotide sequence of the *TCM1* gene (ribosomal protein L3) of *saccharomyces cerevisiae*. J Bacteriol 155:8.
14. Teem J (1983). Ph.D. Thesis, Brandeis University, Waltham, MA.
15. Fried HM, Nam HG, Loechel S, Teem J (1985). Characterization of yeast strains with conditionally expressed variants of ribosomal protein genes *tcm1* and *cyh2*. Molec Cell Biol 5:99.
16. Jones EW (1984). The synthesis and function of proteases in *saccharomyces*: genetic approaches. Ann Rev Genet 18:233.

17. Adams AEM, Pringle JR (1984). Relationship of actin and tubulin distribution to bud growth in wild-type and morphogenetic-mutant *saccharomyces cerevisiae*. J Cell Biol 98:934.
18. Kilmartin JV, Wright B, Milstein C (1982). Rat monoclonal antitubulin antibodies derived by using a new nonsecreting rat cell line. J Cell Biol 93:576.
19. Chou PY, Fasman GD (1977). β-turns in proteins. J Mol Biol 115:135.
20. Chothia C (1984). Principles that determine the structure of proteins. Ann Rev Biochem 53:537.
21. Itakura K, Hirose T, Crea R, Riggs AD, Heyneker HL, Bolivar F, Boyer HW (1977). Expression in *Escherichia coli* of a chemically synthesized gene for the hormone somatastatin. Science 198:1056.
22. Franke WW, Scheer U, Krohne G, Jarasch ED (1981). The nuclear envelope and the architecture of the nuclear periphery. J Cell Biol 91:39s.
23. Feldherr CM, Kallenbach E, Schultz N (1984). Movement a karyophillc protein through the nuclear pores of oocytes. J Cell Biol 99:2216.
24. Wunderlich F (1972). The macronuclear envelope of *Tetrahymena pyriformis* GL in different physiological states. J Membrane Biol 7:220.
25. De Robertis EM (1983). Nucleocytoplasmic segregation of proteins and RNAs. Cell 32:1021.
26. Byers B (1981). Cytology of the yeast life cycle. In Strathern J, Jones E, Broach JR (Eds.): "Molecular Biology of the Yeast *Saccharomyces*," New York: Cold Spring Harbor Press, p. 59.
27. Lanford RE, Butel JS (1984). Construction and characterization of an SV40 mutant defective in nuclear transport of T-antigen. Cell 37:801.
28. Kalderon D, Richardson WD, Markham AF, Smith AE (1984). Sequence requirements for nuclear location of simian virus 40 large-T-antigen. Nature 311:33.
29. Lanford RE, Butel JS (1980). Biochemical characterization of nuclear and cytoplasmic forms of SV40 tumor antigens encoded by parental and transport defective mutant SV40-adenovirus 7 hybrid viruses. Virology 105:314.
30. Warner JR (1979). Distribution of newly formed ribosomal proteins in HeLa cell fractions. J Cell Biol 80:767.
31. Dingwall C, Sharnick SV, Laskey RA (1982). A polypeptide domain that specifies migration of nucleoplasm into the nucleus. Cell 30:449.

ENTRY OF A PROCARYOTIC ENDONUCLEASE INTO THE NUCLEUS OF *Saccharomyces cerevisiae*[1]

Jasper Rine and Georjana Barnes

Department of Biochemistry, University of California, Berkeley, California 94720

ABSTRACT Recent experiments have documented the existence of amino acid sequences, acting as nuclear signal sequences, that result in the accumulation of proteins in the nucleus. In this paper, the issue of whether or not a protein must possess a nuclear signal in order to enter the nucleus is examined. For these experiments, a plasmid capable of expressing EcoRI endonuclease in the yeast Saccharomyces cerevisiae has been constructed and transformed into several yeast strains. Two results demonstrate that this bacterial protein can enter the yeast nucleus: First, yeast cells expressing the endonuclease gene die with kinetics that are proportional to the capacity of the strain to repair double stranded breaks in nuclear DNA. Secondly, the nuclear DNA contains extensive double stranded breaks at EcoRI sites and only at EcoRI sites. Therefore, there is no apparent requirement for a protein to contain a complex nuclear localization signal in order to enter the nucleus. Additional experiments demonstrate that in frame fusion of an open reading frame to the 5' end of the endonuclease structural gene results in synthesis of a hybrid protein that retains endonucleolytic activity. Mutants of the endonuclease are described that allow the activity of the enzyme to be modulated independently of its synthesis.

[1]This research was supported by National Institutes of Health GM31105, and by the University of California Cancer Research Coordinating Committee, and by USPHS Training Grant CM07232.

INTRODUCTION

The nucleus is the largest organelle of eucaryotic cells, yet the mechanisms that operate to ensure that nuclear proteins accumulate in the nucleus and nonnuclear proteins do not are largely unknown. On topological grounds, the localization of proteins to the nucleus may be an example of protein sorting that operates independently of the mechanisms that sort proteins destined for the golgi, vacuole, plasma membrane, or secretion. For example, the nucleus is surrounded by an inner nuclear membrane and an outer nuclear membrane (1). Furthermore, the outer nuclear membrane is contiguous with the endoplasmic reticulum (2). Thus a protein that enters the endoplasmic reticulum, or any other secretory organelle, is in a space topologically equivalent to the space between the inner and outer nuclear membranes and not equivalent to the nucleoplasm.

Nuclear pores span the inner and outer nuclear membranes and offer a potential route for the flow of macromolecules between the nucleus and cytoplasm. The nuclear pore has a complex structure consisting of an octet of subunits forming a ring around a central plug (3). In general, pores display remarkable structural conservation among diverse organisms. The strongest evidence that pores are a route for nuclear-cytoplasmic exchange comes from experiments using electron microscopy. In these experiments, colloidal gold particles were coated with a nuclear protein of Xenopus oocytes that has been shown to contain a nuclear signal sequence (4). Upon injection into the cytoplasm of oocytes, the gold particles become concentrated in and around nuclear pores (5). This result indicates that at least some molecules may enter the nucleus through the pores, but leaves open the possibility that other proteins enter the nucleus by other routes.

Conceptually, the accumulation of a protein within the nucleus has three requirements. First, the protein must reach the nucleus after leaving the ribosome. Second, the protein must pass through the nuclear membranes. Third, the rate of nuclear entry must exceed the rate of nuclear exit. Nuclear accumulation could be achieved by an active, energy requiring process in which nuclear proteins are recognized and transported into the nucleus against a concentration gradient, perhaps by a carrier or shuttle protein. Alternatively, nuclear accumulation could be achieved through passive entry of all proteins into and out of the nucleus with selective retention of those destined to be accumulated in the nucleus. At present, there is insufficient evidence to favor either class of models.

A theme that has emerged from other studies of protein localization is that signals present in the structure of proteins determine their route through the cell. Similarly, nuclear protein contains signals that are sufficient to direct their nuclear localization. Nuclear localization signals are short amino acid sequences that need not occur at any particular place within a protein. Furthermore, when fused to heterologous proteins, the nuclear signals result in the accumulation of the heterologous protein in the nucleus. Nuclear signals appear not to be removed from nuclear proteins during or after nuclear entry (4,6,7,8,9).

The work presented in this paper has two foci. First, the issue of whether or not the presence of a nuclear signal is required for the entry of a protein into the nucleus is addressed. For these experiments, we assume that procaryotic proteins will lack signal sequences for nuclear localization. The second focus will be the development of a genetic selection that may allow the isolation of mutations that block the entry of proteins into the nucleus. These mutations should provide critical information on the route of nuclear entry and on the mechanisms that distinguish nuclear proteins from nonnuclear proteins.

RESULTS

Production of EcoRI Endonuclease in Yeast

Two plasmids have been constructed that contain the EcoRI coding sequences under the transcriptional control of the yeast GAL1 promoter (Figure 1; ref. 10). Only four nucleotides 5' of the endonuclease coding sequence were retained in this plasmid in an effort to prevent fortuitous production of the endonuclease in E. coli. The two plasmids, YCpGal:RIa and YCpGal:RIb, differ only by the presence or absence, respectively, of a unique EcoRI cleavage site. Transcription from the GAL1 promoter is under both positive control, with galactose acting as the inducer, and negative control by catabolite repression when the cells are grown on media containing glucose (11). The YCpGal:HIS3 plasmid consists of the yeast HIS3 gene under the transcriptional control of the GAL1 promoter and serves as a control plasmid in all experiments.

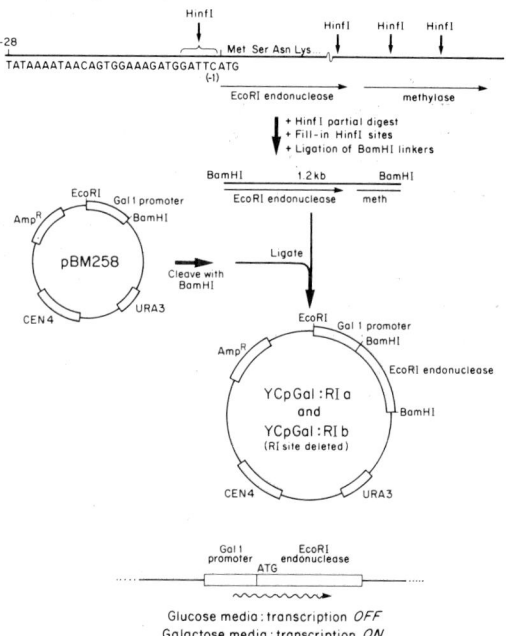

FIGURE 1. Construction of the YCpGal:RIa and -b fusion plasmids. The EcoRI endonuclease coding sequences were placed under the regulation of the GAL1 promoter as shown. The mode of transcriptional regulation of the promoter-gene fusion is shown at the bottom. The wavy arrow indicates the direction of transcription of the YCpGal:RI plasmids. pBM125 is described in reference 10.

In order to evaluate the effects of EcoRI production in yeast, two types of strains were used: rad52 mutants and wild type cells (RAD52). rad52 mutants are an exquisitely sensitive monitor of double stranded breaks in DNA. In these mutants, one double stranded break is a lethal event (12,13). Thus, these mutants should be very sensitive to any EcoRI endonuclease that may enter the yeast nucleus. A RAD52 strain (JRY438) and a rad52 strain (JRY481) were transformed with YCpGal:RIa, YCpGal:RIb, and YCpGal:HIS3 and the transformants tested for the ability to grow on media with inducing or repressing carbon sources (Figure 2). Either

The Nucleus of *Saccharomyces cerevisiae*

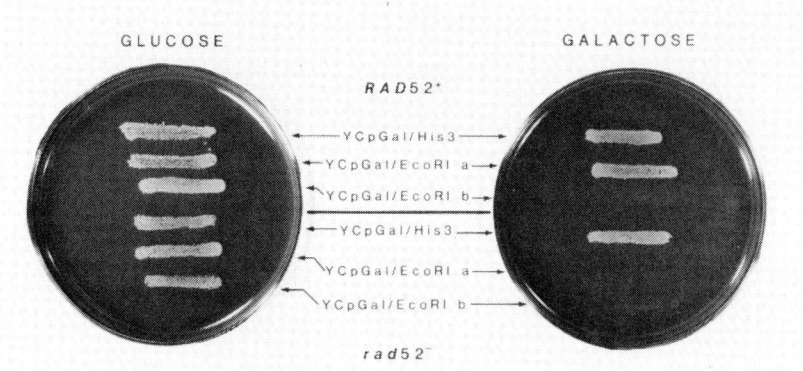

FIGURE 2. Growth of yeast transformants. Yeast strains JRY438 (RAD52) and JRY481 (rad52), transformed with the YCpGal:RIa or -b plasmids or the control YCpGal:HIS3 plasmid, were tested for the ability to grow under conditions inducing or repressing transcription of the GAL1 promoter. (Left) Plate containing YM medium with 2% glucose (repressing conditions). (Right) Plate containing YM medium with 2% galactose (inducing conditions). All cells were grown in glucose medium and then duplicate aliquots were plated. The plates were incubated at 30°C for 2 days before being photographed. Note that JRY481 (rad52) cells are more sensitive to the presence of the YCpGal:RIa plasmid than are JRY438 (RAD52) cells.

RAD52 or rad52 strains grow well on glucose containing media regardless of the plasmid present in the transformants. On galactose containing media, either strain containing the YCpGal:HIS3 plasmid grows well. However, neither strain is capable of growth when it contains the YCpGal:RIb plasmid, indicating that a toxic product is produced in these cells. The difference in sensitivity of RAD52 and rad52 cells to the effect of YCpGAL:RIa suggests that the toxicity is due to double stranded DNA breaks. The viability of RAD52 cells containing the YCpGal:RIa plasmid is somewhat surprising. Since the EcoRI site in this plasmid is near the GAL1 promoter, it is possible that cleavage of this site and repair of the plasmid occur in such a way as to mutate the GAL1 promoter.

Kinetics of Cell Death

In order to determine whether expression of the EcoRI gene merely inhibits growth on media containing galactose or actually causes cell death, the plating efficiency of individual cells was determined at increasing times after galactose induction. At each time point, individual cells were isolated by micromanipulation and transferred to minimal glucose medium to determine if, and when, cells died after induction. Death is defined as the inability to form a colony. The plating efficiency of RAD52 or rad52 cells containing the YCpGal:RIb plasmid grown in either glucose (repressing media) or glycerol (nonrepressing, noninducing) media was uniformly high (Figure 3). With cells pre-grown in glucose containing medium and then shifted to galactose containing medium, the viability of rad52 cells began to decrease 8 hours after induction. The viability of RAD52 cells began to decrease after 10 hours. When the same cells were pregrown in glycerol containing medium and then shifted to galactose medium, the viability of both RAD52 and rad52 cells falls approximately 1.5 to 2 hours after induction. These results demonstrate that galactose induction of the YCpGal:RIb plasmid results in cell death and not simply growth inhibition. The difference in the kinetics of killing in the two experiments is proportional to the time required to induce expression from the GAL1 promoter under these two growth conditions (14).

EcoRI Endonuclease Activity in Yeast Lysates

The results described above predict that, upon induction, cells containing a YCpGal:RI plasmid contain active EcoRI endonuclease. To test this prediction, lysates were prepared from cells containing the YCpGal:RIb plasmid and assayed for endonuclease activity both before and 18 hours after a shift from glucose containing media to galactose containing media. In order to assay for activity, bacteriophage lambda DNA was incubated with aliquots of the lysate and cleavage monitored by agarose gel electrophoresis. As seen in Figure 4, lysates from galactose induced cells contain a substantial amount of EcoRI endonuclease activity. Additional experiments have shown that EcoRI activity is not detectable in cells grown in media containing glycerol as the carbon source nor in cells containing the YCpGal:HIS3

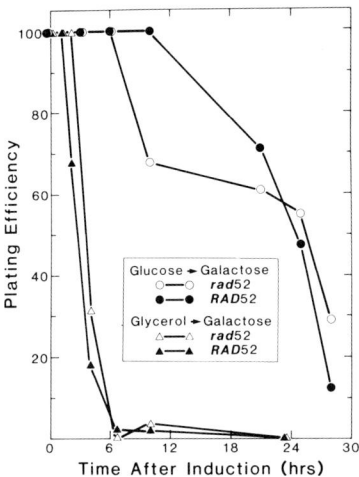

FIGURE 3. Plating efficiency assay. This assay demonstrates the effects on the growth of rad52 and RAD52 cells containing the YCpGal:RIb plasmid after galactose induction. As described in the RESULTS, JRY481 and JRY438 cells transformed with either the control YCpGal:HIS3 or YCpGal:RIb plasmid were grown to mid log phase in either YM glucose or YM glycerol media. The cells were harvested by centrifugation, washed once in YM, and resuspended in YM galactose media (0 hour). Aliquots were removed at 0, 3, 6, 10, 21, 25, and 28 hour time points for glucose to galactose shifted cultures and at 0, 1, 2, 4, 6.5, 10, and 23.5 hour time points for glycerol to galactose shifted cultures. 50 cells were individually isolated by micromanipulation to monitor the viability of specific cells. Plating efficiency (%) was calculated as the number of viable cells with the YCpGal:RIb plasmid/number of viable cells with the YCpGal:HIS3 plasmid. The cells containing the YCpGal:HIS3 plasmids maintained a viability of 90-100% throughout the experiment. The slower response to the shift from glycerol of the JRY481 (rad52) cells is probably due to the longer generation time of these cells. Note that both strains respond to the presence of the EcoRI endonuclease within one generation time.

plasmid grown in media containing galactose. Furthermore, endonuclease activity is detectable at least as early as 6 hours after induction, a time at which the plating efficiency of the same cells is zero.

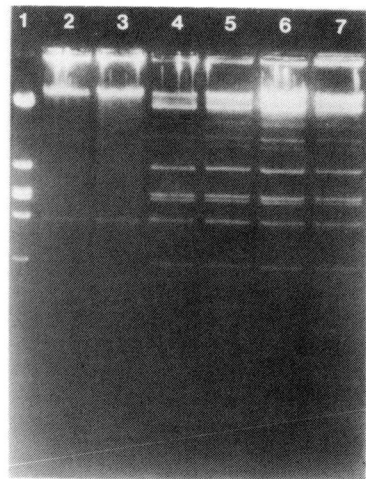

FIGURE 4. EcoRI endonuclease activity in cell lysates. JRY438 and JRY481 cells transformed with YCpGal:RIb plasmid were grown in medium containing glucose, and then half of each culture was shifted to medium containing galactose and incubated for 18 hr. Lysates were prepared by glass-bead disruption and were assayed for EcoRI endonuclease activity by incubating aliquots (50 or 100 µl volumes) with 1 µg of phage λ cI857 S7 DNA for 1 hr at 37°C. The reaction was then electrophoresed on a 1% agarose gel. Lanes: 1, control digest of phage λ cI857 S7 DNA with commercial EcoRI; 2 and 3, assays of 100 µl aliquots of lysates prepared from JRY438 and JRY481 cells with the YCpGal:RIb plasmid grown in glucose; 4 and 5, 50 and 100 µl aliquots, respectively, of lysates from 438 cells with the YCpGal:RIb plasmid, which were induced in galactose; 6 and 7, 50 and 100 µl aliquots, respectively, of lysates from 481 cells with the YCpGal:RIb plasmid induced in galactose.

EcoRI Cleaves Nuclear DNA in vivo

If the lethality caused by EcoRI endonuclease is due to damage to nuclear DNA, then this damage should be evident in a gel transfer hybridization experiment (Figure 5). Genomic DNA was isolated from both RAD52 and rad52 strains containing either the YCpGal:RIa or the YCpGal:RIb plasmid. Samples were taken both before (0 hour lanes) or six hours after a shift from glycerol to galactose containing medium. The DNA

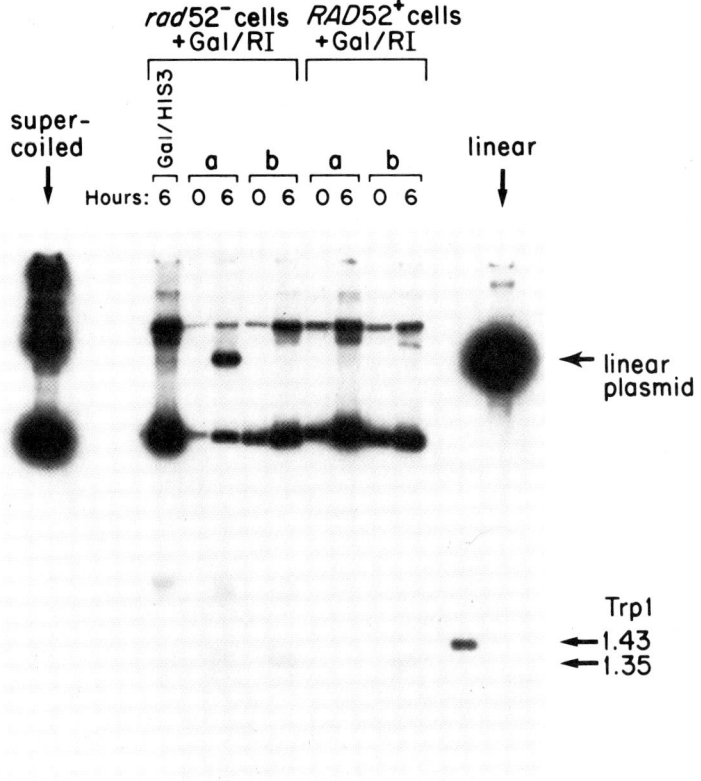

FIGURE 5. Analysis of EcoRI activity in vivo. In a gel transfer hybridization experiment, DNA was isolated from RAD52 and rad52 cells containing the YCpGal:RIa or -b plasmid (lanes designated a or b). The cells were induced for 6 hr in galactose medium or were grown under noninducing conditions in glycerol medium (0 hr). The DNA was probed with radio-labeled YRp7 DNA. The top portion of the gel shows that plasmid DNA isolated from all cells at 0 hr is intact and supercoiled. The yeast TRP1 gene is flanked by EcoRI sites in the chromosome, and cleavage of yeast chromosomal DNA in vitro with EcoRI generates the 1.43 kb TRP1 gene fragment.

was electrophoretically separated on an agarose gel, transferred to nitrocellulose, and hybridized to radiolabelled YRp7 DNA. YRp7 consists of pBR322 plus the 1.43 EcoRI restriction fragment that contains the yeast TRP1 gene (15).

This probe serves to measure the integrity of all plasmids used in this experiment as well as the chromosomal DNA in the vicinity of the TRP1 gene. The outermost lanes in the autoradiogram are controls showing the mobility of either supercoiled YCpGal:RIa plasmid or the mobility of YCpGal:RIa linearized in vitro with purified EcoRI endonuclease. In induced or uninduced RAD52 cultures, both YCpGal:RIa and YCpGal:RIb plasmids have the mobility of supercoiled plasmids. However, in galactose induced rad52 cultures, YCpGal:RIa has the mobility of linearized plasmid, whereas the YCpGal:RIb retains the mobility of supercoiled plasmid. Thus, the EcoRI endonuclease produced in these cells is capable of entering the nucleus and cleaving DNA at EcoRI sites and apparently only at EcoRI sites since the plasmid lacking an EcoRI site remained supercoiled. We infer that the supercoiled state of the YCpGal:RIa plasmid in RAD52 cells and their ability to grow in media containing galactose are both due to alteration of the sequences at and near the EcoRI site such that the EcoRI site is mutated and endonuclease expression is impaired.

The effect of the endonuclease on chromosomal DNA is evident in the lower portion of Figure 5. A fragment is present in the lanes from the induced cells that is approximately 1.35 kilobases in length and hybridizes to the TRP1 gene. As indicated in the control lane containing genomic DNA cleaved in vitro with EcoRI endonuclease, the TRP1 gene is present on a 1.43 kilobase EcoRI restriction fragment. The slightly smaller size and broader appearance of the TRP1 fragment in the induced lanes is probably due to exonuclease digestion in vivo since, with longer induction times, the band is smaller and more diffuse. As expected, this band is absent in the lanes containing DNA prepared from cells containing the YCpGal:HIS3 plasmid and from all uninduced cultures.

Phenotypes of Cells Producing EcoRI

A priori, it is difficult to predict whether the absense of intact chromosomal DNA would have any specific effect on the stage in the cell cycle at which the cells would arrest. In control cultures containing the YCpGal:HIS3 plasmid, all cells had a normal morphology. In contrast, a subset of cells induced for the expression of the endonuclease is shown in Figure 6. These cells with aberrant morphologies represent a total of 10% of the cells in the induced

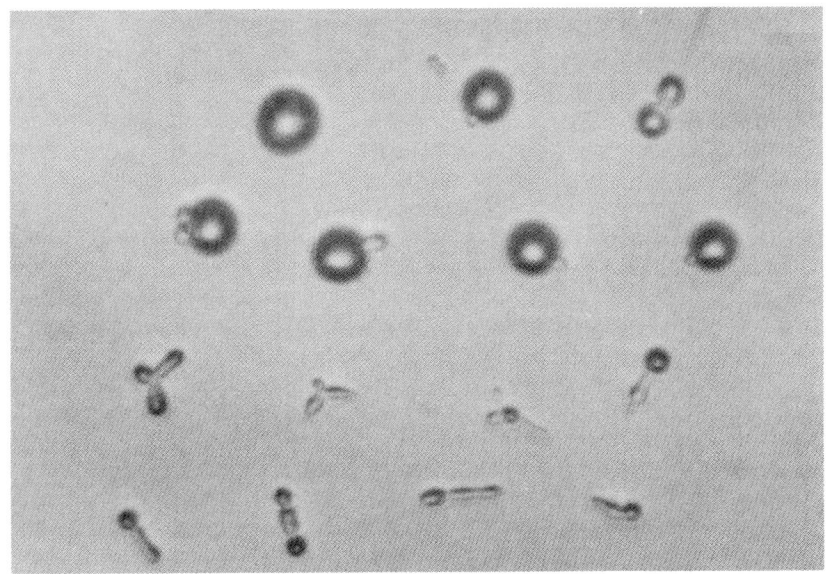

FIGURE 6. The phenotypes of cells expressing EcoRI endonuclease. In cultures expressing EcoRI endonuclease, 10% of the cells exhibited one of two different morphologies shown in this figure. Both novel cell shapes have a volume much larger than the galactose induced control cell shown in the top center of the figure.

culture. Ninety percent of the cells appeared normal and were distributed through the cell cycle similarly to the control culture.

At this point it is unclear why only 10% of the cells exhibit this unusual morphology. Perhaps only cells in one portion of the cell cycle are competent to continue this unusual growth without cell division. The large size of these cells suggests that metabolism may continue in the absence of intact DNA. Perhaps cells with cleaved chromosomal DNA will only produce gene products whose messages are transcribed from intact DNA molecules such as the supercoiled YCpGal:RIb plasmid. Therefore, these cells may be analogous to bacterial maxicells (16) and may prove useful in identification and accumulation of gene products ordinarily made in low abundance.

Hybrid Proteins

The results presented above indicate that the EcoRI endonuclease is able to enter the yeast nucleus. Thus, unless the endonuclease has by chance a sequence equivalent to a nuclear signal, it is unlikely that a nuclear signal is required for nuclear entry per se. However, these results do not bear on the question of whether the route taken by the EcoRI endonuclease into the nucleus is the same route taken by authentic yeast nuclear proteins. In order to determine whether the endonuclease could be used to tag authentic nuclear proteins and assay their nuclear entry, we determined whether hybrid proteins could be formed with the endonuclease that would retain endonuclease activity. In the first experiment, 11 amino acids were added to the amino terminus of EcoRI endonuclease by the construction shown in Figure 7. This construction results in the expression of the hybrid gene under the transcriptional and translational control of the E. coli lac operon. The effect of this construction on E. coli cells is shown in Figure 8. On glucose medium, the lac operon is not expressed and the cells are viable. However, in the presence of IPTG, expression of the hybrid endonuclease protein results in death of the cell. To determine whether the hybrid protein was causing cell death, or whether the hybrid protein was being proteolyzed in vivo and thereby regenerating the active native form of the endonuclease, the integrity of the hybrid endonuclease protein was examined by a western blot using antibody directed against the endonuclease. As can be seen in Figure 9, cells containing a plasmid that encodes authentic EcoRI endonuclease produce a protein that is 31 Kdal as predicted from the sequence of the endonuclease gene (18). Cells containing the plasmid encoding the endonuclease with the 11 amino acid N-terminal extension produce an endonuclease that is approximately one kilodalton larger as predicted. This result indicated that hybrid proteins can be constructed with the endonuclease that still contain endonuclease activity.

Since hybrid endonuclease proteins can retain endonucleolytic activity, a plasmid was constructed that should fuse the aminoterminal portion of the yeast GAL4 protein in frame with the EcoRI endonuclease as shown in Figure 10. Since this portion of GAL4 contains a signal for nuclear localization of the GAL4 protein (7), this hybrid protein should be localized by whatever means bona fide nuclear proteins are localized. The hybrid genes were constructed on a plasmid

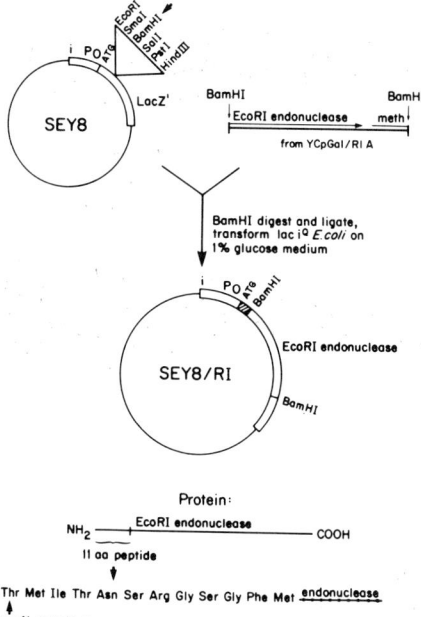

FIGURE 7. Construction of a hybrid endonuclease gene. The EcoRI endonuclease coding sequence (19) was ligated in frame to an open reading frame that includes the amino terminus of β-galactosidase and sequences contributed by the poly linker. In the resulting plasmid, transcription and translation of the hybrid protein are controlled by the regulatory sequences of the lac operon. SEY8 is described in (17).

on which the expression of the hybrid protein is regulated by the transcriptional control of the GAL1 promoter. The data in Figure 10 describe the growth properties of cells containing these plasmids. It appears from the difference in sensitivity of the RAD52 and rad52 cells, that both hybrid proteins retain at least partial EcoRI endonuclease activity. The smaller hybrid, which still contains the GAL4 nuclear signal sequence, may have more endonuclease activity. These results suggest that hybrid proteins between GAL4 and the endonuclease may be useful for studying the route that authentic nuclear proteins take into the nucleus.

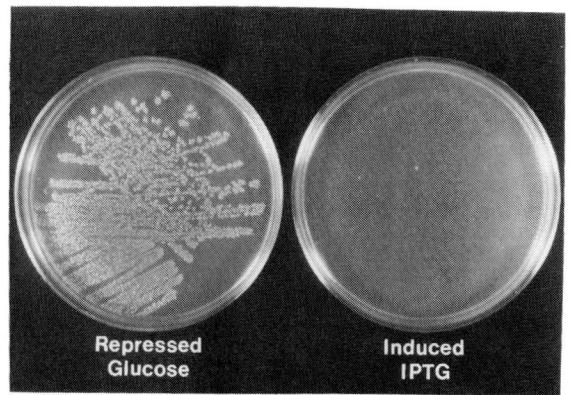

FIGURE 8. A laci^Q E. coli strain (18) was transformed with the SEY8/RI plasmid diagrammed in Figure 7. A transformant was streaked out on an L plate containing 1% glucose and, after growth overnight at 37°C, was replica-plated onto an L plate containing 1% glucose and onto an L plate containing isopropylthio-β-D-galactoside (80 μM). As indicated above, induction of the lac operon with IPTG blocks all growth.

Selection of Mutations in Nuclear Localization

One goal of this research is to identify the route of protein entry into the nucleus by identification of mutations in the genes encoding proteins that function along the route(s). The nuclear specific lethality associated with expression of the EcoRI endonuclease will be used to select for mutations that block nuclear entry of the endonuclease. The import of proteins into the nucleus is, in all probability, an essential process. Therefore, mutations defective in this process will have to be isolated as conditional lethals such as temperature sensitive mutants. A conceptual problem with using conditional expression of the endonuclease to select for conditional mutations in nuclear import is that production of endonuclease can be controlled, but activity of previously synthesized endonuclease cannot. Thus, mutants selected at the high temperature as survivors of endonuclease induced lethality, will not grow at the high temperature due to the inability of bona fide nuclear proteins to enter the nucleus. Furthermore, if the temperature is lowered to allow growth of the mutants, and glucose is added to block further expression of the endonuclease, the mutants may still die

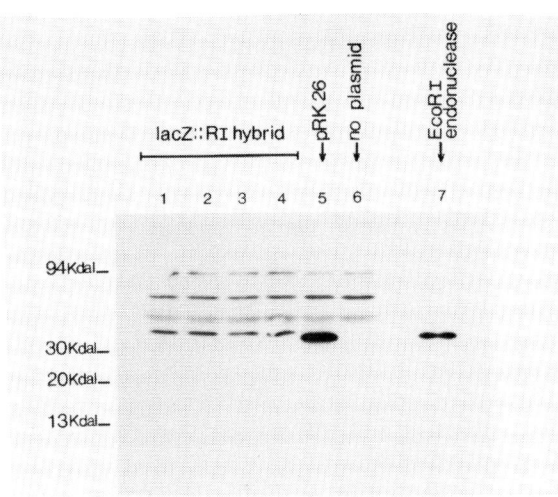

FIGURE 9. Western blot analysis of hybrid EcoRI endonuclease proteins. Lanes 1-4 contain protein extracts from an induced E. coli laci^Q strain containing the SEY8/RI plasmid that is described in Figure 7. Lane 5 contains an extract from the same E. coli strain containing pRK26, a plasmid that contains the EcoRI endonuclease-methylase operon (a gift from Rosalind Kim). Lane 6 contains an extract of E. coli lacking a plasmid. Lane 7 contains 10 units of EcoRI endonuclease obtained from a commercial source. After electroblotting the gel to a nitrocellulose filter, the filter was incubated with antibodies directed against EcoRI endonuclease. Immune complexes were visualized with ^{125}I labelled protein A followed by autoradiography. The three largest protein bands in lanes 1-6 are background. The smallest protein band contains the endonuclease determinant. By comparison of lanes 1-4 with lanes 5 and 7, it is evident that the hybrid endonuclease gene produces a hybrid protein that is larger by the amount predicted by the amino acid sequence.

due to subsequent entry of endonuclease that had accumulated in the cytoplasm. In an effort to circumvent this genetic conundrum, mutants were isolated in the EcoRI endonuclease gene that rendered the activity of the endonuclease sensitive to the osmotic strength of the media that the cells are grown in. In these experiments, approximately 5×10^7 cells containing the YCpGal:RIb plasmid were plated onto yeast minimal

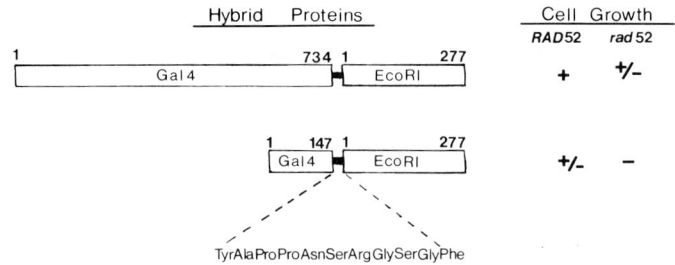

FIGURE 10. A plasmid containing the yeast GAL4 gene under the transcriptional control of the GAL1 promoter was provided by R. Gesteland. In frame fusions between convenient restriction sites in the GAL4 sequence and the endonuclease gene were constructed in this plasmid with the aid of an oligonucleotide linker that encodes the amino acid sequence shown. The numbers refer to the amino acids in the GAL4 and endonuclease sequence that should be present in the resulting hybrid proteins. Plasmids containing the hybrid proteins were transformed into RAD52 and rad52 strains on glucose containing media. The growth characteristics of the transformants on galactose containing media are indicated.

media containing galactose as the carbon source. After three days of incubation, survivors were picked, patched onto a master plate, and after subsequent growth, were replica-plated onto yeast minimal media containing galactose and 1.2 M KCl. Colonies that grew on normal minimal media (low osmolarity media) but not on high osmolarity media were picked as candidates for being osmotic remedial EcoRI endonuclease mutants. Of 13 candidates, only 8 contained mutations that mapped to the plasmid. By resubcloning the endonuclease gene from each of these candidates into the original GAL1 expression vector, it was discovered that in only four plasmids was the osmotic remedial galactose induced lethality due to a mutation in the endonuclease coding sequence. These plasmids with osmotic remedial endonuclease mutations provide essential flexibility in the use of the endonuclease as a selection for mutations in nuclear localization. The GAL1 promoter provides the ability to regulate production of the endonuclease and the osmotic remedial mutation provides an independent means of controlling endonuclease activity. Although osmotic remedial mutants are frequently temperature sensitive as well (20,21), these osmotic remedial endonuclease mutants are not. Thus, the use of temperature as the variable condition for isolation of mutations affecting nuclear localization should still be feasible.

CONCLUSIONS

The experiments described here demonstrate that the entry of a protein into the nucleus has no apparent requirement for a nuclear signal sequence. Although it is possible that the EcoRI endonuclease fortuitously contains a sequence that resembles a nuclear signal sequence, we have found no significant homology between the endonuclease and any of the identified nuclear signal sequences. Our experiments do not indicate how many routes there may be for a protein to enter the nucleus. Nor do these experiments provide information on the rate of the endonuclease entry into the nucleus, nor whether the endonuclease accumulates within the nucleus. The role of nuclear signal sequences may be to affect the extent or rate of nuclear entry, or may enable proteins to enter by routes that are not available to foreign or cytoplasmic proteins.

The EcoRI endonuclease is a soluble protein and is not processed during or after synthesis. Therefore the entry of this protein into the nucleus is unlikely to require passage across a membrane. Since the nuclear pores are the only known structures that allow the cytoplasm and nucleoplasm to be continuous, it is probable that the endonuclease enters the nucleus by passing through the pores. The eventual characterization of mutations that exclude the endonuclease from the nucleus will indicate whether all proteins enter the nucleus by the same route.

The foreign origin of EcoRI endonuclease may be a substantial benefit in studies of nuclear localization. Whereas most of the yeast nuclear proteins that have been studied interact with other proteins in yeast, it is unlikely that the endonuclease location would benefit from or suffer from interactions with other yeast proteins. Therefore, the location of the endonuclease may be free of the complications imposed by interacting proteins and, thus, may reflect all of the locations that an average sized protein can reach if not excluded by a specific mechanism.

Osmotic remedial mutants have been used as conditional mutants in a variety of organisms including yeast (20,21). The osmotic remedial mutants of the endonuclease are an important component in the strategy for isolation of cellular mutations that limit the entry of endonuclease into the nucleus since these mutants allow the synthesis and activity of the enzyme to be controlled separately. Osmotic remedial mutants may be particularly useful in the study of protein localization in general. In addition, since only four out of thirteen osmotic remedial mutations mapped to the endonuclease coding sequence, osmotic remedial mutants may provide an efficient route toward isolation of new types of mutations affecting gene expression.

ACKNOWLEDGMENTS

We thank Scott Emr, Ray Gesteland, Rosalind Kim, and Mark Johnston for various plasmids that were essential for these experiments. We also thank Paul Modrich for generously supplying antibodies against EcoRI endonuclease.

REFERENCES

1. Schatten G, Thoman M (1978). Nuclear surface complex as observed with the high resolution scanning electron microscope. J Cell Biol 77:517.
2. Franke WW, Scheer U, Krohne G, Jarasch ED (1981) The nuclear envelope and the architecture of the nuclear periphery. J Cell Biol 91:39s.
3. Unwin PNT, Milligan RA (1982) A large particle associated with the perimeter of the nuclear pore complex. J Cell Biol 93:63.
4. Dingwall C, Sharnick SV, Laskey RA (1982) A polypeptide domain that specifies migration of nucleoplasmin into the nucleus. Cell 30:449.
5. Feldherr CM, Kallenbach E, Schultz N (1984). Movement of a karyophilic protein through the nuclear pores of oocytes. J Cell Biol 99:2216.
6. Hall MN, Hereford L, Herskowitz I (1984). Targeting of E. coli β galactosidase to the nucleus in yeast. Cell 36:1057.
7. Silver PA, Keegan LP, Ptashne M (1984). Amino terminus of the yeast GAL4 gene product is sufficient for nuclear localization. Proc Natl Acad Sci USA 81:5951.
8. Kalderon D, Richardson WD, Markham AF, Smith AE (1984). Sequence requirements for nuclear localization of Simian virus 40 large-T antigen. Nature 211:33.

9. Kalderon D, Roberts BL, Richardson WD, Smith AE (1984). A short amino acid sequence able to specify nuclear localization. Cell 39:499
10. Barnes G, Rine J (1985). Regulated expression of endonuclease EcoRI in Saccharomyces cerevisiae: Nuclear entry and biological consequences. Proc Natl Acad Sci USA 82:1354.
11. Johnston M, Davis RW (1984). Sequences that regulate the divergent GAL1-GAL10 promoter in Saccharomyces cerevisiae. Mol Cell Biol 4:1440.
12. Ho KSY, Mortimer RK (1973). Induction of dominant lethality by X-rays in a radiosensitive strain of yeast. Mutat Res 20:45.
13. Malone RE, Esposito RE (1980). The RAD52 gene is required for homothallic interconversion of mating types and spontaneous mitotic recombination in yeast. Proc Natl Acad Sci USA 77:503.
14. Adams BG (1972). Induction of galactokinase in Saccharomyces cerevisiae: kinetics of induction and glucose effects. J Bacteriol. 111:308.
15. Stinchcomb DT, Struhl K, Davis RW (1979). Isolation and characterization of a yeast chromosomal replicator. Nature 282:39.
16. Sancar A, Hack AM, Rupp WD (1979). Simple method for identification of plasmid-coded proteins. J Bacteriol 137:692.
17. Schauer I, Emr S, Gross C, Schekman R (1985). Invertase signal and mature sequence substitutions that delay intercompartmental transport of active enzyme. J Cell Biol 100:1664.
18. Müller-Hill B, Crapo L, Gilbert W (1968). Mutants that make more lac repressor. Proc Natl Acad Sci USA 59:1259.
19. Green PJ, Gupta M, Boyer HW, Brown WE, Rosenberg JM (1981). Sequence analysis of the DNA encoding the EcoRI endonuclease and methylase. J Biol Chem 256:2143.
20. Hawthorne DC, Friis J (1964). Osmotic-remedial mutants. A new classification for nutritional mutants in yeast. Genetics 50:829.
21. Kohno T, Roth J (1978). Electrolyte effects on the activity of mutant enzymes in vivo and in vitro. Biochem 18:1386.

NUCLEAR PROTEIN LOCALIZATION IN
SACCHAROMYCES CEREVISIAE[1]

Pamela Silver

Department of Biochemistry & Molecular Biology,
Harvard University, Cambridge, Mass. 02138

Hybrid proteins containing as few as 74 NH_2-terminal amino acids of GAL4, a yeast positive regulatory protein, at the amino terminus of E. coli β-galactosidase accumulate in the cell nucleus. GAL4-linked mutants that display altered localization of the normally nuclear-associated fusion protein have been isolated.

INTRODUCTION

The nucleus, like all organelles, is composed of a unique set of proteins necessary for its structure and function. It is surrounded by a double membrane with pores distributed throughout. The diameter of the nuclear pore has been estimated to be about 90Å (1). The nuclear pore has been proposed for the site of entry of proteins into the nucleus. Only recently has good experimental proof demonstrated that one protein injected into the cytoplasm of the Xenopus oocyte can actually traverse the nuclear pore to gain access to the nuclear interior (2).
Several possible mechanisms have been proposed for how specific proteins are localized to the cell nucleus (reviewed in ref. 3). By one model, nuclear association may be the result of simple diffusion of proteins through the nuclear pore and selective retention of a subset of proteins within the nucleus. By an alternative model, receptors on the nuclear envelope (or even in the

---

[1] This work was supported by NIH grant GM 32308 to Mark Ptashne in whose lab this work was done, and by a postdoctoral fellowship from NIH (GM 08871).

cytoplasm) may recognize specific amino acid sequences or tertiary structures and allow passage of only certain proteins across the nuclear envelope. Specific information for nuclear localization would then reside in the transported protein. This mechanism is distinguished from the first because it suggests selective transport into the nucleus rather than passive retention of a subset of proteins.

EXPERIMENTAL APPROACH AND RESULTS

In order to define the steps of nuclear protein localization, several groups have undertaken studies of the yeast, S. cerevisiae. The yeast nucleus has many structural features in common with that of mammalian cells. By applying a combined genetic and biochemical approach ammenable to yeast, necessary cellular components will hopefully be identified. One now common approach is to use protein fusions to study protein trafficking in yeast. In the case of nuclear protein assembly, chimeric proteins are formed between portions of nuclear proteins and normally non-nuclear proteins. The intracellular location of the hybrid proteins is determined by immunofluorescence and cell fractionation.

The hybrid proteins are the products of gene fusions constructed in vitro. These gene fusions can be easily introduced into yeast on autonomously replicating plasmids. The product of the E. coli lacZ gene, β-galactosidase, is often chosen as the non-nuclear component of the chimeric protein for several reasons. As a soluble protein of the E. coli cytoplasm, β-galactosidase is probably not specifically localized to any subcellular organelle in yeast. Its hydrophilicity would not allow it to pass through membranes, and its size (116kd) might limit its ability to diffuse through the nuclear pore. It has been demonstrated that fusions between yeast genes and lacZ when introduced into yeast, produce active β-galctosidase (4,5). The presence of the β-galactosidase then provides both a biochemical and immunological tag.

The product of the yeast GAL4 gene has been particularly useful for the study of nuclear protein uptake in the yeast S. cerevisiae (6). As a positive activator protein of the yeast genes necessary for galactose metabolism, it interacts directly with specific stretches of DNA (7,8), and is hence a nuclear protein.

Under normal growth conditions the GAL4 protein is not required. However, it is absolutely necessary for cell growth on galactose. These characteristics make the GAL4 protein useful for genetic studies. Finally, the GAL4 gene has been isolated and its nucleotide sequence determined (9,10). The protein has been overproduced in E. coli and shown to bind to specific DNA sequences in vitro (L. Keegan, P. Silver, and M. Ptashne, unpublished).

The approach of using fusion proteins has been used to study the intracellular localization of the GAL4 gene product (6). Gene fusions with various amounts of the GAL4 gene fused to the end of the E. coli lacZ gene encoding the N-terminus of β-galactosidase have been constructed. When these gene fusions are placed into yeast, chimeric proteins are produced that have β-galactosidase activity. The intracellular distribution of these chimeric proteins can be determined by indirect immunofluorescence using anti-β-galactosidase antibody (Figure 1a). By this type of analysis, the first 74 N-terminal amino acids of the GAL4 gene product were shown to be sufficient to localize the GAL4-β-galactosidase chimeric protein to the yeast nucleus. A fusion protein containing only the first 18 amino acids of the URA3 gene product fused to LacZ is, on the other hand, not localized to the nucleus. The immunofluorescent material appears to be excluded from the nucleus, consistent with the notion that β-galactosidase alone cannot gain access to the nucleus (Figure 1c).

Recent studies of nuclear uptake of proteins in yeast and mammalian cells have lent support to the notion that specific nuclear targeting signals lie in the transported protein. A single amino acid change in SV40 T antigen renders the protein non-nuclear in vivo but still capable of specific DNA binding in vitro (11). This finding led to the suggestion that a very short stretch of amino acids with the sequence pro-lys-lys-lys-arg-lys-val could act to direct a protein to the nucleus. Similar experiments with the yeast MATα2 gene product have led Hall et al. (12) to propose that some nuclear proteins contain the specific sequence lys-ile-pro-ile-lys, which is responsible for targeting them to the nucleus. No strong homology to either of these sequences occurs in the first 74 amino acids of GAL4.

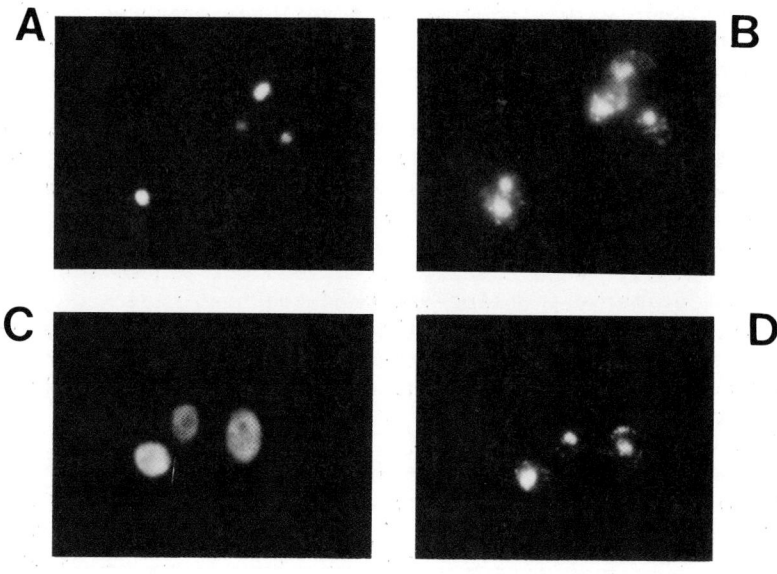

FIGURE 1. Immunofluorescence of cells producing nuclear-associated GAL4-LacZ chimeric proteins (A,B) or cytoplasmic URA3-LacZ chimeric proteins (C,D). Cells were prepared for indirect immunofluorescence and treated with either rabbit anti-β-galactosidase antibody, followed by FITC-conjugated anti-rabbit IgG (A,C) to identify the chimeric proteins, or DAPI (B,D) to identify cell nuclei. A and B, GAL4-LacZ; C and D, URA3-LacZ.

An unexpected finding was that chimeric proteins lacking the nuclear determinant (and apparently excluded from the nucleus) are produced at higher levels when compared to the level of the nuclear associated chimeras (6). On Xgal indicator plates, yeast producing non-nuclear chimeric proteins form bluer colonies than yeast harboring otherwise isogenic gene fusions that encode nuclear proteins. This finding was used as the basis for the isolation of several GAL4-linked mutations that alter the intracellular location of a normally nuclear GAL4-β-galactosidase fusion protein. The mutations are single base substitutions within the GAL4 coding sequence. The amino acid changes are listed in Table 1.

TABLE 1
MUTATIONS AFFECTING GAL4-β-GALACTOSIDASE LOCALIZATION

Mutant Allele	Amino Acid Change
gal4-N1-lacZ	ser 6 to phe
gal4-N2-lacZ	glu 8 to lys
gal4-N3-lacZ	lys 20 to glu
gal4-N4-lacZ	arg 51 to ser

As one can see, the amino acid changes are distributed throughout the entire 74 amino acids and are not localized to any one particular region. Further analysis is required to distiguish mutations that may define a nuclear localization determinant from those that may simply cause the protein to fold incorrectly.

DISCUSSION

The fact that nuclear associated chimeric proteins are present at lower levels than otherwise non-nuclear chimeras encoded by isogenic gene fusions may reflect some fundamental aspect of the nuclear localization process. It is possible that nuclear transport somehow limits the amount of a nuclear protein made by the cell. For example, as suggested above, there may be a limited number of nuclear-envelope associated receptors that recognize some property of the first 74 GAL4 amino acids. Once these sites are fully occupied, newly synthesized protein (presumably in the cytoplasm) that can no longer bind to, or be translocated into the nucleus, may instead be degraded. A mutation might render the protein resistant to degradation. Alternatively, it is possible that the mRNA's encoding the non-nuclear proteins are transcribed or translated at a faster rate.

By the above type of analysis, studies of the yeast MATα2 (12), GAL4 (6), and histones H2B (B. Moreland and L. Hereford, unpublished) have shown that short stretches of NH_2-terminal amino acids are sufficient to target β-galactosidase to the nucleus. These results, in combination with others obtained from studies of mammalian cells (11,13,14), lend support to the notion that specific nuclear targeting signals lie in the transported protein. However, we do not yet know the precise role of these sequences in the transported protein. By extending this genetic analysis, we hope to identify cellular components involved in nuclear protein transport in yeast.

REFERENCES

1. Paine, PL, Moore, LC, Horowitz, SB (1975). Nature 254:110.
2. Feldherr, CM, Kallenbach, E, Schultz, SB (1984). J Cell Biol 99:2216..
3. deRobertis, EM (1983). Cell 32:1021.
4. Guarente, L, Ptashne, M (1981) Proc Natl Acad Sci 78:2199.
5. Rose, M, Casadaban, MJ, and Botstein, D (1981) Proc Natl Acad Sci 78:2460.
6. Silver, P, Keegan, L, Ptashne, M (1984) Proc Natl Acad Sci 81:5951.
7. Braum, RJ, Kornberg, RD, (1985) Proc Natl Acad Sci 82:43.
8. Giniger, E, Varnum, SM, Ptashne M (1985) Cell 40:767.
9. Johnston, SA, Hopper, JE (1982) Proc Natl Acad Sci 79:6971.
10. Laughon, A, Driscoll, R, Wills, N, Gesteland, R (1984) Mol Cell Biol 4:268.
11. Kalderon, D, Richardson, WD, Markham, AF, Smith, A (1984) Nature 311:33.
12. Hall, MN, Hereford, L, Herskowitz, I (1984) Cell 36:1057.
13. Dingwall, C, Sharnick, SV, Laskey, RA (1982) Cell 30:449.
14. Davey, J, Dimmock, NJ, Colman, A (1985) Cell 40:667.

NUCLEAR PROTEIN LOCALIZATION SIGNALS IN YEAST

Michael N. Hall

Department of Biochemistry and Biophysics
University of California
San Francisco, California 94143

Proteins synthesized in the cytoplasm and destined to reside in any one of a variety of organelles, specialized membranes, or the cell exterior contain signals that are responsible for the removal of these proteins from the cytoplasm. Recently, this general rule has been shown to apply to proteins that are localized to the nucleus (1-8). Dingwall et al. (1) have shown that nucleoplasmin, an abundant protein of the Xenopus oocyte nucleus, can be cleaved into two fragments, one of which contains a signal required for entry into the nucleus. Lanford and Butel (3) and Kalderon et al. (4) have shown that mutations within a stretch of five basic amino acids, lys-lys-lys-arg-lys, in SV40 T-antigen block entry of this protein into the nucleus. Despite adhering to the general rule mentioned above, nuclear proteins are providing some surprises. Here I review results which indicate that the yeast nuclear protein α2 (a DNA-binding regulatory protein) has a signal that is capable of mediating nuclear localization but which, unlike previously identified localization signals, is not necessary for localization. These results suggest the unique possibility that the α2 protein might have more than a single localization determinant.

To identify a possible nuclear localization determinant within a nuclear protein, we have taken a genetic approach with the yeast Saccharomyces cerevisiae (2). We

[1]This work was supported by a Research Grant from the Public Health Service to I. Herskowitz. M.N.H. was the recipient of a Helen Hay Whitney Fellowship.

constructed a set of gene fusions that code for hybrid proteins containing varying amounts of the yeast nuclear protein α2 at the amino terminus and a constant, enzymatically active portion of Escherichia coli β-galactosidase at the carboxy terminus. The rationale for identifying a localization determinant made the assumption that should α2 contain such a signal, then fusion of β-galactosidase to the appropriate amount of α2 would convert β-galactosidase from a cytoplasmic protein to a nuclear protein. The amount of α2 required for this conversion would give an indication of the site of the localization determinant within α2. The hybrid proteins contain 3, 13, 25, 67, or all 210 amino acids of α2, and are referred to as 3α2-LacZ, 13α2-LacZ, 25α2-LacZ, 67α2-LacZ, and 210α2-LacZ, respectively.

Indirect immunofluorescence studies with cells containing the hybrid proteins indicate that the amino-terminal thirteen residues of α2 are sufficient for targeting β-galactosidase to the nucleus. Comparison of amino acid sequences of other nuclear proteins with these thirteen amino acids of α2 reveals a sequence that may be necessary for nuclear targeting, lys_3-ile-pro-ile-lys_7 (2).

Is the signal within the amino-terminal thirteen residues of α2 necessary for nuclear localization? A deletion of amino acids three through twenty of α2 does not affect, by the indirect immunofluorescence assay, nuclear localization of hybrid 210α2-LacZ (unpublished data). Thus, the amino-terminal thirteen residues of α2 are sufficient but not necessary for targeting β-galactosidase to the nucleus. This raises the unexpected possibility that α2 has a second localization determinant which, like the first, is sufficient but not necessary.

Should α2 have two nuclear localization determinants, they are not necessarily functionally equivalent (the sequence at the amino teminus is not reiterated). For reasons described below, the signal at the amino terminus is likely a specialized signal directly involved in mediating nuclear localization, perhaps by binding a receptor on the nuclear envelope. The presumed second signal could be 1) a DNA-binding domain, 2) a protein-protein interaction or multimerization domain, or 3) a second specialized localization signal. A commonly invoked mechanism of nuclear localization is that all proteins freely diffuse into the nucleus with selective retention by binding to a non-diffusible intra-nuclear substrate (9,10). Accordingly, the second signal could be a DNA-binding domain mediating retention of α2 in the nucleus. This is consistent with the observation that a deletion of the amino terminus of α2 does not affect DNA binding (unpublished data). The work of Dingwall et al. (1) on the pentameric protein nucleoplasmin suggests that a defective subunit can "piggyback" into the nucleus by multimerization with a functional subunit. In the case of α2, the notion of multimerization and subsequent ride to the nucleus is supported by the observation that the 210α2-LacZ hybrid containing a deletion of the amino-terminal localization signal confers a dominant α2- phenotype i.e., the mutated protein interacts with wild-type α2. Experiments to test directly whether the mutated hybrid protein requires wild-type α2 for nuclear localization have not been possible since the mutated hybrid is, for unknown reasons, toxic in the absence of wild-type α2. Thus, the second localization signal is not necessarily a specialized signal directly involved in mediating translocation into the nucleus. The following evidence suggests that the signal within the amino-terminal thirteen residues is a specialized localization signal. First, this signal can mediate nuclear localization in the absence of wild-type α2. Second, the amino terminus of α2 is neither required nor sufficient for DNA binding, as determined by a non-specific DNA-binding assay with the different hybrid proteins and with the 210α2-LacZ hybrid containing a small amino-terminal deletion (unpublished data).

Where within α2 could the presumed second signal be? Interestingly, the α2 protein contains a homeo box at its carboxy terminus (11,12). A homeo box is a sequence first found common to the protein products of several homeotic genes in Drosophila. Subsequent analysis has shown that a homologous sequence is found in vertebrates as well as in yeast (13). The function of the homeo box in not known; possibilities include the three functions described above which could be ascribed to a second nuclear localization determinant. It is tempting to speculate that the second signal is at the carboxy terminus within the homeo box.

Additional experiments are necessary to clarify the existence, roles, and sites of the possibly multiple nuclear localization signals in α2. Clarification will require not only further analysis of α2 itself but also identification of the cellular components with which the localization signal(s) interact.

REFERENCES

1. Dingwall, C, Sharnick, SV, and Laskey, RA (1982). A polypeptide domain that specifies migration of nucleoplasmin into the nucleus. Cell 30:449-458.

2. Hall, MN, Hereford, L, and Herskowitz, I (1984). Targeting of E coli β-galactosidase to the nucleus in yeast. Cell 36:1057-1065.

3. Lanford, RE, and Butel, JS (1984). Construction and characterization of an SV40 mutant defective in nuclear transport of T antigen. Cell 37:801-813.

4. Kalderon, D, Richardson, WD, Markham, AF, and Smith, AE (1984). Sequence requirements for nuclear location of simian virus 40 large-T antigen. Nature 311:33-38.

5. Kalderon, D Roberts, BL, Richardson, WD, and Smith, AE (1984). A short amino acid sequence is able to specify nuclear location. Cell 39:499-509.

6. Silver, PA, Keegan, LP, and Ptashne, M (1984). Amino terminus of yeast GAL4 gene product is sufficient for nuclear localization. Proc Natl Acad Sci USA 81:5951-5955.

7. Fischer-Fantuzzi, L and Vesco, C (1985). Deletion of 43 amino acids in the amino-terminal half of the large tumor antigen of simian virus 40 results in a non-karyophilic protein capable of transforming established cells. Proc Natl Acad Sci USA 82:1891-1895.

8. Davey, J, Dimmock, NJ, and Colman, A (1985). Identification of the sequence responsible for the nuclear accumulation of the influenza virus nucleoprotein in Xenopus oocytes. Cell 40:667-675.

9. Bonner, WM (1978). Protein migration and accumulation in nuclei. In Busch H (ed): "The Cell Nucleus" Vol 6 New York: Academic Press, p 97-148.

10. De Robertis, EM (1983). Nucleocytoplasmic segregation of proteins and RNAs. Cell 32:1021-1025.

11. Laughon, A and Scott, MP (1984). Sequence of a Drosophila segmentation gene: protein structure homology with DNA-binding proteins. Nature 310:25-31.

12. Shepard, JCW, McGinnis, W, Carrasco, AE, De Robertis, EM, and Gehring, WJ (1984). Fly and frog homoeo domains show homologies with yeast mating type regulatory proteins. Nature 310:70-71.

13. McGinnis, W, Garber, RL, Wirz, J, Kuroiwa, A, and Gehring, WJ (1984). A homologous protein-coding sequence in Drosophila homeotic genes and its conservation in other metazoans. Cell 37:403-408.

VI. MACROMOLECULAR TRAFFIC II: SECRETION, ENDOCYTOSIS, AND THE CELL SURFACE

THE ROLE OF CLATHRIN IN YEAST CELL GROWTH
AND PROTEIN TRANSPORT[1]

Gregory S. Payne and Randy Schekman

Department of Biochemistry, University of California
Berkeley, California 94720

ABSTRACT Clathrin-coated vesicles have been implicated as intermediates in several pathways of intracellular membrane traffic. The identification of coated vesicles in Saccharomyces cerevisiae provides the opportunity to pursue a genetic approach to determine the function of clathrin during intercompartmental transport. Yeast clathrin heavy chain has been purified and characterized. A molecular clone of the gene encoding clathrin heavy chain has been isolated and used to disrupt the heavy chain gene (CHC1) in vivo. Cells harboring a nonfunctional CHC1 allele are viable but grow substantially more slowly than wild-type cells. This result challenges several common conceptions of clathrin heavy chain function.

The complex compartmental organization of eukaryotic cells requires, for its generation and maintenance, mechanisms which sort and distribute newly synthesized proteins to their ultimate residences. The general itinerary of newly synthesized plasma membrane and secreted proteins involves cotranslational segregation into the endoplasmic reticulum (ER), transport to and passage through the Golgi apparatus where these proteins are packaged into secretory

---

[1]This work was supported by grants from the National Institute of General Medical Sciences of the National Institutes of Health and the National Science Foundation. G.P. is a fellow of the Jane Coffin Childs Memorial Fund for Cancer Research.

vesicles and shuttled to the cell surface (reviewed in ref. 1-3). Other routes of intracellular transport also involve transfer of proteins between distinct membrane-enclosed compartments. Newly synthesized proteins, such as proteases, destined for the major hydrolytic compartment (the vacuole in yeast and the lysosome in mammalian cells) are diverted from the secretory pathway within the Golgi body and then transferred to the hydrolytic organelle (2,4). The vacuole or lysosome is also the recipient of material delivered from the cell surface by the process of endocytosis (5,6). Although the major stations along each of these intracellular routes have been defined and characterized, the mechanism of transfer between topologically distinct organelles remains obscure.

Transport Vesicles.

Logical candidates for the role of inter-compartment shuttles are the small vesicles commonly observed budding from organelle membranes (1-3). These vesicles have been proposed to ferry molecules from the site of vesiculation to specific recipient organelles. According to this hypothesis, each vesicle contains the information (perhaps expressed as surface receptors) necessary to direct it to a precise destination. Segregation of molecules destined for different subcellular compartments occurs by sorting the molecules into regions of the organelle generating properly addressed vesicles. Subsequent fusion of the transport vesicle membrane with the target organelle membrane results in delivery of the vesicle contents to the recipient compartment. This cycle of membrane vesiculation and fusion transfers soluble and membrane components from one membrane-enclosed compartment to another without requiring the molecules to leave their particular microenvironment. That is, membrane proteins are transferred from the donor organelle membrane into the vesicle membrane, then delivered to the target compartment membrane without ever passing through the cytoplasm. In an analogous fashion soluble molecules are carried within the vesicle lumen and never cross into or through the lipid bilayer during transport.

The process of vesicle formation may therefore represent a pivotal stage of intracellular transport where molecules are sorted and packaged for delivery to specific subcellular compartments.

The most clearly defined example of vesicular transport occurs in mammalian cells during the early stages of receptor-mediated endocytosis (7-9). Certain molecules destined for internalization are bound by specific receptors at the cell surface. The receptors for these ligands are located at, or migrate upon ligand-binding to, indentations along the plasma membrane termed coated pits. Coated pits derive their name and appearance in the electron microscope from the fuzzy proteinaceous coat displayed on their cytoplasmic surface. The coated pits invaginate and finally vesiculate to form coated vesicles carrying the receptor-bound ligands. After shedding their coats the vesicles fuse with an uncoated, low pH compartment referred to as an endosome (8). An alternative view, however, suggests that although coated pits invaginate, the ultimate product of vesiculation already lacks the coat (9).

Properties of Clathrin-coated Membranes.

Since coated pits appear to represent the precursor of endocytic vesicles, the proteinaceous coat has been implicated in the membrane invagination and fusion which generates transport vesicles (7-11). In accord with these morphological studies, structural and biochemical investigations described below suggest that the proteins which comprise the coat are uniquely suited to function in transport vesicle biogenesis (10,11). Detailed inspection of coated membranes, using the electron microscope, revealed that the coat consists of a polyhedral protein lattice. The principal constituent of the lattice is clathrin. The lattice can be disassembled and clathrin specifically extracted from membranes by subjecting preparations of coated vesicles to incubation in either mildly denaturing buffers (e.g., 2 M urea) or high concentrations of primary amines (e.g. 0.5 M Tris). Clathrin extracted in this fashion forms a distinctive three-legged "triskelion" (12) which is composed of three molecules of clathrin heavy chain (molecular weight of approximately 180 kd) and three molecules of clathrin light chain(s) (either one or two species ranging in size from 30-36 kd) (Figure 1). Purified triskelions can be induced to reassemble into empty baskets (Figure 1) or rebind to coated vesicles previously stripped of their coats (13,14). Additional coated vesicle proteins have been shown to facilitate these _in vitro_ reassembly reactions (13). Conceivably, an analogous assembly of triskelions into a lattice basket _in vivo_ could drive the formation of a coated vesicle from a coated region of mem-

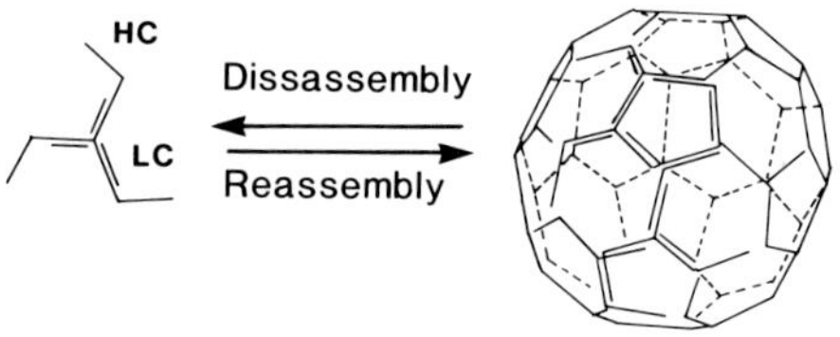

FIGURE 1. Clathrin coats can be disassembled and reassembled in vitro. The product of disassembly is a triskelion composed of three molecules of clathrin heavy chain (HC) and three molecules of clathrin light chain (LC).

brane (10,11). Once formed, the clathrin basket could be removed from the vesicle by an "uncoating ATPase" that has been identified and purified from mammalian cells (15). This protein hydrolyzes ATP and binds stoichiometrically to triskelions to effect release of clathrin from coated membranes. There is, therefore, ample morphological and biochemical evidence that clathrin undergoes a cycle of assembly and disassembly both in vitro and in vivo. If this cycle provides the driving force to generate endocytic vesicles, then it represents a paradigm for the vesiculation events that occur during other stages of intracellular transport. Indeed, coated vesicles have been implicated as intermediates in traffic between the ER and Golgi, through the Golgi, and between the Golgi and the lysosome (10). Although these observations provide compelling circumstantial evidence, there is no direct demonstration, either from in vitro or in vivo systems, that clathrin is required for membrane vesiculation and the formation of transport vesicles.

A Genetic Approach to Clathrin Function.

In an attempt to address directly the role of clathrin during intracellular transport, we have sought mutant cells unable to form or disassemble coated vesicles. The opportunity to pursue this genetic approach arose when Mueller and Branton identified coated vesicles in cell extracts from Saccharomyces cerevisiae (16).

Preliminary attempts to identify coated membranes in yeast by electron microscopic examination of whole cells were hampered by the granular appearance of the cytoplasm which prevented clear visualization of structures along the cytoplasmic face of membranes. Mueller and Branton successfully identified coated vesicles by preparing a membrane fraction from yeast cell extracts and chromatographing the membranes on a Sephacryl S-1000 gel filtration column. Protein profiles of coated vesicle-enriched fractions displayed a prominent 190 kd protein which Mueller and Branton proposed to be yeast clathrin heavy chain. These observations provided the impetus for our genetic evaluation of clathrin function. Our approach consisted of three stages: 1) further characterizing the yeast coated vesicles and obtaining additional evidence that the 190 kd protein is clathrin heavy chain; 2) isolating a molecular clone of the yeast gene encoding clathrin heavy chain; 3) disrupting the gene in vitro, then replacing, in vivo, the intact gene with the disrupted version in order to assess the phenotypic consequences of a clathrin deficiency.

Characterization of Yeast Clathrin.

The definition of mammalian clathrin rests principally on the structural properties outlined earlier (Table 1).

TABLE 1
COMPARISON OF YEAST AND MAMMALIAN CLATHRIN

	Mammals	Yeast
Major protein of coated vesicles	+	+
Extracted by 2 M urea or 0.5 M Tris	+	+
Associated with light chain(s)	+(1 or 2)	+(1)
Heavy and light chain(s) form triskelions	+	+
Molar ratio HC:LC(s) in triskelions	1	1
HC molecular weight	180kd	190kd
LC(s) molecular weight(s)	30-36kd	36kd
Triskelions reassemble into baskets	+	?
Triskelions bind to "stripped" coated vesicles	+	?
Recognized by anti-mammalian clathrin antibody	+	−
Recognized by anti-yeast heavy chain antibody	−	+

In order to establish the identity of the 190 kd yeast clathrin heavy chain, we examined properties characteristic of mammalian heavy chain. The Sephacryl S-1000 column fractions enriched in coated vesicles were pooled and incubated in either 2 M urea or 0.5 M Tris-HCl, pH 7.5. Both treatments specifically extracted the 190 kd, a 55 kd and a 36 kd protein species from the membrane vesicles. The 190 kd and 36 kd proteins cofractionate and are separated from the 55 kd protein when the urea-extracted proteins were chromatographed through a Sepharose 4B gel filtration column. This result suggests that the 190 kd and 36 kd polypeptides form a molecular complex. A densitometer scan of the gel displaying the column fractionated complex indicated that the molar ratio of 190 kd to 36 kd proteins in the complex was one. Tomas Kirchhausen (Harvard University) examined this molecular complex in the electron microscope and observed triskelions which were indistinguishable from mammalian triskelions. These results, summarized in Table 1, argue strongly that the 190 kd protein is clathrin heavy chain and the 36 kd protein is clathrin light chain.

Isolation of a Molecular Clone of the Clathrin Heavy Chain Gene.

While these experiments were in progress, Richard Young in Ron Davis' lab developed a procedure which employed antibody to identify molecular clones of the gene encoding the protein antigen (17). The strategy involved constructing a λ bacteriophage vector, designated λgt11, which carried the E. coli lacZ gene. The 3' end of the lacZ gene harbors an EcoRl recognition site which is unique in λgt11. Random fragments of yeast genomic DNA were inserted into this EcoRl site to create a library representing the entire yeast genome. If yeast DNA carrying an open reading frame is inserted in the proper orientation and proper coding frame, then a β-galactosidase-yeast protein fusion will be created. Young and Davis reported that fusion proteins generated in this fashion could be detected in phage plaques by adsorbing the plaque proteins to nitrocellulose, incubating the filter with antibody, then visualizing the bound antibody with ^{125}I-labelled protein A.

Before pursuing this approach, however, we required antibody which specifically recognized yeast clathrin heavy chain. Daniel Louvard (Institut Pasteur) provided a high titre preparation of antibody specific for mammalian heavy chain but, unfortunately, this antibody failed to recognize yeast heavy chain. Subsequently we raised antibodies in rabbits using yeast clathrin heavy chain as the immunogen. This polyclonal antiserum specifically reacted with yeast heavy chain, but failed to detect mammalian heavy chain. Several quantitative studies using our antibody determined that clathrin heavy chain comprises 0.1% of the mass of total yeast proteins. This number is similar to the amount of clathrin heavy chain measured in a variety of mammalian cells (18).

The antiserum was used to screen a library of yeast DNA carried by λgt11. A number of phage were isolated which produced immunoreactive plaques. Restriction endonuclease and molecular hybridization analyses separated the yeast DNA isolates into nine non-overlapping sets. Several criteria were applied to identify the phage harboring the authentic heavy chain gene. First, since the heavy chain has a molecular weight of 190 kd, we can estimate that the gene must span at least 5,000 base pairs. Thus, DNA encoding the gene should hybridize to a species of RNA at least 5 kb long. Only two classes of inserts detected RNA longer than 5 kb. We then prepared lysates of E. coli cells infected with phage carrying these two insert types. The lysate proteins were affixed to nitrocellulose and tested for their ability to deplete the antiserum of anti-heavy chain antibodies. Only one of the two phage classes fulfilled this criterion. This result implied that the protein produced from the inserted yeast DNA carried most, if not all, of the antigenic determinants recognized by the anti-heavy chain antibodies. Finally, RNA capable of hybridizing to this yeast DNA was purified from a preparation of total yeast RNA. The purified RNA was translated in vitro using a rabbit reticulocyte translation system. The major translation product was a 190 kd protein which comigrated with authentic yeast heavy chain when electrophoresed through polyacrylamide-SDS gels and was specifically recognized by the anti-heavy chain antibody.

We have not further characterized the other eight classes of inserts.

Mutating the Clathrin Heavy Chain Gene.

With at least a part of the heavy chain gene (designated CHC1 for Clathrin Heavy Chain), we prepared to introduce alterations into the gene in vivo. As a first step, restriction endonuclease recognition sites in the CHC1 gene were mapped and positioned with respect to the approximate boundaries of the transcription unit (Figure 2A). Hybridization of the cloned DNA to yeast chromosomal DNA under conditions of lowered stringency revealed only a single copy of CHC1. The following consideration influenced our strategy for disrupting CHC1. Temperature-sensitive, conditionally lethal sec mutations have been isolated which block secretion in yeast at the non-permissive temperature (19). Studies using mutant sec strains have shown that the secretory process is essential for cell growth (19). The expectation that clathrin would prove necessary for intracellular transport implied that a mutation of CHC1 would be lethal to the cell. We thus planned to alter only one of the two homologous CHC1 alleles in a diploid cell. The effect of a heavy chain deficiency on cell growth could then be assessed by sporulating the diploids and evaluating the meiotic products. If clathrin is required for either spore germination or cell growth, then only two of every four sibling spores should produce colonies.

A 600 bp DNA fragment situated approximately 1.5 kb from the 5' end of the CHC1 gene was removed and in its place a fragment of DNA coding for the LEU2 gene was inserted (Figure 2A). A linear DNA fragment encompassing the altered region was then isolated and introduced into diploid leu2 mutant cells. Recombination at both ends of the fragment replaces the wild-type CHC1 gene with the disrupted copy (20). Cells harboring the mutant chc1 allele were identified by their ability to grow in the absence of leucine by virtue of the LEU2 gene present at chc1 (Figure 2B).

These diploids were sporulated and dissected into tetrads. Surprisingly, many diploids gave rise to four viable spores. Each complete tetrad, however, consisted of two wild-type sized colonies and two extremely small colonies; the two large colonies were leu2 and the small colonies were LEU2. This provided genetic evidence that the slow-growing colonies contained a disrupted clathrin heavy chain gene. In the case of two tetrads we confirmed this interpretation by using restriction enzyme analysis, Southern transfer and molecular hybridization to determine the physical structure of the heavy chain gene. The results of three experiments argue that the chc1 disruption eliminates expression of clathrin heavy chain. First, we used immunoblotting procedures to search chc1 cell extracts for protein species capable of reacting with the anti-heavy chain antibody. Neither clathrin heavy chain nor heavy chain fragments were detected. Secondly, vesicle fractions were prepared by applying chc1 cell membranes to a Sephacryl S-1000 column. The protein profiles of vesicle fractions were examined by polyacrylamide-

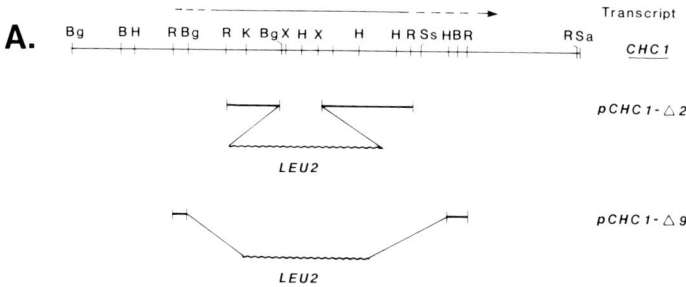

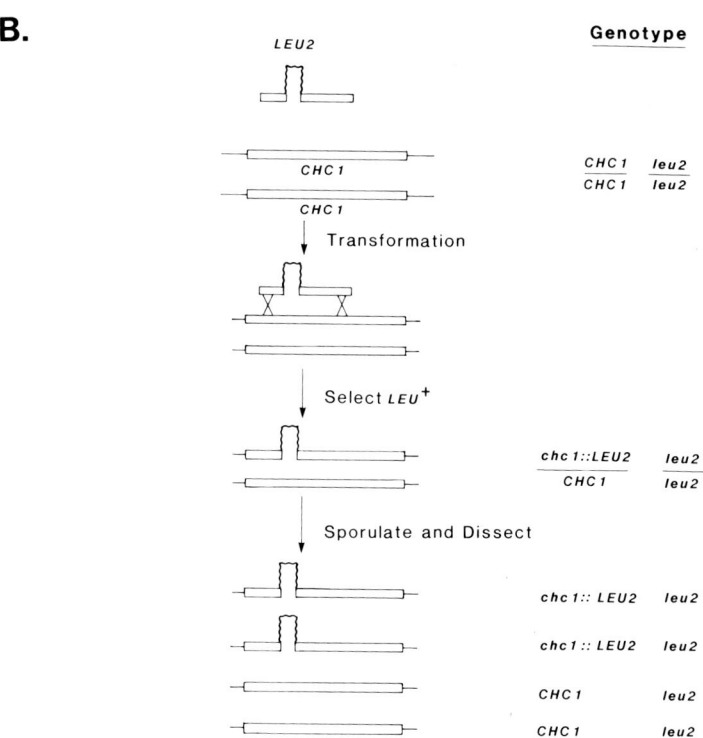

FIGURE 2. Strategy for CHC1 gene disruption. A) Structure of the clathrin gene, the approximate limits of the transcription unit and two gene disruptions constructed in vitro (see text for details). B: BamHl; Bg: BglII; E: EcoRl; H: HindIII; K: KpnI; Sa: SalI; Ss: SstI; X: XbaI. B) Strategy for single step disruption of CHC1 in vivo, using the constructions shown in part A.

SDS gel electrophoresis followed either by Coomassie-blue staining or immunoblotting using the anti-heavy chain antibody. The results of each analysis suggested that mutant cell vesicles are devoid of clathrin heavy chain. In addition, the vesicles also lacked clathrin light chain, almost certainly because light chains associate with vesicles by binding to heavy chains (10). This finding provides further substantiation that the mutant cell vesicles are not coated. Finally, we have constructed another chcl gene disruption in which approximately 4.5 kb have been deleted and replaced by LEU2 DNA. We estimate that this deletion covers at least the first 80% of the CHCl gene. Cells containing this chcl mutation exhibit the same growth properties as the mutant cells with the less extensive chcl alteration described above.

Re-examining Clathrin Heavy Chain Function.

It thus appears that yeast cells grow, albeit slowly, in the absence of clathrin heavy chain. Consequently, our results challenge the model, presented earlier, which postulates necessary coupling between the formation of clathrin-coated vesicles and intracellular transport.

The existence of a protein capable of functionally substituting for clathrin heavy chain would reconcile our data with the model. Since we were unable to detect sequences related to CHCl using lowered stringency hybridization conditions, this putative substitute could display only limited primary sequence similarities. Recently, evidence obtained from mammalian cells is consistent with the existence of other vesicle coats. When cells are incubated at $20^{\circ}C$ the secretory process is blocked within the Golgi apparatus. Golgi-like organelles wearing cytoplasmic coats are observed under these conditions, but the coat proteins fail to react with anti-clathrin antibody (21). A similar observation has been made by Rothman and his colleagues while studying transport through the Golgi apparatus in vitro (22). Golgi membranes which are primed for transport form coated regions of vesiculation that do not bind anti-clathrin antibody (L. Orci and J. E. Rothman, personal communication).

Clathrin heavy chain functional substitutes might be identified in yeast by isolating chcl strains in which the slow growth phenotype is suppressed. One strategy to obtain suppressors of chcl entails introducing into chcl cells a library of yeast DNA carried on a multicopy plasmid vector. This approach relies on the assumption that multiple gene copies of a functional analog would produce sufficient product to cure the chc1 strain's retarded growth. On the other hand, mutations altering the expression or functional properties of a putative heavy chain substitute could alleviate the chc1 growth defect in the absence of exogenously added DNA. These types of genetic approaches should complement the biochemical studies performed on mammalian cells.

Since chc1 cells are viable but exhibit a marked growth phenotype, we are poised to address the precise role of heavy chain during intracellular transport. The principal activity of clathrin may occur during events temporally and spacially distinct from transport vesicle biogenesis. For instance, the clathrin coat has been shown to prevent calcium-mediated fusion of coated vesicles to lysosomes in vitro (23). The clathrin coat might, therefore, function by shielding the vesicle from non-specific fusion with other intracellular membranes encountered during vesicle movement to its targeted destination. This model predicts that chc1 strains will undergo some degree of mis-directed transport. The ability to assess the rate and fidelity of secreted, plasma membrane and vacuolar protein transport will provide a test of this scenario.

As mentioned previously, studies of endocytosis present the most compelling support for the involvement of clathrin in membrane vesiculation. It is conceivable that exocytosis occurs in the absence of clathrin heavy chain but endocytosis is eliminated. Howard Reizman has recently described an endocytic process in yeast (6). He observed that yeast cells will take up a fluorescent membrane-impermeant dye which collects in the vacuole. By applying this assay to chc1 cells, it should be possible to determine the requirement for clathrin in endocytosis.

Conclusion.

The construction and characterization of clathrin heavy chain deficient yeast strains reveal that clathrin heavy chain plays an important but not essential role in cell growth. The genetic and biochemical approaches available in this system promise further elucidation of the precise function of clathrin coats during intracellular transport.

ACKNOWLEDGEMENTS

We gratefully acknowledge discussions with and suggestions from Leslie Berg, Tom St. John and members of our lab, in particular Mitchell Bernstein, Raymond Deshaies and Linda Silveira.

REFERENCES

1. Palade G (1975). Intracellular aspects of the process of protein secretion. Science 189:347.
2. Farquhar MG, Palade GE (1981). The Golgi apparatus complex)-(1954-1981)-from artifact to center stage. J Cell Biol 91:77s.
3. Sabatini DD, Kreibich K, Morimoto T, Adesink M (1982). Mechanisms for the incorporation of proteins in membranes and organelles. J Cell Biol 92:1.
4. Stevens T, Esmon B, Schekman R (1982). Early stages in the yeast secretory pathway are required for transport of carboxypeptidase Y to the vacuole. Cell 30:439.
5. Steinman RM, Mellman IS, Muller WA, Cohen ZA (1983). Endocytosis and the recycling of plasma membrane. J Cell Biol 96:1.
6. Reizman H (1985). Endocytosis in yeast: several of the yeast secretory mutants are defective in endocytosis. Cell 40:1001.
7. Goldstein JL, Anderson RGW, Brown MS (1979). Coated pits, coated vesicles, and receptor-mediated endocytosis. Nature 279:679.
8. Helenius A, Mellman I, Wall D, Hubbard A (1983). Endosomes. Trends Biochem Sci 8:245.
9. Pastan I, Willingham M (1983). Receptor-mediated endocytosis: coated pits, receptosomes and the Golgi. Trends Biochem Sci 8:250.
10. Pearse BMF, Bretscher MS (1981). Membrane recycling by coated vesicles. Ann Rev Biochem 50:85.
11. Harrison SC, Kirchhausen T (1983). Clathrin, cages, and coated vesicles. Cell 33:650.
12. Ungewickell E, Branton D (1981). Assembly units of clathrin coats. Nature 289:420.

13. Zaremba S, Keen JH (1983). Assembly polypeptides from coated vesicles mediate reassembly of unique clathrin coats. J Cell Biol 97:1339.
14. Hanspal M, Luna E, Branton D (1984). The association of clathrin fragments with coated vesicle membranes. J Biol Chem 259:11075.
15. Schlossman DM, Schmid SL, Braell WA, Rothman JE (1984). An enzyme that removes clathrin coats: purification of an uncoating ATPase. J Cell Biol 99:723.
16. Mueller SC, Branton D (1984). Identification of coated vesicles in Saccharomyces cerevisiae. J Cell Biol 98: 341.
17. Young RA, Davis RW (1983). Yeast RNA polymerase II genes: isolation with antibody probes. Science 222:778.
18. Goud B, Huet C, Louvard D (1985). Assembled and unassembled pools of clathrin: a quantitative study using an enzyme immunoassay. J Cell Biol 100:521.
19. Schekman R, Novick P (1982). The secretory process and yeast cell-surface assembly. In Strathern JN, Jones EW, Broach JR (eds): "Molecular Biology of the Yeast Saccharomyces: Metabolism and Gene Expression," Cold Spring Harbor: Cold Spring Harbor Laboratory, p 361.
20. Rothstein R (1983). One step gene disruption in yeast. Meth Enzymol 101:202.
21. Griffiths G, Pfeiffer S, Simons K, Matlin K (1985). Exit of newly-synthesized membrane proteins from the trans cisternae of the Golgi complex to the plasma membrane. J Cell Biol:in press.
22. Balch WE, Glick BS, Rothman JE (1984). Sequential intermediates in the pathway of intercompartmental transport in a cell-free system. Cell 39:525.
23. Altstiel L, Branton D (1983). Fusion of coated vesicles with lysosomes: measurement with a fluorescent assay. Cell 32:921.

ENDOCYTOSIS IN YEAST : RELATIONSHIP TO OTHER CELLULAR PATHWAYS*

Howard Riezman, Yolande Chvatchko and Isabelle Howald

Membrane Unit, Swiss Institute for Experimental Cancer Research, CH-1066 Epalinges, Switzerland

ABSTRACT We have followed endocytosis in the yeast, Saccharomyces cerevisiae, using lucifer yellow CH. Uptake of the fluorescent dye into the yeast vacuole is time-, temperature- and energy-dependent. Several of the yeast mutants conditionally defective for secretion (13,14) are also conditionally defective for endocytosis (17). Studies of these mutants show that endocytosis can be uncoupled from the early stages of secretion. In contrast, the late stages of secretion may be obligatorily coupled to secretion.
We have recently begun isolating mutants defective in endocytosis that secrete normally. Two of these mutants have been characterized. One is deficient in biogenesis of vacuolar carboxypeptidase Y, has morphologically abnormal vacuoles and accumulates invaginations of the plasma membrane. The other mutant is normal for biogenesis of carboxypeptidase Y and accumulates an undefined organelle. Both of these mutants are somewhat deficient in response to alpha factor, a yeast mating pheromone.

---

* This work was supported by a grant from the Fonds National Suisse de la Recherche Scientifique.

Endocytosis is the pathway whereby cells internalize portions of their own plasma membrane. Internalization usually occurs via specialized regions of the membrane enriched in receptors and their ligands. Due to the vesicular nature of this process a certain amount of fluid is entrapped within each budding vesicle. Thus we have come to define two types of endocytosis, receptor-mediated endocytosis and fluid-phase endocytosis. The two processes are both, of course, time-, temperature- and energy-dependent, but one can distinguish the two biochemically as fluid-phase endocytosis is not saturable with respect to ligand concentration. One should emphasize the following; receptor-mediated endocytosis is most likely always accompanied by fluid uptake. The converse may not be true.

Endocytosis probably serves multiple functions though few of these have been proven. In lower eucaryotes, such as slime molds and amebas, endocytosis serves to bring macronutrients into the cell in the form of whole bacteria or yeast (1,2). In the higher eucaryotes, micronutrients such as iron, cholesterol and vitamin B_{12} enter the cell by endocytosis (3). Other functions postulated for endocytosis are control of cell shape and size, a role in cell motility, surface remodeling and membrane conservation (e.g. reutilization of secreted membrane). Perhaps one of the most important functions of endocytosis will be in growth control. In vertebrates many polypeptide hormones are internalized by receptor-mediated endocytosis (3). The precise role of endocytosis in the hormone response is not known, however evidence suggests that the epidermal growth factor receptor is degraded in lysosomes after endocytosis (4) and that endocytosis may need to occur in order to bring the protein kinase activity of the receptor and its substrate into contact (5). Endocytosis is also important in several cellular dysfunctions as some enveloped viruses, toxins and parasitic bacteria enter their target cells by endocytosis (6,7,8).

The mechanism of endocytosis is unknown. Only a few enzymes and proteins have been implicated in the process and their precise role is not yet clear. They are clathrin and its associated polypeptides (9), an uncoating ATPase that removes clathrin coats from coated vesicles (10) and an ATP-dependent proton pump responsible for endosome and coated vesicle acidification (11). This acidification process is important for the dissociation of certain receptors from their ligands (12).

In order to understand the mechanism of endocytosis and to gain new insights and additional evidence for the functions of the process, we decided to study endocytosis in an organism where a genetic approach would be readily available. For this we chose <u>Saccharomyces cerevisiae</u>. This yeast has the additional advantage that many mutants are already available in secretion (13,14) which permits us to study immediately the relationship between the endocytic and exocytic pathays. Our approach was : 1, show that yeast cells endocytose, 2, analyze the known yeast secretory mutants for their ability to accumulate an endocytic marker, 3, isolate and analyze mutants defective in accumulation of endocytic content but normal for secretion. Of course, the availability of many complementation groups of endocytosis mutants will allow us to identify the genes involved in the process. Perhaps most importantly, the cloning of these genes will allow the expression of fusion proteins and the production of antibodies with specificity against endocytic organelles. This could open the door to a more sophisticated biochemical reconstitution of the process in vitro where one can hope to dissect the mechanism of endocytosis.

YEAST CELLS ENDOCYTOSE

In order to demonstrate that yeast cells endocytose we chose as a marker a small fluorescent dye, lucifer yellow CH. This dye has the following advantages; it is small enough that it can readily pass through the yeast cell wall, it is negatively charged at all pH values above 2, it does not cross membranes, its fluorescence is remarkably stable over a wide pH range (pH 2 - pH 10) and is not quenched by concentrated yeast extracts. Lucifer yellow CH has also been used as an endocytic tracer in mammalian cells (15,16). The intracellular accumulation of lucifer yellow is time- and temperature-dependent. We have quantified the amounts of lucifer yellow CH sequestered by the yeast strain AH 216 at 37°C (17). If we assume that the dye is a marker of the fluid-phase we can calculate that the rate of accumulation is approximately $30 nl/10^8$ cells/hr or 0.1 percent of the cell volume per generation.

Morphological examination of the yeast cells after uptake experiments by fluorescence microscopy show that the vast majority of internalized lucifer yellow CH is localized in the vacuole. Uptake of lucifer yellow CH into the vacuole is also inhibited if the cells are deprived of an energy source and is sensitive to inhibitors of energy metabolism such as combinations of azide and fluoride ions. Therefore, uptake is also energy-dependent. By varying the external concentration of lucifer yellow CH and measuring its accumulation by yeast cells in a fixed-time assay we could show that the accumulation of lucifer yellow CH is not saturable. Thus far, all facets of lucifer yellow CH accumulation, time-, temperature-, energy-dependence and non-saturability are consistent with fluid-phase endocytosis as the uptake mechanism.

ENDOCYTOSIS IN THE YEAST SECRETORY MUTANTS

Schekman's group has isolated 25 different complementation groups of yeast secretory (sec) mutants (13,14). As secretion is an essential function of the cell, Schekman's group isolated conditionally-lethal mutants which were screened for secretion at the non-permissive temperature (37°C). Using our assay of the accumulation of lucifer yellow CH as a measure for endocytosis we examined representatives from each of the complementation group of sec mutants at permissive (24°C) and non-permissive (37°C) temperature (17). At 24°C, all of these mutants could accumulate lucifer yellow CH in the vacuole, although several of them less than the parent strain (X2180-1A). At 37°C, many of the sec mutants were defective in accumulation of lucifer yellow in the vacuole. Of those sec mutants which accumulate secretory proteins in the endoplasmic reticulum, some are blocked in lucifer yellow CH accumulation, others are not. Those mutants which are blocked in secretion at the endoplasmic reticulum, but which endocytose normally allow us to conclude that endocytosis can be uncoupled from net secretion and that transport of proteins from the endoplasmic reticulum to the Golgi is not necessary for endocytosis. The most likely explanation for the mutants which are blocked in endocytosis is that these gene products also act at later stages of the pathway or perhaps directly on the endocytic pathway. There are two mutants which accumulate secretory proteins in the Golgi apparatus, sec 7 and sec 14. Sec 7 is blocked in endocytosis, sec 14 is not. Thus, the SEC 14 step is not necessary for endocytosis. At least some portion of protein transport through the Golgi can be uncoupled from endocytosis.

All of the ten complementation groups of secretory mutants which accumulate secretory proteins in secretory vesicles are blocked in endocytosis. This suggests, but does not prove, that endocytosis may be coupled with the later stages of the secretory pathway. One way to explain this would be that the endocytosis-recycling pathway and the latter portions of the secretory pathway form a circular pathway. We are currently investigating this possibility at the ultrastructural level. Another possibility is that due to the presence of the turgor pressure of the yeast cell, the plasma membrane would be pushed against the cell wall with such force that only a secretory vesicle freshly fused with the plasma membrane could be an intermediate for the pinching off of the initial endocytic organelle. Our studies of endocytosis in the secretory mutants provide evidence that endocytosis can occur in the absence of net secretion in yeast and suggest the possibility that portions of the endocytic and secretory pathways are coupled with each other.

ISOLATION OF MUTANTS DEFECTIVE IN ACCUMULATION OF ENDOCYTIC CONTENT

In order to find mutants that are defective in accumulation of endocytic content we first constructed a collection of temperature-sensitive lethal strains. The rationale behind this was that if endocytosis is necessary for cell growth we could find our mutant in this bank. If endocytosis is not necessary for cell growth, we could still find mutants in this heavily mutagenized population, but upon genetic characterization the endocytosis defect would segregate away from temperature-sensitive lethality. Each of 750 temperature-sensitive clones was screened for its ability to accumulate lucifer yellow CH in the vacuole by examination in the fluorescence microscope. Those which did not accumulate endocytic content were subsequently screened for their ability to incorporate $^{35}SO_4^=$ into protein. Those mutants which did not accumulate endocytic content, but which synthesized protein at near normal rates were next tested for their ability to secrete invertase. We found 2 mutants which were defective in accumulation of lucifer yellow CH and positive for invertase secretion. We have called these genes end 1 and end 2. Through several generations the trait of temperature-sensitive lethality cosegregates by tetrad analysis with the inability to accumulate endocytic content at non-permissive temperature. We were very surprised then to find out that these mutants are also defective for endocytosis at permissive temperature. There are several ways to explain this observation, but we prefer to await further characterization of the nature of these mutations before we do so.

End 1 and end 2 have been examined for their ability to synthesize carboxypeptidase Y (CPY), a vacuolar enzyme. The mutants were pulse labeled for 10 minutes and chased for 40 minutes. The cells were then lysed and CPY was immunoprecipitated and analyzed by SDS-gel electrophoresis and autoradiography. End 1 produces a CPY which comigrates with the form produced in pep 4 strains. Pep 4 mutants are unable to activate CPY by proteolytic processing. This is the case at both temperatures. We have recently found that, in steady state, proCPY is secreted into the medium. Thus, this mutant shows a similarity to the vpl mutants isolated by Steven's group (this volume) and the vac mutants isolated by Emr's group (this volume). End 2 is normal for CPY biosynthesis and targetting.

We have examined these mutants by electron microscopy using improved techniques for membrane enhancement (18) and have come up with the following results. End 1 is devoid of a normal vacuole at permissive temperature. Instead it has many small vesicles of the same apparent morphology as a normal vacuole and accumulates invaginations of the plasma membrane. At 37°C (non-permissive temperature) these vacuole-like structures lose their electron density with time and the invaginations of the plasma membrane disappear. At 23°C, end 2 appears to accumulate a membrane-bounded organelle whose function has not yet been discovered in yeast cells. This organelle appears also in wild-type cells but is encountered less frequently. It has a fairly rough morphology and its interior is slightly more dense than the cytoplasm under our conditions of sample preparation. The vacuole appears nearly normal. At 37°C, this newly described organelle loses its internal electron density and with time the vacuole also appears to break into pieces with irregular morphology. As we have no data on the function of this new organelle, but we do know that it apparently forms "holes" inside the yeast cell, we have called it the "Emmenthaler body".

PHEROMONE RESPONSE IN THE END MUTANTS

As many polypeptide hormones are internalized by endocytosis in mammalian cells, we wondered whether our mutants, defective in accumulation of endocytic content, would be insensitive or supersensitive to the yeast mating pheromones. These pheromones are in many ways analogous to peptide hormones in mammalian cells. Therefore, we performed dose-response curves for alpha factor effect on a cells containing the end 1 and end 2 mutations. To judge the response we measured agglutination, cell-cycle arrest and projectile formation. Both mutants are less sensitive to alpha factor than the wild-type parent. End 2, which is somewhat leaky in its endocytosis phenotype, is approximately 10 times less sensitive whereas end 1 is greater than 500 times less sensitive to alpha factor. For end 1, this is confirmed by a 10-30 fold drop in mating efficiency in either mating type background.

A trivial explanation for these results could be that the end mutants are secreting an inhibitor of alpha factor. In order to test this possibility, we mixed mutant a cells with wild-type a cells and treated them with alpha factor. If the mutants secrete an inhibitor of alpha factor they should be able to protect the wild-type from the action of alpha factor. This is not the case. In mixed populations, the wild-type cells respond, the mutants do not. These results suggest, but do not prove, that endocytosis may play a role in pheromone response.

CONCLUSIONS AND FUTURE PROSPECTS

It now seems that the yeast cell may be an advantageous system for the study of endocytosis. It is clear that yeast cells endocytose (see also work by Makarow, ref. 19) and the availability of secretory mutants has and will allow us to perform very interesting experiments concerning the relationship between the two pathways. In addition, we have been able to isolate mutants defective in accumulation of endocytic content and the study of these mutants may provide us with some clue about the role of endocytosis in vacuole biogenesis and pheromone response. Certain properties of these mutants have recently allowed us to develop a highly efficient screening procedure to find new mutants. The obtention of a large number of endocytosis mutants should allow us to pursue our goal of identifying the genes and steps involved in endocytosis, cloning these genes and analyzing their protein products in the hopes of obtaining antibodies against endocytic organelles. These antibodies could be of great use in eventual *in vitro* reconstitution experiments of the individual steps of endocytosis.

REFERENCES

1. Weisman, R.A. and Korn, E.D. (1967) Biochem. 6, 485-497.
2. Loomis, W.F. (1975) Dictyostelium discoideum : A development System (New York Academic Press).
3. Goldstein, J.L., Anderson, R.G.W. and Brown, M.S. (1979) Nature 279, 679-685.
4. Carpenter, G. and Cohen, S. (1976) J.Cell Biol. 71, 159-171.
5. Sawyer, S.T. and Cohen, S. (1985) J. Biol. Chem. 260, 8233-8236.
6. Helenius, A., Marsh, M. and White, J. (1980) Trends Biochem. Sci. 5, 104-106.
7. Sandvig, K. and Olsnes, S. (1981) J. Biol. Chem. 256, 9068-9076.
8. Smith, H. and Fitzgeorge, R.B. (1964) Brit. J. Exp. Path. 45, 174-186.
9. Pearse, B.M.F. and Bretscher, M.S. (1981) Ann. Rev. Biochem. 50, 85-101.
10. Schlossman, D.M., Schmid, S.L., Braell, W.A. and Rothman, J.E. (1984) J. Cell Biol. 99, 723-733.
11. Galloway, C.J., Dean, G.E., Marsh, M., Rudnick, G. and Mellman, I. (1983) Proc. Natl. Acad.Sci. USA 80, 3334-3338.
12. Helenius, A., Mellman, I., Wall, D. and Hubbard, A. (1983) Trends Biochem. Sci. 8, 245-250.
13. Novick, P., Field, C. and Schekman,R. (1980) Cell 21, 205-215.
14. Ferro-Novick, S., Novick, P., Field, C. and Schekman, R. (1984) J. Cell Biol. 98, 35-43.
15. Miller, D.K., Griffiths, E., Lenard, J. and Firestone, R.A. (1983) J. Cell Biol. 97, 1841-1851.
16. Swanson, J.A., Yirinec, B.D. and Silverstein, S.C. (1985) J. Cell Biol. 100, 851-859.
17. Riezman, H. (1985) Cell 40, 1001-1009.
18. Willingham, M.C. and Pastan, I. (1983) Proc. Natl. Acad. Sci. USA 80, 5617-5621.
19. Makarow, M. (1985) Eur. Mol. Biol. Org. J. 4, 1861-1866.

ENDOCYTOSIS IN *SACCHAROMYCES CEREVISIAE*
Internalization of soluble and particulate
markers into cells and spheroplasts

Marja Makarow

Recombinant DNA Laboratory, University
of Helsinki, Valimotie 7, SF-00380 Helsinki,
Finland

ABSTRACT *S. cerevisiae* appears to be able to internalize components from the growth medium. Enveloped viruses are established probes in research of endocytosis in the mammalian cell system. One of them, Vesicular stomatitis virus (VSV), became irreversibly associated with spheroplasts upon incubation at 37C. It could not be removed by techniques which detach surface-bound viruses, or which destroy extracellular particles. Instead, all virus components could be recovered from spheroplast lysates. This suggests that the virus particles were internalized by the spheroplasts. A soluble enzyme, α-amylase, was also internalized by spheroplasts, as well as by intact cells. Subcellular fractionation showed that internalization of virus and α-amylase occured into intracellular organelles. One of these compartments comigrated in density gradients with vacuolar markers. Internalization into the vacuole was visualized directly with fluorescent dextran. Uptake of all three markers was dependent on the concentration of the marker, as well as time and temperature of incubation.

INTRODUCTION

Eukaryotic cells differ from prokaryotic ones in that they are compartmentalized. Communication between many compartments occurs by vesicular membrane traffic (1). Membrane traffic in the mammalian cells includes exocytosis and endocytosis. Exocytosis takes polypeptides from their site of synthesis at the endoplasmic reticulum via the Golgi complex to the exterior of the cell (1). Endocytosis captures material from the extracellular fluid and directs it e.g. to the lysosomes (2). In *S. cerevisiae* exocytosis appears to operate according to similar principles as in the mammalian cell (3). However, it has not been clear whether the yeast cell is capable of endocytosis. Here enveloped viruses were chosen as markers to study endocytosis in yeast spheroplasts. To study internalization into intact cells a soluble enzyme marker, α-amylase, and a fluorescent marker, FITC-conjugated dextran, were used.

RESULTS

Uptake of Virus

$[^{35}S]$ Methionine labeled VSV was incubated with *S. cerevisiae* (strain S13, derivative of S288C) spheroplasts at 37C in growth medium. At different time points samples were taken, free virus was washed off and the spheroplast-associated viral radioactivity was determined. There was a linear increase of association of virus with the spheroplasts for 90-120 min (Fig. 1). Surface-bound virions can be detached by proteinase K digestion in the cold, which degrades the envelope glycoprotein (4). After incubation with $[^{35}S]$ VSV the spheroplasts were again washed and then treated with proteinase K in the cold. A similar association curve was obtained as before after washing, but now with a lower background (Fig. 1). Thus, the radioactivity which was removed by proteinase K probably represented virus which was only superficially bound to the spheroplast surface. In contrast, the protease-resistant radioactivity probably represented internalized particles. This conclusion was substantiated by further experiments with Triton X-100. After incubation with VSV at 37C the spheroplasts were treated for 5 min with 0.1% Triton X-100. The treatment solubilizes the virus envelope completely releasing the nucleocapsid, but under these conditions the spheroplasts retain a spherical morphology and can still be surface-stained with FITC-concanavalin A. After the detergent treatment the spheroplast-associated viral radioactivity was the same as after proteinase K digestion (Fig. 1). SDS-PAGE analysis of the Triton X-100 treated spheroplasts showed that all viral polypeptides were present in the spheroplasts (4).

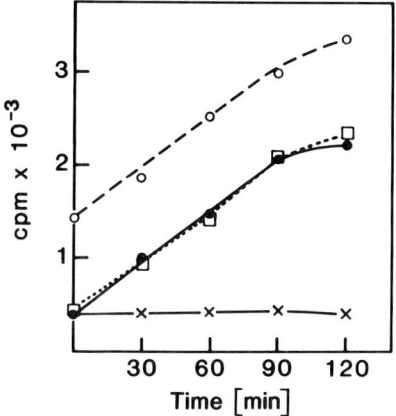

FIGURE 1. Uptake of VSV into spheroplasts. Spheroplasts (1.5×10^9/ml) were incubated with [^{35}S] VSV (5×10^5 cpm/ml) in growth medium supplemented with 1.2 M sorbitol at 37C. Samples of 100 μl were removed, washed with PBS-sorbitol and counted for viral radioactivity (O). Parallel samples were digested with proteinase K in the cold (●) or treated with Triton X-100 (□) prior to counting. Another set of parallel samples were incubated at 10C and treated with proteinase K (X). For experimental details see (4).

Uptake of α-Amylase

A soluble enzyme, α-amylase, was used to see whether also intact cells would internalize extracellular markers. α-Amylase was found to be bound in the cold to the cell surface. Proteinase K digestion in the cold removed the majority of the bound enzyme. When α-amylase was incubated with cells at 37C, increasing amounts of the marker became proteinase K-resistant (Fig. 2), suggesting uptake of the marker into the cells. Similar uptake curves were obtained for spheroplasts, but the level of uptake was 4-fold higher than into intact cells.

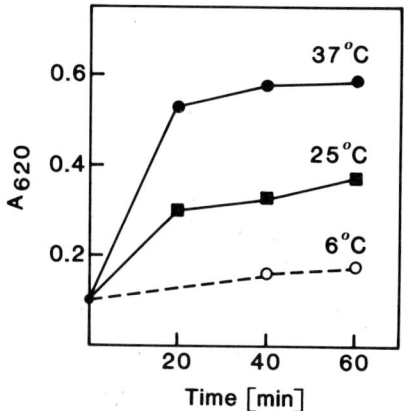

FIGURE 2. Uptake of α-amylase into cells. Cells (1.5×10^9/ml) were incubated with α-amylase (90 µg/ml) in growth medium at different temperatures. Samples of 100 µl were washed with PBS, digested in the cold with proteinase K to remove extracellular α-amylase and then subjected to zymolyase digestion and lysis for determination of α-amylase activity (5).

Subcellular Fractionation

In endocytosis internalized material is taken into intracellular organelles. An efficient homogenization procedure was adapted for yeast, which allows the use of high (0.8 M) sorbitol concentrations (4). Under these conditions at least the vacuole appears to be well preserved. Subcellular fractionation in Percoll density gradients revealed internalized VSV in three compartments (Fig. 3). About 4 to 18% of the viral radioactivity comigrated with the vacuole markers. About one third migrated in fractions 19 to 24 and about 60% very close to the plasma membrane marker. Most of the internalized α-amylase cosedimented with vacuole markers, the rest floating at the top of the gradient (Fig. 3).

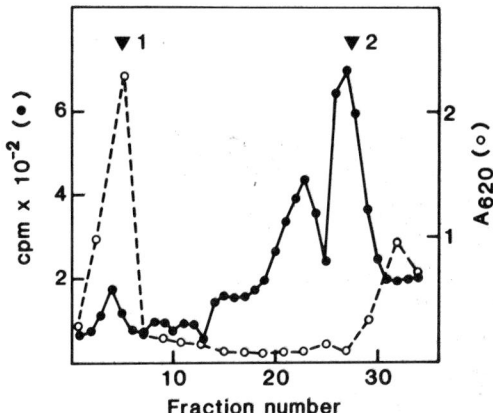

FIGURE 3. Subcellular distribution of internalized markers. Spheroplasts were allowed to internalize at 37C [^{35}S] VSV for 90 min (●) or α-amylase for 30 min (○). Extracellular markers were removed by proteinase K digestion in the cold. The spheroplasts were lysed and subjected to fractionation in 20% Percoll density gradients. Radioactivity and α-amylase activity were determined from the fractions. The position of the vacuole (arrow 1) was determined by assaying for α-mannosidase and non-specific protease activity.

The position of the plasma membrane (arrow 2) was determined by binding [^{3}H] concanavalin A to the spheroplast surface in the cold prior to homogenization (4).

Uptake of FITC-Dextran

Since subcellular fractionation suggested that both VSV and α-amylase were found in the vacuole after internalization, a morphological marker, FITC-dextran, was used to visualize directly whether there was uptake into the vacuole. When cells were incubated with the marker at 37C, a bright fluorescence was observed inside the cells coinciding with the vacuole, which was revealed by Nomarski optics (Fig. 4A, B). Similar staining of the vacuoles was observed in the absence and presence of 1.2 M sorbitol in the incubation mixture. In the cold there was no staining of the cells (Fig. 4C). Dead cells were stained over the whole cytoplasm (Fig. 4D).

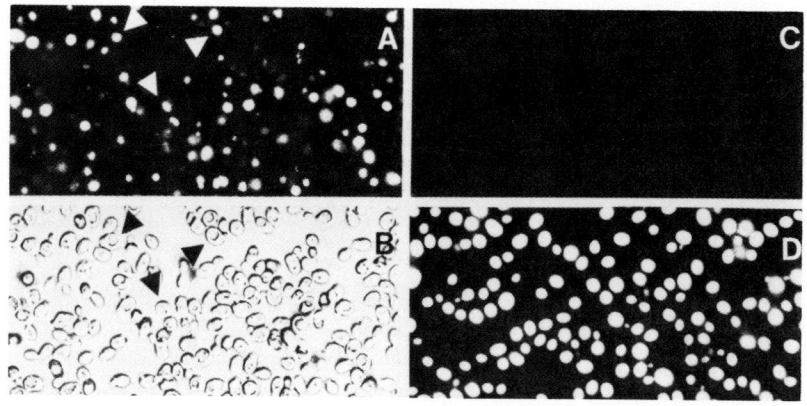

FIGURE 4. Staining of the vacuoles with FITC-dextran. Cells were incubated for 30 min in 100 μl of PBS containing 100 mg/ml of FITC-dextran at 37 C (A, B) or at 0C (C). In D the cells were killed by heating prior to staining. The cells were prepared for microscopy and viewed through fluorescence filters (A, C and D) or by Nomarski

optics (B). In A and B the arrow heads point to
the same cells. Magnification 620X.

DISCUSSION

Incubation of VSV with yeast spheroplasts
appeared to result in internalization of virus
particles. The spheroplast-associated virus could
not be removed by proteolytic digestion which
degrades the envelope glycoproteins, or by solubilization of the viral envelope with nonionic
detergent. α-Amylase was also taken up into
spheroplasts, as well as into intact cells. Uptake
of both markers was temperature-dependent, ceasing
at 10C.

Internalized α-amylase was found by subcellular fractionation mostly with the vacuoles.
This suggests that there is a transport route in
S. cerevisiae leading from the plasma membrane to
the vacuole. Such a route should involve intermediary organelles operating between these two
compartments. The intermediary organelles would
escape detection if the half life of internalized
material in them would be very short. This is
the case for secretory proteins in the yeast
Golgi complex (3).

In contrast, only a minor fraction of internalized VSV was found with the vacuoles. Most of
the virus was found in two as yet undefined compartments. In the mammalian cell enveloped
animal viruses are internalized into endosomes,
where the low internal pH activates the fusogeni-
city of the viruses. This leads to fusion of the
viral envelope with the membrane of the endosome,
resulting in introduction of the nucleocapsid of
the virus into the cytoplasm to start infection
(6). VSV can be fused at low pH with the plasma
membrane of *S. cerevisiae* spheroplasts (Makarow
and Sareneva, submitted). Thus, if yeast also
would have an endosome-like, acidic compartment,
VSV could be expected to fuse with its membrane.
This might cause the accumulation of VSV, which
is seen in subcellular fractionation (Fig. 3).
Interaction of VSV with the intermediary compart-

ment could also explain the different kinetics of intracellular accumulation of VSV and α-amylase. The former leveled off in 90-120 min and the latter in 20-40 min. Part of internalized α-amylase, but not virus, appeared to recycle back to the growth medium (Makarow, unpublished). Recycling of pinocytosed fluid has been described in the mammalian cell system (7).

Fluorescent dextran was accumulated specifically into the vacuole. Again, the accumulation was time-, temperature- and concentration-dependent. If the cells were killed, no specific staining of any organelles could be observed, but the whole cells were stained.

Uptake of FITC-dextran to the vacuoles was similar in the presence and absence of 1.2 M sorbitol in the incubation medium. This indicates that endocytosis in yeast is independent of turgor pressure. Coated vesicles have been identified in yeast (8). It will be interesting to learn whether they have any function in endocytosis in this organism.

The mechanisms of membrane dynamics and correct targetting of vesicular traffic are as yet unsolved. The yeast system should provide great potential to study these questions, because useful mutants can be generated. Many temperature-sensitive exocytosis-deficient mutants isolated by Schekman *et al.* (3) have recently been shown to be temperature-sensitive also for endocytosis (5, 9).

REFERENCES

1. Palade GE (1983). Membrane biogenesis: an overview. Meth Enzymol 96:XXIX.
2. Steinman RM, Mellman IS, Muller WA and Cohn Z (1983). Endocytosis and the recycling of plasma membrane. J Cell Biol 96:1.
3. Schekman R and Novick P (1982). The secretory process and yeast cell-surface assembly. In Strathern JN, Jones EW and Broach JR (eds): "The Molecular Biology of the Yeast *Saccharomyces*: Metabolism and Gene Expression," New York: Cold Spring Harbor Laboratory, p 361.

4. Makarow M (1985a). Endocytosis in *Saccharomyces cerevisiae*: Internalization of enveloped viruses into spheroplasts. EMBO J 4: 1855.
5. Makarow M (1985b). Endocytosis in *Saccharomyces cerevisiae*: Internalization of α-amylase and fluoresent dextran into cells. EMBO J 4: 1861.
6. Helenius A and Marsh M (1982). Endocytosis of enveloped animal viruses. Ciba Sym 92:59.
7. Besterman J M, Airhart J A, Woodworth RC and Low RB (1981). Exocytosis of pinocytosed fluid in cultured cells: kinetic evidence for rapid turn over and compartmentation. J Cell Biol 91:716.
8. Mueller SC and Branton D (1984). Identification of coated vesicles in *Saccharomyces cerevisiae*. J Cell Biol 98:341.
9. Riezman H (1985). Endocytosis in yeast: several of the yeast secretory mutants are defective in endocytosis. Cell 40:1001.

POST-TRANSLATIONAL PROCESSING EVENTS IN THE MATURATION OF YEAST PHEROMONE PRECURSORS[1]

Robert Fuller[#], Anthony Brake[+], Rachel Sterne[#], Riyo Kunisawa[#], Debra Barnes[#,2], Monica Flessel[#] and Jeremy Thorner[#]

[#]Department of Biochemistry, University of California, Berkeley, California 94720 and [+]Chiron Research Laboratories, Emeryville, California 94608

ABSTRACT The mating pheromones of yeast are peptide hormones that are excised from larger precursor polypeptides by precise proteolytic cleavages and other types of post-translational covalent modification. Identification of mutants deficient in pheromone production has permitted the cloning of genes that encode several of these specific processing enzymes and, in turn, has allowed detailed characterization of their catalytic and structural features.

INTRODUCTION

Intercellular signalling by means of hormones, growth factors, and neurotransmitters that are peptides is a common feature of the growth and differentiation of all eukaryotic cells, especially in multicellular organisms. Despite its deceptively simple lifestyle as a unicellular eukaryote, the yeast Saccharomyces cerevisiae is no exception to this rule and produces peptide hormone-like molecules that are responsible for triggering the events required for mating between haploid yeast cells to form diploids (1). Given their ap-

---

[1]Our work was supported by NIH Grant GM21841 (to JT), by a Helen Hay Whitney Postdoctoral Fellowship (to RF), and by USPHS Predoctoral Traineeships GM07232 (to DB, MF, and RS).
[2]Present address: Molecular Parasitology Group, School of Public Health, University of California, Berkeley, California 94720 and Naval Biosciences Laboratory, Naval Supply Center, Oakland, California 94625.

parent biological function and the fact that they are secreted into the culture medium, these molecules are called mating pheromones (2).

The production of these pheromones is one of the primary characteristics that distinguish the three yeast cell types (3). The pheromone secreted by haploids of the MATα cell type is called α-factor and is a 13-residue peptide (4). This molecule acts only on haploids of the other mating type, MATa cells. The pheromone produced by MATa cells is called a-factor, and it only affects MATα cells. This molecule was originally reported to be an 11-residue peptide (5), but may be larger (as discussed later here). Diploid (MATa/MATα) cells do not produce, or respond to, either of the pheromones.

In all eukaryotes in which it has been studied, a common strategy is used for the biosynthesis of secreted bioactive peptides. All such molecules are derived by excision from larger precursor polypeptides. The phylogenetic ubiquity of this mechanism suggests that there are important reasons for its use. As will be described in this review, the ability to define and characterize specific biochemical steps required for the maturation of the yeast pheromones has contributed to our understanding of the biological relevance of precursors and has elucidated the properties of the processing enzymes responsible for these universally encountered events.

BIOSYNTHESIS OF α-FACTOR

Structure of α-Factor Precursors.

Using oligonucleotide probes corresponding to the amino acid sequence of mature α-factor, two different genes that encode α-factor precursors were isolated (6,7). One of these genes, MFα1, is the major gene in the sense that it is expressed at a much higher level and, hence, is responsible for the majority (99% or greater) of the α-factor produced (8,9). The predicted protein encoded by the MFα1 gene is 165 amino acids long. This molecule begins with a hydrophobic signal (or pre-) sequence of about 20 residues, followed by a segment of 60 or so very polar amino acids that includes three sites for the addition of asparagine-linked oligosaccharide. Attachment of carbohydrate is often diagnostic of proteins destined for secretion. The carboxy-terminal half of the molecule contains four tandem repeats of the mature pheromone sequence separated from one another, and from the leader segment, by spacer regions consisting of the dipeptide -LysArg-

followed by two or three repeats of -GluAla- (or -AspAla-) dipeptides. The product encoded by the minor gene, MFα2, is similar in overall structure and shares considerable homology; however, it is shorter (because it contains only two α-factor repeats), has a different placement of its three glycosylation sites, and specifies, in one of the two pheromone repeats, a variant form of α-factor containing two conservative amino acid replacements (Asn for Gln and Arg for Lys). This variant does not appear to perform some special function because deletion of the MFα2 locus has no discernible effect on the mating efficiency of MATα cells or on the level of α-factor activity that they produce (9).

The fact that there are two functional genes for producing the pheromone probably explains why no structural gene mutations in these loci were discovered among the mating-deficient mutants isolated previously (3). Even in a strain deleted for the MFα1 locus, sufficient α-factor is produced from the MFα2 gene to permit MATα cells to mate, at only a somewhat reduced efficiency (10-25% of normal) (9,10). Defects in other genes required for mating can cause mating efficiency to drop to 10^{-5}-10^{-6}. Conversely, the fact that there are two very similar pheromone precursors dictates that the only mutations that can cause a pronounced pheromone-deficiency are lesions in genes whose products define functions common to the processing of both precursor molecules (11).

The overall structural motif of the α-factor precursor, namely pheromone repeats separated by spacer segments of the generic sequence, $-LysArg-(X-Ala)_n-$, has been found in all the Saccharomyces species examined (8) and even in the α pheromone gene of a quite distantly related yeast, Kluyveromyces lactis (A. Brake, unpublished results). In fact, in the precursors of all peptide hormones and neurotransmitters examined to date, from marine invertebrates to man, the amino acids corresponding to the mature bioactive peptides are always flanked on either side by doublets of basic amino acids (most often -LysArg-, but also -ArgArg-, -ArgLys-, and, more rarely, -LysLys-) (12).

Intermediates in Prepro-α-Factor Processing.

The overall pathway of α-factor biosynthesis was elucidated by an approach applicable to the dissection of any biosynthetic pathway. First, an assay was needed to detect α-factor-related molecules. For this purpose, specific antibodies directed against the mature pheromone were generated (13). Second, using in vitro translation of mRNA selected by virtue of its specific hybridization to MFα1 DNA, it was demonstrated that these antibodies were able to efficiently immunoprecipitate the entire 165-residue precursor (8). Third, various means to block the generation of α-factor were employed in order to accumulate in the cell intermediates that might be diagnostic of different stages in the processing of the precursor. Such methods included the use of temperature-sensitive secretion-defective (sec) mutants that at the non-permissive temperature halt the translocation of all exported proteins in discrete compartments of the secretory apparatus (14), and the use of specific drugs, like the glycosylation inhibitor, tunicamycin. These studies showed (15) that the primary translation product made in vivo is the 165-residue molecule; that the precursor receives three N-linked core oligosaccharides in the endoplasmic reticulum; that the carbohydrate chains are elongated by the addition of mannose-rich outer chains early in the Golgi; that initial endoproteolytic scission of the precursor occurs later in the Golgi; and that final N- and C-terminal exoproteolytic maturation of the pro-pheromone units may take place in secretory vesicles en route to the cell surface.

While these results outlined the progression of α-factor maturation, it was the examination of mutants specifically defective in pheromone production that proved most incisive for defining the specific genes and gene products required for prepro-α-factor processing.

KEX2 Gene Product is an Endoprotease Specific for Pairs of Basic Residues.

Because production of mature α-factor is essential for mating, mutations that affect the mating ability of MATα cells, but not MATa cells-- so-called "α-specific sterile" mutations --are candidates for genetic lesions that prevent prepro-α-factor processing. The kex2 mutation is one such α-specific sterile defect. The kex2 mutation was originally identified as a lesion blocking production of another small protein, called "killer toxin", released by certain yeast strains (16). MATα kex2 mutants are sterile because they also are unable to produce biologically active α-factor. Like α-factor, it has been shown that killer toxin is derived by scission of a larger precursor protein which contains -LysArg-doublets at several presumptive processing sites (17).

We found that prepro-α-factor is not processed in MATα kex2 mutants and is secreted intact into the culture medium (18). This finding suggested that kex2 mutants are unable to initiate the entire maturation process and, hence, might lack the specific endopeptidase responsible for making the first incisions of the precursor at its -LysArg- sites. Indeed, we demonstrated that kex2 mutants are deficient in an endopeptidase specific for cleaving on the carboxyl side of substrates containing -LysArg- and -ArgArg- residues (18,19).

Many different types of enzymes in animal cell systems have been proposed to perform this kind of processing step on the basis of in vitro experiments. In fact, most of the proteases that display specificity toward basic residues have been invoked as the true "prohormone convertase", including trypsin-like and kallikrein-like serine proteases and cathepsin B-like thiol proteases. Only in the case of yeast, however, has it been possible through the use of a mutation in vivo to identify the enzyme required for processing precursors at pairs of basic residues. For this reason, it was of interest to examine in detail the catalytic and structural features of the yeast KEX2 gene product.

For this purpose, we cloned the KEX2 gene on the basis of its ability to correct all the phenotypic characteristics of a MATα kex2 mutant (18), determined its entire nucleotide sequence and used expression vectors containing the galactose-inducible GAL1 promoter to overproduce the KEX2 protein 200-500-fold in yeast (19,20). The enzyme has been purified over 100-fold from such a source. Among the most striking features of the enzyme revealed to date are its relatively large molecular weight (~100,000), its tight membrane association, and the absolute dependence of its catalytic activity on Ca^{2+}. After removal of Ca^{2+} with chelator, readdition of Ca^{2+} (and no other divalent metal ion tested) at micromolar concentration is sufficient to fully reactivate the enzyme. Although the KEX2 enzyme appears to be a neutral thiol protease with certain similarities to the Ca^{2+}-dependent neutral thiol proteases of animal cells (so-called calpains), it can be distinguished from calpains by its resistance to certain inhibitors and by its lack of cross-reaction with anti-calpain antibodies (R. Fuller, D. Croall and G. DeMartino, unpublished results). Thus, the KEX2 enzyme represents a novel, and previously unrecognized, class of protease.

The primary structure of the 814-residue KEX2 polypeptide deduced from the DNA sequence includes two markedly hydrophobic regions (a potential signal sequence at the N-terminus and a putative transmembrane domain near the C-terminus), a cysteine-rich region, several sites for the attachment of Asn-linked oligosaccharide, a Ser/Thr-rich domain, and potential Ca^{2+}-binding sites. To follow the polypeptide in various kinds of experiments, it can be tagged covalently by its reaction with [^{125}I]Tyr-Ala-Lys-Arg-chloromethylketone. Examination of tunicamycin-treated cells, and digestion in vitro with endoglycosidase H (which can remove N-linked oligosaccharides from glycoproteins), indicate that the native enzyme does carry N-linked carbohydrate chains. A series of C-terminal deletions of the gene that remove as many as 200 amino acids indicates that the carboxyl-end of the molecule is not required for catalytic activity, but is important for retaining the enzyme within the secretory system in the cell.

Given the pleiotropic role of the KEX2 enzyme in precursor processing, it is perhaps surprising that cells which carry a kex2 deletion that removes all but the 120 amino-terminal residues, and eliminates catalytic function, are still viable. Nonetheless, the properties of the KEX2 protease may be of general significance if the enzyme is indeed a prototype for the prohormone convertase(s) of higher eukaryotes. For example, its membrane localization may serve to order its activity in the secretory system, both temporally and spatially. The low concentration of Ca^{2+} required for enzyme activity suggests that substrate cleavage in vivo could be regulated by modulation of the intracompartmental Ca^{2+} concentration. Finally, if the KEX2 enzyme is an integral component of the Golgi body, this protein might provide a probe for examining the biogenesis of, and "traffic" through, this organelle.

We are currently analyzing further the structure-function relationships of the domains of the KEX2 enzyme, and attempting to further understand the features of the polypeptide that direct its proper subcellular localization, by several means. First, we are generating antibodies against different segments of the molecule by using as antigens lacZ-KEX2 fusion proteins. Second, we intend to create various single amino acid replacements and small deletions within the protein by using in vitro site-directed mutagenesis of the gene.

STE13 Gene Product is a Membrane-Bound Dipeptidyl Aminopeptidase.

Another of the α-specific sterile defects is the ste13 mutation (21). We found that MATα ste13 mutants do not produce mature α-factor, but release instead a collection of incompletely processed forms of markedly reduced specific biological activity (22). Digestion with carboxypeptidases A and Y indicated that these species had the same C-terminus as the mature pheromone. However, amino-terminal analysis of these peptides indicated that essentially all the molecules possessed extra N-terminal residues, the majority of which had the sequence H$_2$N-GluAlaGluAla-α-factor or H$_2$N-AspAlaGluAla-α-factor. These findings suggested that N-terminal maturation of the pro-pheromone repeats excised from the precursor is defective in ste13 mutants.

Exoproteases that attack the free N-termini of polypeptide chains consisting of repeating -X-Ala- (or -X-Pro-) sequences, and cleave so as to release X-Ala (and X-Pro) dipeptides as the products, had been detected in vitro in a variety of other systems. These observations suggested that ste13 mutants might lack such a dipeptidyl aminopeptidase enzyme. Using synthetic substrates that made it possible to follow this kind of cleavage by a colorimetric assay, we were able to demonstrate that yeast cells carrying ste13 mutations lacked one of at least two such dipeptidyl aminopeptidase activities that are present in normal yeast cells (22). The STE13 gene was cloned on the basis of its ability to correct the phenotypic defects of a MATα ste13 mutant; and, cells carrying the STE13 gene on a multi-copy plasmid greatly overproduced this same dipeptidyl aminopeptidase.

Two distinguishing characteristics of the dipeptidyl aminopeptidase encoded by the STE13 gene are its tight membrane association and its resistance to inactivation by heat. Another much more thermosensitive dipeptidyl aminopeptidase activity is localized in the vacuole (23) and probably serves a degradative, rather than biosynthetic, function. The STE13 gene encodes a transcript of 2.9 kb suggesting that its polypeptide product could be as large as 100,000 MW; and, membrane fractions that should be enriched in the STE13 gene product have elevated levels of a protein species in this molecular weight range (D. Julius and A. Brake, unpublished results). Nucleotide sequence analysis of the entire gene is currently in progress.

Since ste13 mutants secrete processed pheromone molecules that are immature only at their amino-terminal end, all the other steps involved in excision and secretion of the pheromone repeats can occur, suggesting that processing by the dipeptidyl aminopeptidase is a late step in pheromone synthesis, in agreement with our overall view of the precursor processing pathway (15). Moreover, the removal of the N-terminal -X-Ala- repeats appears to be the most rate-limiting step in α-factor biosynthesis because normal cells carrying the MFα1 gene on a multi-copy plasmid produce mainly the same incompletely processed pro-pheromone species as made by ste13 mutants and very little mature pheromone. In support of this conclusion, introduction of multiple copies of both the MFα1 gene and the STE13 gene on the same plasmid permits normal cells to produce a greatly elevated level of the mature pheromone (10).

The effect of ste13 mutations on α-factor processing provided the first in vivo demonstration in any organism that a dipeptidyl aminopeptidase has an essential role in the maturation of a secreted peptide. It is noteworthy in this regard that the pro-protein segments that are removed from the N-terminal side of secreted peptides in other eukaryotic organisms are often sequences of repeating -X-Ala-, -X-Pro-, or -X-Gly- dipeptides that may be trimmed away by the action of an enzyme similar to the STE13 gene product.

A Carboxypeptidase B-like Activity is Required for Pheromone Maturation.

Because the KEX2 endopeptidase cleaves the precursor on the carboxyl side of the -LysArg- sites in the spacer regions, three of the four pheromone units so released will possess two basic residues on their carboxyl end (Fig. 1). Hence, in addition to the N-terminal residues that must be removed from each pro-pheromone unit by the STE13 gene product, the basic residues also must be removed from the C-terminus in order to produce fully mature α-factor.

Several classes of exoproteases with the appropriate specificity for this kind of reaction have been described in animal cells, including carboxypeptidase B, carboxypeptidase N, cathepsin B2, and the Co^{2+}-stimulated enzyme involved in removing basic residues from the C-terminus of proenkephalin molecules in the chromaffin granules of the adrenal medulla (24). In yeast, neither the soluble vacuolar serine protease, carboxypeptidase Y, nor the vacuolar membrane-associated metalloprotease, carboxypeptidase S, are required for α-factor biosynthesis because a mutant lacking these activities (MATα pep4 prc1 cps1) produces a normal level of mature pheromone (L. Blair and D. Julius, unpublished results). Although no mutations have yet been identified that result in production of inactive pheromone molecules that retain basic residues on their C-terminus, and hence might represent lesions in the specific enzyme required for their removal, a membrane-bound carboxypeptidase activity with this precise specificity has been detected in extracts (25).

Our current understanding of the generation of mature α-factor pheromone from its precursor, prepro-α-factor, is summarized in the following diagram (Fig. 1).

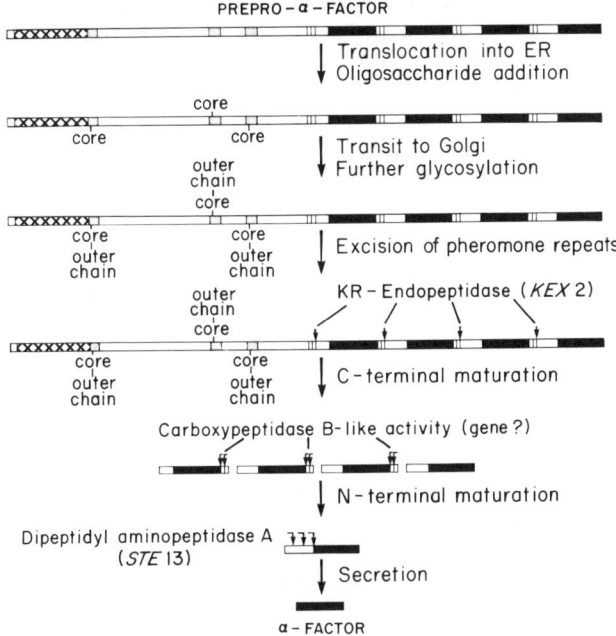

FIGURE 1. Pathway of posttranslational processing of the prepro-α-factor precursor in yeast.

BIOSYNTHESIS OF a-FACTOR

Structure of a-Factor Precursors.

It was reported originally that MATa cells secrete two nearly identical, 11 amino acid-long pheromones, that differ only in the residue at position 6 (one molecule had Val and the other Leu) (5). It might have been reasonable to assume that the explanation for the production of both the Val- and Leu-containing peptides by a single MATa strain was that the a-factor precursor(s) bear resemblance to the α-factor precursors and contain at least two tandem repeats of the mature

a-factor sequence that differ by a single base pair change in
the codon corresponding to position 6. However, more careful
reflection suggests why a-factor must be made by a very different route than that responsible for the generation of α-
factor. The main reason for this inference is that the enzyme deficiencies that prevent α-factor production (namely,
ste13 and kex2 mutations) are MATα cell-specific, and have no
effect on the ability of MATa cells to produce a-factor. Because the STE13 and KEX2 genes are transcribed (and the encoded proteases are present) in all three yeast cell types, the
α-specific nature of these functions must mean that they are
only required to process prepro-α-factor, but are not required for any step in the conversion of an a-factor precursor
to the mature a-factor pheromone. Because the a-factor precursors are not substrates for the action of these processing
enzymes, they must have a structure that differs dramatically
from prepro-α-factor, as indeed turns out to be the case.

Using oligonucleotide probes corresponding to the purported amino acid sequence of mature a-factor (5), two different
genes that encode a-factor precursors were isolated (26,27).
The product encoded by one gene, MFa1, is only 36 amino acids
long and contains within it a single copy of the Val-containing a-factor bracketed on both its amino-terminal and carboxyterminal sides by additional residues. The product specified
by the second gene, MFa2, is very similar, but is 38 amino
acids long and contains a single copy of the Leu-containing
a-factor. These precursors are highly unusual in several respects. First, they are quite short. Second, the amino-terminal extensions lack the striking hydrophobicity characteristic of the signal sequences of other secreted proteins.
Third, there are no sites for the attachment of Asn-linked
oligosaccharide.

Nonetheless, two pieces of evidence (27) support the conclusion that both of these genes are expressed and functional.
Using the DNAs as probes for analysis of mRNAs, corresponding
transcripts (340 bases for MFa1 and 420 bases for MFa2) are
found in MATa cells, but not in MATα or in MATa/MATα cells.
Second, when carried on multi-copy plasmids, the presence of
either MFa1 or MFa2 causes MATa haploids to overproduce a-
factor activity, as judged by bioassay.

Potential Intermediates in a-Factor Precursor Processing.

Both a-factor precursors show considerable homology
(60%) in their amino-terminal extensions and are identical,
except for the Val-to-Leu difference, in the stretch of 22

residues that correspond to their carboxy-terminal halves.
Just preceding the a-factor sequence in the carboxy-terminal
half is a -LysLys- doublet, which might represent a site for
endoproteolytic processing (Fig. 2). (It should be recalled
that the KEX2 endoprotease is highly specific for pairs of
basic residues that contain at least one Arg.) If cleavage
occurred here, an -Asp-Asn- dipeptide would remain at the
amino-terminal end of the a-factor residues. Removal of
these amino acids to produce the mature N-terminus of a-factor
could be carried out by either an aminopeptidase or a dipeptidyl aminopeptidase.

Difficulties in determining the amino acid sequence of
the carboxyl end of purified a-factor (5, 28), and the solu-
bility characteristics and behavior during purification of the
pheromone isolated from MATa culture medium (29), indicate to
us that mature a-factor may bear some non-polar posttranslational modification at its carboxy-terminus. Several hetero-
basidiomycetous yeast species produce lipopeptide mating pheromones (30) that carry a farnesyl moiety in thioether linkage to a C-terminal Cys residue. Posttranslational addition
of isoprenoids onto specific cell proteins has recently been
detected in animal cells (31). It may be significant, therefore, that the putative a-factor sequence is immediately followed in the precursors by a Cys residue and three additional
hydrophobic amino acids (-Val-Ile-Ala-COOH). It is interesting to note that at the C-terminus of the yeast RAS gene products (and their higher cell counterparts) a nearly identical
sequence is present (32) and that these proteins are known to
be acylated on the Cys with palmitate (or myristate) (33).

Approaches for Investigating a-Factor Biosynthesis.

To address these questions, we have prepared the 12-resi-
due molecule that contains the additional C-terminal Cys by
solid phase peptide synthesis and found that it possesses low,
but detectable, biological activity (34). In contrast, the
11-residue a-factor-related sequence lacking the Cys also has
been prepared by chemical synthesis, but has no detectable
biological activity (J. Becker, personal communication). We
intend to modify the Cys residue in vitro to determine if
attachment of aliphatic substituents greatly enhances the
biological activity of the molecule.

We have used the synthetic 12-residue a-factor-related
peptide as an immunogen for preparing anti-a-factor antibodies. When MATa cells are labeled with [^{35}S]SO$_4^{2-}$, the
antiserum immunoprecipitates two radioactive species (R.

Sterne, unpublished results). Both molecules are a-factor-related because their precipitation is specifically blocked when excess synthetic a-factor is added. One of the a-factor-related molecules migrates similarly to the synthetic dodecapeptide upon electrophoresis in SDS-polyacrylamide gels; the other species is larger, and of appropriate apparent molecular size to be the full-length precursor. Both of these molecules are overproduced in MATa cells transformed with the MFa1 gene on a multi-copy plasmid, providing further support for the view that the cloned DNA represents a fully functional a-factor structural gene. Only the smaller species is found extracellularly. Because overproduction of this secreted molecule correlates with an increase in biological activity, the smaller species may represent the final mature pheromone.

Given that the only sulfur-containing amino acids in the entire a-factor precursor are the initiator methionine and the internal Cys, the fact that the small extracellular a-factor-related peptide is labelled with $[^{35}S]SO_4^{2-}$ indicates that at least a fraction of secreted a-factor does contain the Cys residue predicted by the nucleotide sequence of the gene.

STE6 and STE14 Gene Products Are Required for Post-Translational Processing of the a-Factor Precursors.

Two MATa cell-specific sterile mutations, ste6 (35) and ste14 (36), appear to prevent the mating of MATa cells only because they prevent the ability of MATa cells to produce a sufficient amount of, or a sufficiently active form of, a-factor (11). One formal possibility is that STE6 and STE14 products might be required for expression of both MFa1 and MFa2; however, mRNAs for both of the pheromone structural genes are produced at normal levels in both ste6 and ste14 mutants (27; S. Michaelis and I. Herskowitz, personal communication). Again, therefore, because the yeast genome contains two expressed genes for a-factor precursors, the STE6 and STE14 genes are candidates for functions that are required for steps common to the processing of both precursors.

It is not yet clear what posttranslational modifications of the a-factor precursors are defective in ste6 and ste14 mutants. On the other hand, in preliminary experiments, we have detected intracellular accumulation of a species larger than the major a-factor peptide secreted by normal MATa cells in both MATa ste6 and MATa ste14 mutants (R. Sterne, unpublished results).

Our current speculation concerning the processing of the

a-factor precursors is given below (Fig. 2).

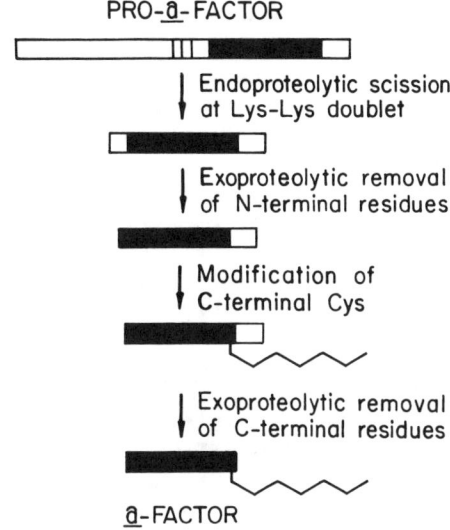

FIGURE 2. Hypothetical processing pathway for the maturation of a-factor from its precursor.

REFERENCES

1. Thorner, J (1980). Intercellular interactions of the yeast Saccharomyces cerevisiae. In Leighton, TJ, Loomis, WF (eds): "The Molecular Genetics of Development," New York: Academic Press, Inc., p. 119.
2. Thorner, J (1981). Pheromonal regulation of development in Saccharomyces cerevisiae. In Strathern, JN, Jones, EW, Broach, JR (eds): "Molecular Biology of the Yeast Saccharomyces: Life Cycle and Inheritance," Cold Spring Harbor: Cold Spring Harbor Laboratory, p. 143.
3. Sprague, GF jr, Blair, LC, Thorner, J (1983). Cell interactions and regulation of cell type in the yeast Saccharomyces cerevisiae. Ann. Rev. Microbiol. 37: 623.
4. Stotzler, D, Kiltz, H-H, Duntze, W (1976). Primary structure of α-factor peptides from Saccharomyces cerevisiae. Eur. J. Biochem. 69: 397.
5. Betz, R, Manney, TR, Duntze, W (1981). Hormonal control of gametogenesis in the yeast Saccharomyces cerevisiae. Gamete Res. 4: 571.
6. Kurjan, J, Herskowitz, I (1982). Structure of a yeast pheromone gene (MFα): A putative α-factor precursor con-

tains four tandem repeats of mature α-factor. Cell 30: 933.
7. Singh, A, Chen, EY, Lugovoy, J, Chang, CN, Hitzeman, R, Seeburg, PH (1983). Saccharomyces cerevisiae contains two discrete genes coding for the α-factor pheromone. Nucl. Acids Res. 11: 4049.
8. Brake, AJ, Julius, DJ, Thorner, J (1983). A functional prepro-α-factor gene in Saccharomyces yeasts can contain three, four, or five repeats of the mature pheromone sequence. Molec. Cellul. Biol. 3: 1440.
9. Kurjan, J (1985). α-Factor structural gene mutations in Saccharomyces cerevisiae: Effects on α-factor production and mating. Molec. Cellul. Biol. 5: 787.
10. Barnes, DA (1985). Isolation and characterization of the LYS2 gene and its use for genetic manipulation of the yeast Saccharomyces cerevisiae. Ph.D. Thesis, University of California, Berkeley.
11. Chan, RK, Melnick, LM, Blair, LC, Thorner, J (1983). Extracellular suppression allows mating by pheromone-deficient sterile mutants of Saccharomyces cerevisiae. J. Bacteriol. 155: 903.
12. Schwartz, TW, Wittels, B, Tager, HS (1983). Hormone precursor processing in the pancreatic islet. In Hruby, VJ, Rich, DH (eds): "Peptides: Structure and Function (Proceedings of the 8th American Peptide Symposium," Rockford, Illinois: Pierce Chemical Co., p. 229.
13. Jones-Brown, YR, Thorner, J (1978). Radioimmunoassay for α-factor, the oligopeptide pheromone released by Saccharomyces cerevisiae cells of mating-type α. In Lawrence, C, Prakash, L, Prakash, S, Sherman, F (eds) "9th International Conference on Yeast Genetics and Molecular Biology," Rochester, New York, p. 89.
14. Schekman, R (1985). Protein localization and membrane traffic in yeast. Ann. Rev. Cell Biol. 1: 115.
15. Julius, D, Schekman, R, Thorner, J (1984). Glycosylation and processing of prepro-α-factor through the yeast secretory pathway. Cell 36: 309.
16. Wickner, RB, Leibowitz, MJ (1976). Two chromosomal genes required for killing expression in killer strains of Saccharomyces cerevisiae. Genetics 82: 429.
17. Bostian, KA, Elliot, Q, Bussey, H, Burn, V, Smith, A, Tipper, DJ (1984). Sequence of the prepro-toxin dsRNA gene of type I killer yeast: Multiple processing events produce a two-component toxin. Cell 36: 741.
18. Julius, D, Brake, A, Blair, L, Kunisawa, R, Thorner, J (1984). Isolation of the putative structural gene for

the lysine-arginine-cleaving endopeptidase required for processing of yeast prepro-α-factor. Cell 37: 1075.
19. Fuller, RS, Brake, AJ, Julius, DJ, Thorner, J (1985). The KEX2 gene product required for processing of yeast prepro-α-factor is a calpain-like endopeptidase specific for cleaving at pairs of basic residues. In Gething, M-J (ed) "Protein Transport and Secretion," Cold Spring Harbor: Cold Spring Harbor Laboratory, p. 97.
20. Fuller, RS, Brake, A, Thorner, J (1986). The yeast KEX2 gene, required for processing prepro-α-factor, encodes a calcium-dependent endopeptidase that cleaves after Lys-Arg and Arg-Arg sequences. In Leive, L. (ed) "Microbiology-1986," Washington, DC: American Society for Microbiology, in press.
21. Sprague, GF jr, Rine, J, Herskowitz, I (1981). Control of yeast cell type by the mating type locus II: Genetic interactions between MATα and unlinked α-specific STE genes. J. Molec. Biol. 153: 323.
22. Julius, D, Blair, L, Brake, A, Sprague, G, Thorner, J (1983). Yeast α-factor is processed from a larger precursor polypeptide: The essential role of a membrane-bound dipeptidyl aminopeptidase. Cell 32: 839.
23. Bordallo, C, Schwencke, J, Suarez-Rendueles, M (1984). Localization of the thermosensitive X-prolyl dipeptidyl aminopeptidase in the vacuolar membrane of Saccharomyces cerevisiae. FEBS Letters 173: 199.
24. Fricker, LD, Snyder, SH (1982). Enkephalin convertase: Purification and characterization of a specific enkephalin-synthesizing carboxypeptidase localized to adrenal chromaffin granules. Proc. Natl. Acad. Sci. USA 79: 3886.
25. Achstetter, T, Wolf, DH (1985). Hormone processing and membrane-bound proteinases in yeast. EMBO J. 4: 173.
26. Brake, A, Merryweather, J, Najarian, R, Thorner, J (1983). Identification and characterization of a structural gene for the yeast peptide mating pheromone, a-factor. J. Cellul. Biochem. Suppl. 7B: 375.
27. Brake, AJ, Brenner, C, Najarian, R, Laybourn, P, Merryweather, J (1985). Structure of genes encoding precursors of the yeast peptide mating pheromone a-factor. In Gething, M-J (ed) "Protein Transport and Secretion," Cold Spring Harbor: Cold Spring Harbor Laboratory, p. 103.
28. Betz, R, Duntze, W (1979). Purification and partial characterization of a-factor, a mating hormone produced by mating-type-a cells from Saccharomyces cerevisiae. Eur. J. Biochem. 95: 469.
29. Strazdis, JR, MacKay, VL (1982). Reproducible and rapid

methods for the isolation and assay of a-factor, a yeast mating hormone. J. Bacteriol. 151: 1153.
30. Ishibashi, Y, Sakagumi, Y, Isogai, A, Suzuki, A (1984). Structure of tremerogens A-9291-I and A-9291-VIII: Peptidyl sex hormones of Tremella brasiliensis. Biochemistry 23: 1399.
31. Schmidt, RA, Schneider, CJ, Glomset, JA (1984). Evidence for post-translational incorporation of a product of mevalonic acid into Swiss 3T3 cell proteins. J. Biol. Chem. 259: 10175.
32. Powers, S, Kataoka, T, Fasano, O, Goldfarb, M, Strathern, J, Broach, J, Wigler, M (1984). Genes in S. cerevisiae encoding proteins with domains homologous to the mammalian ras proteins. Cell 36: 607.
33. Chen, Z-Q, Ulsh, LS, DuBois, G, Shih, TY (1985). Post-translational processing of p21 ras proteins involves palmitylation of the C-terminal tetrapeptide containing cysteine. J. Virol. 56: 607.
34. Sterne, RE, Thorner, J (1985). Biogenesis of the yeast peptide mating pheromone, a-factor, in Saccharomyces cerevisiae. Symp. Amer. Soc. Protein Chemists, San Diego, CA, p. 38 (Abstr).
35. Wilson, KL, Herskowitz, I (1984). Negative regulation of STE6 gene expression by the α2 product of Saccharomyces cerevisiae. Molec. Cellul. Biol. 4: 2420.
36. Blair, LC (1979). Genetic analysis of mating type switching in yeast. Ph.D. Thesis, University of Oregon, Eugene.

CALMODULIN AND OTHER CALCIUM-BINDING PROTEINS IN YEAST[1]

Trisha N. Davis and Jeremy Thorner

Department of Biochemistry, University of California
Berkeley, California 94720

ABSTRACT Regulation of cellular processes by changes in calcium ion concentration occurs in all eukaryotic cells. Calcium regulation is thought to be largely mediated by binding of the ion to small proteins that act as calcium receptors and, in turn, interact specifically with, and control the activity of, a variety of target enzymes. Cloning and characterization of genes for two calcium-binding proteins, calmodulin and the CDC31 gene product, have demonstrated that both of these proteins perform vital, but different, functions in the yeast cell.

INTRODUCTION

Calcium is the fifth most abundant element in the Earth's crust. In the milieu in which eukaryotic cells grow (for example, in blood plasma for the propagation of animal cells or in standard minimal salts medium for the culturing of yeast cells), calcium ion is present at relatively high concentration (about 1 mM). However, because Ca^{2+} can combine with inorganic phosphate to form very insoluble salts (e.g. hydroxylapatite), the intracellular concentration of calcium must be kept quite low so that ATP breakdown and resynthesis are not compromised. In other words, one problem for cells, in general, is not how to acquire sufficient Ca^{2+}, but rather how to exclude the ion and only admit it under the appropriate circumstances or stimulus. Given the low steady state level of the ion (about 0.1 μM), small changes in Ca^{2+} flux can have

---

[1] Our work was supported by NIH Grant GM21841 (to JT) and by Anna Fuller Fund Postdoctoral Fellowship #584 and USPHS Postdoctoral Traineeship CA09041 (to TD).

a dramatic effect on the absolute concentration of Ca^{2+} inside the cell. In this sense, living systems are poised to respond to changes in Ca^{2+} level. Ca^{2+} also is an appropriate choice for an intracellular signal molecule in other respects.

Probably the most important of these attributes are the chemical properties of Ca^{2+}, which are well suited for both tight and specific binding to proteins. First, Ca^{2+} is a divalent ion, and therefore possesses a higher affinity for polyanionic sites on proteins than monovalent cations, like sodium or potassium. Second, Ca^{2+} has a moderate ionic radius (0.99Å) that is compatible in atomic dimension with the size of pockets that are created on the surface of proteins by the folding of polypeptide backbones and the extension of amino acid side chains. Third, because of the configuration of its valence orbitals, Ca^{2+} can ligand to as many as eight electron donors (usually oxygen atoms on a protein). In this respect, Ca^{2+} is superior to magnesium, which can bind to only six electron donors (1). Furthermore, because Mg^{2+} has a smaller ionic radius (0.65Å) and cannot completely fill the kind of cavities available on proteins, often water molecules will occupy one or more of the bonding positions and thereby greatly weaken the interaction between Mg^{2+} and the protein. These considerations presumably explain, in part, how Ca^{2+} can bind specifically to a protein in the presence of the large excess of Mg^{2+} (1 mM) that exists inside cells.

In highly differentiated cells of multicellular organisms, changes in intracellular Ca^{2+} level trigger a variety of complex phenomena, including muscle contraction, hormone secretion and neurotransmitter release, cytoskeletal restructuring in the fertilized egg, etc. The changes in Ca^{2+} level that elicit these responses are thought to be generated by one of two basic mechanisms. In some cells, chemical signals (like binding of a hormone to its receptor) or electrical signals (like the voltage change upon depolarization of nerve cells) transiently open Ca^{2+}-specific protein channels (2) in the plasma membrane, thus harnessing the very large Ca^{2+} gradient across the cell membrane to drive Ca^{2+} entry. In other cells, activation of cell surface receptors triggers hydrolysis of membrane phosphoinositide lipids thereby releasing into the cytosol inositol-1,4,5-tris-phosphate (IP_3), a compound that mobilizes bound or stored Ca^{2+} from internal reserves (3).

The rise in intracellular Ca^{2+} level (measurements utilizing trapped fluorescent indicator dyes [4] or Ca^{2+}-specific electrodes indicate that the free cytosolic Ca^{2+} can reach 5-10 μM, depending on the cell type and the stimulus) is "sensed" by Ca^{2+}-binding proteins that act as the transducers of this event.

Paramount among the Ca^{2+}-binding proteins is calmodulin, an unusual acidic protein of 16,000 MW, that has been found in every eukaryotic system that has been examined (5). At least as measured in vitro with the purified protein, the affinity of calmodulin for Ca^{2+} ($K_d \approx 1\ \mu M$) is such that under normal resting conditions most molecules would be unoccupied; but, after the rise in Ca^{2+} level, most of the calmodulin molecules would be saturated with the ion. This view is undoubtedly an oversimplification because each calmodulin molecule has four Ca^{2+} binding sites and their affinity for Ca^{2+} will be influenced by pH, ionic strength, occupancy of the other sites, association with other proteins, etc.

Upon binding of Ca^{2+}, calmodulin changes conformation, revealing hydrophobic sites that permit the molecule to interact with other proteins. In this way, calmodulin appears to mediate cellular responses to changes in intracellular Ca^{2+} level. Many roles in the cell have been ascribed to calmodulin, based on its interactions in vitro and effects on activity with a variety of enzymes, on its subcellular distribution (located by immunocytochemistry with anti-calmodulin antibodies), and on the perturbations of cell physiology and morphology caused by calmodulin-directed drugs.

Calmodulin/Ca^{2+} complex interacts with certain specific protein kinases, for example phosphorylase kinase (6), and with several apparently multi-functional protein kinases (6a). Ca^{2+} itself stimulates other protein kinases, for example the membrane-bound protein kinase C (7). Hence, at least one immediate effect of a change in Ca^{2+} level would be a change in the pattern of protein phosphorylation. Calmodulin/Ca^{2+} also interacts with enzymes that control the metabolism of adenosine 3',5'-cyclic monophosphate, including cyclic nucleotide phosphodiesterase (8) and, in some cells, adenylate cyclase (9). Thus, a concomitant result of an alteration in intracellular Ca^{2+} concentration is presumably a change in cAMP level. Therefore, protein phosphorylation by the catalytic subunit of cAMP-dependent protein kinase (10), and perhaps gene expression resulting from the putative topoisomerase activity of the regulatory subunit of cAMP-dependent protein kinase (11), may also be affected. In addition, quite a number of cytoskeletal elements and other cellular polypeptides have been identified on the basis of their behavior as "calmodulin-binding proteins".

Because Ca^{2+} is a metal ion that cannot be created or destroyed by biological processes, a cell can only return to its normal resting level of Ca^{2+} by actively extruding Ca^{2+} or by re-sequestering the ion inside the cell. For this purpose, most of the membranes that enclose cellular compartments, especially the plasma membrane, possess "pumps" (Ca^{2+}-ATPases) and other mechanisms (e.g. Na^+/Ca^{2+} exchangers) for transporting Ca^{2+} away from the cytosol. It is to be expected that the process of Ca^{2+} removal should be as tightly controlled as the regulation of its entry. It is not surprising, therefore, that the plasma membrane Ca^{2+}-pumping Mg^{2+}-ATPase in animal cells seems to be stimulated by binding of calmodulin/Ca^{2+} complex and by phosphorylation via both the cAMP-dependent and a calmodulin/Ca^{2+}-dependent protein kinase (12). This kind of autoregulatory loop would help to ensure that any rise in Ca^{2+} level would only be transient.

Although yeast cells are not highly specialized in the sense that muscle cells are for contraction, that neurons are for synaptic transmission, or that spermatozoa are for motility, they nevertheless carry out life processes, and must respond to cues in their environment, in ways (and presumably by mechanisms) that are quite similar to those used by their higher cell counterparts. For example, yeast cells must coordinate their metabolism, growth, and development to the availability of nutrients (13). In addition, haploid yeasts secrete hormones and are hormonally-responsive cells (14). For these reasons, yeast cells represent a relatively uncomlicated and attractive system in which to use a molecular genetic approach to dissect Ca^{2+} regulation, to analyze the structure and function of calcium-binding proteins, and to determine the spectrum of <u>in vivo</u> targets for the action of these agents.

In this review, we will summarize what is known about Ca^{2+} and its metabolism in yeast in three areas: (a) transport of Ca^{2+} and transmembrane signalling; (b) calcium-binding proteins, in particular our own work on yeast calmodulin; and, (c) the potential targets of Ca^{2+} regulation in this organism.

CONTROL OF CYTOSOLIC CALCIUM CONCENTRATION IN YEAST

Entry of Calcium.

We performed a simple experiment that serves to illustrate the importance of Ca^{2+} for yeast growth and the fact that yeast cells must be able to control their intracellular Ca^{2+} level (Fig. 1). Even at the highest concentration of Ca^{2+} added (10 <u>mM</u>), the final yield of cells was unaffected. However, if Ca^{2+} was removed by titration with a chelator (EGTA), or if unregulated entry (or exit) of Ca^{2+} was caused by the presence of A23187 (an antibiotic that is a Ca^{2+}-specific ionophore), yeast cell growth was severely compromised.

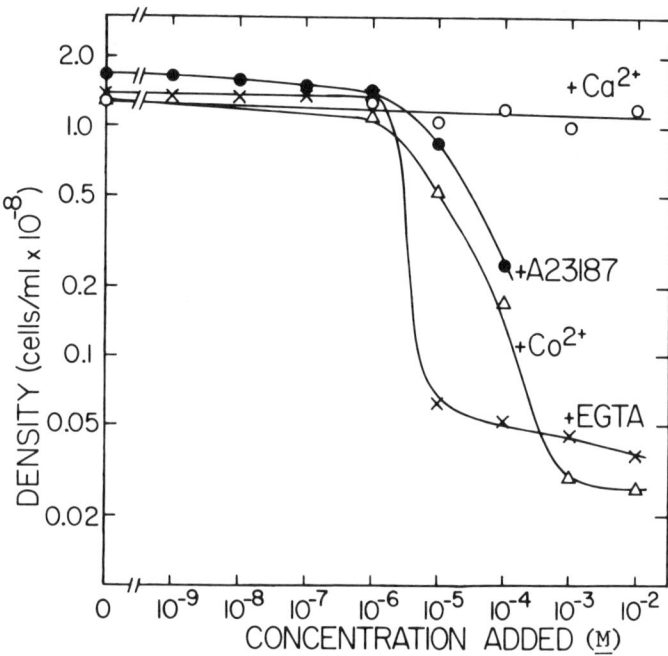

Figure 1. Effect of calcium concentration, calcium ionophore, calcium chelator, and calcium analog on yeast cell growth. A MATa strain was grown overnight at 30°C in a succinate-buffered, low phosphate, synthetic medium, containing 1 μM added Ca^{2+} and 2 mM Mg^{2+}, in the absence or in the presence of the indicated concentrations of $CaCl_2$, A23187, EGTA, or $CoCl_2$.

In other eukaryotic cells, and presumably in yeast, passage of Ca^{2+} through the plasma membrane is conducted by Na^+/Ca^{2+} exchangers and/or by Ca^{2+} specific channels. To date, no electrophysiological measurements or other evidence for the existence in yeast of such carriers or channels has been reported, although yeast cells may be of sufficient size to apply the patch-clamp method to determine if the cytoplasmic membrane of protoplasts possesses gated Ca^{2+}-specific pores similar to those described in animal cells (W.G. Owen, personal communication). In higher cells, a variety of drugs (for example, verapamil or nifedipine) and other ions (for example, Co^{2+}; see also, Fig. 1) appear to be relatively specific Ca^{2+} channel "blockers". Hence, another approach to this question would be to attempt to isolate yeast mutants that are either hypersensitive or more resistant to such compounds. Alternatively, yeast mutants that have more (or more efficient) Ca^{2+} channels might be selected by their ability to grow on medium containing a limiting concentration of Ca^{2+} or containing EGTA. Conversely, if entry of Ca^{2+} requires such a mechanism, yeast mutants with defective channels might require greatly elevated external Ca^{2+} concentration for growth and, therefore, might be isolated as Ca^{2+}-dependent mutants. Indeed, two groups have identified temperature-sensitive conditionally lethal mutations in two different complementation groups, cal1 (15) and fsr2 (K. Matsumoto, personal communication), whose effects are suppressed by growth of the cells on rich medium containing 0.1 $\underline{M}$ Ca^{2+} (but not on medium containing equivalent concentrations of Mg^{2+} or K^+). Cloning and nucleotide sequence analysis of the FSR2 gene, isolated by its ability to restore growth of an fsr2 mutant at the non-permissive temperature on normal medium, indicates that it is a very large protein with several markedly hydrophobic domains, consistent with its being a membrane protein (K. Matsumoto, personal communication). The FSR2 gene is on chromosome XI.

Effect of Mating Pheromones on Intracellular Calcium.

Haploid yeast cells secrete peptide hormones, called mating pheromones, that are responsible for triggering the conjugation process that leads to the formation of diploid cells (14). MATα haploids release α-factor which specifically affects the MATa haploids; MATa haploids produce a-factor that acts only on the MATα haploids. Among the effects induced in their respective target cells by the presence of these peptides are: changes in gene expression; cell division cycle arrest in the G1 phase; appearance of cell surface agglutinins; and, pronounced changes in cell shape and cell wall structure (for review, see ref. 16).

Like other peptide hormones, the yeast pheromones act through specific cell surface receptors. Direct binding measurements using radioactive α-factor and non-mating ("sterile") MATa mutants strongly support the conclusion that the product of the STE2 gene is (or is a component of) the receptor for α-factor (17). This view is further supported by the features of the protein predicted by the nucleotide sequence of the

cloned STE2 gene (18,19). Similarly, the STE3 gene product
is probably the receptor for a-factor on MATα cells (18,20).
Hydropathy plots suggest that both of these predicted polypeptides could be capable of spanning a lipid bilayer seven times
(similar to what has been observed for the subunits of acetylcholine receptor and for the visual pigment, rhodopsin) and
hence could serve as pheromone-regulated ion channels. Alternatively, because a number of oligopeptide hormones and neurotransmitters of animal cells that are similar in length and
and properties to the yeast pheromones act through receptors
that are coupled to the hydrolysis of phosphatidylinositol-
4,5-bis-phosphate (PIP_2), α-factor and a-factor may influence
the intracellular Ca^{2+} pool through mobilization of the ion
from cellular stores via generation of IP_3.

To determine if the yeast pheromones affect the level of
either PIP_2, or its hydrolysis product IP_3, we labelled MATa
cells with radioactive precursors before and after exposure
to a physiologically reasonable level of α-factor. To follow
PIP_2, cells were pre-labelled exhaustively with $[^{32}P]PO_4^{2-}$
prior to pheromone treatment, and then solubilized with organic solvents known to preferentially extract acidic phospholipids. To follow IP_3, an inositol-requiring (inol) MATa
mutant was pre-labelled exhaustively with $[^3H]$inositol prior
to pheromone treatment, and then extracted under conditions
that preferentially recover soluble molecules. The ^{32}P-lipids were examined by thin layer chromatography and the soluble ^3H-inositol-containing compounds were examined by ion
exchange chromatography. At present, our conclusion from preliminary studies of this type is that α-factor does not have
a substantial effect on the level of either PIP_2 or IP_3 (J.
DeClue, G.S. Martin and J. Thorner, unpublished results).

To explore the alternative possibility that pheromone
binding directly affects Ca^{2+} level, we intend to determine
if the intracellular free Ca^{2+} concentration in MATa cells
changes significantly immediately following their exposure
to α-factor by using trapped fluorescent indicator dyes (4).
We are currently attempting to find conditions that overcome
the notorious impermeability of the yeast envelope and permit
cells to be reproducibly pre-loaded with sufficient dye to
allow a reliable measurement (M. Poenie, R. Tsien and J.
Thorner, unpublished results). It has been reported, however,
that within 30-40 minutes after exposure of MATa cells to a
relatively high dose of α-factor (3 μM or higher) $^{45}Ca^{2+}$ uptake is specifically induced and Ca^{2+} is accumulated (21).
It is unlikely that these changes in Ca^{2+} flux represent an
initial response to the pheromone because the dose dependence

and kinetics of these effects do not correspond well with the low level of pheromone (10^{-10}-10^{-9} M) and brief time of exposure (a few minutes) required to cause cell cycle arrest or induce agglutinin expression. Rather, the high dose dependence and relatively slow kinetics of appearance of the increased Ca^{2+} influx correlate reasonably well with the onset of the morphological changes elicited by α-factor. To be sustained, the elevated Ca^{2+} influx requires both the continuous presence of α-factor and de novo protein synthesis (21). Taken together, these findings suggest (a) that the protein responsible for the increase in Ca^{2+} uptake accumulates to the appropriate level only after pheromone induction, (b) that it turns over rapidly and needs to be constantly replenished, and (c) that, possibly, it requires the presence of the pheromone to be functional. All of these characteristics are features of the α-factor receptor itself. First, STE2 mRNA is accumulated to a markedly higher level after exposure of MATa cells to α-factor (S. Van Arsdell and J. Thorner, unpublished results). Second, bound radioactive α-factor molecules, and presumably the receptors to which they are complexed, undergo rapid endocytosis (D. Jenness, personal communication; H. Riezman, personal communication). Third, both kinetic analysis of the response to normal α-factor (22,23) and the differential responses elicited by various synthetic α-factor analogs (24) suggest that the receptor may have a dual role depending, in part, on its level of occupancy by pheromone. Hence, it is possible that at a relatively high level of receptor in the membrane and at a relatively high level of occupancy (and/or after some modification of the receptor has been induced following the initial exposure of cells to α-factor), individual receptor molecules may oligomerize (and/or interact with other membrane proteins) to form a Ca^{2+} channel or carrier. In a mating pair, the MATa cell will be exposed to the highest α-factor concentration at the point where it is immediately juxtaposed to the MATα partner. Local formation of Ca^{2+} channels, and subsequent Ca^{2+} influx, might be responsible for establishing the polarized cell surface growth required for cell fusion. Eggs of the brown algae Fucus and Pelvetia, when exposed to a gradient of calcium ionophore, send out their rhizoidal processes from the side of the cell exposed to the highest concentration of ionophore (and, therefore, of Ca^{2+}) (25).

Even if occupied pheromone receptors eventually aggregate to form Ca^{2+} channels, the primary signal elicited when a pheromone first binds to its receptor is still unknown.

Efflux of Calcium.

Saccharomyces cerevisiae (26) [and other yeasts (27) and fungi (28)] possess a plasma membrane-bound Mg^{2+}-dependent ATPase that functions as a proton pump to establish the electrochemical gradient that is used to drive transport of many nutrients into the cell. The enzyme can be distinguished from the mitochondrial ATPase by its resistance to oligomycin, by its apparent MW (~100,000), by its sensitivity to inhibition by vanadate (VO_4^{3-}), and by the fact that it forms a phosphoenzyme intermediate (like the Na^+,K^+-ATPase of animal cell membranes). Everted plasma membrane vesicles from Neurospora crassa catalyze the Mg^{2+}- and ATP-dependent uptake of $^{45}Ca^{2+}$ and can accumulate an internal concentration 40-fold higher than that added externally initially (100 µM) (28). Because these vesicles are in inverted orientation, these results indicate that the plasma membrane Mg^{2+}-ATPase provides the energy for pumping Ca^{2+} out of the intact cell. (In N. crassa, the actual carrier appears to be a H^+/Ca^{2+} antiporter, rather than the ATPase itself or a K^+ or Na^+/Ca^{2+} exchanger.)

The N. crassa enzyme shares many properties in common with the yeast plasma membrane ATPase. In fact, antibody against the purified N. crassa ATPase cross-reacts with the purified yeast ATPase and immunoprecipitates a 100,000 MW species from the products made by in vitro translation of S. cerevisiae mRNA (G. Willsky, personal communication). Because of this close relatedness, the yeast enzyme probably also functions to provide the energy (in the form of the H^+ gradient established by the hydrolysis of ATP) to drive Ca^{2+} export. Hence, yeast cells, like their higher cell counterparts, are likely to have a plasma membrane-based mechanism for expelling Ca^{2+}.

In other eukaryotic cells, internal membranes also have the capability of transporting Ca^{2+}. In muscle cells, the membrane of the sarcoplasmic reticulum possesses an ATPase that can actively transport Ca^{2+} from the cytoplasm into the intracisternal spaces of the tubules. In other animal cells, the endoplasmic reticulum and the mitochondrion also appear to actively sequester Ca^{2+}. In yeast, however, the mitochondrion does not seem to be involved in the active removal of cytosolic Ca^{2+} (28a). In contrast, it has been shown by elegant experiments that the yeast vacuolar membrane possesses a Mg^{2+}-ATPase that can efficiently drive Ca^{2+} uptake by a H^+/Ca^{2+} antiport mechanism, such that a 150-fold concentration gradient can be achieved (29). The vacuolar ATPase can be distinguished from the plasma membrane enzyme on the basis of its polypeptide composition and its spectrum of sensitivity to inhibitors (30). In vivo a significant fraction of the Ca^{2+} sequestered in the vacuole presumably complexes with the large amounts of polyphosphates that are also stored there. The apparent K_m values for transport of Ca^{2+} of both the vacuolar and plasma membrane systems are very similar (about 100 µM), as are their V_{max} values (in terms of nmoles Ca^{2+}/mg membrane protein/min); hence, the two systems may be equally responsible for removal of Ca^{2+} from the yeast cytosol and, therefore, for maintaining Ca^{2+} homeostasis (31,32).

If it is essential for the cell to avoid too high a Ca^{2+} concentration in the cytoplasm, then one might expect that mutants defective in either the plasma membrane or vacuolar Ca^{2+} export systems might have a Ca^{2+}-sensitive phenotype. Such Ca^{2+}-sensitive mutants (cls1-cls18) have been isolated (33). It is not known if any of these loci correspond to functions involved in expulsion of Ca^{2+} from the cytosol. Interestingly, however, mutations conferring Ca^{2+}-sensitivity at the cls4 locus are also temperature-sensitive for growth and are allelic to the cdc24 locus, which appears to encode a function required for bud emergence (13,33).

CALCIUM-BINDING PROTEINS OF YEAST

Calmodulin.

Convincing evidence for the presence of calmodulins very similar in properties to the calmodulins originally isolated from vertebrates have been obtained for a broad range of eukaryotic microorganisms (Table 1). Although several groups initially reported their inability to detect a calmodulin-like protein in S. cerevisiae (34-36), more recent preliminary communications (37,38) suggested that calmodulin could be obtained from S. cerevisiae if measures against non-specific proteolysis were taken. By including a battery of protease inhibitors in the buffers, or by using as a source of protein strains carrying a mutation (pep4) that greatly reduces the level of the major vacuole-associated proteases (39), proteolysis in yeast extracts can be avoided. By incorporating such strategies, we were able to purify yeast calmodulin to homogeneity using modifications of a procedure developed for bovine brain calmodulin (51). Following removal of the bulk of cell protein by heat treatment (which also may inactivate any remaining proteolytic activities), 2000-fold purification was achieved by hydrophobic chromatography in the presence of Ca^{2+}. Because calmodulin is considerably less abundant in yeast than

TABLE 1
EUKARYOTIC MICROORGANISMS FROM WHICH CALMODULINS
HAVE BEEN ISOLATED

Yeasts
    Candida albicans (37)
    Saccharomyces cerevisiae (38,40)
Filamentous fungi
    Neurospora crassa (41)
    Phycomyces blakesleeanus (42)
Water mold
    Blastocladiella emersonii (43)
Slime mold
    Dictyostelium discoideum (44)
Higher fungi
    Agaricus bisporus (45)
    Agaricus campestris (38)
    Coprinus lagopus (38)
Amoeba
    Amoeba proteus (46)
Ciliated protozoa
    Tetrahymena pyriformis (47)
    Paramecium tetraurelia (48)
Flagellated protozoa
    Trypanosoma brucei (49)
    Euglena gracilis (46)
Green alga
    Chlamydomonas reinhardtii (50)

in brain tissue, two additional steps (anion exchange chromatography and gel filtration) were necessary to complete the purification.

Attempts to obtain an amino-terminal sequence of the purified protein were unsuccessful, suggesting that yeast calmodulin, like other calmodulins, has a blocked amino terminus. Overall the amino acid composition of the yeast protein resembled other calmodulins, in particular by the absence of Trp and Cys and the low molar ratio of His and Pro. Peptide fragments of the purified protein were prepared by trypsin digestion. Two of the tryptic peptides were purified by HPLC and their amino acid sequences were determined by automated Edman degradation (40). Sequences homologous to the two peptides could be found in the sequence of mammalian calmodulin.

For each of the two peptides, two kinds of oligonucleotide probes (a completely degenerate 20-mer and a longer unique probe that was based on assumptions about yeast codon usage bias) were synthesized (40). All four probes were used to screen a yeast genomic library in a centromere vector by colony hybridization. Two clones were identified that were homologous to both of the degenerate probes and to one of the unique probes. These clones contained overlapping segments of the yeast genome, as determined by restriction endonuclease cleavage site mapping. Nucleotide sequence analysis of this region revealed an open reading frame corresponding to a 147-residue polypeptide that is 60% identical to vertebrate calmodulin (Fig. 2). The yeast protein is even more similar to calmodulins (up to 96% homology), depending on the extent to which conservative amino acid substitutions are allowed. The predicted protein contained perfect matches to the amino acid sequences of both of the tryptic fragments; hence, the gene cloned encodes the protein that was purified.

Hybridization with sub-clones containing the entire coding region to chromosomal DNA digested with two different restriction enzymes indicates that only a single locus exists in the yeast genome. Because the nucleotide sequence data demonstrate that only a single copy of the coding region is present in the cloned DNA, the gene encoding yeast calmodulin (which we have designated CAM1) is a unique single-copy gene. To determine if calmodulin is essential for the growth and viability of yeast cells, we disrupted the CAM1 gene in vitro by insertion of the cloned URA3 gene and introduced the disrupted gene in place of the normal locus by integrative gene replacement using DNA-mediated transformation of appropriate diploid recipients. The resulting diploid transformants, which carry one functional CAM1 gene and one non-functional (cam1::URA3) copy, were induced to sporulate to yield asci containing four spores that represent the four products of a meiotic division. In every tetrad examined (total of 40), two of the spores were viable, and two of the spores never grew to form a visible colony. All of the viable spores lacked the URA3 allele and hence contained the intact CAM1 gene. Therefore, the dead spores were those that harbored the disrupted CAM1 gene. Our results (40) are the first direct demonstration in any organism that calmodulin performs an essential function in eukaryotic cells.

A number of drugs, especially the phenothiazines, are known to be reasonably specific inhibitors of calmodulin, and such compounds attached to solid supports are often effective means for purifying the protein by affinity chromatography (45). In fact, yeast calmodulin binds in a Ca^{2+}-dependent manner to Chlorpromazine-Sepharose and can be eluted by buf-

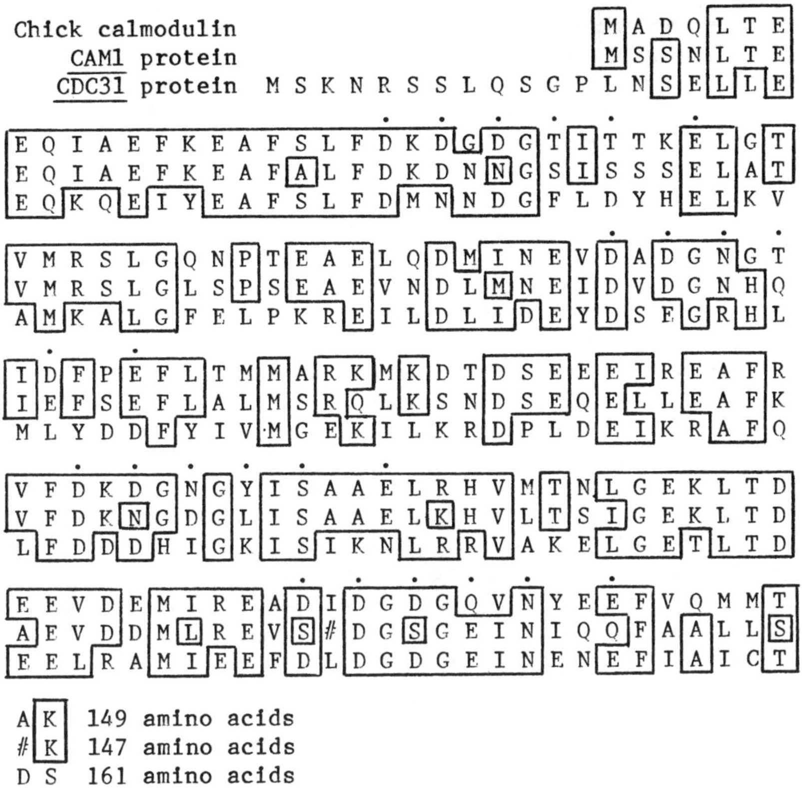

```
Chick calmodulin                                    M A D Q L T E
   CAM1 protein                                     M S S N L T E
  CDC31 protein  M S K N R S S L Q S G P L N S E L L E
```

```
E Q I A E F K E A F S L F D K D G D G T I T T K E L G T
E Q I A E F K E A F A L F D K D N N G S I S S S E L A T
E Q K Q E I Y E A F S L F D M N N D G F L D Y H E L K V

V M R S L G Q N P T E A E L Q D M I N E V D A D G N G T
V M R S L G L S P S E A E V N D L M N E I D V D G N H Q
A M K A L G F E L P K R E I L D L I D E Y D S F G R H L

I D F P E F L T M M A R K M K D T D S E E E I R E A F R
I E F S E F L A L M S R Q L K S N D S E Q E L L E A F K
M L Y D D F Y I V M G E K I L K R D P L D E I K R A F Q

V F D K D G N G Y I S A A E L R H V M T N L G E K L T D
V F D K N G D G L I S A A E L K H V L T S I G E K L T D
L F D D D H I G K I S I K N L R R V A K E L G E T L T D

E E V D E M I R E A D I D G D G Q V N Y E E F V Q M M T
A E V D D M L R E V S # D G S G E I N I Q Q F A A L L S
E E L R A M I E E F D L D G D G E I N E N E F I A I C T
```

```
A K    149 amino acids
# K    147 amino acids
D S    161 amino acids
```

Figure 2. Comparison of the primary sequence of vertebrate calmodulin (52) with the predicted sequences of the products of two yeast genes, CAM1 (40) and CDC31 (53). Only identical residues are boxed. The # symbols indicate two single-residue gaps inserted to maximize the alignment. Residues (.) known to be involved in Ca^{2+} binding by X-ray analysis of vertebrate calmodulin are shown (54).

fers containing EGTA (G. Stetler and J. Thorner, unpublished results). Furthermore, yeast cells carrying the CAM1 gene on a multi-copy plasmid show elevated resistance to growth inhibition by three other calmodulin-directed drugs, trifluoperazine, calmidazolium, and W13 (40). Therefore, it should be possible to isolate calmodulin mutants by selecting for resistance to such agents. Indeed, one class of fluphenazine-resistant mutants (fsr1) that is temperature-sensitive for growth, and not suppressed by a high exogenous Ca^{2+} concentration, maps on chromosome II (K. Matsumoto, personal communication). Chromosome blotting (55) indicates that the CAM1 gene is carried on chromosome II (40). Thus, fsrl mutations may already define one type of temperature-sensitive lesion that can occur in the yeast calmodulin structural gene.

CDC31 protein.

The spindle pole body embedded in the nuclear membrane is the microtubule-organizing center of the yeast cell (56). In normal cells, the spindle pole body duplicates just prior to bud emergence and the initiation of chromosomal DNA replication, analogous to the duplication of centrioles that occurs at the beginning of the mammalian cell cycle. Hence, spindle pole body duplication has been considered a landmark for the transition between the G1 and S phases of the cell division cycle in yeast (13). Cells bearing temperature-sensitive mutations in the CDC31 gene are specifically defective in duplication of this structure at the non-permissive temperature (57). As a result, the spindle pole body becomes unusually large and establishes a monopolar spindle that is unable to properly partition chromosomes (58). The process of spindle pole body duplication, and its regulation, are clearly pivotal for proper mitosis. Meiosis also fails at the restrictive temperature in a homozygous (cdc31/cdc31) diploid, although it is in the second, not the first, meiotic division that the spindle pole body does not duplicate.

The CDC31 gene was cloned by selecting for plasmids that restored temperature-resistance to a cdc31 mutant (53). Subcloning, transcript mapping, and complementation tests localized the gene to a region of the cloned DNA that contains an open reading frame which encodes a 161-residue polypeptide, revealed by nucleotide sequence analysis (53). The predicted CDC31 product is clearly a calmodulin-like protein (Fig. 2). However, it is no more closely related to yeast calmodulin (38% identity) than it is to human or Tetrahymena calmodulin (42% and 41% identity, respectively). While the similarity between CDC31 and calmodulins is not confined strictly to the putative Ca^{2+}-binding sites (Fig. 2), there are several features that clearly distinguish CDC31 protein from calmodulin. First, an invariant lysine residue found in every calmodulin sequenced to date (and that is post-translationally modified to trimethyl-lysine in many calmodulins), the so-called "Lys 115", is absent in CDC31 protein and replaced by a Thr. Second, the X-ray structure of vertebrate calmodulin indicates that the molecule consists of two globular lobes connected by a long exposed α-helix. In the CDC31 product, but not in yeast calmodulin (CAM1 product), the amino acids that should comprise the long helical region include two residues, Gly and Pro, that are normally considered "α-helix breakers". Third, at the C-terminus, or penultimate thereto, all calmodulins characterized to date possess a basic residue. In contrast, CDC31 protein lacks this feature and instead has an Asp penultimate to the C-terminal residue. Fourth, most calmodulins, including the yeast CAM1 gene product, lack Cys residues (except those from higher plants which have a Cys within the first Ca^{2+}-binding domain near the amino terminus); CDC31 protein has a Cys three residues from its C-terminus. Therefore, while it is likely that the CDC31 product binds Ca^{2+} and serves as a Ca^{2+}-dependent switch in some manner, in other respects its structure and function cannot be very similar to calmodulin. Hence, the CDC31 gene product probably represents a novel member of the Ca^{2+}-binding regulatory protein family.

Ca^{2+}-binding proteins in animal cells are often localized exclusively at their sites of action by their assembly into larger supramolecular structures, for example troponin C is complexed with the regulatory light chain of myosin in muscle fibers. Similarly, the CDC31 product may itself be a component of the spindle pole body. The simplest way to address this issue would be by immunocytochemical examination using specific antibodies directed against the CDC31 protein. Affinity-purified antibodies against an acidic, ~20,000 MW, Ca^{2+}-binding protein isolated from algal flagellar rootlets also cross-reacts with a component in mammalian mitotic spindle poles (58a). The yeast CDC31 gene product may be a homologous protein.

In animal cells there is abundant evidence that fluxes of Ca^{2+} play a regulatory role in mitogenesis and in progression through mitosis (59). Several lines of investigation indicate an involvement of Ca^{2+} in cell cycle control in yeast. First, it has been reported that in synchronized cells uptake of Ca^{2+} increases three-fold before bud emergence and then decreases as the cells enter S phase (60). This observation is consistent with a role for Ca^{2+} in passage through the G1/S boundary, where CDC31 function is apparently involved. Second, other studies implicate Ca^{2+}-dependent functions in the G2 phase. The temperature-sensitive cal1 mutants mentioned earlier (that can only grow at the non-permissive temperature in medium containing a high Ca^{2+} concentration) eventually arrest with a small bud and contain a single nucleus with twice the normal DNA content when cells are shifted at the restrictive temperature to medium containing a low Ca^{2+} concentration (15). This terminal morphology is distinct from that caused by cdc31 mutations (57), suggesting that Ca^{2+}-dependent functions other than those requiring the CDC31 product are essential for cell cycle progression. Perhaps these other activities are mediated by calmodulin.

Other calcium-binding proteins.

In yeast, there have been described a variety of cellular functions that are affected by Ca^{2+} itself, and do not require the intermediacy of calmodulin (or any other small Ca^{2+}-dependent regulatory protein). In most of these cases, it appears that Ca^{2+} has direct effects on the conformation, stability, or activity of the protein. Hence, in this sense, these molecules must also be considered Ca^{2+}-binding proteins. For example, the specific endoprotease (KEX2 gene product) that is required for processing the prepro-α-factor and prepro-killer toxin precursors in the secretory system is a Ca^{2+}-dependent enzyme (61). After removal of Ca^{2+} with chelator, readdition of Ca^{2+} (and no other divalent metal ion tested) at micromolar concentration is sufficient to fully reactivate the enzyme. Examination of the predicted amino acid sequence of the KEX2 protease deduced from its nucleotide sequence indicates potential Ca^{2+}-binding sites, but these domains do not strictly resemble the highly conserved so-called "EF-hand" sequence (62) found in calmodulin (see Fig. 2), troponin C, and parvalbumin. Yeast transketolase appears to contain two tightly bound Ca^{2+} ions that are required for maximal activity of the enzyme (63). Polymerization of highly purified yeast actin in vitro is greatly stimulated by 100 μM Ca^{2+} (64); and, disassembly of yeast tubulin in vitro is promoted by the presence of Ca^{2+} at the same concentration (65). Purified type I DNA topoisomerase from yeast does not require a divalent metal ion for catalysis; however, activity is stimulated 10-20-fold by millimolar Ca^{2+} (although this is presumably a non-physiological level of the ion and may reflect effects on the supercoiled DNA substrate rather than on the enzyme itself) (66). Flocculation ("non-specific" cell-cell agglomeration) of yeast strains is promoted specifically by extracellular Ca^{2+} in the millimolar range and mediated by the FLO1 gene product (67), which may be a Ca^{2+}-binding lectin, like concanavalin A.

POTENTIAL TARGETS FOR CALMODULIN ACTION IN YEAST

As discussed earlier, quite a number of functions have been ascribed to calmodulin in other systems. Because of this very complexity, a genetic analysis might actually pinpoint the more vital roles of calmodulin in vivo. This task might be hopeless if this ubiquitous calcium-dependent regulatory protein truly interacts with a plethora of cellular target proteins and interacts with all of its targets in precisely the same manner. However, recent evidence suggests that this is not the case and hence the targets of calmodulin action may be accessible to a combined genetic and biochemical approach. First, certain proteolytic fragments of brain calmodulin are able to activate some of its target enzymes (e.g. phosphorylase kinase), but not others (e.g. cAMP phosphodiesterase) (68). This finding, and the results of additional studies involving chemical modification of calmodulin, suggest that the domains by which calmodulin associates with its target enzymes are not always the same. Second, as already discussed in detail above, at least two clearly calmodulin-like proteins, the CAM1 and CDC31 gene products, have been identified in yeast (40,53). Because the CDC31 locus was originally defined by temperature-sensitive conditionally lethal mutations, the CDC31 protein, like the CAM1 protein, must have an essential function and, furthermore, that role cannot be performed by calmodulin itself. This situation is in contrast to the RAS1 and RAS2 gene products which appear to be largely interchangeable (69,70). Third, hybridization of both CAM1 and CDC31 to genomic blots at low stringency indicates that there is at least one (and possibly another) related gene in the yeast genome. Hence, it is possible that the essential functions of calmodulin are limited to only a few of its targets. To identify these targets, we intend to use the approach of selecting allele-specific second-site suppressors (pseudorevertants) of calmodulin mutants. Of course, to initiate such a study, one first needs defined calmodulin mutations. We are currently in the process of generating a collection of temperature-sensitive mutations in yeast calmodulin by site-directed in vitro mutagenesis of the CAM1 gene.

Biochemical evidence obtained in vitro in yeast and in other lower eukaryotic systems provides some indication of the classes of target proteins for calmodulin that might be anticipated. We found that partially purified yeast calmodulin was able to stimulate the activity of calmodulin-depleted bovine brain cyclic nucleotide phosphodiesterase, but only two-fold (T. Davis and J. Thorner, unpublished results), in agreement with an earlier report of 2.5-fold stimulation of brain phosphodiesterase using a crude preparation of S. cerevisiae calmodulin (37). Recently, stimulation of the brain enzyme by the purified yeast protein as high as 6-fold has been observed; furthermore, the activation was Ca^{2+}-dependent and EGTA- and fluphenazine-inhibitable (Y. Ohya, I. Uno, T. Ishikawa and Y. Anraku, personal communication). It has also been reported that fractions containing yeast calmodulin stimulate other calmodulin-regulated enzymes of higher cells, including the Ca^{2+},Mg^{2+}-ATPase of the red blood cell membrane (37), the myosin light chain kinase of rabbit muscle (38), and the NAD kinase of pea seedling (38; Y. Ohya, I. Uno, T. Ishikawa, and Y. Anraku, personal communication). Importantly, for maximum activation of plant NAD kinase, a ten-fold higher concentration of yeast calmodulin is required than when wheat germ or mammalian calmodulin is used, and the maximum activity achieved is significantly lower (10-50%). Conversely, animal calmodulin does not detectably stimulate the NAD kinase activity of yeast (38). In the same regard, only Tetrahymena calmodulin, and no other calmodulin tested, activates Tetrahymena guanylate cyclase in vitro (47); and, when microinjected, only Paramecium calmodulin, and no other calmodulin tested, will restore function to Paramecium mutants that have defective Ca^{2+}-dependent potassium channels (71). These latter results serve to illustrate dramatically why it is crucial in examining the properties and functions of a calmodulin to use potential target enzymes and proteins from the same or a closely related organism.

For some of the types of enzymes and proteins that should be responsive to Ca^{2+}/calmodulin complex, there is only limited information available for yeast and other fungal systems. At least one protein kinase potentially responsive to Ca^{2+}/calmodulin, phosphorylase kinase, has been partially purified from S. cerevisiae (72). Ca^{2+} weakly stimulates an apparent threonine/serine-specific protein kinase activity associated with the CDC28 gene product (73), another function that, like CDC31 protein, is required for transit of the G1/S boundary (13). Evidence for a more general Ca^{2+}/calmodulin-stimulated protein kinase activity has been obtained in the filamentous fungus Neurospora crassa (74). A myosin heavy chain kinase is inhibited by Ca^{2+}/calmodulin in the slime mold, Dictyostelium discoideum (75); unlike the non-motile S. cerevisiae, however, slime mold cells are highly differentiated for amoeboid movement.

S. cerevisiae possesses two cyclic nucleotide phosphodiesterases: a "low K_m" (~0.2 μM) enzyme (76) and a "high K_m"

(100-200 μM) enzyme (77). Both enzymes have been highly purified. The high K_m species is apparently a zinc metalloprotein, but is not inhibited by EDTA even at concentrations as high as 25 mM. Mutants (pdel) apparently lacking the high K_m enzyme have been isolated as suppressors of mutations (CYR3) that raise the K_a of the regulatory subunit of cAMP-dependent protein kinase (78). Yeast cells carrying the pdel mutation alone are viable, and have elevated levels (4-5-fold) of cAMP, suggesting that the high K_m enzyme plays some role in regulating cellular cAMP content. This finding is perhaps surprising for several reasons. First, measurements of the steady state level of intracellular cAMP in yeast are in the range 0.01 to 1 μM, far below the apparent K_m (79). Second, although yeast cells have undetectably low levels of cGMP, the enzyme hydrolyzes cGMP (at about one-half the rate of cAMP). Third, under physiological conditions (30°C, pH 6.4) and at a physiologically reasonably cAMP concentration (0.25 μM), the specific activity of cAMP hydrolysis in whole cell extracts attributable to the high K_m enzyme is only one-tenth that due to the low K_m activity.

The low K_m enzyme also appears to be a zinc metalloprotein, but requires added divalent cation for activity (Mg^{2+} is optimal, but Ca^{2+} has not yet been critically examined). Unlike the other phosphodiesterase, the affinity of the low K_m enzyme is clearly in the physiological range. In addition the low K_m enzyme is highly specific for cAMP. Mutants deficient in the low K_m phosphodiesterase (sra5) have been isolated as suppressors of the poor growth of ras2 mutants on nonfermentable carbon sources (J. Cannon, K. Tatchell, J. Gibbs, I. Sigal and E. M. Scolnick, personal communication). The RAS2 gene is required for full activity of adenylate cyclase (80). The fact that a defect in the low K_m enzyme can spare the cAMP deficiency of the ras2 mutant indicates that the low K_m enzyme must also have a physiologically important function in regulating the intracellular cAMP concentration. In crude extracts the low K_m activity is not stimulated by added mammalian brain calmodulin nor inhibited by Chlorpromazine (H. Liao and J. Thorner, unpublished results). However, these parameters should be reexamined with the purified enzyme and with purified yeast calmodulin. Using the cloned CAM1 gene, it should be possible to over-produce yeast calmodulin from expression vectors. If so, sufficient material may be obtained to prepare yeast calmodulin-affinity columns. Such columns have been extremely effective tools for purifying calmodulin-responsive cAMP phosphodiesterases from mammalian tissues (81) and such a strategy may also work for identifying which, if any, of the two yeast phosphodiesterases may be calmodulin-regulated. A Ca^{2+}/calmodulin-dependent cAMP phosphodiesterase has been identified in N. crassa (82).

Yeast adenylate cyclase activity in crude membrane preparations is not stimulated by added Ca^{2+} (100 μM) (83); however, the fraction examined may have already been saturated with the ion because membranes were prepared using concanavalin A binding in buffer containing excess Ca^{2+} prior to spheroplast lysis. Because the structural gene for the catalytic subunit of yeast adenylate cyclase has recently been cloned (84,85), it should be possible to examine the involvement of yeast calmodulin in yeast adenylate cyclase activity with purified components in vitro. Again, in animal cells, the catalytic subunit of adenylate cyclase can be purified by affinity chromatography on calmodulin-agarose columns (86). In N. crassa, Ca^{2+}/calmodulin stimulates solubilized adenylate cyclase (87).

Based on the effects of phenothiazine drugs, the plasma membrane Mg^{2+}-ATPase of S. cerevisiae required for Ca^{2+} extrusion may also be a Ca^{2+}/calmodulin-dependent activity (31). The effects of these same calmodulin-directed compounds on the vacuolar Mg^{2+}-ATPase involved in Ca^{2+} pumping have not been reported.

Finally, among the most effective inhibitors in vitro of calmodulin action are amphipathic peptides that are both hydrophobic and net positively charged. For example, the 26-residue honey bee venom peptide, mellitin, has a K_d for calmodulin lower than 3 nM, and a 14-residue peptide from wasp venom, mastoparan, has a $K_d \sim 0.3$ nM (88). A variety of peptide hormones also have been reported to show reasonably tight binding to calmodulin in vitro (K_d values= 50 nM - 3 μM), including ACTH, β-endorphin, substance P, glucagon, dynorphin 1-13 and dynorphin 1-17, secretin, gastric inhibitory peptide, and vasoactive intestinal peptide (88). In addition to these natural peptides, calmodulin also binds tightly to synthetic amphiphilic peptides that are positively charged and that can assume an α-helical conformation (89). Because the yeast mating pheromones are both hydrophobic and net positively charged, it is possible that one aspect of their mode of action may be the result of their direct interaction with yeast calmodulin, assuming that the pheromones can enter the cytoplasm (for example, by receptor-mediated endocytosis and subsequent release from the endosomal compartment).

These hypotheses and others mentioned earlier in this review are testable both genetically and biochemically, now that the yeast calmodulin structural gene has been isolated.

ACKNOWLEDGEMENTS

We are grateful for the helpful comments of, and stimulating discussions with, W. Zacheus Cande, Howard Schulman, Roger Y. Tsien, and Robert S. Zucker. We also thank Peter R. Baum and Breck F. Byers, and Kunihiro Matsumoto, for the communication of results prior to publication.

REFERENCES

1. Carafoli, E, Penniston, JT (1985). The calcium signal. Sci. Amer. 253 (5): 70.
2. Tsien, RW (1983). Calcium channels in excitable cell membranes. Ann. Rev. Physiol. 45: 341.
3. Hokin, LE (1985). Receptors and phosphoinositide-generated second messengers. Ann. Rev. Biochem. 54: 205.
4. Tsien, RY, Pozzan, T, Rink, TJ (1984). Measuring and manipulating cytosolic Ca^{2+} with trapped indicators. Trends in Biochem. Sci. 9: 263.
5. Klee, CB, Crouch, TH, Richman, PG (1980). Calmodulin. Ann. Rev. Biochem. 49: 489.
6. Cohen, P (1982). The role of protein phosphorylation in neural and hormonal control of cellular activity. Nature 296: 613.
6a. Schulman, H (1984). Calcium-dependent protein kinases and neuronal function. Trends in Pharmacol. Sci. 5: 188.
7. Nishizuka, Y (1984). Protein kinases in signal transduction. Trends in Biochem. Sci. 9: 163.
8. Lin, YM, Cheung, WY (1980). Ca^{2+}-dependent cyclic nucleotide phosphodiesterase. In Cheung, WY (ed): "Calcium and Cell Function", Orlando, FL: Academic Press, Vol. I, p. 79.
9. Kopf, GS, Vacquier, VD (1984). Characterization of a calmodulin-stimulated adenylate cyclase from abalone spermatozoa. J. Biol. Chem. 259: 7590.
10. Builder, SE, Beavo, JA, Krebs, EG (1981). Mechanism of activation and inactivation of the cAMP-dependent protein kinase. In Rosen, OM, Krebs, EG (eds): "Protein Phosphorylation", Cold Spring Harbor, NY: Cold Spring Harbor Laboratory, Book A, p. 33.
11. Constantinou, AI, Squinto, SP, Jungmann, RA (1985). The phospho-form of the regulatory subunit RII of cyclic AMP-dependent protein kinase possesses intrinsic topoisomerase activity. Cell 42: 429.
12. Carafoli, E (1984). Calcium-transporting systems of plasma membranes, with special attention to their regula-

tion. Adv. Cyclic Nucl. Prot. Phosph. Res. 17: 543.
13. Pringle, JR, Hartwell, LH (1981). The Saccharomyces cerevisiae cell cycle. In Strathern, JN, Jones, EW, Broach, JR (eds): "Molecular Biology of the Yeast Saccharomyces cerevisiae", Cold Spring Harbor, NY: Cold Spring Harbor Laboratory, Vol. I, p. 97.
14. Thorner, J (1981). Pheromonal regulation of development in Saccharomyces cerevisiae. ibid, p. 143.
15. Ohya, Y, Ohsumi, Y, Anraku, Y (1984). Genetic study of the role of calcium ions in the cell division cycle of Saccharomyces cerevisiae: A calcium-dependent mutant and its trifluoperazine-dependent pseudorevertants. Molec. Gen. Genet. 193: 389.
16. Sprague, GF, jr, Blair, LC, Thorner, J (1983). Cell interactions and regulation of cell type in the yeast Saccharomyces cerevisiae. Ann. Rev. Microbiol. 37: 623.
17. Jenness, DD, Burkholder, AC, Hartwell, LH (1983). Binding of α-factor pheromone to yeast a cells: Chemical and genetic evidence for an α-factor receptor. Cell 35: 521.
18. Nakayama, N, Miyajima, A, Arai, K (1985). Nucleotide sequences of STE2 and STE3, cell type-specific sterile genes from Saccharomyces cerevisiae. EMBO J. 4: 2643.
19. Burkholder, AC and Hartwell, LH (1985). The yeast α-factor receptor: Structural properties deduced from the sequence of the STE2 gene. Nucl. Acid Res. 13: 8463.
20. Hagen, DC, McCaffrey, G, Sprague, GF, jr (1985). Evidence that the yeast STE3 gene encodes a receptor for the peptide pheromone, a-factor: Gene sequence and implications for the structure of the presumed receptor. Proc. Natl. Acad. Sci. USA, in press.
21. Ohsumi, Y, Anraku, Y (1985). Specific induction of Ca^{2+} transport activity in MATa cells of Saccharomyces cerevisiae by a mating pheromone, α-factor. J. Biol. Chem. 260: 10482.
22. Moore, SA (1983). Comparison of dose-response curves for α-factor-induced cell division arrest, agglutination, and projection formation of yeast cells: Implications for the mechanism of α-factor action. J. Biol. Chem. 258: 13849.
23. Moore, SA (1984). Yeast cells recover from mating pheromone α-factor-induced division arrest by desensitization in the absence of α-factor destruction. J. Biol. Chem. 259: 1004.
24. Baffi, RA, Shenbagamurthi, P, Terrance, K, Becker, JM, Naider, F, Lipke, PN (1984). Different structure-function relationships for α-factor-induced morphogenesis and agglutination in Saccharomyces cerevisiae.

J. Bacteriol. 158: 1152.
25. Robinson, KR, Cone, R (1980). Polarization of fucoid eggs by a calcium ionophore gradient. Science 207: 77.
26. Willsky, GR (1979). Characterization of the plasma membrane Mg^{2+}-ATPase from the yeast, Saccharomyces cerevisiae. J. Biol. Chem. 254: 3326.
27. Amory, A, Foury, F, Goffeau, A (1980). The purified plasma membrane ATPase of the yeast Schizosaccharomyces pombe forms a phosphorylated intermediate. J. Biol. Chem. 255: 9353.
28. Stroobant, P, Scarborough, GA (1979). Active transport of calcium in Neurospora plasma membrane vesicles. Proc. Natl. Acad. Sci. USA 76: 3102.
28a. Carafoli, E, Balcavage, WX, Lehninger, AL, Mattoon, JR (1970). Ca^{2+} metabolism in yeast cells and mitochondria. Biochim. Biophys. A. 205: 18.
29. Ohsumi, Y, Anraku, Y (1983). Calcium transport driven by as proton motive force in vacuolar membrane vesicles of Saccharomyces cerevisiae. J. Biol. Chem. 258: 5614.
30. Uchida, E, Ohsumi, Y, Anraku, Y (1985). Purification and properties of H^+-translocating, Mg^{2+}-adenosine triphosphatase from vacuolar membranes of Saccharomyces cerevisiae. J. Biol. Chem. 260: 1090.
31. Eilam, Y (1984). Effect of phenothiazines on inhibition of plasma membrane ATPase and hyperpolarization of cell membranes in the yeast Saccharomyces cerevisiae. Bio-. chim. Biophys. A. 769: 601.
32. Eilam, Y, Lavi, H, Grossowicz, N (1985). Cytoplasmic Ca^{2+} homeostasis maintained by a vacuolar Ca^{2+} transport system in the yeast Saccharomyces cerevisiae. J. Gen. Microbiol. 131: 623.
33. Ohya, Y, Miyamoto, S, Ohsumi, Y, Anraku, Y (1986). Calcium-sensitive cls4 mutant of Saccharomyces cerevisiae with a defect in bud formation. J. Bacteriol. 165: 28.
34. Jamieson, GA, Bronson, DD, Schachat, FH, Vanaman, TC (1980). Structure and function relationships among calmodulins and troponin C-like proteins from divergent eukaryotic organisms. Ann. NY Acad. Sci. 356: 1.
35. Grand, RJA, Nairn, AC, Perry, SV (1980). The preparation of calmodulins from barley (Hordeum sp.) and basidiomycete fungi. Biochem. J. 185: 755.
36. Clarke, M, Bazari, WL, Kayman, SC (1980). Isolation and properties of calmodulin from Dictyostelium discoideum. J. Bacteriol. 141: 397.
37. Hubbard, M, Bradley, M, Sullivan, P, Shepherd, M, Forrester, I (1982). Evidence for the occurrence of calmod-

ulin in the yeasts Candida albicans and Saccharomyces cerevisiae. FEBS Letts. 137: 85.
38. Nakamura, T, Fujita, K, Eguchi, Y, Yazawa, M (1984). Properties of calcium-dependent regulatory proteins from fungi and yeast. J. Biochem. (Tokyo) 95: 1551.
39. Jones, EW (1984). The synthesis and function of proteases in Saccharomyces: Genetic approaches. Ann. Rev. Genet. 18: 233.
40. Davis, TN, Urdea, MS, Masiarz, FR, Thorner, J (1986). Isolation and characterization of the yeast calmodulin gene: Calmodulin is an essential protein. Cell, submitted for publication.
41. Cox, JA, Ferraz, C, Demaille, JG, Perez, RO, van Tuinen, D, Marme', D (1982). Calmodulin from Neurospora crassa: General properties and conformational changes. J. Biol. Chem. 257: 10694.
42. Martinez-Cadena, G, Lucas, M, Goberna, R (1982). Biological properties of a calcium-modulator protein from Phycomyces blakesleeanus. Comp. Biochem. Physiol. 71: 515.
43. Gomes, SL, Mennucci, L, Maia, JCC (1979). A calcium-dependent protein activator of mammalian cAMP phosphodiesterase from Blastocladiella emersonii. FEBS Letts. 99: 39.
44. Marshak, DR, Clarke, M, Roberts, DM, Watterson, DM (1984). Structural and functional properties of calmodulin from the eukaryotic microorganism Dictyostelium discoideum. Biochemistry 23: 2891.
45. Charbonneau, H, Cormier, MJ (1979). Purification of plant calmodulin by fluphenazine-Sepharose affinity chromatography. Biochem. Biophys. Res. Commun. 90: 1039.
46. Kuźnicki, J, Kuźnicki, L, Drabikowski, W (1979). Ca^{2+}-binding modulator proteins in protozoa and myxomycete. Cell Biol. Intl. Rep. 3: 17.
47. Yazawa, M, Yagi, K, Toda, H, Kondo, K, Narita, K, Yamazaki, R, Sobue, K, Kakiuchi, S, Nagao, S, Nozawa, Y (1981). The amino acid sequence of the Tetrahymena calmodulin which specifically interacts with guanylate cyclase. Biochem. Biophys. Res. Commun. 99: 1051.
48. Rauh, JJ, Nelson, DL (1981). Calmodulin is a major component of extruded trichocysts from Paramecium tetraurelia. J. Cell Biol. 91: 860.
49. Ruben, L, Egwuagu, C, Patton, CL (1983). African trypanosomes contain calmodulin which is distinct from host calmodulin. Biochim. Biophys. A. 758: 104.
50. Lukas, TJ, Wiggins, ME, Watterson, DM (1985). Amino acid sequence of a novel calmodulin from the unicellular green alga Chlamydomonas. Plant Physiol. 78: 477.

51. Gopalakrishna, R, Anderson, WB (1982). Ca^{2+}-induced hydrophobic site on calmodulin: Application for purification by phenyl-Sepharose affinity chromatography. Biochem. Biophys. Res. Commun. 104: 830.
52. Putkey, JA, Ts'ui, KF, Tanaka, T, Lagacè, L, Stein, JP, Lai, EC, Means, AR (1983). Chicken calmodulin genes: A species comparison of cDNA sequences and isolation of a genomic clone. J. Biol. Chem. 258: 11864.
53. Baum, PR, Furlong, C, Byers, B (1985). Yeast gene required for spindle pole body duplication: Homology of its product with Ca^{2+}-binding proteins. Proc. Natl. Acad. Sci. USA, submitted for publication.
54. Babu, YS, Sack, JS, Greenhough, TJ, Bugg, CE, Means, AR, Cook, WJ (1985). Three-dimensional structure of calmodulin. Nature 315: 37.
55. Carle, GF, Olson, MV (1985). An electrophoretic karyotype for yeast. Proc. Natl. Acad. Sci. USA 82: 3756.
56. Byers, B. (1981). Cytology of the yeast life cycle. In Strathern, JN, Jones, EW, Broach, JR (eds): "Molecular Biology of the Yeast Saccharomyces", Cold Spring Harbor, NY: Cold Spring Harbor Laboratory, Vol. I, p. 59.
57. Byers, B. (1981). Multiple roles of the spindle pole bodies in the life cycle of Saccharomyces cerevisiae. In von Wettstein, D, Friis, J, Kielland-Brandt, M, Stenderup, A (eds): "Molecular Genetics in Yeast (Alfred Benzon Symp.)", Copenhagen, Denmark: Munksgaard, Vol. 17, p. 119.
58. Schild, D, Ananthaswarmy, HN, Mortimer, RK (1981). An endomitotic effect of a cell cycle mutation of Saccharomyces cerevisiae. Genetics 97: 551.
58a. Salisbury, JL (1985) Striated flagellar roots and pericentriolar satellite material: Calcium-sensitive contractile organelles. J. Cell Biol. 101: 26a.
59. Harris, PJ (1981). Calcium regulation of cell cycle events. In Zimmerman, AM, Forer, A (eds): "Mitosis/Cytokinesis", New York, NY: Academic Press, Inc., p. 29.
60. Saavedra-Molina, A, Villalobos, R, Borbolla, M (1983). Calcium uptake during the cell cycle of Saccharomyces cerevisiae. FEBS Letts. 160: 195.
61. Fuller, RS, Brake, AJ, Thorner, J (1986). The yeast KEX2 gene, required for processing prepro-α-factor, encodes a calcium-dependent endopeptidase that cleaves after Lys-Arg and Arg-Arg sequences. In Leive, L. (ed): "Microbiology-1986", Washington, DC: Amer. Soc. for Microbiol., in press.
62. Kretsinger, RH (1980). Structure and evolution of cal-

cium-modulated proteins. CRC Crit. Rev. Biochem. 8: 119.
63. Kochetov, GA, Philippov, PP (1970). Calcium: Cofactor of transketolase from baker's yeast. Biochem. Biophys. Res. Commun. 38: 930.
64. Greer, C, Schekman, R (1982). Calcium control of Saccharomyces actin assembly. Molec. Cellul. Biol. 2: 1279.
65. Baum, PR (1981). Identification and characterization of the tubulin proteins from the yeast S. cerevisiae. Ph.D. thesis, University of California, Berkeley, 209pp.
66. Goto, T, Laipis, P, Wang, JC (1984). The purification and characterization of DNA topoisomerases I and II of the yeast Saccharomyces cerevisiae. J. Biol. Chem. 259: 10422.
67. Miki, BLA, Poon, NH, James, AP, Seligy, VL (1982). Possible mechanism for flocculation interactions governed by gene FLO1 in Saccharomyces cerevisiae. J. Bacteriol. 150: 878.
68. Newton, DL, Oldewurtel, MD, Krinks, MH, Shiloach, J, Klee, CB (1984). Agonist and antagonist properties of calmodulin fragments. J. Biol. Chem. 259: 4419.
69. Tatchell, K, Chaleff, DT, DeFeo-Jones, D, Scolnick, EM (1984). Requirement for either of a pair of ras-related genes of Saccharomyces cerevisiae for spore viability. Nature 309: 523.
70. Kataoka, T, Powers, S, McGill, C, Fasano, O, Strathern, J, Broach, J, Wigler, M (1984). Genetic analysis of yeast RAS1 and RAS2 genes. Cell 37: 437.
71. Hinrichsen, RD, Burgess-Cassler, A, Soltvedt, BC, Hennessey, T, Kung, C (1985). Calmodulin restores a defective Ca^{2+}-dependent K^+ current in a mutant of Paramecium. Nature, in press.
72. Fosset, M, Muir, LW, Nielsen, LD, Fischer, EH (1971). Purification and properties of yeast phosphorylase a and b. Biochemistry 10: 4105.
73. Reed, SI, Hadwiger, JA, Lörincz, AT (1985). Protein kinase activity associated with the product of the yeast cell division cycle gene CDC28. Proc. Natl. Acad. Sci. USA 82: 4055.
74. van Tuinen, D, Perez, RD, Marmè, D, Turian, G (1984). Calcium/calmodulin-dependent protein phosphorylation in Neurospora crassa. FEBS Letts. 176: 317.
75. Maruta, H, Baltes, W, Dieter, P, Marmè, D, Gerisch, G (1983). Myosin heavy chain kinase inactivated by Ca^{2+}/calmodulin from aggregating cells of Dictyostelium discoideum. EMBO J. 2: 535.
76. Souranta, K, Londesborough, J (1984). Purification

of intact and nicked forms of a zinc-containing, Mg^{2+}-dependent, low K_m cyclic AMP phosphodiesterase from baker's yeast. J. Biol. Chem. 259: 6964.
77. Londesborough, J, Souranta, K (1983). The zinc-containing high K_m cyclic nucleotide phosphodiesterase of baker's yeast. J. Biol. Chem. 258: 2966.
78. Uno, I, Matsumoto, K, Ishikawa, T (1983). Characterization of a cyclic nucleotide phosphodiesterase-deficient mutant in yeast. J. Biol. Chem. 258: 3539.
79. Liao, HH, Thorner, J (1981). Adenosine 3',5'-phosphate phosphodiesterase and pheromone response in the yeast Saccharomyces cerevisiae. J. Bacteriol. 148: 919.
80. Broek, D, Samiy, N, Fasano, O, Fujiyama, A, Tamanoi, F, Northup, J, Wigler, M (1985). Differential activation of yeast adenylate cyclase by wild-type and mutant RAS proteins. Cell 41: 763.
81. Kincaid, RL, Vaughn, M (1979). Sequential adsorption-electrophoresis: Combined procedure for purification of a calcium-dependent cyclic nucleotide phosphodiesterase. Proc. Natl. Acad. Sci. USA 76: 4903.
82. Ortega-Perez, R, van Tuinen, D, Marmè, D, Turian, G (1983). Calmodulin-stimulated cyclic nucleotide phosphodiesterase from Neurospora crassa. Biochim. Biophys. A. 758: 84.
83. Liao, H, Thorner, J (1980). Yeast mating pheromone α-factor inhibits adenylate cyclase. Proc. Natl. Acad. Sci. USA 77: 1898.
84. Casperson, GF, Walker, N, Bourne, HR (1985). Isolation of the gene encoding adenylate cyclase in Saccharomyces cerevisiae. Proc. Natl. Acad. Sci. USA 82: 5060.
85. Kataoka, T, Broek, D, Wigler, M (1985). DNA sequence and characterization of the S. cerevisiae gene encoding adenylate cyclase. Cell 43: 493.
86. Coussen, F, Haiech, J, D'Alayer, JD, Monneron, A (1985). Identification of the catalytic subunit of brain adenylate cyclase: A calmodulin-binding protein of 125 kDa. Proc. Natl. Acad. Sci. USA 82: 6736.
87. Reis, JA, Tellez-Inon, MT, Flawia, MM, Torres, HN (1984). Activation of Neurospora crassa soluble adenylate cyclase by calmodulin. Biochem. J. 221: 541.
88. Malencik, DA, Anderson, SR (1984). Peptide binding by calmodulin and its proteolytic fragments and by troponin C. Biochemistry 23: 2420.
89. Cox, JA, Comte, M, Fitton, JE, DeGrado, WF (1985). The interaction of calmodulin with amphiphilic peptides. J. Biol. Chem. 260: 2527.

GENETICS OF VACUOLAR PROTEASES

E.W. Jones, C. Moehle, M. Kolodny, M. Aynardi, F. Park, L. Daniels and S. Garlow

Dept. of Biological Sciences, Carnegie-Mellon University
Pittsburgh, Pennsylvania 15213

ABSTRACT Expression of protease B activity was examined in strains bearing plasmids containing the protease structural gene PRB1 and/or the processing gene PEP4. The level of protease B activity found is a function of the stage of growth of the cells and of the plasmid(s) present in the strain. The YEp(PRB1) plasmid can substitute for the PEP4 gene if given a "running start". Mutations in the PEP7 locus are pleiotropic and result in low levels of vacuolar hydrolases. Low levels of antigen corresponding to protease A and carboxypeptidase Y are found in the mutants and the CPY-like antigens present show some variation in glycosidic side chain components. The alg1-alg6 mutations of Huffaker (1-3) affect the structure of CPY antigens, but all complement pep7 mutations.

INTRODUCTION

The yeast vacuole is a major cellular repository for amino acids and polyphosphate and contains a number of the major hydrolases of the cell (4-6). Levels of vacuolar hydrolases vary with growth stage (7) and energy and nitrogen sources (8). The vacuolar hydrolases are glycoproteins and are synthesized as inactive precursors (see 9 for review). Vacuolar hydrolase precursors share part of the secretion pathway with secreted proteins, passing through the endoplasmic reticulum and Golgi membranes before being sorted from secretory proteins along a route which leads to the vacuole (10). Unlike secreted invertase (11), removal of a signal sequence from the vacuolar hydrolase carboxypeptidase Y (CPY) appears not to occur upon transport into the endoplasmic reticulum (12). Two kinetic precursors to mature CPY and protease A(PRA) have been detected, which differ in the composition of their carbohydrate side chains. These correspond to species found in the endoplasmic reticulum (p1) and Golgi (p2) (10, 13, 14). The final step in maturation of the hydrolases appears to be removal of a peptide, known to be N-terminal in the case of CPY (15-17). This final maturation step, known to occur very late in the processing pathway, possibly even in the vacuole itself, fails to occur in pep4 mutants (14, 15, 18).

To facilitate investigation of the mechanisms of regulation of vacuolar hydrolases and of the pathways of processing and localization of these enzymes, we have isolated genes which encode the vacuolar hydrolases and gene products required for processing. We have continued to study other pleitropic mutants defective in these processes. I report here on the PEP4 and PRB1 (protease B structural) genes and on the pep7 mutant.

RESULTS

We recovered plasmids capable of complementing the pep4-3 mutation from the YEp24 bank (19). Three of these resulted in synthesis of active, mature CPY. Segregation analysis of crosses involving strains in which one such plasmid (CBZ1B1) had integrated revealed tight linkage between the PEP4 locus and the plasmid borne URA3 gene, thus proving that the yeast DNA insert derived from the PEP4 region of chromosome 16L (data not shown). A preliminary restriction map of the DNA from this region is given in Figure 1. A plasmid bearing the 1.1 kb RI-XhoI fragment complements the pep4-3 mutation.

Plasmids capable of complementing the prb1-1122 mutation were isolated form the YEp13 bank (20) and the YEp24 bank (19). Segregational analysis of crosses involving integrants revealed tight linkage between LEU2 or URA3, respectively, and the CAN1 and PRB1 loci on chromosome 5L, thus proving that the insert DNAs originated from the region which encodes protease B (data not shown). A preliminary restriction map of DNA from this region is given in Figure 1. The rightmost end of the region which encodes protease B is to the left of the XhoI site. The leftmost end is to the right of the Hind III site.

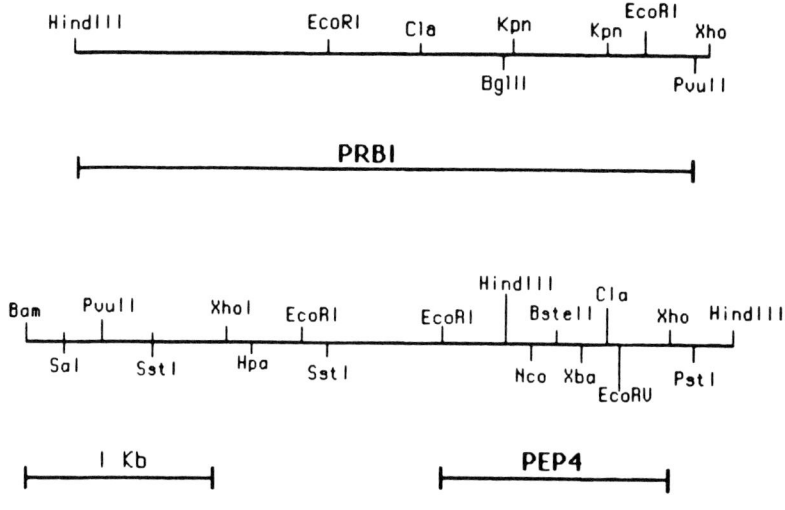

FIGURE 1. Preliminary restriction maps for DNA fragments which carry the structural gene for protease B (top) and the PEP4 gene (bottom).

As part of our initial characterization of the YEp(PRB1) clones, we determined levels of protease B in strains bearing the YEp(PRB1) plasmids. Enzyme levels were only elevated 2-4 fold rather than the 15-20 fold we expected on the basis of copy number. Because protease B levels were reported to rise in late stages of growth (7) and because levels were low in pep4/+ heterozygotes (21), we determined the effects of growth stage and the YEp(PEP4) plasmid on expression of protease B levels. The results are given in Figure 2. For the wild type strain, enzyme levels derepress more than 300 fold as the cells leave the diauxic plateau. If the cells bear the YEp(PEP4) plasmid, the same level of derepression is achieved but derepression occurs somewhat earlier in growth of the

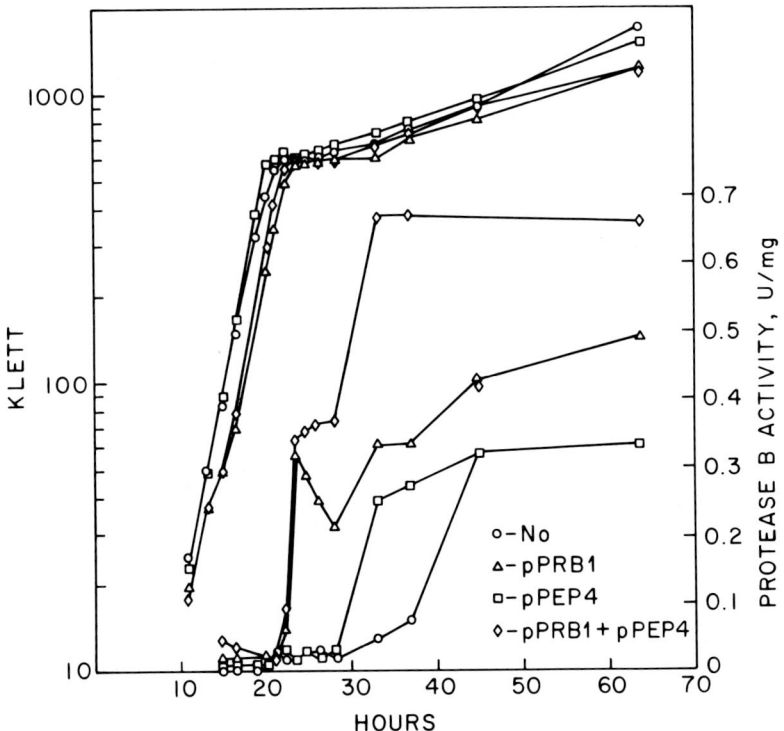

FIGURE 2. Expression of protease B activity as a function of growth stage in cells bearing YEp plasmids containing the PRB1 or PEP4 genes. The chromosomal genotype of the strain was trp1 ura3-52 leu2. Plasmid YEp13:PRB8R is a derivative of YEp13 bearing PRB1 in the yeast DNA insert. Plasmid CBZ1 is a derivative of YEp24 bearing PEP4 in the yeast DNA insert. Strains were cultured overnight with selection for plasmid retention in minimal medium to which needed nutrients were added. These inocula were diluted 1/100 into YEPD. Growth was monitored as turbidity. Samples were withdrawn at intervals; medium and cell-free extracts were assayed for protease B activity (22). Specific activities are plotted without correction for plasmid loss.

culture. If cells bear the YEp(PRB1) plasmid, enzyme levels derepress just after the plateau is reached. Early derepression and the levels of protease B expected for cells bearing a YEp plasmid are only achieved if the cells bear both the YEp(PRB1) and YEp(PEP4) plasmids. Presumably under this condition there is sufficient processing activity to process most of the protease B precursor present in the precursor pool. After correction for plasmid loss, the relative specific activities are 1, 2.3, 3.6 and 14.7 for cells bearing no, PEP4, PRB1 and both plasmids. In some, but not all, cultures of the strain bearing both plasmids, substantial levels of protease B are found in the medium. Preliminary indications (from assays of α-glucosidase acitivity, Coomassie staining of electropherograms of concentrated medium samples and methylene blue staining of cell populations) are that some fraction of the cells is lysing.

During the course of preparing strains for the above determinations, we made the observation that the PRB1 gene in high copy number can substitute for the PEP4 gene so long as it is given a "running start". These observations are summarized in Table 1. In crosses involving pep4 mutations, there is a substantial phenotypic lag in expression of the mutation, which is eliminated by mutations in PRB1 (23). We inferred that the protease B present in the spores was activating its own and other vacuolar hydrolase precursors, but that the level of protease B present was below some threshold level needed for maintenance of the wild type phenotype. Direct transformation of the YEp(PRB1) plasmid into a pep4 strain has no effect on expression of the pep4 mutation (line 4, Table 1). If cells of the identical genotype are generated via an intermediate strain which carries both the YEp(PEP4) and YEp(PRB1) plasmids, followed by loss of the YEp(PEP4) plasmid, the cells are permanently CPY$^+$. We infer that the high levels of protease B generated in the strain bearing both plasmids are above the threshold level of protease activity needed to process hydrolase precursors in a self sustaining fashion.

TABLE 1

THE PRB1 GENE IN HIGH COPY NUMBER CAN SUPPRESS THE pep4-3 MUTATION[a]

	Genotype		Mode of Construction	Phenotype
	Chromosome	Plasmid		
1)	pep4	none	mitosis	CPY⁻
2)	pep4	none	meiosis (pep4/+)[b]	CPY⁻, after phenotypic lag
3)	pep4 prb1	none	meiosis (pep4 +/+ prb1)	CPY⁻, no phenotypic lag
4)	pep4	YEp(PRB1)	direct transformation	CPY⁻
5)	pep4	YEp(PRB1)+ YEp(PEP4)	meiosis (pep4 +; pPEP4, pPRB1)/+ prb1	CPY⁺
6)	pep4	YEp(PRB1)+ YEp(PEP4)	sequential transformation	CPY⁺
7)	pep4	YEp(PRB1)	mitosis; loss of p(PEP4) from genotypes 5 or 6	permanently CPY⁺

a) The pep4-3 mutation is followed as CPY deficiency.
b) In parenthesis is genotype of parent diploid.

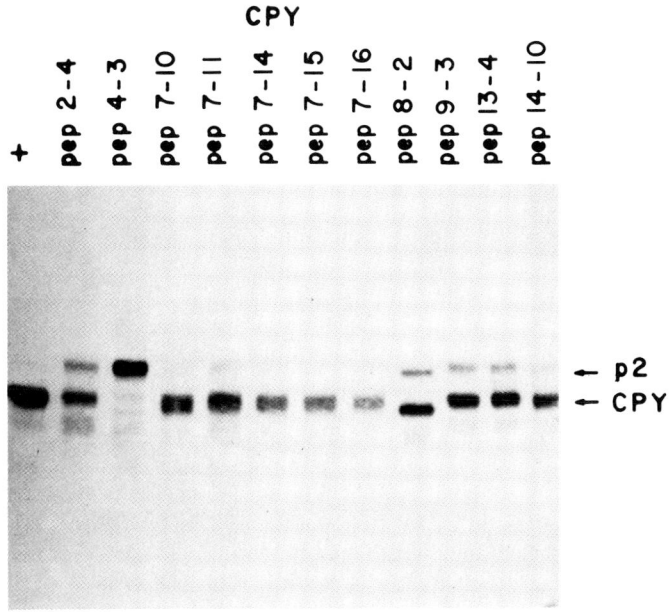

FIGURE 3. Carboxypeptidase Y-like antigens present in various CPY-deficient mutants. Cells were incubated with ^{35}S sulfate overnight and immunoprecipitates and fluorograms made according to the procedure in (10). Arrows point to positions of mature CPY(CPY) and the larger, fully glycosylated precursor of CPY(p2). Equal numbers of counts were loaded in each lane, hence the band intensity needn't reflect intracellular concentration.

The pep4 mutation is only one of several pleiotropic mutations which affect the levels of activity of vacuolar hydrolases (24, 25, unpublished). Mutations in the PEP7 locus result in reduced levels of activity for the vacuolar hydrolases protease A, protease B, CPY, alkaline phosphatase and RNase(s) and in low levels of CPY-like and PRA-like antigens, at least [24, 25, and unpublished; the allele designated pep17-1 in (25) is pep7-1]. The CPY-like antigen(s) present in pep7 mutants are seen as a diffuse band with mobility near that of mature CPY (Figure 3).

In some preparations the diffuse band can be seen to be a doublet. The diffuse or doublet band collapses into a singlet of normal band width when the samples are treated with EndoH (data not shown). Thus the antigen species differ from one another in their carbohydrate components. CPY contains four asparagine-linked glycosidic side chains whose synthesis is dolichol-mediated (9, 26, 27). The glycolipid donor for synthesis is depicted in Figure 4 (28). After transfer of the $(GlcNAc)_2 Man_9 Glc_3$ unit to

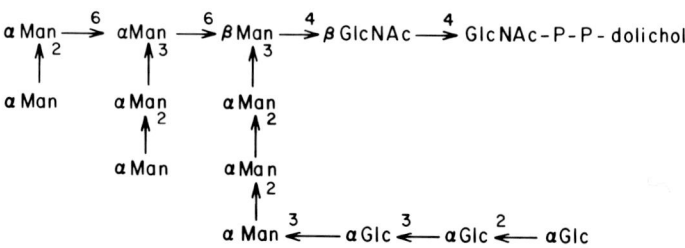

FIGURE 4. Structure of the oligosaccharide-lipid donor molecule. In alg5 and alg6 mutants this glycolypid donor lacks the three Glc residues (28). GlcNAc: N-Acetylglucosamine; Glc: glucose; Man: mannose.

protein, the three glucose units are removed as is one mannose unit (29) and several α-1-3 linked mannose units are added (30). It has been shown that gls1 mutants fail to remove the three glucose units from CPY (31) and that mnn1 mutants fail to add α-1-3-linked mannose units to the core carbohydrate of CPY (30).

Were a fraction of the CPY molecules to receive one too few carbohydrate units, a fuzzy or doublet band might result. Because the unglucosylated glycolipid is transferred to protein slowly in vitro (32), it seemed possible that pep7 mutants might have a defect in synthesis of the glycolipid donor, and, in particular, in glucosylation of the glycolipid. To investigate this possibility and to determine whether all steps in synthesis of the glycolipid donor are common for secreted and vacuolar enzymes, we analyzed the CPY-like antigens present in the alg mutants isolated by Tim Huffaker (1-3) and in crosses between the alg mutants and the pep7 mutant. We were especially interested in the alg5 and alg6 mutants, since both make unglucosylated glycolipid (3). The results can be seen in Figure 5, in which CPY-like antigens detected in Western immunoblots are presented for mutants grown at 30°. All of the alg mutants have

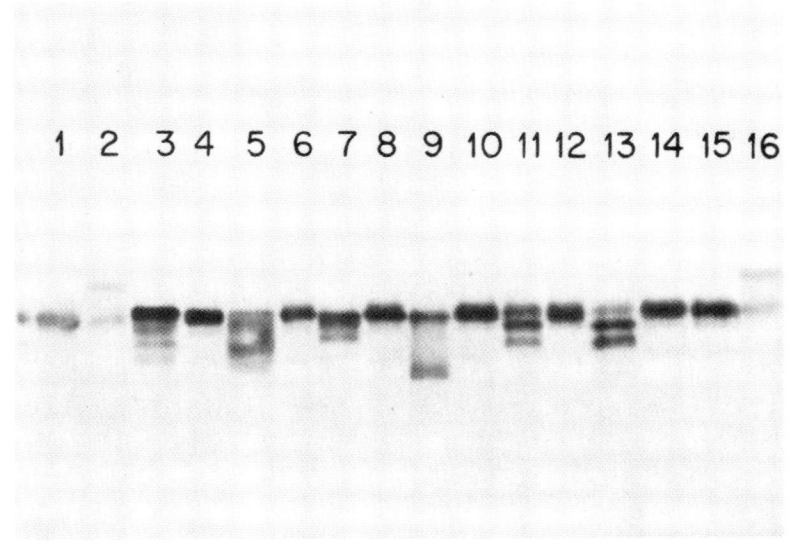

FIGURE 5. Carboxypeptidase Y-like antigens present in alg1-alg6 and pep7 mutants and in selected doubly heterozygous diploids. Strains were grown at 30° to stationary phase in YEPD. Harvested cells were boiled in SDS. Equal amounts of proteins were loaded for electrophoresis. Western blots were made according to (33) with visualization of immune complexes via the avidin-biotin-linked peroxidase activity in the Vectastain kit (Vector Labs). Genotypes of strains in lanes are 1 +; 2 pep7-10; 3 alg1; 4 pep7-10 +/+ alg1; 5 alg2; 6 pep7-10 +/+ alg2; 7 alg3; 8 pep7-10 +/+ alg3; 9 alg4; 10 pep7-10 +/+ alg4; 11 alg5; 12 pep7-10 +/+ alg5; 13 alg6; 14 pep7-10 +/+ alg6; 15 +; 16 pep7-10.

altered as well as normal forms of CPY antigens present, a not unexpected result. In the alg1 mutant four bands are seen, presumably corresponding to species bearing one, two, three or four carbohydrate chains. A band with greater mobility yet, presumably corresponding to the carbohydrate free but processed protein, is seen in the alg4 (sec53) mutant that is defective in protein translocation into the endoplasmic reticulum at the restrictive temperature (34, 35). The alg5 and alg6 mutants show three bands, presumably corresponding to two, three or four carbohydrate chains/molecule, in line with the in vitro results. The relative proportions of the three species in the alg5 and alg6 mutants are reproducible and characteristic. The pep7 mutation complements all six alg mutations, for the six diploids have CPY antigen of mature size only. It is worth noting that the alg mutations appear to have little effect on levels of CPY-like antigen, in contrast to the pep7 mutations. These results imply that the primary effect of the pep7 mutations is not on synthesis of the glycolipid. As the pep7 mutations have no effect on synthesis or mobility of invertase activity (data not shown) it seems unlikely that pep7 mutants are defective per se in transfer of the glycosidic side chains to recipient proteins.

DISCUSSION

The timing and extent of derepression of protease B activity depends not only on the number of copies of the structural gene but also on the amount of processing activity available for acting upon the protease B precursor. The amount of processing activity available for processing protease B may depend on the number of copies of the PEP4 gene, the state of derepression of the PEP4 gene, and the size and composition of the pool of precursor proteins which must be processed. Moreover, it is not farfetched to think that the ultimate product of the PEP4 gene may be a vacuole protein, synthesized via an inactive precursor, which must itself be processed by the same machinery which processes all other hydrolase precursors.

Simple explanations can be adduced for the changes in level and/or timing of derepression of protease B seen in Figure 2 which invoke sequential derepression of the PEP4 gene and the PRB1 gene, partial substitution of protease B activity for the PEP4 gene product, and increased levels of proPRB and/or the precursor to PEP4 product in the precursor pool leading to elevated levels of product by improved competition for processing. The relation of simple explanation to truth awaits investigation.

The finding that the PRB1 gene in high copy number can substitute for the PEP4 gene inevitably reinforces our bias that the PEP4 gene product is the protease responsible for processing all of the vacuolar hydrolase precursors. Of more interest, possibly, is the observation that cells of identical genotype can have different phenotypes. The simplest explanation is that the protease activity encoded by PEP4 (see above) processes the large amount of proPRB present in cells bearing both plasmids to active protease B and that the resulting high levels of protease B are sufficient to process its own precursor as well as the other hydrolase precursors at rates which are self sustaining. Proteolytic cleavages have always had appeal as possible determinative events since they are largely irreversible. It is interesting in this regard that raising the level of a protease (protease B, artificially, by activating more precursor) can commit the cell to a different phenotype, despite the genetic identity of product and ancestral cell.

The pep7 mutants produce CPY-like antigens with altered carbohydrate side chains or/and content, but the mutants seem not to be defective in synthesis of the glycolipid or in transfer of the glycolipid to proteins per se. Our unpublished experiments suggest that the pep7 mutations have a profound effect on the kinetics of maturation of CPY, slowing down the rate of processing for all steps by a factor of about five.

ACKNOWLEDGEMENTS

This work was supported by NIH research grants AM18090 and GM29713 to E.W. J., by an NIH training grant (CM), and by an NIH postdoctoral fellowship (LD). The skilled assistance of Guy Bennardo is gratefully acknowledged.

REFERENCES

1. Huffaker T, Robbins P (1982). Temperature-sensitive yeast mutants deficient in asparagine-linked glycosylation. J Biol Chem 257:3203.

2. Huffaker T, Robbins P (1983). Yeast mutants deficient in protein glycosylation. Proc Natl Acad Sci 80:7466.
3. Runge K, Huffaker T, Robbins P (1984). Two yeast mutations in glucosylation steps of the asparagine glycosylation pathway. J Biol Chem 259:412.
4. Wiemken A, Durr M (1974). Characterization of amino acid pools in the vacuolar compartment of Saccharomyces cerevisiae. Arch Microbiol 101:45.
5. Wiemken A, Schellenberg M, Urech K (1979). Vacuoles: The sole compartments of digestive enzymes in yeast (Saccharomyces cerevisiae)? Arch Microbiol 123:35.
6. Messenguy F, Colin D, Ten Have J (1980). Regulation of compartmentation of amino acid pools in Saccharomyces cerevisiae and its effects on metabolic control. Eur J Biochem 22:277.
7. Saheki T, Holzer H (1975). Proteolytic activities in yeast. Biochem Biophys Acta 384:203.
8. Hansen R, Switzer R, Hinze H, Holzer H (1977). Effects of glucose and nitrogen source on the levels of proteinases, peptidases and proteinase inhibitors in yeast. Biochem Biophys Acta 196:103.
9. Jones E (1984). The synthesis and function of proteases in Saccharomyces: Genetic approaches. Ann Rev Genetics 18:233.
10. Stevens T, Esmon B, Schekman R (1982). Early stages in the yeast secretory pathway are required for transport of carboxypeptidase Y to the vacuole. Cell 30:439.
11. Emr S, Schekman R, Flessel M, Thorner J (1983). In $MF\alpha 1$-SUC2 (α-factor-invertase) gene fusion for study of protein localization and gene expression in yeast. Proc Natl Acad Sci 80:7080.
12. Distel B, Al R, Tabak H, Jones E (1983). Synthesis and maturation of the yeast vacuolar enzymes carboxypeptidase Y and aminopeptidase I. Biochem Biophys Acta 741:128.
13. Müller M, Müller H (1981). Synthesis and processing of in vitro and in vivo precursors of the vacuolar yeast enzyme carboxypeptidase Y. J Biol Chem 256:11962.
14. Zubenko G, Park F, Jones E (1983). Mutations in the PEP4 locus of Saccharomyces cerevisiae block the final step in the maturation of two vacuolar hydrolases. Proc Natl Acad Sci 80:510.
15. Hemmings B, Zubenko G, Hasilik A, Jones E (1981). Mutant defective in processing of an enzyme located in the lysozyme-like vacuole of Saccharomyces cerevisiae. Proc Natl Acad Sci 78:435.

16. Hasilik A, Tanner W (1978). Biosynthesis of the vacuolar yeast glycoprotein carboxypeptidase Y. Conversion of precursor into enzyme. Eur J Biochem 91:567.
17. Mechler B, Muller M, Muller H, Muessdoerffer F, Wolf D (1982). In vivo biosynthesis of the vacuolar proteinases A and B in the yeast Saccharomyces cerevisiae. J Biol Chem 257:11203.
18. Mechler B, Müller M, Müller H, Wolf D (1982). In vivo biosynthesis of vacuolar proteinases in proteinase mutants of Saccharomyces cerevisiae. Biochem Biophys Res Comm 107:770.
19. Carlson M, Botstein D (1982). Two differentially regulated mRNAs with different 5' ends encode secreted and intracellular forms of yeast invertase. Cell 28:145.
20. Nasmyth K, Reed S (1980). Isolation of genes by complementation in yeast. Proc Natl Acad Sci 77:2119.
21. Jones E, Zubenko G, Parker R (1982). PEP4 gene function is required for the expression of several vacuolar hydrolases in Saccharomyces cerevisiae. Genetics 102:665.
22. Zubenko G, Mitchell A, Jones E (1979). Septum formation, cell division, and sporulation in mutants of yeast deficient in proteinase B. Proc Natl Acad Sci 76:2395.
23. Zubenko G, Park F, Jones E (1982). Genetic properties of mutations at the PEP4 locus in Saccharomyces cerevisiae. Genetics 102:679.
24. Jones E (1977). Proteinase mutants of Saccharomyces cerevisiae. Genetics 85:23.
25. Jones E, Zubenko G, Parker R, Hemmings B, Hasilik A (1981). Pleiotropic mutations of S. cerevisiae which cause deficiency for proteinases and other vacuole enzymes. In von Wettstein D, Friis J, Kielland-Brandt M, Stenderup A (eds): Alfred Benzon Symposium 16:183, Copenhagen: Munksgaard.
26. Hasilik A, Tanner W (1978). Carbohydrate moiety of carboxypeptidase Y and perturbation of its synthesis. Eur J Biochem 91:567.
27. Trimble R, Maley F (1977). The use of endo-β-N-acetyl--glucosaminidase H in characterizing the structure and function of glycoproteins. Biochem Biophys Res Comm 78:935.

28. Li E, Tabas I Kornfeld S (1978). The synthesis of complex-type oligosaccharides I. The structure of the lipid-linked oligosaccharide precursor of the complex-type oligosaccharides of the vesicular stomatitis virus G protein. J Biol Chem 253:7762.
29. Byrd J, Tarentino A, Maley F, Atkinson P, Trimble R (1982). Glycoprotein synthesis in yeast. J Biol Chem 257:14657.
30. Isai P-K, Frevert J, Ballou C (1984). Carbohydrate structure of Saccharomyces cerevisiae mnn9 mannoprotein. J Biol Chem 259:3805.
31. Esmon B, Esmon P, Schekman R (1984). Early steps in processing of yeast glycoproteins. J Biol Chem 259:10322.
32. Lehle L (1980). Biosynthesis of the core region of yeast mannoproteins. Eur J Biochem 109:589.
33. Burnette W (1981). "Western blotting": Electrophoretic transfer of proteins from sodium dodecyl sulfate-polyacrylamide gels to unmodified nitrocellulose and radiographic detection with antibody and radioiodinated protein A. Anal Bioch 112:195.
34. Ferro-Novick S, Novick P, Field C, Schekman R (1984). Yeast secretory mutants that block the formation of active cell surface enzymes. J Cell Biol 98:35.
35. Ferro-Novick S, Hansen W, Schauer I, Schekman R (1984). Genes required for completion of import of protein into the endoplasmic reticulum of yeast. J Cell Biol 98:44.

TRANSLOCATION, SORTING AND TRANSPORT OF YEAST VACUOLAR GLYCOPROTEINS[1]

Tom H. Stevens, Elizabeth G. Blachly, Craig P. Hunter Joel H. Rothman, and Luis A. Valls

Institute of Molecular Biology, University of Oregon, Eugene, Oregon 97403

ABSTRACT The PRC1 gene encoding the vacuolar glycoprotein carboxypeptidase Y (CPY) has been cloned and sequenced. Sequence data indicate that CPY is synthesized as a precursor 111 aa larger than mature CPY. Deletion analysis within the 111 aa precursor propeptide suggests that aa 25-31 are essential for sorting of the CPY precursor and correct localization of CPY to the vacuole. A new class of mutant (VPL) has been isolated whose phenotype is the pleiotropic secretion of vacuolar glycoproteins. A number of these vpl mutants appear to be defective in the sorting of vacuolar proteins from secretory proteins along the secretory pathway.

INTRODUCTION

The secretory pathway in yeast has been identified and characterized genetically by isolating conditional lethal mutants that block secretion at discrete steps. The sec mutants accumulate secretory glycoproteins in the endoplasmic reticulum (ER), Golgi body, or secretory vesicles at the non-permissive temperature (1,2). These accumulated proteins proceed to subsequent stages in the pathway when cells are returned to the permissive temperature. Analysis of the sec mutants indicates that these mutational blocks define a linear pathway for secre-

[1]This work was supported by a grant from the NIH, GM 32448, The Chicago Community Trust/Searle Scholars Program, and the Oregon Affiliate of the American Heart Association.

tory glycoproteins from their site of synthesis at the ER to their site of discharge at the plasma membrane.

The secretory pathway in yeast is required for localization of vacuolar glycoproteins (3, 4), just as the secretory pathway of mammalian cells is necessary for transport of proteins to the lysosome (5,6). The yeast sec mutants blocked in transport from the ER or from the Golgi body fail to transport the vacuolar glycoprotein carboxypeptidase Y (CPY) to the vacuole (3). Precursor forms of CPY accumulate at the restrictive temperature in these sec mutant cells and transport of CPY to the vacuole is restored when cells are returned to the permissive temperature. sec mutations that block secretory protein transport after the Golgi step do not affect the ability of the cell to correctly localize CPY. These results indicate that sorting of vacuolar and secretory proteins occurs at or prior to the Golgi apparatus.

The recognition signals that facilitate sorting of vacuolar and secretory proteins in yeast remain to be elucidated. In mammalian cells a mannose-6-phosphate determinant has been identified that directs transport of some lysosomal glycoproteins from the Golgi body to the lysosome (7). Although yeast phosphorylate mannose residues of secretory glycoproteins and CPY (8,9) during transit through the secretory pathway, this phosphorylation step appears to be unnecessary for localization of CPY to the vacuole. Yeast cells treated with tunicamycin, a drug that blocks the synthesis of high-mannose oligosaccharides, properly sort and transport CPY to the vacuole (3,10) and localize the secretory glycoprotein invertase to the periplasm (11). It is likely then, that the determinant for vacuole localization resides within the protein structure of the CPY polypeptide. In such an event, it should be possible to determine the nature and location of this sorting determinant within the amino acid sequence of CPY.

To understand the protein sequences or domains responsible for targeting CPY to the vacuole we have cloned the CPY structural gene, PRC1. Early experiments with this PRC1 gene clone indicated that the ability of yeast cells to localize CPY exclusively to the vacuole depends on the amount of CPY synthesized by the cells (12). The presence of PRC1 on a multiple copy 2μ plasmid causes yeast to overproduce CPY about 3 to 4-fold, resulting in the secretion of about 10% of the total CPY through the late secretory pathway. CPY is overproduced 5 to 10-fold when the PRC1 gene is placed under the transcriptional regulation of the phosphate-repressible PHO5 promoter. Under these conditions, greater than 50% of the total CPY synthesized is mislocalized to the cell surface, strongly suggesting that the sorting of CPY is a saturable reaction (12). The phenomenon may be general for all vacuolar proteins: we have recently cloned the gene encoding proteinase A using a novel immunochemical approach and have found that on a multiple copy plasmid this PRA1-containing DNA fragment causes proteinase A to appear at the cell surface (13).

When CPY is secreted due to overexpression of PRC1 it is the 69 kd proCPY form which appears in the periplasm. This 69 kd proCPY species co-migrates by SDS·PAGE with the proCPY form delivered to the vacuole (12) indicating that no further glycosyl modification occurs during transit through the final stages of the secretory pathway. Whereas proCPY is proteolytically activated to the 61 kd mature CPY species in the vacuole under control of the PEP4 locus (14), secreted proCPY is processed to the active 61 kd protein independent of the allelic state of the PEP4 locus (12). These observations have provided us with a powerful selection procedure for obtaining yeast mutants that fail to sort, segregate or transport proCPY to the vacuole, by selecting mutants that secrete and activate proCPY to CPY in a pep4 background.

Although the nature of the sorting recognition determinant is unclear, work from a number of laboratories allows us to rule out some hypotheses about it. The tunicamycin studies mentioned above show that it is not a carbohydrate determinant. Mutations in the PEP4 gene isolated by E. Jones and coworkers cause yeast cells to accumulate inactive precursor forms of a number of vacuolar glycoproteins (14,15). Since these precursors accumulate in the vacuole (3) in a pep4 mutant, we conclude that the PEP4 gene product is required for proenzyme cleavage but not vacuolar localization. Interestingly, the precursors for CPY, proteinase A and proteinase B are all larger than their mature forms by 8-10 kd (14,16-18). We are pursuing the possibility that the 8-10 kd propeptides of these enzymes not only function to maintain the enzymes in an inactive form during transit through the secretory pathway, but also share a functional domain which binds to a sorting component in the Golgi.

In this paper we describe our progress in uncovering the localization signals on vacuolar proteins and identifying the genes that code for vacuolar protein sorting functions.

RESULTS

<u>PRC1</u> Gene

The partial amino acid sequence of mature CPY has been obtained by conventional amino acid sequencing techniques (19,20). However, to determine the sequence of the transiently attached propeptide portion of proCPY we cloned and sequenced the CPY structural gene, <u>PRC1</u> (12) by complementing a <u>prc1</u>-1 mutation. A <u>prc1</u>-1 <u>leu2</u>-3 <u>leu2</u>-112 yeast strain was transformed with a YEp13 (<u>LEU2</u>) yeast genomic bank and <u>LEU2</u> transformants were selected (12). The <u>LEU2</u> transformants were screened for production of CPY by a CPY plate stain (15). We have constructed fusion proteins using the complementing fragment and the <u>lacZ</u> gene. These hybrid proteins are immunoprecipitable from E. coli cells using anti-CPY antibody (TH Stevens and SD Emr, unpublished observations). This serves as proof that the complementing fragment indeed encodes CPY.

The restriction map for <u>PRC1</u> is shown in Figure 1. The direction of transcription was determined by analysis

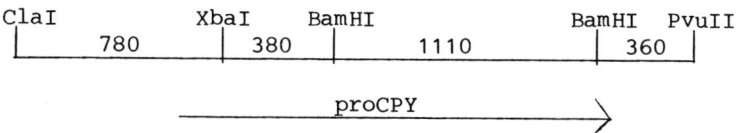

Figure 1. Restriction map of <u>PRC1</u> illustrating protein coding region.

of hybrid proteins resulting from <u>PRC1</u>-<u>LacZ</u> gene fusions (TH Stevens and S Emr, unpublished observations).

Both strands of the <u>PRC1</u> gene have been sequenced from the ClaI site to the PvuII site at the 3' end of the gene (21). Comparison of the sequence of this 2.6 kb <u>PRC1</u> DNA fragment with the published amino acid sequence of CPY indicates that the CPY precursor, proCPY, is synthesized as a 532 aa protein. Ambiguities in amino acid sequence and uncertainty in the full size of mature CPY (19,20) were resolved using the DNA sequence (21). Mature enzymatically active CPY is 421 aa. Within the mature protein there are four sites of asparagine-linked oligosaccharide addition, asn-x-thr or asn-x-ser. These asparagine residues are at amino acid positions 13, 87, 168 and 368 counting from the first amino acid of the mature protein. In addition, there are no asparagine residues that can serve as oligosaccharide addition sites in the 111 aa propeptide region of proCPY, indicating that N-linked glycosylation is restricted to the mature protein.

The 111 aa propeptide of PRC1 has a number of unusual features (Figure 2). The first 18 aa of the precursor contain a stretch of 16 uncharged and hydrophobic residues (aa 3-18) that probably serve as a signal sequence to direct nascent CPY peptides to the ER. It is unlikely that cleavage of the signal sequence occurs since it has been observed that <u>in vitro</u> translated proCPY co-migrates by SDS·PAGE with proCPY synthesized by tunicamycin-treated yeast cells (22). Furthermore, this N-terminal stretch of amino acids does not contain a consensus signal sequence cleavage site (23). However, definitive proof concerning signal sequence cleavage awaits N-terminal amino acid

StuI
Met·Lys·Ala·Phe·Thr·Ser·Leu·Leu·Cys·Gly$_{10}$·Leu·Gly·Leu·Ser·

Thr·Thr·Leu·Ala·Lys·Ala ·Ile·Ser·Leu·Gln·Arg·Pro·Leu·Gly·
XbaI   20

Leu·Asp$_{30}$·Lys·Asp·Val·Leu·Leu·Gln·Ala·Ala·Glu·Lys$_{40}$·Phe·

Gly·Leu·Asp·Leu·Asp·Leu·Asp·His·Leu$_{50}$·Leu·Lys·Glu·Leu·Asp·
                    ApaI
Ser·Asn·Val·Leu·Asp$_{60}$·Ala·Trp·Ala·Gln·Ile·Glu·His·Leu·Tyr·

Pro$_{70}$·Asn·Gln·Val·Met·Ser·Leu·Glu·Thr·Ser·Thr$_{80}$·Lys·Pro·

Lys·Phe·Pro·Glu·Ala·Ile·Lys·Thr$_{90}$·Lys·Lys·Asp·Trp·Asp·

Phe·Val·Val·Lys·Asn$_{100}$·Asp·Ala·Ile·Glu·Asn·Tyr·Gln·Leu·

Arg·Val$_{110}$·<u>Asn:Lys</u>·Ile

Figure 2. Amino acid sequence of the propeptide of proCPY. Lys·ile are the first two amino acids of mature CPY.

analysis of the proCPY species isolated a from pep4 strain. Our laboratory is proceeding with this analysis. The next 93 amino acids of the 111 aa propeptide contain a high proportion of charged residues: approximately 35% charged residues compared with about 15% in mature CPY. Finally, cleavage of proCPY to CPY occurs between the asn and the lys in the stretch of amino acids val_{110}·asn·lys·ile (Figure 2).

PRC1 signal sequence mutations. The role of the 20 aa putative proCPY signal sequence can be ascertained by obtaining mutations in this region and analyzing their phenotypes. The DNA sequence revealed a unique StuI restriction site at amino acid 12 (Figure 2) in the middle of the putative signal sequence. We constructed deletions by cutting the PRC1 gene at StuI, treating the DNA with the exonuclease Bal31 for short intervals (24) and then transforming the ligated DNA into a yeast strain that had been deleted for the entire PRC1 coding region (12). We then screened for the transformants carrying PRC1 genes with in-frame deletions by assaying via western blot analysis for the synthesis of CPY (24). Western analysis also revealed whether CPY accumulated in cells as the glycosylated proCPY species (70 kd), glycosylated mature CPY (61 kd) or as unglycosylated proCPY (59 kd). Transformants that produced the 59 kd species were tested for the failure to glycosylate CPY by immunoprecipitation of CPY from yeast cells labeled with ^{35}S-met in the presence or absence of tunicamycin. A number of transformants produced a 59 kd CPY species on SDS PAGE gels even in the absence of tunicamycin, indicating a failure to translocate the unglycosylated cytoplasmic proCPY across the ER membrane (24). DNA from four of these transformants was isolated and sequenced to determine the precise boundaries and sizes of the deletions (24). Table 1 summarizes the results of the signal sequence deletion analysis. Comparison of Table 1 with the wild-type sequence in Figure 2 suggests that a reduction in length of the uncharged stretch of amino acids at the N-terminus results in the failure to translocate and glycosylate the mutant proCPY species. These observations are consistent with results from signal sequence deletions in E. coli secretory proteins which result in a failure of the mutant precursor protein to be translocated across the E. coli inner membrane (25).

Mutations in the PRC1 propeptide. The propeptide of CPY has at least two known functions. We have demon-

TABLE 1
SIGNAL SEQUENCE DELETIONS

	Size of deletion	Amino acids deleted	Location of intracellular CPY
ssΔ1	22 aa	aa 9-30	cytoplasm
ssΔ2	6 aa	aa 7-12	cytoplasm
ssΔ3	11 aa	aa 13-23	cytoplasm
ssΔ4	5 aa	aa 8-12	cytoplasm
wt	-----	-----	vacuole

strated that the N-terminal 20 aa of proCPY are required for efficient translocation across the ER membrane. A second function of the propeptide is to render CPY enzymatically inactive during its transit through the secretory pathway, since PEP4-dependent processing occurs either at a very late transport step or upon delivery to the vacuole (3,18). In addition, the 111 aa propeptide may carry the recognition signal for targeting CPY to the vacuole. Hybrid proteins have been constructed in which 5' coding regions of the PRC1 gene are fused to the structural gene for the secretory protein invertase, SUC2. Preliminary studies with these fusions indicate that the N-terminal 150 amino acids of PRC1 are sufficient to direct the fusion protein to the vacuole (SD Emr, personal communication). These results strongly suggest that the N-terminal 150 aa of proCPY contain a protein domain which is recognized by the cell's sorting machinery.

To investigate in more detail the vacuolar targeting signals contained in the CPY propeptide we have constructed mutations in this region of PRC1. The PRC1 gene was mutagenized in vitro by treatment with Bal31 exonuclease after cutting the DNA at either the XbaI or ApaI restriction sites within the coding region. The ligated plasmid DNA was transformed into E. coli cells and 5,000-10,000 individual transformants were combined. The combined transformants were grown selectively for about 10 generations and plasmid DNA was isolated. This in vitro mutagenized CEN4-URA3-PRC1 plasmid DNA was used to transform a Δprc1 ura3-52 yeast strain to Ura$^+$. Yeast transformants were screened for aberrantly secreted CPY by immunoblotting with anti-CPY antibody (26). This immunoblot procedure takes advantage of the fact that CPY secreted to the periplasm gets through the cell wall (12), allowing replica-plated colonies to deposit secreted CPY onto a nitrocellulose membrane. Secreted CPY can be detected even if the proCPY cannot be processed or if the protein is not enzymatically active.

Yeast transformants found to be secreting CPY by the immunoblot technique were isolated and tested quantitatively for the extent of CPY secretion (26). This was accomplished by labeling yeast cells with ^{35}S-sulfate and immunoprecipitating CPY from the medium, periplasm and an intracellular fraction, as described in reference 12. Transformants secreting a higher percentage of CPY than wild-type ($\cong$1%) were used as sources of plasmid DNA for sequencing of PRC1 to determine the precise amino acids deleted.

Table 2 summarizes data for two deletions near the XbaI site at amino acid 30 of the propeptide. The 7 aa deletion in PRC1 ΔX1 secretes almost all of its CPY to the

TABLE 2
PHENOTYPES OF PRC1 PROPEPTIDE DELETIONS

	Size of deletion	Amino acids deleted	CPY location in cell
ΔX1	7 aa	aa 25-31	>80% secreted
ΔX2	12 aa	aa 20-31	15-20% secreted
ΔXA[a]	~34-38 aa	~aa 29-63	100% intracellular

[a]The ΔXA deletion has not yet been sequenced.

cell surface (26), even though the ΔX1 PRC1 allele is maintained at single copy on a centromere plasmid. The 10-20% of the ΔX1 CPY that is intracellular appears not to be in the vacuole as it accumulates as the proCPY species in PEP4 cells. Either this small amount of intracellular proCPY is in transit to the cell surface or it is in the vacuole and resistant to PEP4-dependent cleavage. Interestingly, a larger deletion of 12 aa ΔX2 with the same 3' deletion end point has a much less dramatic secretion phenotype: only about 20% of the CPY is secreted, and the intracellular CPY accumulates as mature CPY, indicating vacuolar localization (26). A fraction of the proCPY secreted in both the ΔX1 and ΔX2-containing cells is processed in the periplasm to active CPY. Thus far, in vitro Bal31 deletions created at the ApaI site have not yielded PRC1 alleles coding for mislocalized CPY.

A deletion was also constructed in the PRC1 propeptide that extended from the XbaI site to the ApaI site, by cutting at these unique restriction sites, followed by trimming and filling in with T4 DNA polymerase to create an in-frame deletion (26). This approximately 35 aa deletion, ΔXA, results in the synthesis of a proCPY species that is retained 100% within the cell. However, ΔXA CPY accumulates as a proCPY species in a PEP4 strain indicating that either it is blocked in transit to the vacuole or it cannot be processed upon delivery to the vacuole. Isolation of vacuoles from yeast cells carrying the ΔXA PRC1 allele will allow us to resolve this question.

Isolation of Vacuole Protein Localization (VPL) Mutants

Identifying the genes that encode vacuolar protein sorting functions will allow us to more clearly define the sorting pathway and eventually to isolate the various gene products that mediate these sorting functions. To obtain mutants that are defective in proper sorting of CPY we have devised a powerful selection scheme. This selection procedure allows us to rapidly identify mutant cells that aberrantly localize CPY activity to the periplasm. The selection is based on the observation that leucine auxotrophs lacking CPY activity due to a pep4-3 mutation cannot grow on medium containing the N-blocked dipeptide CBZ•pheleu as the sole leucine source. Because PEP4-independent proteolytic activation of proCPY to active CPY can occur in the periplasm (12), we reasoned that one class of mutants arising on the dipeptide medium would be those that aberrantly secrete CPY. After plating mutagenized pep4-3 leu2-3 leu2-112 cells on the dipeptide medium we find that indeed most of the mutants capable of growing under the selective conditions exhibit extracellular CPY activity (27). Such putative CPY sorting mutants are readily screened from a collection of growers on dipeptide medium by using a plate filter assay that detects secreted CPY activity.

TABLE 3
VPL COMPLEMENTATION GROUPS

VPL	Number of alleles	Number of temperature-sensitive alleles
1	26	22
2	4	0
3	12	3
4	8	1
5	5	1
6	15	8
7	7	5
8	10	4
Dominants	24	---
Others	61	---
	172	

The phenotypes of pairwise combinations of mutations in diploid cells were examined using the CPY activity filter assay. This has allowed us to classify the mutants that aberrantly secrete CPY into eight complementation groups (27). Each of the eight vacuole protein localization (VPL) complementation groups contains multiple alleles. In addition, there are a number of mutants that fall into none of the eight groups suggesting that many more than eight genes can cause a vpl phenotype. Many of the vpl mutants are temperature-sensitive for growth; in the case of vpl1-1, we have shown that temperature sensitivity and CPY secretion are tightly linked as seen in an outcross of a haploid vpl1-1 mutant with a wild-type cell (Table 4). Approximately fifteen percent of the selected vpl mutations are dominant. We have determined that none of these dominant mutations are linked to the PRC1 locus by crossing various vpl mutants with a haploid strain marked with $URA3^+$ adjacent to the PRC1 locus (27). In such crosses, the Ura^+ and vpl phenotypes segregate independently in over 100 dominants analyzed.

The cell surface CPY activity detectable in vpl mutants is a result of mislocalization of the vacuolar enzyme to the periplasm. SDS•PAGE analysis of immunoprecipitates from ^{35}S-sufate-labeled medium, periplasmic and intracellular fractions indicates that a large fraction of CPY antigen (from 10% to greater than 90%) resides in the extracellular fractions (27) following a 30 minute pulse and 60 minute chase period (Table 4). The appearance of CPY at the cell surface is unlikely to be a result of cell lysis since only a small fraction of the normally cytoplasmic glucose-6-phosphate dehydrogenase activity and immunoreactive phosphoglycerate kinase are detectable at the cell surface in vpl mutants. That vpl mutations truly cause mis-sorting that results in the passage of CPY into the late secretory pathway is supported by the observation that late-acting sec mutants (i.e., those that accumulate secretory vesicles) block the aberrant secretion of CPY in vpl1 sec double mutants (27).

Cells mutant in some of the VPL complementation groups show an intracellular accumulation of the unprocessed form of CPY (proCPY) suggesting that delivery of the zymogen to the vacuole is restricted or blocked. This observation may reflect sequestering of CPY to a pre-vacuolar compartment in these vpl mutants. In light of this finding, it may be significant that electron micrographs of certain vpl mutant cells exhibit an accumulation of aberrant non-vacuolar membrane-enclosed organelles (S Huestis and JH Rothman, unpublished observations). It is conceivable that such aberrant structures are related to an intermediate in vacuole assembly.

If vacuolar proteins utilize a common localization pathway to the vacuole, then one might expect that vpl mutants would aberrantly secrete a number of vacuolar proteins. Using antibody to the vacuolar glycoprotein proteinase A we have shown that this enzyme is also mislocalized to the cell surface in vpl mutant complementation groups 1,3,4,6,7 and 8 (27). In addition, assays of alkaline phosphatase indicate the presence of this vacuolar enzyme activity at the cell surface in a number of the vpl mutants (27). A summary of phenotypes seen in the various vpl complementation groups is shown in Table 5.

By complementing the temperature-sensitive lethal defect of vpl1-1 we were able to rapidly clone the VPL1 gene. The restriction map of this gene is shown in Figure 3. Using various subclones we have localized the VPL1 complementing region to a 4.2 kb DNA fragment. Using the method of Orr-Weaver et al. (28) we have integrated a plasmid containing this fragment and the URA3 marker into VPL cells. In sporulated diploids heterozygous for a vpl1-1 mutation and the integrated plasmid, URA3 segregates away from the vpl1-1 mutation in all tetrads analyzed indicating that the cloned DNA fragment maps to the VPL1 locus. A deletion of the VPL1 gene resulting from the replacement of the 1.5 kb BamHI-XhoI segment of VPL1 with a 2.6 kb XhoI-BglII LEU2 fragment (Figure 3) followed by transplacement of the inserted gene into the genome of a haploid yeast cell (29) causes a vpl1 phenotype (27). However, such Δvpl1 cells exhibit a greatly reduced growth rate relative to wild-type cells. Therefore, VPL1 appears to be a non-essential but important gene for haploid vegetative growth.

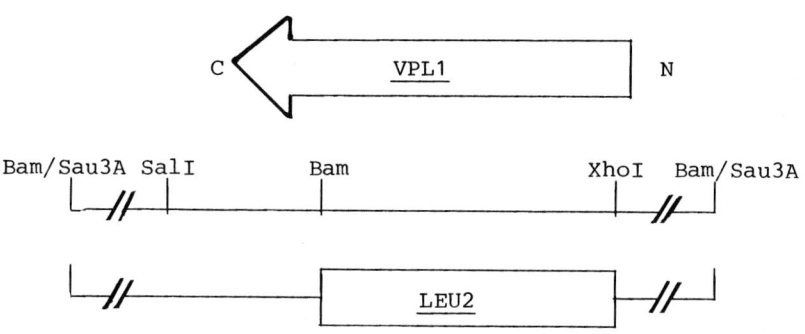

Figure 3. Restriction map of VPL1.

We have determined the direction of transcription of the VPL1 gene by constructing protein fusions with β-galactosidase. This fusion was constructed by ligating the 5' end of lacZ onto the left of the central BamHI site shown in Figure 3. The gene fusion encodes a protein of approximately 1400 amino acids indicating that the VPL1 gene product is at least 400 amino acids long. We are currently determining whether the VPL1-β-galactosidase fusion protein is membrane-bound, and are generating antibody to the fusion protein in order to determine the intracellular location of the VPL1 gene product.

To select against cells that aberrantly secrete CPY, we have devised a negative selection scheme. This selection uses medium containing CBZ·pheF$_3$leu, which we have shown is toxic to CPY-secreting leucine prototrophs. This selective condition prevents growth of vpl pep4 cells but not VPL pep4 cells and will thus prove useful in selecting for clones that complement vpl mutations and second site suppressors of vpl and PRC1-linked mis-sorting mutations.

TABLE 5
SUMMARY OF VPL PHENOTYPES

VPL	Intracellular CPY Activity[a]	Immunoreactive CPY Secreted[a]	Immunoreactive Proteinase A secreted[b]
WT	+ + +	−	−
1	+	+ + +	+
2	−	+ +	−
3	−	+ +	+
4	+	+ +	+
5	+ +	+	−
6	−	+ + +	+
7	−	+ + +	+
8	+	+ + +	+

[a]Degree of phenotype: (−) is <5%, (+) is 10-40%, (++) is 50-70%, (+++) is 80-100%.
[b]Proteinase A is scored (+) or (−) only.

Discussion

We are studying two aspects of the vacuolar protein localization process using both genetic and biochemical approaches. To understand the earliest molecular events that occur during localization of CPY, we have altered the signal sequence of this protein, and are currently probing the interactions of these altered signal sequences with the protein translocation apparatus of yeast cells. The second major localization process we are studying is the mechanism of sorting of vacuolar proteins from other polypeptides traversing the secretory pathway. We believe that both processes require the presence of localization determinants encoded within the CPY structural gene, as well as components (SRP and a CPY sorting receptor) that recognize these determinants and effect ER translocation, packaging and delivery of the protein to its final destination. It is our ultimate goal to understand these complex cellular processes by defining the detailed series of events that occur during localization, and by identifying and isolating the molecular components involved in these processes.

Regions of the N-terminus that are critical for translocation across the ER membrane and subsequent glycosylation of the protein have been identified by in vitro mutagenesis of the putative signal sequence of CPY. It is of great interest to us to identify the yeast analogue of the signal recognition particle (SRP), which has been isolated and studied in mammalian systems (reviewed in 30). Our approach to this problem has been to select second-site suppressor mutations that allow efficient translocation of the CPY polypeptides carrying signal sequence deletions. It is reasonable that such suppressor mutations might reside within the genes encoding SRP analogue subunits or within genes encoding functions that interact with these subunits. We have recently obtained suppressors of CPY signal sequence deletions and are currently determining the relationship of the mutant genes to the protein translocation process. If we are indeed able to identify genes encoding SRP subunits, it will allow us to manipulate genetically the SRP particle. Such an approach should complement the biochemical studies that have been performed on SRP (30).

Because carbohydrate is apparently not important as a localization determinant on CPY, we have undertaken a mutational study of the vacuolar localization signal(s) on the protein. Our present data indicate that a determinant located within amino acids 25-31 of the CPY propeptide is required for efficient delivery of CPY to the vacuole. The results of an XbaI to ApaI in-frame deletion of the propeptide suggest further that the region including amino acids 31 to 65 is also important for vacuolar localization, or alternatively that this region is required for proteolytic processing of the proenzyme. Since no Bal31-generated deletions at ApaI in the propeptide-encoding region have yielded PRC1-linked missorting mutations, this region of the propeptide may not be included in the vacuolar localization domain(s). However, to more precisely define the size and distribution of sites required for proper sorting of CPY, we are obtaining random point mutations in PRC1 that lead to CPY secretion.

To avoid localization artifacts resulting from mutations causing a non-native conformation of CPY, we are studying only those mutants that exhibit some CPY activity in the vacuole or at the cell surface. In this way, we hope to identify sequences that are specifically associated with the recognition and sorting process.

Some of the VPL genes are likely to be intimately involved in the vacuolar protein sorting process. However, since a large number of mutant vpl complementation groups can lead to mis-sorting of CPY, it is possible that many of the vpl mutations prevent correct vacuolar localization in a non-specific manner. For example, mutations that affect the integrity of structural components of the Golgi apparatus might disrupt proper sorting. To resolve this issue, we are presently isolating unlinked suppressors of PRC1-linked sorting mutations. Such unlinked suppressors may provide a means of directly identifying the vpl genes coding for components that specifically recognize the CPY sorting determinant(s). If we can identify this presumptive sorting receptor, it may be possible to ascertain which vpl genes interact with the receptor, thereby allowing us to define the presently unknown reactions involved in vacuolar protein sorting.

The various classes of VPL complementation groups indicate that we may have isolated lesions in different steps of the vacuolar sorting pathway. For example, mutations in a sorting receptor might lead to mis-sorting at a late ER or Golgi step, and completely block packaging of proteins into organelles bound for the vacuole. Lesions in genes important for later steps in the sorting pathway might cause an accumulation of pre-vacuolar transitory vesicles. In these latter mutants one might expect CPY to reach a vacuole-like compartment in which some PEP4-dependent processing can occur. Our biochemical and electron microscopic evidence is consistent with different vpls falling into both classes of events. By constructing double mutants of various vpl complementation groups, and examining epistatic interactions, we hope to describe a sequential pathway of vacuolar protein sorting.

Different classes of vacuolar proteins might be sorted and delivered via different pathways. The fact that proteinase A, CPY and alkaline phosphatase are all mislocalized in certain vpl mutants suggests that the three proteins follow a common pathway of localization. Further work with both pleiotropic and non-pleiotropic vpl mutants will help to substantiate whether vacuolar localization is a single linear or branched pathway.

Isolating the gene products that carry out vacuolar protein sorting reactions will greatly enhance our understanding of these processes. Our VPL1-β-galactosidase fusion protein will be useful in generating antibody to the VPL1 gene product which will allow us to localize and isolate the VPL1 polypeptide. The negative selection described in the "Results" section will allow us to clone other VPL genes as well. In this way we hope ultimately to identify and understand the function of products encoded by the various VPL genes and to greatly increase our understanding of the complex mechanisms of protein localization in eukaryotic cells.

ACKNOWLEDGEMENTS

We are grateful to Elizabeth Jones for suggesting the use of CBZ•pheleu in our selection procedure, and Scott Emr for helpful discussions and for sharing preliminary observations on SUC2-PRC1 gene fusions. In addition, we greatly appreciate the expert and efficient assistance of Elizabeth Cooksey in the preparation of this manuscript.

REFERENCES

1. Novick P, Schekman R (1979). Secretion and cell surface growth are blocked in a temperature-sensitive mutant of Saccharomyces cerevisiae. Proc Nat Acad Sci 76:1858.
2. Novick P, Field C, Schekman R (1980). Identification of 23 complementation groups required for post-translational events in the yeast secretory pathway. Cell 21:205.
3. Stevens T, Esmon B, Schekman R (1982). Early stages in the yeast secretory pathway are required for transport of carboxypeptidase Y to the vacuole. Cell 30:439.
4. Jones EW (1983). Genetic approaches to the study of protease function and proteolysis in Saccharomyces cerevisiae. In Spencer JFT, Spencer DM, Smith ARW (eds): "Yeast Genetics: Fundamental and Applied Aspects," New York: Springer-Verlag, p 167.

5. Erickson AH, Blobel G (1979). Early events in the biosynthesis of the lysosomal enzyme, cathepsin D. J Biol Chem 254:11771.
6. Rome LH, Garbin AJ, Allietta MM, Neufeld EF (1979). Two species of lysosomal organelles in cultured human fibroblasts. Cell 17:143.
7. Hasilik A, Neufeld E (1980). Biosynthesis of lysosomal enzymes in fibroblasts: phosphorylation of mannose residues. J Biol Chem 255:4946.
8. Ballou, CE (1976). Structure and biosynthesis of the mannan component of the yeast cell envelope. Adv Microbiol Physiol 14:93.
9. Hashimoto C, Cohen RE, Zhang W, Ballou CE (1981). Carbohydrate chains on yeast carboxypeptidase Y are phosphorylated. Proc Nat Acad Sci USA 78:2244.
10. Schwaiger H, Haslik A, von Figura K, Wiemken A, Tanner W (1982). Carbohydrate-free carboxypeptidase Y is transferred into the lysosome-like yeast vacuole. Bioch Biophys Res Commun 104:950.
11. Ferro-Novick S, Hansen W, Schauer I, Schekman R (1984). Genes required for completion of import of proteins into the endoplasmic reticulum in yeast. J Cell Biol 98:44.
12. Stevens T, Payne G, Rothman JH, Schekman R (1985). Carboxypeptidase Y is secreted when a component involved in transport to the vacuole is overloaded. Submitted to J Cell Biol.
13. Rothman JH, Hunter CP, Valls LA, Stevens TH. Cloning of the gene for the vacuolar enzyme proteinaseA using a novel immunochemical technique. In preparation.
14. Hemmings BA, Zubenko GS, Hasilik A, Jones EW (1981). Mutant defective in processing of an enzyme located in the lysosome-like vacuole of Saccharomyces cerevisiae. Proc Nat Acad Sci USA 78:435.
15. Jones EW (1977). Proteinase mutants of Saccharomyces cerevisiae. Genetics 85:25.
16. Zubenko GS, Park FJ, Jones EW (1983). Mutations in PEP4 locus of Saccharomyces cerevisiae block final step in maturation of two vacuolar hydrolases. Proc Nat Acad Sci USA 80:510.
17. Mechler B, Muller M, Muller H, Wolf DH (1982). In vivo biosynthesis of vacuolar proteinases in proteinase mutants of Saccharomyces cerevisiae. Bioch Biophys Res Commun 107:770.

18. Jones EW (1984). The synthesis and function of proteases in Saccharomyces: Genetic approaches. In Roman HL, Campbell A, Sandler LM (eds): "Annual Review of Genetics" Vol. 18, Palo Alto: Annual Reviews Inc., p 233.
19. Martin B, Svendsen I, Viswanatha T, Johansen JT (1982). Amino acid sequence of carboxypeptidase Y. I. Peptides from cleavage with cyanogenbromide. Carlsberg Res Commun 47:1.
20. Svendsen I, Martin BM, Viswanatha T, Johansen JT (1982). Amino acid sequence of carboxypeptidase Y. II. Peptides form enzymatic cleavages. Carlsberg Res Comun 47:15.
21. Valls LA, Stevens TH. DNA sequence of the PRC1 gene reveals a novel precursor structure for proCPY. In preparation.
22. Distel B, Al R, Tabak H, Jones E (1983). Synthesis and maturation of the yeast vacuolar enzymes carbopeptidase Y and aminopeptidase I. Biochim Biophys Acta 741:128.
23. von Heijne G (1983). Patterns of amino acids near signal-sequence cleavage sites. Eur J Biochem 133:17.
24. Blachly E, Stevens TH. CPY signal sequence mutations block translocation of proCPY across the endoplasmic reticulum membrane. In preparation.
25. Emr SD, Silhavy TJ (1982). Molecular components of the signal sequence that function in the initiation of protein export. J Cell Biol 95:689.
26. Valls LA, Hunter CP, Rothman JH, Stevens TH. Carboxypeptidase Y recognition signal necessary for vacuolar localization maps to a N-terminal protein domain. In preparation.
27. Rothman JH, Stevens TH. Identification of eight complementation groups (VPL) required for vacuole protein localization. In preparation.
28. Orr-Weaver TL, Szostak JW, Rothstein RJ (1981). Yeast transformation: A model system for the study of recombination. Proc Nat Acad Sci USA 78:6354.
29. Rothstein RJ (1983). In Wu R, Grossman C, Moldane K (eds) "Methods in Enzymology Vol 101, Recombinant DNA part C". New York: Academy Press p202.
30. Walter P, Gilmore R, Blobel G (1984). Protein translocation across the endoplasmic reticulum. Cell 38:5.

MATURATION AND SECRETION OF THE M_1-dsRNA ENCODED KILLER TOXIN IN S. CEREVISIAE

S. L. STURLEY, S. D. HANES, V. BURN, AND K. A. BOSTIAN

Division of Biology and Medicine
Section of Biochemistry
Brown University
Providence, Rhode Island 02912

ABSTRACT We present here our recent studies on the structural organization, maturation and secretion of the M_1-dsRNA encoded yeast preprotoxin. The transformation of sensitive yeast by a PHO5 (promoter and leader) preprotoxin cDNA gene fusion yielded phosphate repressible, immune, killer strains, indicating that toxin and immunity are encoded within the same open reading frame. The expression of these constructs indicate that during maturation of wild type preprotoxin, the natural leader peptide is not co-translationally removed in the endoplasmic reticulum. The factors controlling such proteolytic cleavage events are also considered.

INTRODUCTION

The type 1 killer phenotype of S. cerevisiae is characterized by the secretion of a low molecular weight polypeptide toxin that kills sensitive cells of yeast by disruption of cytoplasmic membrane functions (1). In general, strains that secrete this toxin are also resistant to its effects. Toxin production and immunity are both determined by a 1.9 kilobase pair (kb) M_1-dsRNA plasmid which is encapsidated in virus-like particles (VLPs), ScV-M_1 (2). Encapsidation and maintenance of the M_1-dsRNA are both dependent on functions provided by another dsRNA species of molecular weight 4.5 kb (L_{1A}-dsRNA) which is also separately encapsidated in ScV-L_{1A} VLPs (3).

The secreted toxin has been shown to be composed of two dissimilar 9.5 and 9.0 kilodalton (kd) disulfide linked subunits, designated α and β respectively (4). The intracellular precursor to this dimer is a 43 kd glycosylated protoxin. Since the carbohydrate moiety of this molecule can be removed by endoglycosidase-H cleavage, it is linked to asparagine in the protoxin via di-N-acetylchitobiose (5). Both protoxin and toxin are derived from a preprotoxin, M_1-P1, the 35 kd in vitro translation product of denatured M_1-dsRNA (6). The biosynthesis of these intermediates and the ultimate secretion of toxin follows the normal yeast secretory pathway (see Table 1), involving the endoplasmic reticulum (ER), the Golgi body, and secretory vesicles as defined by the sec18, sec7, and sec1 mutations respectively (7,8). Other host determined functions are also required for secretion of protoxin and probably represent proteolytic cleavage events. These events could be performed by the KEX1, KEX2, and REX1 gene products (9,10) and may also involve enzymes sensitive to the chymotrypsin inhibitor tosyl-L-phenylalanylchloromethyl ketone (TPCK) (8).

Comparison of the N-terminal sequences of the α and β subunits with the nucleotide sequence of a cDNA clone derived from an in vivo M_1-dsRNA transcript has enabled us to identify the coding domains for the toxin subunits (4). This 316 amino acid open reading frame initiates 14 bp from the 5' terminus of the plus strand of M_1-dsRNA and encodes a 34.8 kd product. The predicted N terminal sequence is identical to that of M_1-P1. This preprotoxin gene comprises 4 protein domains. The α toxin component (86 amino acids) is preceded by the 44 amino acid δ region which possesses a typical hydrophobic leader sequence (residues 12-27). Following α is a 103 amino acid toxin component, γ, which contains all 3 asparagine glycosylation sites presumed to be in protoxin. The 83 amino acid β toxin component comprises the C terminal portion of the preprotoxin gene (4).

Recent work in this laboratory has involved the subcloning of a cDNA copy of the preprotoxin gene into the yeast shuttle vector p1A1, placing it under the control of the repressible yeast acid phosphatase PHO5 promoter (11). Combined with a reappraisal of the effect of the kex1 and rex1 mutations, and of TPCK, on preprotoxin processing, we have begun to ascribe roles for the component domains of preprotoxin in efficient toxin secretion and immunity.

EXPERIMENTAL PROCEDURES

Yeast Strains and Media

Non-killer yeast strains S6 (a/α) and GG100-14D (α his3 ura3 trp1 pho5 pho3) lack M_1-dsRNA. K12-1 (α ade2 arg) is a standard type K1 killer as is the diploid strain T158C/S14a and both have been used in previous studies (4,8). Killer and resistance expression mutants SS005 (a ade2 thr1 kex1) and SS001 (a ade2 rex1) were derived from the kex1 and rex1 strains described by Wickner and co-workers (9,10), and possess the M_1-dsRNA genome of K12-1. Low and high phosphate media and the methionine free medium used in labeling experiments have been described previously (12,8).

Construction and Expression of Recombinant pSH-GB Plasmids

The PHO5-killer cDNA fusions were performed using standard recombinant DNA techniques. Yeast transformations and killer and immunity expression assays were performed by published procedures (12,13).

Isolation and In Vitro Translation of Total Cellular RNAs

Total cellular RNAs were isolated from cells grown in low and high phosphate media, as described previously (5), and then translated in a wheat germ system using L-[^{35}S]-methionine. Translation products were analyzed by immunoprecipitation and SDS-polyacrylamide gel electrophoresis using published procedures (4,6).

Labeling and Immunoprecipitation of In Vivo Protoxin

Cells, grown on the appropriate media to a cell density of 2.5×10^7 were labeled with L-[^{35}S]-methionine for 12 min., followed by a chase of variable duration. Cell extracts, obtained in the presence of Triton X-100 as described previously (5), were analyzed by immunoprecipitation and gel electrophoresis as above. TPCK inhibition experiments were performed as described by Bussey et al. (8), using a TPCK concentration of 1.0 mM, with cold methionine chase periods of 2.5, 10, 20, or 40 minutes.

RESULTS AND DISCUSSION

PHO5 Regulation of a cDNA Copy of the Preprotoxin Gene in Yeast

The PHO5-M_1-cDNA fusion was constructed by ligating the 5' portion of the acid phosphatase PHO5 gene (14), containing the promoter, translation start site and 17 amino acid APase leader peptide sequence, to the unique PstI site within the δ region of the preprotoxin gene. Synthetic BamHI linkers were used to maintain a proper translational reading frame. The hybrid PHO5-preprotoxin gene contained the APase ala-gly leader peptidase cleavage site at residues 17-18 and from residue 20 onwards the normal preprotoxin sequence from residue 34 (Fig. 1A). The pSH-GB shuttle plasmids so constructed were used to transform non-killer yeast strain GG100-14D. The killer and immunity phenotypes were both found to be under PHO5 control. Thus, under derepressed conditions (low-Pi media), transformed cells were immune to the action of the toxin and able to kill sensitive strains. On high-Pi medium neither phenotype was expressed. The in vitro translation product of wild type M_1-dsRNA after immunoprecipitation with antibodies comigrated with M_1-P1 (Fig. 2A, lane d). The cross-reactive fusion gene translation product derived from transformed cell RNAs grown under derepressed conditions (M_1-P1$_{hy}$) migrated as a 33 kd species (Fig. 2A, lane f). This is consistent with the 14 amino acid reduction in size of the N terminal sequence. M_1-P1$_{hy}$ was not detected in translation products of RNA extracted from transformants grown in high-Pi medium.

Northern analysis of the same RNAs using a preprotoxin cDNA probe confirmed the regulated expression of these hybrid RNA transcripts. Transcripts of 1.9 kb were detected in derepressed cells and since they were at a concentration proportional to the in vitro translational activity of M_1-P1 mRNA this suggests that these fusion plasmid transcripts are efficiently processed and transported out of the nucleus.

Intracellular Protoxin and Secreted Toxin in the Fusion Gene Transformants

Immunoprecipitation of L-[^{35}S]-methionine pulse-labeled transformed cell extracts with anti-toxin indicated that immunoreactive species were only found if the transformant

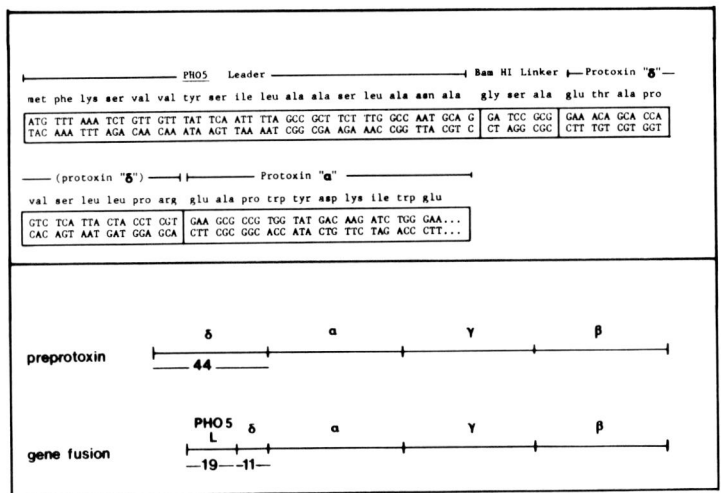

Figure 1. The <u>PHO5</u> preprotoxin cDNA gene fusion.
  A. DNA sequence of the fusion region of the pSH-GB series of expression plasmids: The gene fusion joins the 17 amino acid APase leader peptide sequence (under the control of the <u>PHO5</u> promoter) to the Pst1 site of the preprotoxin cDNA. The BamH1 linker aligns the fusion in frame and recreates the ala-gly leader peptide cleavage site. The third residue encoded by the linker (ala) is equivalent to residue 34 of the wild type preprotoxin molecule.
  B. Functional domains of wild-type preprotoxin (above) and the hybrid gene product (below): Numbers refer to the amino acids preceding the a toxin component. The <u>predicted</u> wild type preprotoxin leader peptide cleavage site lies between residues 26 and 27 within δ. "PHO5 L" includes the 17 amino acid APase leader plus 2 residues encoded by the BamH1 linker.

had been grown in low phosphate medium (Fig. 2B, lane h). These proteins, presumably the hybrid gene preprotoxin (prehy) and the glycosylated hybrid protoxin (prohy) had molecular weights of 33 and 40 kd respectively. The specificity of immunoprecipitation was confirmed by competition experiments using wild type toxin (data

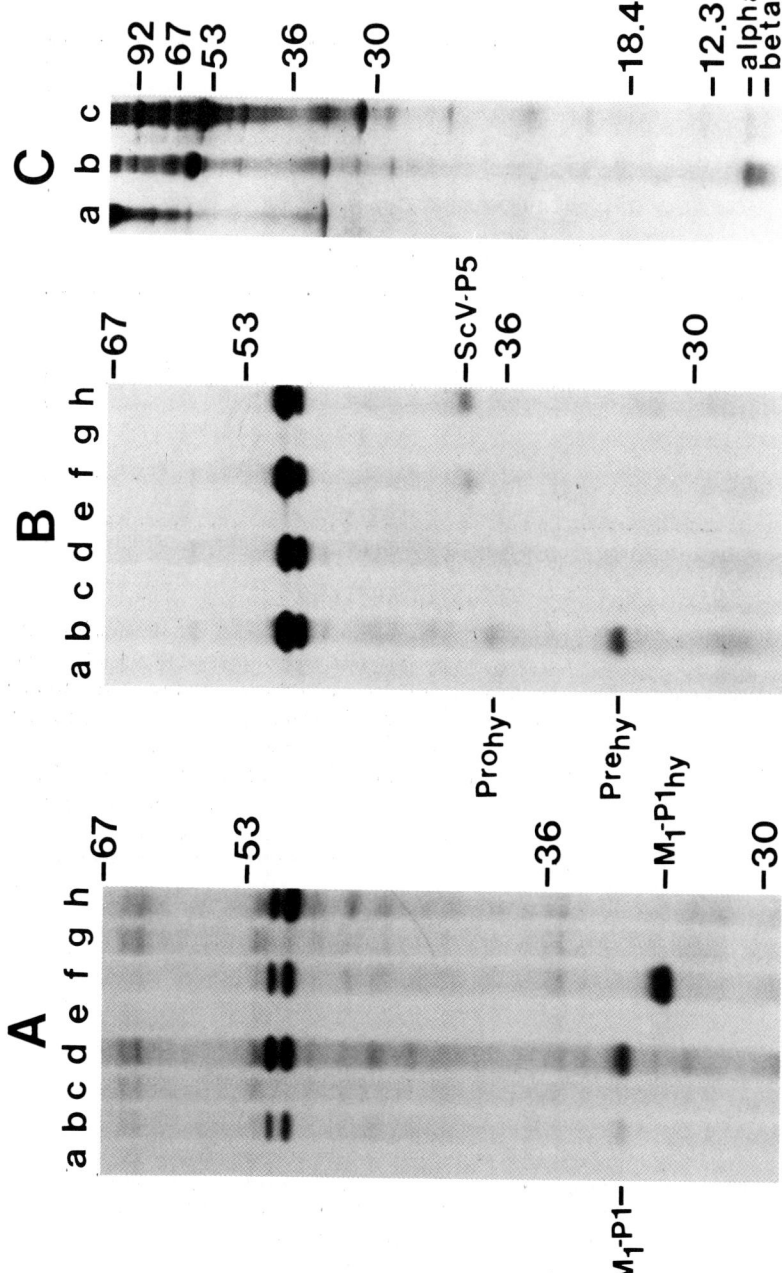

Figure 2. In Vitro and In Vivo products of the PHO5-preprotoxin cDNA gene fusion.
   A. In Vitro translation of total RNA derived from pSH-GB transformants: Cells of wild type killer K12-1 (lanes a-d) and a pSH-GB transformant (lanes e-h) were grown in low-Pi (lanes a, b, e, f) or high-Pi (lanes c, d, g, h) media. Total RNA was extracted, translated and the products analyzed by immunoprecipitation as described in the text. Lanes a, c, e and g received pre-immune serum in place of killer toxin IgG. The positions of the 34.8 kd preprotoxin (M_1-P1) and the 33 kd fusion gene product (M_1-P1$_{hy}$) are indicated.
   B. In Vivo labeling of the fusion gene product: Extracts of pulse labelled cells of K12-1 (lanes e-h) and a pSH-GB transformant (lanes a-d), grown on low-Pi (lanes a, b, e, f) or high-Pi (lanes c, d, g, h) media were analysed by immunoprecipitation with anti-toxin IgG (lanes b, d, f, h) and pre-immune serum (lanes a, c, e, g). ScV-P5 is the 43 kd glycosylated protoxin produced in wild type killer cells and Pro$_{hy}$ (lane b) is the equivalent glycosylated precursor formed in transformants. Pre$_{hy}$ (lane b), a 33 kd protein co-migrates with in vitro synthesised M_1-P1$_{hy}$.
   C. Secretion of normal toxin by pSH-GB transformants: α and β toxin components were detected amongst the total secreted proteins of a pSH-GB transformant. Culture supernatants of K12-1 (lane c) and a transformant (lane b) grown on low-Pi medium were concentrated by ultrafiltration. No such proteins were obtained when the transformant was grown on high-Pi medium (lane a).

not shown). The secreted proteins from the control killer strain K12-1 and the pSH-GB yeast transformants were also compared by SDS polyacrylamide gel electrophoresis. The α and β subunits of the toxin appear to be identical in these two strains (Fig. 2C).

Cleavage of the Preprotoxin Leader Peptide

   Previous experiments involving an in vitro system (dog pancreas microsomal membranes; 5, 6) suggested that cleavage of the preprotoxin leader sequence occurs during translation. The present experiments however indicate this may not be the situation in vivo. If we assume that both the fusion gene product and wild type preprotoxin leaders are cleaved at the predicted sites (the carboxyl side of amino acids 17 and 26 respectively; 15,4) the resulting proteins should consist of a common region (the α, γ and β domains) preceded by 13 and 18 amino acids respectively. This 5 amino acid difference would yield proteins differing by approximately 0.5 kd. In the unlikely event that neither leader peptide is cleaved, then a size difference of about 1.5 kd (13 amino acids, see Fig.1B) would occur. In fact, protoxin and the processed hybrid gene product (Prohy) differ by 3 kd (Fig.2B, lanes h,b) which is a close approximation to the difference between cleaved prehy and uncleaved preprotoxin (31 amino acids). Alternatively, the lack of a complete δ sequence in the gene fusion could lead to a lesser degree of glycosylation of preprotoxin$_{hy}$ and therefore a smaller protoxin$_{hy}$. Endoglycosidase H deglycosylation of natural protoxin and hybrid protoxin indicated that this was not the case since a 3 kd difference persisted between these proteins.

The predominance of a 33 kd hybrid preprotoxin in derepressed pulse-labeled transformed cells seems to indicate somewhat incomplete processing of this precursor. In wild-type cells the preprotoxin band is not visible when labeled under the same conditions. Furthermore, the derepressed PHO5 promoter expressed hybrid preprotoxin mRNA at much higher levels than the native ScV-M_1 plasmid and yet the levels of secreted toxin are similar in wild-type killer and transformants. This could indicate the inability of a PHO5 leader sequence to efficiently direct secretion of certain polypeptides, a situation also found when this construct has been used to secrete certain heterologous proteins (D. Rogers, personal communication). Alternatively an essential membrane spanning function of the preprotoxin δ sequence could be envisaged that is lacking in the gene fusion tested here. The retention of the natural leader sequence hypothesized earlier could therefore be a prerequisite for this process. Experiments in progress to address this issue include the synthesis of the natural killer leader sequence and the coupling of this "complete" killer system to the PHO5 promoter.

The Secretion Pathway

The processing of protoxin in wild-type killer cells is governed by the secretory SEC genes described by Novick and Schekman (7,8; Table 1). Temperature sensitive mutations at these loci result in accumulation of endoplasmic reticulum (sec18), the Golgi body (sec7), or secretory vesicles (sec1) and do not secrete toxin. These strains all accumulate the 43 kd protoxin at the restrictive temperature (37 C) indicating that core glycosylation occurs only in the ER, without the Golgi body associated elongation of the carbohydrate chains common to some secreted yeast proteins. In a sec18 strain, protoxin is stable but in sec1 mutants, it can be chased out of labeled cells with a half-life of 15 minutes. The stability of protoxin in a sec7 strain was intermediate. It seems likely therefore, that the majority of protoxin processing occurs in the secretory vesicles or in the later compartments of the Golgi body. This was substantiated by the treatment of yeast strains with TPCK, followed by the extraction of pulse-labeled proteins. This protease inhibitor blocked toxin secretion and caused the stable accumulation of protoxin (after a 40 minute chase), in normal, sec7, and sec1 mutants at any temperature (7).

We have made a closer examination of the effects of TPCK inhibition on protoxin, primarily by varying the period of chase and it appears that the situation may be more involved. After a brief chase of 2.5 minutes, a ladder of 4 toxin immunoprecipitable proteins are present in TPCK treated labeled cell extracts (Fig. 3A). These have approximate molecular weights of 42, 40, 38 and 36 kd. Since the two extremes closely resemble the sizes of glycosylated protoxin and non-glycosylated preprotoxin, it is likely that the other bands represent glycosylation intermediates, possessing 1 and 2 core units respectively. This was confirmed by endoglycosidase H treatment of the immunoprecipitates since all the intermediates were converted to a single species of a molecular weight consistent with that of such deglycosylation of protoxin. This may be a phenocopy of the sec59 mutation which at 37 C accumulates partial glycosylation intermediates of α-factor (16). This mutant as yet, has not been investigated with respect to preprotoxin maturation. The protoxin precursors in TPCK treated cells are unstable since after a 40 minute chase, the protoxin predominates. This could represent the situation described previously (8). It appears therefore that two TPCK inhibitable processes are involved in preprotoxin maturation, a post sec18 proteolytic cleavage event as well as a step prior to the sec18 regulated exit from the ER. At the concentrations of TPCK used here, the latter seems to be only a partial inhibition and may involve the translocation of preprotoxin through this organelle, as is the case for the sec59 mutation. If this is so, the

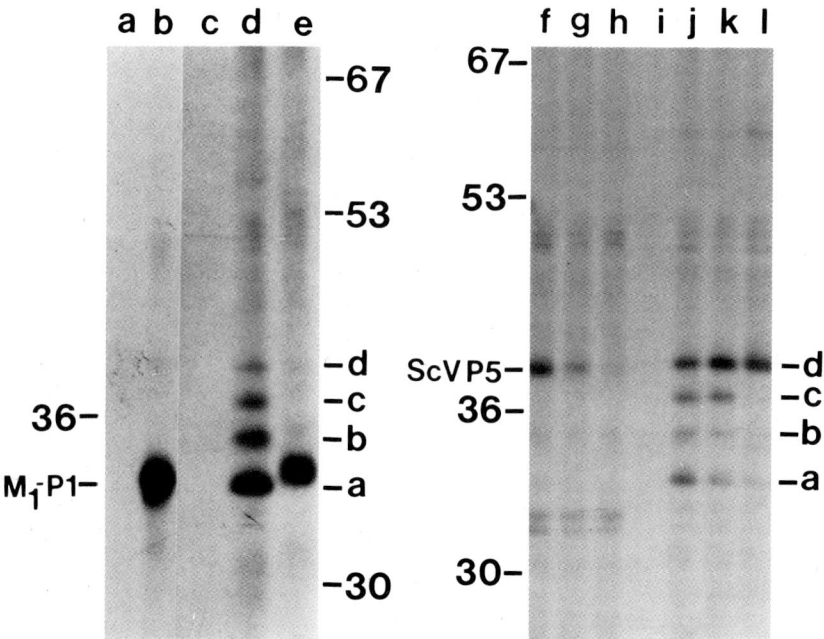

Figure 3. The effect of TPCK on wild-type preprotoxin maturation.

Pulse-labeled cells of strain T158C/S14a (described previously, 8) were treated with TPCK (lanes c-e, i-l) as decribed in the text. Proteins were extracted after cold methionine chases of 2.5 (lanes c-e, f, i, j), 10 (lanes g, k) and 40 (lanes h, l) minutes and analyzed by immunoprecipitation with anti-toxin IgG (lanes b, d-h, j-l) or preimmune serum (lanes a, c, i). Species a, from TPCK treated cells co-migrates with the primary in vitro M_1-dsRNA translation product, M_1-P1 (lane b) and species d has the same mobility as wild-type protoxin, ScV-P5. These proteins and intermediates differ only in the relative degree of glycosylation as shown by endoglycosidase H treatment (lane e).

presence of these intermediates that differ only in the degree of glycosylation from the primary M_1-dsRNA product, M_1-P1, is further evidence for the lack of preprotoxin leader cleavage in the ER. As might be anticipated, TPCK partial inhibition of glycosylation is not restricted to preprotoxin since glycosylation of acid phosphatase was also

TABLE 1
FACTORS AFFECTING PREPROTOXIN MATURATION

Factor	Phenotype	Intracellular Preprotoxin Components	Ref
sec18 (ER exit)	—	43 kd protoxin accumulation, stable. ($t_{1/2}$ = 80 minutes)	7,8
sec7 (Golgi)	—	43 kd protoxin accumulation, moderately stable. ($t_{1/2}$ = 40 minutes)	7,8
sec1 (secretory vesicles)	—	43 kd protoxin accumulation, unstable. ($t_{1/2}$ = 15 minutes)	7,8
sec59 (ER transit)	—	Not investigated. Expect accumulation of preprotoxin (35 kd) and glycosylation intermediates.	16
kex1	R^+K^-	43 kd protoxin accumulation, stable.	8,9
kex2	R^+K^-	43 kd protoxin accumulation, stable.	8,9
rex1	R^-K^+	Not investigated. Might expect secretion of larger toxin.	10
TPCK	$R^?K^-$	Protoxin accumulates after 40 minute chase Glycosylation intermediates apparent after 2.5 minute chase. sec7, sec1 phenotypes stabilized.	8

retarded (data not shown).

Yeast strains harboring M_1-dsRNA as well as killer and resistance expression chromosomal mutations, are currently being investigated using the pulse-chase techniques employed for the TPCK experiments. Preliminary results indicate that rex1 strains (SS001) possess normal levels of intracellular protoxin and generally secrete larger proteins. If abnormally large toxin immunoprecipitable proteins are secreted by such strains, it would be consistent with the theory that this mutation might produce a lesion in an endopeptidase that cleaves at the border between α and γ. Thus instead of being directed to its putative site of action, i.e. the cell membrane, the immunity determinant might be secreted with toxin and so the cells could become sensitive, killers or "suicidal" strains. Similar experiments with a kex1 strain (SS005) are underway.

CONCLUSIONS

The finding that some mutations in M_1-dsRNA lead to loss of either immunity or killer properties (1,10) has led to the hypothesis that the glycosylated portion of protoxin confers immunity (4). Our findings verify that production of, and resistance to killer toxin is encoded by the same transcriptional unit, within one open reading frame, in this case, under the control of a single promoter, PHO5. Since the majority of δ is absent in our construct this renders γ, or alternatively precursor, α or β, domains the most likely causative agent of immunity. To test this, experiments are in progress to make direct fusions of the various coding regions, including γ, to the PHO5 leader and promoter. Protein fusions are also being made to β-galactosidase in order to ultimately produce antibody to the immunity determinant.

The functional organization of preprotoxin and its subsequent processing and secretion bears some resemblance to that of the α-mating factor precursor. Both have a typical N-terminal leader that appears to persist during transit through the ER; a large central region that contains three asparagine residues that are glycosylated only in the ER; extensive proteolytic processing occurring late in the secretory pathway; and a secreted, non-glycosylated active fragment derived from the C-terminal region of the precursor. The dependence of α-factor maturation on the KEX2 lysine-arginine cleaving endopeptidase also indicates the similarities between these proteins. Disparities do occur however; α-factor is still secreted in kex2 strains but in an inactive, immature, highly glycosylated form whereas killer toxin secretion is blocked by this mutation. Furthermore, the kex1 and rex1 mutations have not been implicated in α-factor processing. Their roles in toxin production and immunity are currently being investigated.

ACKNOWLEDGMENTS

The work was supported in part by an NIH Biomedical Research Support Grant, RR05644-15 (K.A.B.), a grant from Biotechnology General (K.A.B), and an NIH grant GM20755 (D.J.T). We thank Quentin Elliott for technical assistance and Donald J. Tipper for his invaluable support and criticism.

REFERENCES

1. Bussey H (1981). Physiology of killer factor in yeast. In Rose AH and Morris G (eds): "Advances in Microbial Physiology," New York: Academic Press, Vol 22, p 93.

2. Wickner RB (1979). The killer double stranded RNA plasmids of yeast. Plasmid 2:303.

3. Bostian KA, Sturgeon JA, and Tipper DJ (1980). Encapsidation of yeast: killer double stranded ribonucleic acids: dependence of M on L. J Bacteriol 143: 463.

4. Bostian KA, Elliot Q, Bussey H, Burn V, Smith A, and Tipper DJ (1984). Sequence of the preprotoxin ds-RNA gene of type 1 killer yeast: multiple processing events produce a two component toxin. Cell 36:741.

5. Bostian KA, Jayachandran S, and Tipper DJ (1983). A glycosylated protoxin in killer yeast: models for its structure and maturation. Cell 32:169.

6. Bostian KA, Hopper JE, Rogers DT, and Tipper DJ (1980). Translational analysis of the killer associated virus like particle ds-RNA genome of S.cerevisiae: M ds-RNA encodes toxin. Cell 19:403.

7. Novick P, and Schekman R (1979). Secretion and cell surface growth are blocked in a temperature sensitive mutant of Saccharomyces cerevisiae. Proc Natl Acad Sci USA 76:1858.

8. Bussey H, Saville D, Greene D, Tipper DJ, and Bostian KA (1983). Secretion of Saccharomyces cerevisiae killer toxin: processing of the glycosylated precursor. Mol Cell Biol 3:1362.

9. Wickner RB, and Leibowitz MJ (1976). Two chromosomal genes required for killing expression in killer strains of Saccharomyces cerevisiae. Genetics 82:429.

10. Wickner RB (1974). Chromosomal and non-chomosomal mutations affecting the "killer character" of Saccharomyces cerevisiae. Genetics 76:423.

11. Rogers DT, Lemire JM, and Bostian KA (1982). Acid phosphatase polypeptides in Saccharomyces cerevisiae are encoded by a differentially regulated multigene family. Proc Natl Acad Sci USA 79:2157.

12. Hinnen A, Hicks JB, and Fink GR (1978). Transformation of yeast. Proc Natl Acad Sci USA 75:1929.

13. Woods DR, and Bevan EA (1968). Studies on the nature of the killer factor produced by Saccharomyces cerevisiae. J Gen Microbiol 51:115.

14. Thill GP, Kramer RA, Turner KJ, and Bostian KA (1983). Comparative analysis of the 5' end regions of two acid phosphatase genes in Saccharomyces cerevisiae. Mol Cell Biol 3:570.

15. Perlman D, and Halvorson HO (1983). A putative signal peptidase recognition site and sequence in eukaryotic and prokaryotic signal peptides. J Mol Biol 167:391.

16. Julius D, Schekman R, and Thorner J (1984). Glycosylation and processing of prepro α-factor through the yeast secretory pathway. Cell 36:309.

COORDINATE REGULATION OF PHOSPHOLIPID SYNTHESIS IN YEAST

Brenda Loewy, Jeanne Hirsch, Margaret Johnson and Susan Henry[1]

Departments of Genetics and Molecular Biology
Albert Einstein College of Medicine
Bronx, New York   10461

ABSTRACT: In Saccharomyces cerevisiae, a number of phospholipid biosynthetic enzymes are regulated coordinately (10). We have studied one of the coordinately regulated enzymes, inositol-1-phosphate synthase, which is encoded by the INO1 gene. The amino acid composition of inositol-1-phosphate synthase was determined and the cloned INO1 gene was used as a probe to study expression of INO1 transcripts. Two polyadenylated transcripts homologous to the cloned INO1 DNA were detected; one is approximately 1.8 kb and the other is less than 1 kb in length. The steady state level of the larger transcript was found to be substantially reduced in cells grown in the presence of inositol. The smaller transcript did not appear to be regulated. The larger transcript is believed to be the message encoding inositol-1-phosphate synthase; while it is not yet clear what role is played by the smaller transcript. Recessive mutations (ino2 and ino4) at two unlinked regulatory loci, pleiotropically prevent derepression of the INO1 gene product, as well as several other phospholipid biosynthetic enzymes (15). An extensive genetic analysis of inositol independent revertants of an ino2, ino4, double mutant strain, revealed the existance of a third regulatory locus, OPI5. Dominant mutations at the OPI5 locus free the INO1 locus from control by ino2 and ino4.

[1]. This work was supported by NIH grants GM19629, GM11301 and CA09060.

INTRODUCTION

Biosynthesis of phospholipids in Saccharomyces cerevisiae is subject to coordinate regulation involving a number of enzymatic activities (10). Among the coordinately regulated enzymes are several membrane associated activities. Also subject to the coordinate control, is the cytoplasmic enzyme, inositol-1-phosphate synthase. The regulated reactions are illustrated in Figure 1 and Table 1 presents the relative levels of the various enzymatic activities in wild type cells grown in the presence or absence of inositol and/or choline. Although each of the regulated enzymes responds in a distinct fashion to inositol and choline, an overall pattern of regulation is apparent. None of the enzymes, for example, is regulated in response to choline alone, and all of the enzymes show maximum repression in the presence of the combination of inositol and choline (Table 1). Also, each of the regulated enzymes responds in some fashion to the presence of inositol alone. For example, inositol-1-phosphate synthase is fully repressed if inositol alone is present, while several other activities are partially repressed. On the other hand, the phospholipid N-methyltransferases, which convert phosphatidylethanolamine to phosphatidylcholine (Figure 1), are actually stimulated by inositol. In contrast, the level of activity of phosphatidylinositol synthase (Table 1) is unaffected by the presence of inositol or choline. This enzyme also fails to respond to the regulatory mutations (to be discussed below) which affect the coordinately regulated activities. Therefore, phosphatidylinositol synthase is not believed to be involved in the coordinate regulation.

The enzymatic activities which respond coordinately to the presence of inositol and choline are also controlled by a single set of positive and negative regulatory genes. For example, the recessive ino2 and ino4 mutations were originally isolated as inositol auxotrophs and they are unable to derepress inositol-1-phosphate synthase, gene product of the INO1 locus (6). The ino2 and ino4 mutants are pleiotropic, in the sense that they also express the repressed levels of the phospholipid N-methyltransferases ([15]). These observations suggest that the INO2 and INO4 genes encode positive regulatory factors required for maximal expression of a number of the coordinately regulated enzymes of phospholipid biosynthesis. Another

TABLE 1

Regulation of Phospholipid Biosynthetic Enzymes
in Response to Soluble Precursors, Inositol
and Choline.

Enzyme[a] Activity	% activity wild type cells medium supplement[b]				Reference
	None	C	I	I + C	
I-1-PS	100%	100%	2%	2%	4,6
PMTFS	100%	100%	200%	10-20%	10,22,24
PSS	100%	100%	60%	25%	2,14
CDP-DGS	100%	100%	70%	40%	11
PSD	–	–	100%	25%	3
PIS	100%	100%	100%	100%	14

Legend: [a] Enzyme abbreviations: I-1-PS, inositol-1-phosphate synthase; PMTFS, phospholipid N-methyltransferases; PSS, phosphatidylserine synthase; CDP-DGS, cytidine diphosphate diacylglycerol synthase; PSD, phosphatidylserine decarboxylase; PIS, phosphatidylinositol synthase.

[b] Activity is normalized to the activity in wild type cells grown in the absence of supplements (ie. first column, none). Supplements; C, choline 1mM; I, inositol 50µM. c. Data for phosphatidylserine decarboxylase is available only for cells grown in presence of inositol plus or minus choline.

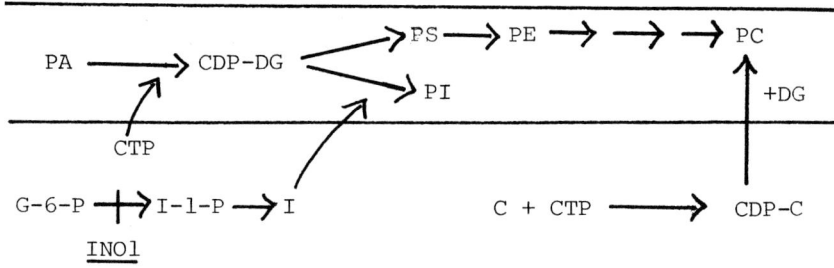

FIGURE 1. Biosynthesis of phospholipids in S. cerevisiae. Membrane associated reactions are shown between the double line. Abbreviations: PA, phosphatidic acid; CDP-DG, cytidine diphosphate-diacylglycerol; CTP, cytidine triphosphate; C, choline; DG, diacylglycerol; I, inositol; I-1-P, inositol-1-phosphate; G-6-P, glucose 6-phosphate; PI, phosphatidylinositol; PS phosphatidylserine; PE, phosphatidylethanolamine; PC, phosphatidylcholine. The position of the step catalysed by the INO1 gene product is indicated.

set of regulatory mutants (called opi for overproduction of inositol) was isolated on the basis of overproduction and constitutive expression of inositol-1-phosphate synthase (9). One of these recessive constitutive mutants (opi1) has recently been shown, also, to be constitutive for the phospholipid N-methyltransferases, phosphatidylserine synthase, the phospholipid N-methyltransferases (14), and CDP-DG synthase (11). Thus, the gene product of the OPI1 gene, like the gene products of INO2 and INO4 loci, appears to be involved in the coordinate regulation of phospholipid synthesis.

In order to further dissect the complex regulatory phenomena described above, we have analyzed the INO1 gene and its gene product in detail. Inositol-1-phosphate synthase is found ubiquitously in eukaryotic organisms as are the inositol containing phospholipids (23). The enzyme has been purified to homogeneity from yeast (6) and rat testis (17). The native enzyme in yeast has a molecular weight in excess of 240,000 daltons and is a tetramer composed of identical subunits of 62,000 daltons (6). However, nothing is known of its amino acid composition, sequence or structure. Furthermore, the precise mechanism of the complex reaction it catalyses remains to be elucidated (7). Inositol-1-phosphate synthase is in some regards atypical of the coordinately controlled enzymes of phospholipid synthesis. First it is a cytoplasmic enzyme and, second, it is repressed by inositol alone. The other coordinately regulated enzymes are membrane associated and require both choline and inositol for full repression. However, the INO1 gene was chosen for extensive analysis because its gene product, inositol-1-phosphate synthase, had been purified to homogeneity (6) and its structural gene had been isolated (13). Furthermore, techniques had been developed for the isolation of a wide variety of regulatory mutants which affect expression of its gene product (9 , 15) and all of these mutants have proven to be defective in the coordinate regulation (11, 15) discussed above. We report here further biochemical analysis of the INO1 gene product and the use of the cloned INO1 gene as a probe to study expression of its transcripts. In addition we report a genetic dissection of the coordinate regulation as it impinges upon the INO1 gene.

MATERIALS AND METHODS

Inositol-1-phosphate synthase was purified using the method of Donahue and Henry (6) and its amino acid composition was determined using the method of Steinman (20). Anti-inositol-1-phosphate synthase antibody was prepared and used in immunoprecipitation studies as previously described (6). The isolation of the INO1 gene from a library of yeast genomic DNA was described by Klig and Henry (13). For use in the present study fragments of the cloned INO1 DNA were subcloned into plasmid pUC9 by standard procedures using the restriction map shown in Figure 2 as a guide. RNA was isolated from yeast by the method of Elion and Warner (8). The poly(A)$^+$ fraction was prepared by oligo (dT)-cellulose chromatography (1) and fractionated on denaturing formaldehyde agarose gels (19). It was transferred to nitrocellulose (21) and hybridized with ^{32}P-labelled nick-translated probes as previously described (18). A probe for the ribosomal protein gene, TCM, was kindly provided by Dr. J. Warner. The phenotypes of the yeast ino2 and ino4 mutants have been previously described (^{15}). Standard methods of yeast genetic analysis were employed.

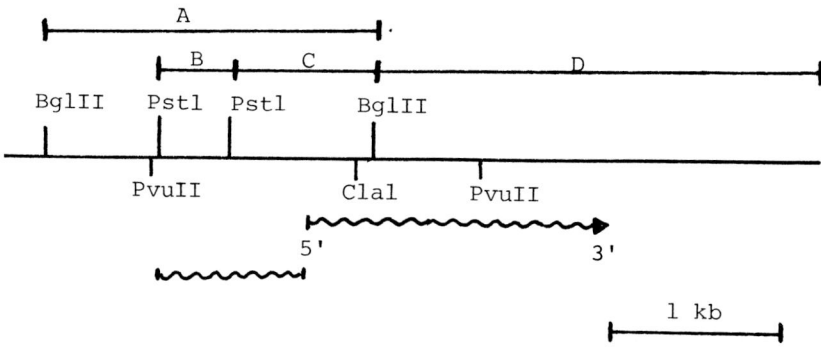

FIGURE 2. Partial Restriction Map of the INO1 gene. Fragments used in subclones described in the text are diagrammed above the map. Fragment A (BglII-BglII) Fragment B (Pst1-Pst1); Fragment C (Pst1-BglII). Fragment D (BglII-to end of cloned sequence, to a site in the vector). Wavy lines below the restriction map indicate the probable positions of the two transcripts.

RESULTS

The amino acid composition of purified inositol-1-phosphate synthase is reported in Table 2. Because the enzyme is regulated in common with membrane associated proteins and because it showed an unusual affinity for hydrophobic resins, a property useful in its purification (6) we have examined the proportion of hydrophobic residues in its composition. Using the formula of Dayhoff et al (5) we calculate that 27% of the amino acid composition is composed of hydrophobic residues.

We employed the cloned inositol-1-phosphate synthase structural gene, INO1, as a probe to study its transcripts. A restriction map of the cloned INO1 DNA is shown in Figure 2. Northern blot analysis was performed using a series of subclones (A,B,C and D - see Figure 2), derived from adjacent fragments in the cloned INO1 DNA. These experiments identified two polyadenylated transcripts (Figure 3). The larger transcript is approximately 1.8 kb in length, while the smaller one is less than 1 kb. Subclone B (see Figure 2) hybridized only to the smaller transcript, while subclones derived from fragments A and C hybridized to both transcripts. The subclone derived from

TABLE 2

Amino Acid Composition of Inositol-1-Phosphate Synthase Purified from Yeast.

Amino Acid	% total n moles	Amino Acid	% total n moles
Histidine	1.8	Glycine	6.8
Lysine	7.8	Alanine	7.6
Arginine	3.0	Half Cystine	0.9
Aspartic Acid	12.9	Valine	7.6
Threonine	6.0	Methionine	1.5
Serine	6.8	Isoleucine	5.6
Glutamic Acid	11.0	Leucine	8.6
Proline	5.4	Tyrosine	3.5
		Phenylalanine	4.0
TOTAL			99.99

Fragment D (Figure 2) hybridized only to the larger transcript. While it is not yet clear whether the transcripts are homologous to any overlapping sequence in the cloned DNA, they must lie in close proximity. A diagram of the probable arrangement of the transcripts is shown in Figure 2. Preliminary mapping of the region by S1 nuclease digestion of RNA-DNA heteroduplexes suggests that the 5' and 3' ends of the larger transcript are arranged as diagrammed (12). The direction of transcription of the smaller transcript is under investigation.

Regulation of expression of the transcripts was examined by Northern blot analysis of polyadenylated RNA derived from wild type yeast cells grown in the presence and absence of inositol and choline. These studies suggest that the presence of exogenous inositol leads to a reduction in the steady state level of the larger transcript (Figure 3). The level of the transcript is lowest when a high concentration of inositol (ie. 75μM) is present in the growth medium, but the presence of even 10μM inositol leads to a detectable decrease in the level of the transcript. The addition of choline to a culture grown in 75μM inositol leads to no further reduction in the level of the tran-

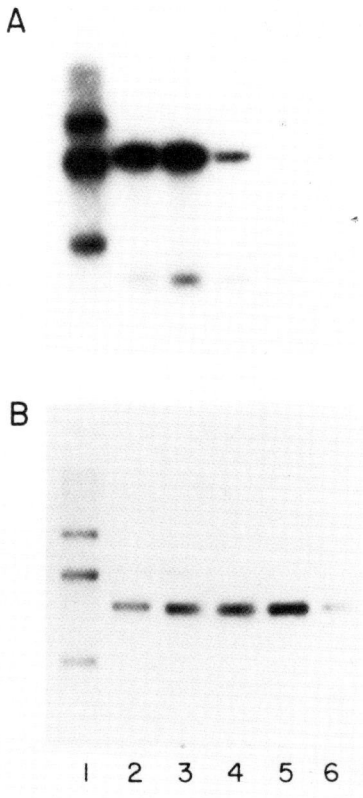

FIGURE 3. Northern blot of INO1 transcripts. Lane 1: size standards, DNA from plasmid pUC9 (PvuI and PvII. digest). Polyadenylated RNA,(.4µg, each lane), was isolated from wild type yeast cells grown under the following conditions; Lane 2: synthetic complete medium; no supplement. Lane 3: 1mM choline, no inositol; Lane 4: 10µM inositol; no choline. Lane 5: 75µM inositol, no choline. Lane 6: 75µM inositol, 1mM choline. Panel A shows the blot probed with INO1 DNA (fragment A, Figure 2). Panel B, shows the same blot probed with ribosomal protein gene TCM as a control for RNA loading. (Lane 6 is somewhat underloaded relative to the other lanes).

script. The smaller mRNA did not appear to be regulated in response to inositol and choline (Figure 3). The variation in the intensity of the smaller mRNA seen in Figure 3 appears to reflect primarily RNA loading.

The regulation of INO1 has also been further analysed on a genetic level. As discussed in the Introduction, the ino2 and ino4 mutants fail to derepress inositol-1-phosphate synthase and they also fail to derepress the phospholipid methyltransferases (15). These mutants exhibit an abnormal phospholipid composition because of the lower levels of the phospholipid methyltransferases but they exhibit no auxotrophy for choline (15). They are, however, inositol auxotrophs due to the failure to express inositol-1-phosphate synthase. A haploid strain of genotype ino2, ino4 was constructed and spontaneous inositol prototrophs were selected against the ino2, ino4 genetic background. Two classes of ino$^+$ revertants were isolated. The class of revertants represented by the mutation INO1-B24 (Table 3) are inositol prototrophs but, in the ino2, ino4 genetic background, they retain the defect in phospholipid methylation. These mutations confer constitutive synthesis of inositol-1-phosphate synthase and are linked to the INO1 locus. Mutants of the INO1-B24 class were found to have major DNA rearrangements at or near the INO1 locus (16), a result consistent with the insertion of a transposable element. The selection for ino$^+$ revertants in the ino2, ino4 genetic background also resulted in the isolation of another class of regulatory mutants which identified a new locus, OPI5, unlinked to the INO2 and INO4 loci. In the ino2, ino4 genetic background these mutants (OPI5$^-$) retain the phospholipid methylation defect characteristic of the ino2, ino4 genetic background (Table 3). These mutants also override the inositol mediated repression of INO1 and confer constitutive synthesis of inositol-1-phosphate synthase. This new class of mutants (OPI5$^-$) are different in two respects from the constitutive mutants such as opi1 described by Greenberg et al., (9). First, OPI5$^-$ mutations are epistatic to ino2 and ino4. The ino2 OPI5$^-$ and ino4 OPI5$^-$ double mutants all have an INO$^+$ phenotype and are constitutive for inositol-1-phosphate synthase (Table 3). On the other hand the ino4 and ino2 mutations are epistatic to the other opi mutants (such as opi1 - see Table 3). All ino4 opi$^-$ and ino2 opi$^-$ double mutant combinations are ino$^-$ in phenotype. Furthermore, the other opi mutants have a recessive constitutive phenotype while

TABLE 3

Phenotype of Regulatory Mutants

Genotype			PMTFS[a]	Expression of enzymes I-1-PS[b] Growth Condition	
				I^+	I^-
wild type			+	−	+
ino2			−	−	−
ino4			−	−	−
opi1			+	+	+
ino2	opi1		−	−	−
ino4	opi1		−	−	−
ino2	ino4		−	−	−
ino2	ino4	INO1-B24	−	+	+
INO2	INO4	INO1-B24	+	+	+
ino2	ino4	OPI5⁻	−	+	+
INO2	INO4	OPI5⁻	+	+	+

[a]. PMTFS is the abbreviation for phospholipid N-methyltransferases. A minus (−) indicates that the strain expresses the low level seen in ino2 and ino4 strains (15).

[b]. I-1-PS is the abbreviation for inositol-1-phosphate synthase. The two growth conditions under which the expression of the enzyme was monitored by immunoprecipitation are: I^+ (repressing-50μM inositol) and I^- (derepressing-no inositol or 10μM inositol). A minus in the body of the Table indicates the repressed level and a plus indicates the derepressed level of the enzyme.

the newly identified OPI5⁻ mutants have a dominant
constitutive phenotype. The phenotypes of these individual
mutants and, as well as the double mutant combinations,
are given in Table 3.

DISCUSSION

Inositol-l-phosphate synthase is the only cytoplasmic
enzyme known to be subject to the coordinate regulation
which controls phospholipid biosynthesis in yeast (10).
The amino acid sequence has never been reported for this
enzyme, from any biological source, and the mechanism of
the reaction it catalyses has never been verified (7).
The isolation of the structural gene for this enzyme from
yeast, therefore, will permit, for the first time, an
analysis of the amino acid sequence and structure of this
essential enzyme. Sequencing of the cloned gene is in
progress as is amino acid sequencing of the N-terminus of
its gene product. We have reported the amino acid
composition of the enzyme subunit; the first such analysis
for this enzyme from any biological source. The enzyme
subunit has a high proportion of hydrophobic residues, but
it is not altogether atypical of soluble proteins in its
size range (5). A high proportion of hydrophobic residues
might explain the affinity this enzyme has for hydrophobic
resins, a property which was useful in its purification
(6).

The cloning of the INO1 structural gene also provides
an opportunity to examine regulation of the gene at the
level of expression of its transcripts. An earlier,
preliminary study of the INO1 transcripts suggested that
there existed a larger transcript from which the regulated
1.8 kb message was processed posttranscriptionally (12 ;
reviewed in 10). The data presented here do not support
the existence of any transcript larger than the regulated
1.8 kb message. The data presented here suggest that the
smaller mRNA is encoded by a sequence in close proximity
to the region where the 5'end of the larger transcript
starts. The arrangement of the two transcripts relative
to each other is under investigation, as is the relationship of the smaller transcript to inositol biosynthesis
and its regulation. The expression of the smaller mRNA did
not appear to be regulated by inositol or choline. However,
the level of the 1.8 kb mRNA varied in response to
exogenous inositol in a fashion consistent with the

reported expression of inositol-1-phosphate synthase (6). The 1.8 kb mRNA was maximally expressed in the total absence of inositol. The presence of high levels of inositol (75µM) in the growth medium led to a substantial reduction in the steady state levels of the larger transcript (Figure 3) but even 10µM inositol appeared to have some effect upon the steady state level of the transcript. The repressing effect of the lower concentration of inositol on the expression of the 1.8 kb INO1 transcript was unexpected, since this concentration of inositol does not appear to reduce the expression of its gene product (4). The addition of choline to the growth medium containing inositol resulted in no detectable further decrease in the level of the larger transcript (Figure 3). This result is entirely consistent with the reported lack of effect of choline upon the expression of the inositol-1-phosphate synthase subunit (14).

A large number of regulatory genes are known to affect expression of inositol-1-phosphate synthase and other coordinately regulated enzymes of phospholipid synthesis in yeast (10). A major goal of our analysis is to establish the mechanisms of interaction of these positive and negative regulatory factors. The regulation of inositol-1 phosphate synthase has been examined in more detail than some of the other regulated enzymes in part because it is cytoplasmic, and because specific antiserum directed against the enzyme subunit is available (6). However, this enzyme is unique among those listed in Table 1, in that it is fully repressed by the addition of inositol alone, to the growth medium. We wished to determine whether any gene (or genes) was more directly responsible for INO1 expression than the already identified pleiotropic regulatory genes, INO2 and INO4. The ino2 and ino4 mutants express repressed levels of inositol-1-phosphate synthase and the phospholipid N-methyltransferases (15). By selecting for inositol prototrophs in an ino2, ino4 genetic background, we selected for expression of the INO1 gene, without necessarily requiring expression of the phospholipid N-methyltransferases. All inositol prototrophs selected in this fashion were found to be linked to the INO1 gene or to the OPI5 locus. The mutations linked to the INO1 gene were all the result of major rearrangements of the DNA in the vicinity of INO1, (16) presumably freeing the gene from its normal regulation. However, the mutations at OPI5 identify a new regulatory gene,

presumably a specific regulator of INO1 since these
mutations, which lead to constitutive INO1 expression, have
no effect upon the phospholipid N-methyltransferases.
Because the OPI5⁻ mutants are dominant it is possible that
they identify a positive regulator of INO1 that has become
altered such that it no longer interacts with an antagonist
or no longer requires an activator. Since the OPI5⁻
mutations free INO1 gene expression from the requirement for
the INO2 and INO4 gene products, both of which play a
positive regulatory role (15) the latter possibility seems
more likely. This hypothesis, however, suggests that
absence of OPI5 function should lead to lack of INO1
expression. In other words it should be possible if our
hypothesis is correct, to isolate recessive ino⁻ mutations
which map to the OPI5 locus. Work presently in progress
will test these ideas.

ACKNOWLEDGEMENTS

Susan Henry was the recipient of an Irma T. Hirschl
Faculty Award.

REFERENCES

1. Aviv H and Leder P (1972). Purification of biologically
   active globin messenger RNA by chromatography on
   oligothymidylic acid-cellulose. Proc Nat Acad Sci USA
   69:1408.
2. Carson M, Atkinson KD and Waechter CJ (1982). Properties
   of particulate and solubilized phosphatidylserine
   synthase activity from Saccharomyces cerevisiae:
   Inhibitory effect of choline in the growth medium. J
   Biol Chem 257:8115.
3. Carson MA, Emala M, Hogsten P and Waechter CJ (1984).
   Coordinate regulation of phosphatidylserine decarboxylase activity and phospholipid methylation in yeast. J
   Biol Chem 259:6-67.
4. Culbertson MR, Donahue TF and Henry SA (1976). Control
   of inositol biosynthesis in Saccharomyces cerevisiae.
   I. Properties of a repressible enzyme system in extracts
   of wild type (Ino⁺) cells. J Bacteriol 26:232.

5. Dayhoff MO, Dayhoff RE, Hunt LT (1976). "Atlas of Protein Sequence and Structure" The National Biochemical Research Foundation, Maryland:p 345.
6. Donahue TF and Henry SA (1981). Myoinositol-1-phosphate synthase: Characteristics of the enzyme and identification of its structural gene in yeast. J Biol Chem 256: 7077.
7. Eisenberg F (1978). Intermediates in the myoinositol-1 phosphate synthase reaction. In:Cyclitols and phosphoinositides (eds. Wells W and Eisenberg F). Academic Press, New York. p.269.
8. Elion EA and Warner JR (1984). The major promotor element of rRNA transcription in yeast lies 2 kb upstream. Cell 39:663.
9. Greenberg M, Reiner B and Henry SA (1982). Regulatory mutations of inositol biosynthesis in Yeast: Isolation of inositol excreting mutants. Genetics 100:19.
10. Henry SA, Klig LS and Loewy BS (1984). The Genetic regulation and coordination of biosynthetic pathways in yeast: Amino Acid and Phospholipid Synthesis. Ann Rev Genet 18:207.
11. Homann MJ, Kelley MJ, Henry SA and Carman GM (1985). Regulation of CDP-diacylglycerol synthase activity in yeast. Submitted Manuscript.
12. Klig LS (1983). Genetic and molecular regulation of inositol biosynthesis and its coordination with phospholipid biosynthesis in Saccharomyces cerevisiae. Ph.D. thesis. Albert Einstein College of Medicine, Bronx, New York.
13. Klig, LS and Henry SA (1984). Isolation of the yeast INO1 gene: Located on an autonomously replicating plasmid, the gene is fully regulated. Proc Nat Acad Sci USA 81:3816.
14. Klig LS, Homann MJ, Carman GM and Henry SA (1985). Coordinate regulation of phospholipid biosynthesis in yeast: The opi1 mutant is pleiotropically constitutive. J Bacteriol In Press.
15. Loewy BS and Henry SA (1984). The INO2 and INO4 loci of yeast are pleiotropic regulatory genes. Mol Cell Biol 4:2479.
16. Loewy B (1985). Coordinate regulation of phospholipid biosynthesis in Saccharomyces cerevisiae. Ph.D Thesis. Albert Einstein College of Medicine, Bronx, New York.

17. Maeda T and Eisenberg F (1980) Purification, structure and catalytic properties of L-myoinositol-1 phosphate synthase from rat testis. J Biol Chem 255: 8458.
18. Rigby PWJ, Dieckmann M, Rhodes C and Berg P (1977). Labeling deoxyribonucleic acid to high specific activity In Vitro by nick translation with DNA polymerase I. J Mol Biol 113:237.
19. Rozek CE and Davidson N (1983). Drosophila has one myosin heavy-chain gene with three developmentally regulated transcripts. Cell 32:23.
20. Steinman HM (1978). The amino acid sequence of mangano superoxide dismutase from Escherichia coli B. J Biol Chem 253:8708.
21. Thomas PS (1980). Hybridization of denatured RNA and small DNA fragments transferred to nitrocellulose. Proc Nat Acad Sci USA 77:(9) 5201.
22. Waechter CJ and Lester RL (1973). Differential regulation of the N-methyltransferases responsible for phosphatidylcholine synthesis in Saccharomyces cerevisiae. Arch Biochem Biophys 158:401.
23. Wells W and Eisenberg, eds (1978). Cyclitols and phosphoinositides. Academic Press, New York.
24. Yamashita S, Oshima A, Nikawa J, Hosaka K (1982). Regulation of the phosphatidylethanolamine methylation pathway in Saccharomyces cerevisiae. Eur J Biochem 128:589.

VII. MACROMOLECULAR TRAFFIC III: IMPORT OF PROTEINS INTO MITOCHONDRIA

SORTING OF CYTOPLASMIC PROTEINS FOR ASSEMBLY IN MITOCHONDRIA

Michael G. Douglas

Department of Biochemistry, University of Texas Health Science Center San Antonio, Texas 78284

The foundations of Cell Biology are rooted in the compartmentalization of the eukaryotic cell. Classic cell fractionation, structural and biochemical studies established the unique composition of different subcellular structures. More recently, advances in the field have given us insights into the mechanisms responsible for this organelle specialization. The mitochondrian occupies a unique position in the biogenesis of intracellular membranes. The two membranes of this organelle, which we designate outer and inner membrane, must assemble their protein and lipid components by a mechanism that is apparently distinct from that required to form all the other intracellular membranes. Indeed, with the notable exception of mitochondria, all other membranes of the cell are synthesized by a process involving intracellular membrane flow to specifically localize cellular components which are initially synthesized in the rough endoplasmic reticulum (1). In addition, a distinct mechanism must exist to specifically sort proteins from the cytoplasm and localize them to the two soluble spaces of mitochondria defined by the outer and inner membrane. How is this distinct organelle assembly process programmed by the cell?

Considerable information on the biogenesis of mitochondria has been provided in recent years primarily through genetic and biochemical studies in yeast. This brief review will describe recent advances in our understanding of the cell biology of mitochondrial development in yeast. I shall concentrate on the role of molecular techniques and mutant analysis which set the stage for elucidation of the distinct mechanisms involved.

Mitochondria are Assembled from Components Synthesized in the Cytoplasm

In yeast, mitochondria which constitute about 10% of the cells protein contain 200-250 unique proteins. Although mitochondria contain a distinct circular genome, it only encodes on the order of 12 polypeptides (2) which remain associated with the organelle. All the remaining proteins of mitochondria, which constitute about 93% of its mass, are encoded in the nucleus and synthesized in the cytoplasm. Early studies documented the presence of specific classes of cytoplasmic ribosomes bound to the mitochondrial outer membrane which were enriched in mitochondrial proteins. However, in vivo pulse-chase labeling studies as well as the demonstration that completed polypeptides of a cell free translation lysate could be sequestered into isolated mitochondria strongly supported a post translational mechanism for mitochondrial delivery (3). The sensitivity of techniques utilized thus far do not favor either mechanism for in vivo mitochondrial delivery during the steady state growth of the cell, however, it is clear that proteins can be incorporated into mitochondria after completion of their synthesis.

The in vitro mitochondrial import model has provided an essential tool in the elucidation of the different steps during transport of a completed protein into the organelle. In each case, a complementary in vivo result has been demonstrated (4). Early studies show that some mitochondrial proteins are made as higher molecular weight precursors containing an additional sequence. For those precursors which have been characterized thus far, the additional sequence is present as an extended amino terminal polypeptide (5). In subsequent studies, it was demonstrated that specific proteolytic processing enzymes exist within mitochondria for the maturation of the delivered precursor (5).

Early proposals that the extensions which are removed during maturation of the protein are "signals" directing import has been confirmed in recent studies (see below). There are exceptions, however, of precursor proteins which are targeted into each of the mitochondrial compartments yet do not exhibit an apparent transient "signal" peptide. It is proposed that these proteins contain (an) internal signal(s) which direct their mitochondrial delivery (6). If these precursors are delivered in a post translational manner from the cytoplasm, what prevents them from assembling with their partner proteins in the wrong location? In the case of transient presequence containing precursors, one function of the additional sequence may be to maintain the precursor in a conformation to both prevent premature assembly outside the mitochondria as well as to direct import. For several proteins which have been shown to be synthesized as the "mature sized" form, internal signals are proposed to maintain the precursor in a conformation compatable with mitochondrial binding. These precursors are triggered for assembly by either a covalent or non-covalent modification which occurs upon binding to the mitochondrial membrane (7).

The Import Apparatus

The specificity of the sorting process implies a receptor mediated process. Since the outer mitochondrial membrane is in direct contact with the cytoplasm, this bilayer has been the focus of the studies to define components which specifically bind mitochondrial precursors. Different studies have defined protease sensitive components on the mitochondrial surface, which exhibit specific and saturable binding of specific mitochondrial precursors (8-10). These binding sites appear to be specific for the precursor and not the mature form of a given mitochondrial protein and represent a true intermediate in the import pathway. To date, however, it has not been possible to demonstrate competition between given mitochondrial precursors for receptor-mediated import.

Although this would indicate that specific receptors for each precursor may be initiating mitochondrial import, the limited number of potential protein candidates available in the outer membrane (only about 20 proteins) suggests that a more involved mechanism may be required for the initial binding. Indeed, recent studies have shown that (a) soluble factor(s) present in the homologous cell free lysates, which can be substituted by a yeast cytosol fraction (11), stimulates the import of a precursor into mitochondria. Whether, in fact, this component directly interacts with soluble precursor in the cytoplasm or with the mitochondria at the "port of import" remains to be determined. However, it is tempting to speculate that cytoplasmic protein components may directly participate in the sorting of precursor proteins for mitochondrial delivery.

Once bound to the mitochondrial surface, little is known about the mechanism of transport across the different mitochondrial membranes. Precursor proteins destined for the mitochondrial matrix must cross two bilayers, whereas, some proteins must cross the outer membrane but not the inner membrane. It is generally believed that the import of precursors past the mitochondrial outer membrane may occur at "points of contact" between the inner and outer membrane. Although such junctions are readily observed by electron microscopy at regions of the outer membrane containing associated cytoplasmic ribosomes (12), no compelling experimental evidence has been provided to document their function as import sites for precursor protein uptake.

Catalyzing Mitochondrial Compartmentation

The *in vitro* and *in vivo* studies have characterized a role for at least two organelle endoproteases and a membrane potential across the mitochondrial inner membrane in the localization of precursors to distinct mitochondrial compartments (3). Briefly, an electrochemical gradient of protons across the mitochondrial inner membrane generated by either electron transfer along the respiratory chain, ATP hydrolysis via the ATPase complex or electrogenic adenine nucleotide exchange, is required for movement of precursors past the outer membrane. In the absence of a membrane potential, due to mutations or membrane potential dissipating antibiotics, precursors bind to the mitochondrial exterior but do not enter. Restoration of a membrane potential across the inner membrane allows import of the precursor past the outer membrane. Further, transport of the precursor to the matrix is with few exceptions accompanied by the proteolytic removal of the presequence which is catalyzed by a metallo-protease located within the matrix. Remarkably, the delivery of proteins to the inter membrane space between the inner and outer membrane requires this same matrix localized metallo protease as the first of two processing steps. Again, there are exceptions, however, the combination of *in vivo* and *in vitro* studies convincingly support a model for the inter membrane space localization of some proteins in which the unprocessed form of the precursor is delivered to the mitochondrial inner membrane where, at least, the amino terminus of the molecule is accessible for processing on the matrix face of this membrane. This generates a membrane bound intermediate which is presumably released from or localized to the membrane cytoplasmic face as the mature protein by an inner membrane associated endoprotease. The reader is referred to an excellent review by Ried (5) detailing the use of yeast mutants and *in vitro* biochemistry to unravel this unique localiztion pathway. In general, the compartmentalization of precursor proteins in yeast mitochondria following delivery from the cytoplasm can be summarized as follows with respect to proteolytic processing and energy dependence:

<u>Outer Membrane</u> - direct insertion - no apparent processing and no energy requirement

<u>Inter Membrane Space</u> - matrix routed localization - two processing steps and an energy requirement for localization, notable exception cytochrome c (7)

<u>Inner Membrane</u> - variable routes, zero to two processing steps and an energy requirement for import

<u>Matrix Space</u> - one processing step, notable exception 2-isopropylmalate synthase (6), energy requirement for import.

Molecular Genetic Analysis

The well described cell biology of the mitochondrial import pathway in yeast provides the foundation for focusing various genetic techniques. This combined approach is already beginning to define the molecular basis for cytoplasmic sorting and the organelle compartmentalization of mitochondrial precursors.

The information directing the cellular fate of a cytoplasmically synthesized mitochondrial precursor is programmed into its primary sequence. In order to define the sequences involved a number of nuclear genes encoding components of mitochondria have been isolated and characterized. The selection of these genes has utilized a number of genetic and physical techniques (13). These genes have been initially utilized to construct gene fusions. This approach allows the yeast cell biologist to define the sequences which participate in the targeting of proteins from the cytoplasm and program their localization within the organelle. The validity of this approach to define the targeting signals for mitochondrial delivery in vivo was established by the observations that gene fusions encoding a mitochondrial protein at the amino terminus could import hybrid proteins consisting of a non-mitochondrial protein at the carboxy terminus (14,15). By limiting the size of mitochondrial precursor sequences expressed at the amino terminus of the hybrid molecule, the targeting information could be localized to the extreme amino terminus (16-18). For the precursor proteins which contain a presequence at their amino terminus, it would appear that sufficient information is present within this transient sequence for import to the mitochondrial matrix.

A comparison of the targeting signals defined in these studies, as well as the sequences at the extreme amino terminus of other characterized mitochondrial precursors (5), however, reveals no apparent homologous targeting element of the type defined for specific protein delivery to the nucleus (19). The mitochondrial targeting elements can best be described as non-secretory signals containing positively charged residues on the average every 5-8 residues. With few exceptions acidic residues are much less abundant or absent. Thus, a paradox is presented by analysis of mitochondrial targeting sequences at this point. How do these short (less than 12-15 residues) non homologous regions specify specific mitochondrial delivery? This dilemma will be resolved by the construction, expression and analysis of targeting sequence mutations in yeast and the selection of compensating host mutants. Thus, armed with a genetic approach, the yeast cell biologist will be able to dissect in some detail the important molecular features of the mitochondrial "addressing sequence" and uncover additional cellular participants of the import/localization pathway.

Another exciting parallel approach to unravel the import mechanism is the selection and characterization of conditional mutants blocked in mitochondrial assembly. For these studies, recessive mutants have been isolated which are temperature-sensitive for growth on any carbon source and which accumulate mitochondrial precursor proteins at the restrictive temperature (20). At present two mutants have been described which represent true lesions in import rather than temperature dependent modifications of a membrane potential (see above). The characterization of these mutants will surely take advantage of yeast technology (21) to select the plasmid encoded genes which restore growth of the mutant at the non-permissive temperature and complement the assembly defect. It is anticipated that mutants of this type will prove useful in defining components at all stages of the import pathway.

Coordination of Mitochondrial Assembly

The assembly of mitochondrial proteins from the cytoplasm is a controlled process (22). Mechanisms must exist to not only regulate cytoplasmic precursor protein synthesis but also to coordinate the rate of mitochondrial growth with that of the cell. To date, little is known about the role mitochondria play in regulating the synthesis of their precursors. It is known that the levels of various energy transducing enzymes and cytochromes of mitochondria are controlled by various physiological signals at the level of transcription (23,24). Do mitochondria "export" regulatory signals to regulate the level of precursors? This question has been answered in large part by the observation that heme (a mitochondrial product) regulates the transcription of cytochrome c (25). Further, trans acting regulatory mutants have been described which block the heme-dependent control of cytochrome c expression (26). Thus, in this one example, a mitochondrial product appears to modulate the synthesis of one of its cytochromes.

How is the expression and coordination of other mitochondrial precursors regulated? What other signals do mitochondria export to regulate its developent? Analysis of mechanism of mitochondrial communication with the nucleus will most likely prove to be the next adventure in the cell biology of mitochondrial assembly.

REFERENCES

1. Palade, G., Farquhar, M. (1981) Cell Biology, in the Biological Principles of Disease (Smith, L. and Thier, S., eds.) pp 1-56, Saunders, Philadelphia

2. Dujon, B. (1981) Mitochondrial Genetics and Functions, in The Molecular Biology of the Yeast Saccharomyces: Life Cycle and Inheritance (Strathern, J., Jones, E. and Broach, J., eds.) p 505-635, Cold Spring Harbor, New York

3. Hay, R., Bohni, P,. Gasser, S. (1984) How mitochondria import proteins. Biochim. Biophys. Acta 779: 65-87

4. Schatz, G., Butow, R. (1983) How are proteins imported into mitochondria. Cell 32: 316-318

5. Reid, G. (1985) Transport of Proteins into Mitochondria, in Current Topics in Membranes and Transport Vol. 24. (Bronner, F., ed.) pp 295-336, Academic Press, New York

6. Gasser, S., Schatz, G. (1983) Import of proteins into mitochondria: in vitro studies on the biogenesis of the outer membrane. J. Biol. Chem. 258: 3427-3430

7. Teintze, M., Neupert, W. (1983) Biosynthesis and assembly of mitochondrial proteins, in Cell Membranes: Methods and Reviews Vol 1 (Elson, E., Frazier, W., Glaser, L., eds.) pp 89-225, Plenum, New York

8. Hennig, B., Koehler, H., Neupert, W. (1983) Receptor sites involved in post translational transport of apocytochrome c into mitochondria. Proc. Natl. Acad. Sci. 80: 4963-4967

9. Riezman, H., Hay, R., Witte, C., Nelson, N., Schatz, G. (1983) Yeast mitochondrial outer membrane specifically binds cytoplasmically-synthesized precursors of mitochondrial proteins. EMBO J. 2: 1113-1118

10. Zwizinski, C., Schleyer, M., Neupert, W. (1983) Precursor to the ADP/ATP carrier binds to receptor sites on isolated mitochondria. J. Biol. Chem. 258: 4071-4084

11. Ohta, S., Schatz, G. (1984) A purified precursor polypeptide requires a cytosolic protein fraction for import into mitochondria. EMBO J. 3: 651-657

12. Kellems, R., Allison, V., Butow, R. (1974) Cytoplasmic type 80S ribosomes associated with yeast mitochondria: Evidence for the association of cytoplasmic ribosomes with outer mitochondrial membrane in situ. J. Biol. Chem. 249: 3297-3303

13. Douglas, M., Takeda, M. (1985) Nuclear genes encoding mitochondrial proteins in yeast. Trends in Biochem. Sci. 10: 192-194

14. Douglas, M., Geller, B., Emr, S. (1984) Intracellular targeting and import of an F_1-ATPase β-subunit β-galactosidase hybrid protein into yeast mitochondria. Proc. Natl. Acad. Sci. 81: 3983-3987

15. Hase, T., Muller, U., Riezman, H., Schatz, G. (1984) A 70 kd protein of the yeast mitochondrial outer membrane is targeted and anchored via its extreme amino terminus. EMBO J. 3: 3157-3164

16. Hurt, E., Presold-Hurt, B., Schatz, G. (1984) The amino terminal region of an imported mitochondrial precursor polypeptide can direct cytoplasmic dihydrofolate reductase into the mitochondrial matrix. EMBO J. 3: 3149-3156

17. Vassarotti, A., Smagula, C., Douglas, M. (1985) In vivo targeting and assembly of F_1-ATPase β-subunit deletions, in Yeast Cell Biology (Hicks, J., ed.) this issue

18. Emr, S., Vassarotti, A., Garrett, J., Geller, B., Takeda, M., Douglas, M. (1985) The amino terminus of the yeast F_1-ATPase β-subunit precursor functions as a mitochondrial import signal, in press

19. Hall, M., Hereford, L., Herskowitz, I. (1984) Targeting of E. coli β-galactosidase to the nucleus in yeast. Cell 36: 1057-1065

20. Yaffe, M., Schatz, G. (1984) Two nuclear mutations that block mitochondrial protein import in yeast. Proc. Natl. Acad. Sci. 81: 4819-4825

21. Struhl, K. (1983) The New Yeast Genetics. Nature 305:391-397

22. Yaffe, M., Schatz, G. (1984) The future of mitochondrial research. Trends in Biochem. Sci. 9: 179-181

23. Zitomer, R., Montgomery, D., Nichols, D., Hall, B.(1979) Transcriptional regulation of the yeast cytochrome c gene. Proc. Natl. Acad. Sci. 76: 3627-3631

24. Szekely, E., Montgomery, D. (1984) Glucose represses transcription of S. cerevisiae nuclear genes that encode mitochondrial components. Mol. Cell. Biol. 4: 939-946

25. Guarente, L., Mason, T. (1983) Heme regulates transcription of the CYC1 gene of S. cerevisiae via an upstream activation site. Cell 32: 1279-1286

26. Guarente, L., Lalonde, B., Gifford, P., Alani, E. (1984) Distinctly regulated tandem upstream actication sites mediate catabolite repression of CYC1 gene in S. cerevisiae. Cell 36: 503-511

HYBRID PROTEIN DETECTION AND SUBCELLULAR LOCALIZATION FOR TARGETING ANALYSIS DURING MITOCHONDRIAL BIOGENESIS[1]

Linda Marshall-Carlson, Jerry Lynn Allen[2] and Michael G. Douglas[2]

Department of Biochemistry, University of Texas Health Science Center, San Antonio, Texas 78284.

ABSTRACT We are developing a combined approach of genetics, biochemistry, and immunocytochemistry to investigate import and assembly of nuclear-encoded mitochondrial proteins. Using the yeast F_1-ATPase β-subunit as our model, a family of plasmids was constructed bearing various 5' portions of this gene fused to the E. coli lacZ gene. β-galactosidase activity was monitored to determine subcellular targeting of the hybrid protein in host cells. Subcellular fractionation, immunofluorescence microscopy and immunoelectron microscopy reveal that efficiency of chimeric protein targeting and import appears to be a function of specific polypeptide regions of F_1-β. While levels of synthesis of the various hybrid proteins are similar, mitochondrial targeting and import are strikingly different. The degrees of sequestration to mitochondria of the longest and shortest fusion proteins are compared here using different techniques.

INTRODUCTION

Most of the proteins constituting the mitochondrion are encoded by nuclear genes, translated on free cytoplas-

[1]This work was supported by NIH Grant GM26713 and R.A. Welch Grant AQ814.
[2]Present address: Department of Biochemistry, University of Texas Health Science Center, Dallas, Texas 75235.

mic ribosomes and are subsequently translocated across one or both mitochondrial membranes. While much is known about the structure, composition and metabolic properties of mitochondria, only recently has progress been made regarding the molecular mechanisms underlying the coordinate control of synthesis, targeting, binding and import of the mitochondrially-destined precursor proteins. The information for correct targeting of these proteins is apparently contained within the amino acid sequence of the nascent precursor proteins. In order to determine which polypeptide regions specifically mediate protein import into mitochondria, a gene fusion study was initiated using as our model the β-subunit of the F_1-ATPase. $F_1\beta$ is encoded by the nuclear ATP2 gene and is synthesized as a 509 amino acid precursor which undergoes N-terminal processing during maturation, to remove a peptide of about 20 residues (1). A family of plasmids was constructed containing a series of overlapping segments from the 5' end of ATP2 fused to the lacZ gene from E. coli. By monitoring the β-galactosidase activity encoded in the lacZ portion of the hybrid proteins, we have followed cellular targeting and subcellular distribution of these proteins (2).

Biochemical subfractionation studies are potentially subject to proteolysis and other artifacts that may arise during organelle isolation and compartment separation. Therefore, a complementary approach has been developed using immunofluorescence microscopy and fracture-label electron microscopy to detect hybrid proteins in previously fixed cells, hence, providing additional evidence for the location in vivo (4).

METHODS

Cell Culture

All procedures were performed at 4°C and in the presence of protease inhibitors, unless otherwise stated. Saccharomyces cerevisiae strain DB745 was chosen as the host organism since spheroplasts could be generated rapidly at room temperature, hence minimizing proteolysis. Cultures were grown in appropriate selective medium with sodium lactate as the carbon source. This strain was able to grow on a nonfermentable carbon source in the presence of any of the gene fusions employed. Three hours prior to harvesting, cultures were supplemented with YP lactate, and 10 minutes prior to harvesting phenylmethyl sulfonyl fluoride (PMSF) in ethanol was added to a final concentration of 0.5 mM.

Hybrid Proteins

In all cases, the genes for the hybrid proteins were expressed in vivo from yeast-E. coli shuttle vectors maintained in yeast by ura$^+$ selection. The hybrid genes expressed in this study were pCβZ1 and pCβZ15, encoding 350 and 39 residues of the $F_1\beta$ subunit ATP2 gene respectively, fused to lacZ (2); pKB41, encoding 125 residues of the put2 gene product Δ'-pyrroline-5-carboxylate dehydrogenase, fused to lacZ (a gift from M. Brandriss) and p649-19 encoding 61 aa of the 70 kd outer mitochondrial membrane protein (a gift from G. Schatz).

Antibodies

Purified β subunit protein from the F_1-ATPase complex (3), and commercially purified β-galactosidase (Bethesda Research Laboratories, Inc.) were used as antigens to prepare antisera as described elsewhere (4). The antibodies were tested for specificity by Ouchterlony immune diffusion, Western blotting against cell lysates and against the original antigen, immune precipitation and fluorescence microscopy. Antibody against citrate synthase was kindly donated by P. Srere.

Fluorescence Microscopy

Cells were fixed using 4% paraformaldehyde/0.1% glutaraldehyde in 0.1M sodium cacodylate pH 7.2 and washed extensively using 0.1M glycine in phosphate buffered saline pH 7.4, (PBS). Spheroplasts were generated by zymolyase digestion, were washed and rendered permeable to antibody or fluorescent probes using 0.1% Triton X-100 in PBS. The cells were washed in PBS, incubated for 2 hours at room temperature with a 200 fold dilution of primary antiserum, pre-immune serum or fluorescent probe and washed twice with PBS. Cells incubated with primary serum were further incubated with fluorescein-conjugated goat-anti-rabbit IgG (FITC-IgG, Sigma) diluted 640 fold, for one hour, and washed. Stained cells were observed using a Zeiss Universal fluorescence microscope equipped for epifluorescence illumination, and were photographed on Kodak Tri-Pan X film.

Freeze Fracture

Fixed spheroplasts were embedded in 30% bovine serum albumin (BSA), glycerol impregnated, frozen and fractured, using a modification of the method of Pinto da Silva et al, (5).

Cytochemical Labeling

Deglycerinated fragments were incubated for 2 hours at room temperature with 200 fold diluted primary antiserum, and controls with pre-immune serum or secondary probe. After washing, the serum-incubated samples were incubated with secondary (electron dense) probe overnight at 4°C and washed. Fragments were treated with 0.2% glutaraldehyde in 0.1M sodium cacodylate pH 7.2, followed by 1% osmium tetroxide in the same buffer, then dehydrated and embedded in Spurr's resin. After polymerization, desired fracture face orientation was determined by light microscopy of thick sections. Appropriate thin sections were obtained using a diamond knife and were observed unstained at 60 KV, in a Jeol Jem 100-cx electron microscope.

Fluorescence Microscopy

Fixed, detergent-treated spheroplasts of yeast cultures harboring various lacZ hybrid protein expression vectors were subjected to protein localization studies using primary antibodies against β-galactosidase, the F_1 ATPase, and citrate synthase, with FITC-IgG used to detect the primary probe. Cells were counterstained using 4'6-diamidine-2-phenyl indole dihydrochloride (DAPI), and photographed under conditions of epifluorescence illumination or phase contrast. For all experiments, controls were performed in parallel using pre-immune serum. No crossover illumination occurred between the FITC and DAPI fluorescence spectra.

RESULTS

Localization of Hybrid Proteins

Fixed permeabilized spheroplasts were incubated with individual antisera raised in rabbits against β-galactosidase, F_1-ATPase β-subunit and citrate synthase.

The bound antibody was detected in each case by fluorescence microscopy, using fluorescein-conjugated goat anti-rabbit IgG. Cells were also incubated with DAPI, which binds to DNA and hence reveals the positions of the nuclear and mitochondrial DNA inside the cell.

Initially, cells were examined by phase contrast light microscopy to check for digestion of the cell wall, the presence of refractile internal organelles and normal morphology, (Fig. 1,a,d,g,j,m) since lysed spheroplasts often maintained their physical configuration despite partial loss of cell contents. When observed under conditions of DAPI fluorescence, (Fig. 1,b,e,h,k,n), the positions of nuclei and mitochondria could clearly be distinguished; nuclei appeared much larger than mitochondria by virtue of their greater DNA content.

Finally, each sample was observed using appropriate excitation radiation and filters for FITC fluorescence, to reveal indirectly the distribution of the β-galactosidase portions of the hybrid proteins (Fig. 1,c,f,i,l,o). When the fluorescence images of DAPI and FITC are compared for the various hybrid protein-expressing cultures, a marked difference can be seen in the intensity of staining of the mitochondria relative to the cytoplasmic staining. The hybrid proteins are synthesized on cytoplasmic ribosomes, so a background of precursor protein in the cytoplasm is always present, but in the cases of pKB41.put2-lacZ124 (Fig. 1,g-i), p649-19 (outer membrane localized) (Fig. 1, j-l) and pCβZ1 (Fig. 1,a-c) a clear correlation can be observed between the hybrid protein and mitochondrial DNA fluorescence. In contrast to this, for pCβZ15, a very small degree of mitochondrial localization can be seen Fig. 1, d-f) but a high cytoplasmic background fluorescence persists.

That this distribution of the βZ15 protein is not merely a reflection of inaccessibility to antibody, is revealed by the control experiment in which antibodies directed against a normal mitochondrial protein component, namely citrate synthase (Fig. 1,j-l) can be clearly seen to be localized to mitochondria. Further controls to test the specificity of the observed fluorescence (Fig. 1,m-o) confirmed that there is no native protein in the wild-type strain that cross reacts either with the anti-β-galactosidase antibody, or with the FITC-linked second antibody. No fluorescence was detected when pre-immune rabbit serum was substituted for the primary antiserum (data not shown). These results indicate that the antisera

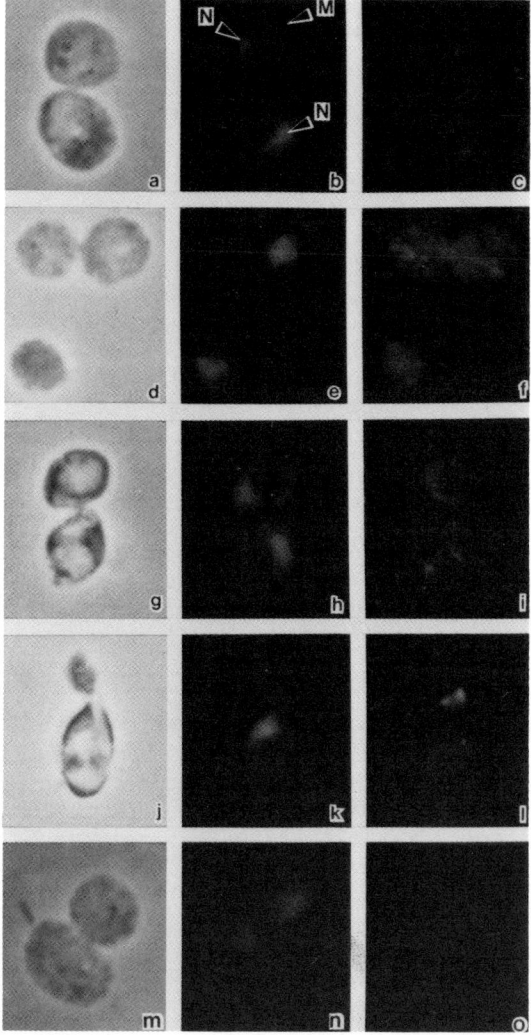

Figure 1   Visualization of hybrid protein location in fixed yeast cells.

Spheroplasts harboring various lacZ fusion protein expression vectors were subjected to fluorescence-localization techniques and photographed. Left hand panels:-appearance of cells by phase-contrast transmitted light; center panels:-DAPI fluorescence revealing the positions of nuclear and mitochondrial DNA; right hand panels:-FITC fluorescence indicating hybrid protein subcellular distribution.

a-c Cells expressing pCβZ1-lacZ, the longest fusion protein, show bright FITC fluorescence, (panel c) which is co-localized with DAPI mitochondrial fluorescence, panel b. N=nucleus, M=mitochondrion.

d-f Distribution of FITC fluorescence shows that the shortest fusion protein, cβZ15-lacZ, does localize to some extent in mitochondria, but also exhibits a high cytoplasmic pool level.

g-i A matrix-localized lacZ fusion protein, put-2-lacZ can clearly be detected with FITC-IgG; cytoplasmic levels are low for the hybrid protein.

j-l LacZ fused to 61 residues of 70 kD OMP (fusion 649-19-lacZ) becomes localized to mitochondria with low cytoplasmic background.

Comparison of panels c,f,i and l suggests that the shortest fusion protein (panel f) is less efficiently targeted to mitochondria than the other fusions examined, reflected in a high cytoplasmic fluorescence.

m-o Control experiments with wild type DB745 spheroplasts failed to bind any detectable levels of FITC-IgG, panel c, after prior probing with anti β-galactosidase antiserum. This reduces the possibility of non-specific binding artefacts being responsible for the FITC localization in the hybrid protein expressing strains.

****

are indeed specific for the proteins against which they were raised.

Fracture Label

Thick sections cut from blocks of embedded freeze-fractured cell-BSA gel suspensions were examined by light microscopy, and spheroplast morphology evaluated. In every case, overall conformation and general preservation appeared good. The cut margins of the original gel blocks could be easily distinguished from fracture profiles by the

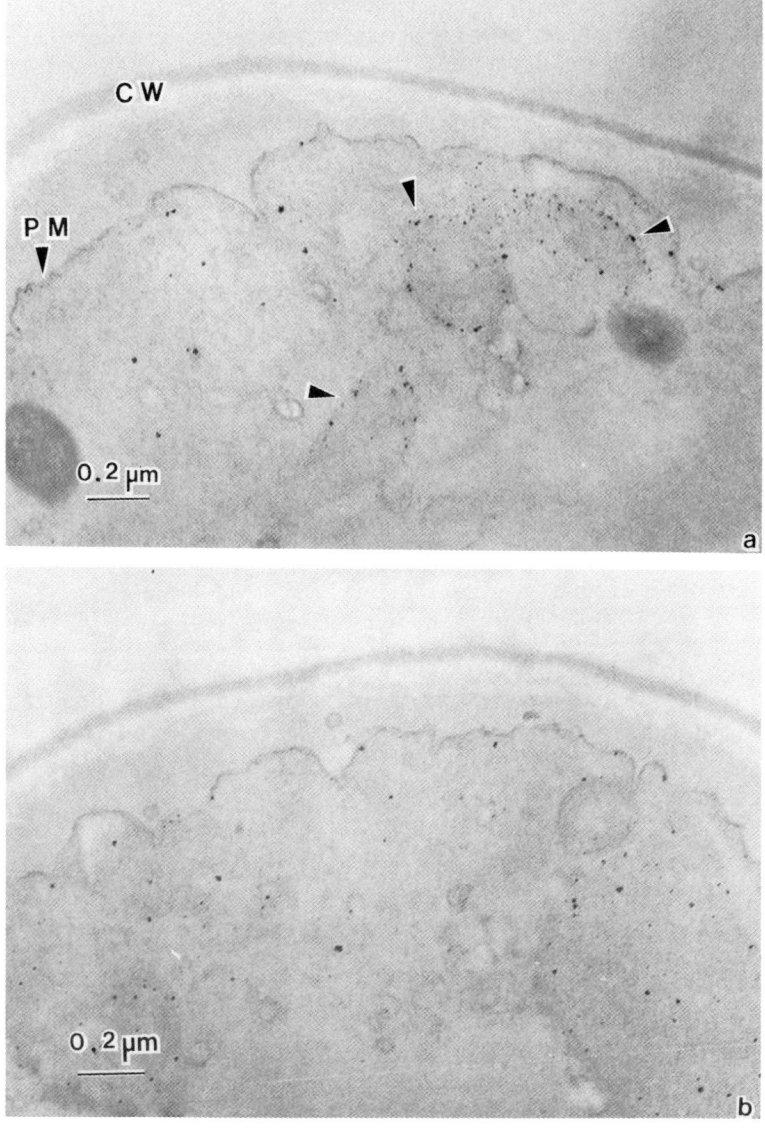

Figure 2   Fracture-label immuno-electron microscopy.

Thin sections are in the plane of fracture of BSA-embedded yeast spheroplasts. a. ferritin particles (arrowed) are localized at the periphery of membrane-bound subcellular organelles, possibly mitochondria, in a cell expressing pC$_\beta$Z1, the longest hybrid protein. b. Ferritin particles reveal that hybrid protein distribution in pC$_\beta$Z15 plasmids bearing cell appears to be predominantly cytoplasmic with low membrane association. CW=residual cell wall, PM=plasma membrane.

****

frayed appearance of the former with many cells pulled out from the matrix, and the absence of cross-fractured spheroplasts. In contrast, the fracture margins appeared sharp and regular with frequent cross-fractured cells and with occasional cracks from incomplete fracturing.
 In preliminary experiments using the fracture-label technique, hybrid protein was detected by primary anti-β-galactosidase antibody, and was visualized with ferritin-conjugated anti-rabbit antibody. Ultrastructural examination of thin sections revealed pC$_\beta$Z1 hybrid protein to be predominantly localized at the periphery of organelle profiles (Fig. 2a), with a low cytoplasmic background. In contrast, for pC$_\beta$Z15 the secondary probe exhibited a high degree of cytoplasmic association as well as some membrane targeting (Fig. 2b). Control experiments, using pre-immune serum in place of the primary antibody (data not shown), or using non-plasmid-bearing cells (Fig 1, m-o), did not show secondary probe labeling.

DISCUSSION

The information for efficient targeting of proteins synthesized on cytoplasmic ribosomes is inherent within the sequence and probably the secondary structure of the nascent precursor. Our subcellular fractionation studies demonstrate that hybrid proteins containing different N-terminal portions of the F_1 ATPase β-subunit exhibit different degrees of targeting to the mitochondrion. Comparisons with a matrix-directed fusion protein, and an outer membrane-directed hybrid protein both fused to an active β-galactosidase protein, indicate that the difference in targeting detected by immunofluorescence and freeze-fracture cytochemistry reflects a true distribution difference, not merely altered degrees of antibody accessibility to the antigen. Controls using antibodies against the F_1 ATPase β-subunit (wild type) and citrate synthase (present in the mitochondrial matrix) support this finding. Whether the lowered efficiency of mitochondrial targeting of pCβZ15 compared with pCβZ1 results directly from the absence of a specific portion of the β-subunit protein sequence, or whether the truncated polypeptide experiences a simple physical impediment, for example, direct blocking or masking of crucial targeting recognition sequences by the β-galactosidase portion of the hybrid, remains to be determined. The difference in distribution of the anti-β-galactosidase antiserum detected in the pCβZ15 transformed strain, compared with the other transformed strains examined could result from a differential accessibility of the antigen, rather than a truly altered distribution. While such possibilities are difficult to exclude, they are unlikely to be the cause of the observed differences, since distribution of citrate synthase and the F_1 β subunit is revealed with essentially equal intensity in all the strains, using the same technique. Our results demonstrate that immunocytochemical techniques applied to fixed cells, at the levels of fluorescence, and electron microscopy, provide a useful complementary visualization approach to the classical biochemical subfractionation analyses.

ACKNOWLEDGEMENTS

We thank Laura Vallier and Marjorie Britten for expert technical support and Janet Kendall for skillful editorial assistance.

REFERENCES

1. Takeda M, Vassarotti A, Douglas M (submitted). Nuclear genes coding the yeast mitochondrial adenosine triphosphatase complex: the catalytic subunits of the F_1-ATPase are imported independently.
2. Emr S, Vassarotti A, Garrett J, Geller B, Takeda M, Douglas M (submitted). The amino terminus of the yeast F_1-ATPase β-subunit precursor functions as a mitochondrial import signal.
3. Douglas M, Koh Y, Dockter M, Schatz G (1977). Aurovertin binds to the β-subunit of the yeast mitochondria ATPase. J Biol Chem 252:8333-8335.

4. Marshall-Carlson L, Vallier L and Douglas M Hybrid protein targeting analysis in mitochondrial biogenesis (in preparation).
5. Pinto da Silva P, Parkison C, Dwyer N (1981). Fracture-Label: Cytochemistry of freeze-fracture faces in the erythrocyte membrane. Proc Natl Acad Sci USA 78:343-347.

IN VIVO TARGETING AND ASSEMBLY OF F_1 ATPase β-SUBUNIT DELETIONS

Alessio Vassarotti[2], Cynthia Smagula[3] and Michael G. Douglas[3]

Department of Biochemistry, University of Texas Health Science Center
San Antonio, Texas 78284

ABSTRACT Nested deletions centered on codon 34 of the ATP2 gene encoding a set of internally deleted mitochondrial F_1-ATPase β subunit proteins have been constructed and analyzed. These constructs delete in some cases a portion or all of the F_1 β presequence. Surprisingly large deletions in this region of the F_1 β protein will be tolerated by the cell for its in vivo delivery to mitochondria and assembly into a functional complex. Only 15 amino terminal residues of the ATP2 presequence are required for in vivo assembly into the functional mitochondrial complex. Deletions from codon 34 to codon 47 and beyond will target but may not assemble the functional enzyme within the mitochondria.

[1]This work was supported by the USPHS Grant GM26713 and Grant AQ-814 from the Robert A. Welch Foundation
 [2]Present Address: Laboratoire del'Heredite Cytoplasmique des Plantes Cultivees, Universite de Louvain, Belgium
 [3]Present address: Department of Biochemistry UT Health Science Center, Dallas, Texas, 75235

INTRODUCTION

Most mitochondrial proteins are encoded in the nucleus and synthesized in the cytoplasm. Recent studies have focused on the genetic and biochemical analysis of various steps in the delivery and import of these proteins into mitochondria (1). The F_1-ATPase β subunit has been defined as one of the best models to examine proteins which are targeted to the mitochondrial matrix. Its location within mitochondria in addition to the cellular and energy requirements for its import into the organelle are the most thoroughly documented for any mitochondrial protein.

The ATP2 gene encoding the $F_1\beta$ subunit (2) has been utilized to construct gene fusions to genes encoding non-mitochondrial proteins in order to monitor $F_1\beta$-protein sequence-dependent targeting in vivo (3). These and additional studies have defined the mitochondrial targeting determinants for its in vivo delivery to the first 27 residues of the cytoplasmic precursor subunit (4).

In the present studies, a set of plasmid encoded ATP2 genes harboring deletions within the first 50 codons has been constructed and expressed in the atp2⁻ deletion host to define the determinants for both the targeting and functional assembly of the protein.

METHODS

Construction of a Chromosomal atp2⁻::LEU2 Mutant

A yeast chromosomal mutant was constructed using the single step gene replacement method (5). The wild type strain SEY2102 MATα ura3 leu2 his4 (3) was transformed with a linear fragment of DNA containing the yeast LEU2 gene flanked by ATP2 DNA. The fragment was generated such that 350 bp of DNA 5' to ATP2 as well as the first 300 bp of ATP2 coding DNA was replaced by the LEU2 gene and its adjacent sequences. The resulting pet mutant, AVY4/1, contained LEU2 in place of ATP2. It lacked the expression of any detectable ATP2 gene product and was unable to grow on a mitochondrial dependent carbon source. This nonreverting atp2⁻::LEU2 ura3⁻ host served as a recipient for analysis of various plasmid encoded ATP2 constructions transformed into it (see below).

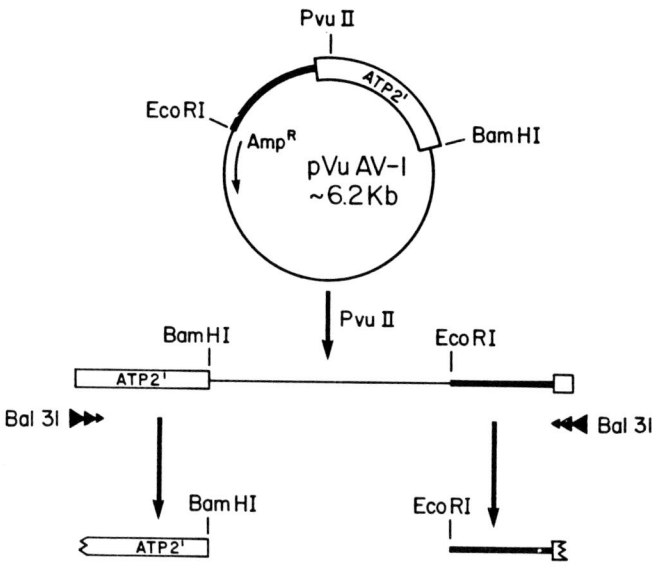

Figure 1. 5' Deletions in ATP2

Construction of Deletions in the 5' Region of ATP2

The ATP2 gene contains a unique PvuII site at codon 34. A pBR322 derivative containing an Eco-Bam fragment of yeast DNA encoding the 5' region and 380 of the 509 codons of ATP2 was opened at the PvuII site. Following a brief digestion with Bal31 exonuclease, the subsequently generated fragments (Figure 1) were religated into the starting construction to replace either the wild type Eco-Pvu or Pvu-Bam fragments. This strategy generated a set of nested 5' and 3' deletions centered on codon 34 which were subsequently screened for correct reading frame by ligating the deleted Eco-Bam fragment into the yeast-E. coli shuttle plasmid pSEY101 (Figure 2) and scoring for ATP2 dependent β-galactosidase expression in yeast (3).

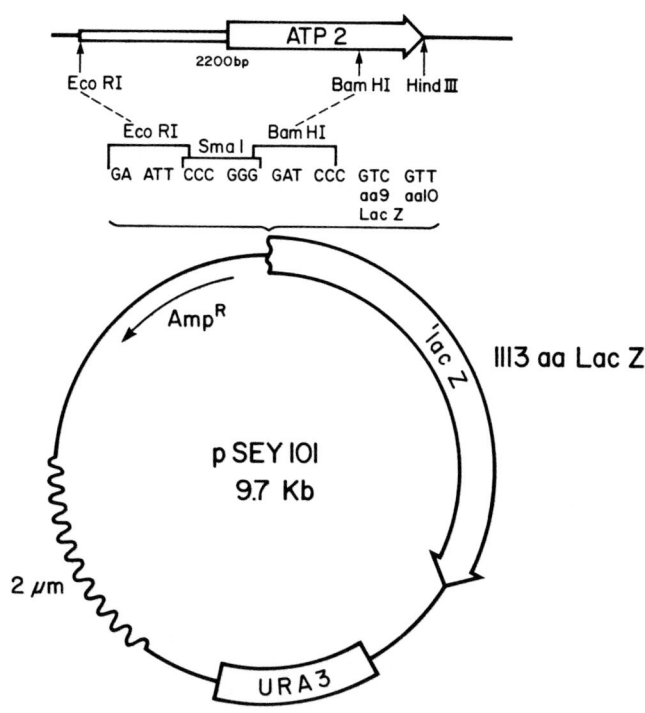

Figure 2. lacZ Fusion Plasmid

An intact plasmid encoded ATP2 gene containing the deletions was constructed in a derivative of the plasmid pSEY101 which contained coding DNA for ATP2 codons 380 to 509 plus 280 bp of the 3' noncoding region in place of lacZ. This URA3, 2 micron plasmid, pAVO-10 retained the Eco and Bam sites of pSEY101 for a one step insertion of the modified Eco-Bam fragments of ATP2.

RESULTS

The deletions in Table I define the limits of ATP2 for in vivo delivery of the F_1 β subunit to mitochondria and its assembly into a functional energy transducing complex. When the mitochondrial delivery of

TABLE 1
SUMMARY OF DELETIONS WITHIN THE FIRST 46 CODONS OF ATP2

ATP2 Deletions	ATP2-lacZ plasmid[a] ATP2 leu2 Host		ATP2 plasmid[b] atp2::LEU2 Host
	Mitochondrial Delivery	Glycerol Phenotype	Glycerol Phenotype
complete	+	−	+
* − 35	−	+	−
15 − 35	−	+	+
16 − 35	+/−	+/−	+
24 − 35	+/−	+/−	+
27 − 35	+	−	+
33−38	+	−	+
33−42	+	−	+
33−47	+	−	−

[a] ATP2 constructs expressed in pSEY101
[b] ATP2 constructs expressed in pAVO-10
*a non ATP2 sequence expressed in place of the first 34 codons

the hybrid pβZ1 product (380 aa of the β subunit-fused to β-galactosidase) harboring the deletions was determined by subcellular fractionation and hybrid protein-dependent phenotype (3), those which exhibit less efficient targeting in vivo were deletions which extend into the region of the precursor protein near the processing site for the matrix localized chelator-sensitive protease (6). A control gene fusion which replaced the first 34 codons of ATP2 with another ATG start codon, blocked delivery of the hybrid lacZ gene product to mitochondria. Therefore, the determinants for efficient in vivo delivery of the F_1 β subunit are present within the amino terminal 15-16 residues. This was confirmed by analysis of ATP2 reconstructants harboring the deletions in the nonreverting atp2⁻ host. Only the deletion which completely removed the presequence (*-35) was unable to import and assemble a functional complex.

All deletions between codon 16 and 34 will support growth of AVY4/1 on a nonfermentable carbon source. Deletions 3' of codon 34 define sequences which are required for expression of a functional enzyme in vivo. Expression of the ATP2 33-42 construction will support mitochondrial dependent growth of AVY4/1 whereas the 33-47 will not. Studies to determine whether the imported subunit is either not assembled or assembled into a nonfunctional enzyme are currently in progress.

DISCUSSION

The present studies have further defined the limits of the F_1-ATPase subunit required for its in vivo delivery to mitochondria and assembly into a functional complex. Deletions within the first 42 codons of ATP2 which leave the first 15 residues of the F_1 precursor intact will still express, correctly localize and assemble the protein. This observation is in accord with in vitro analysis of another matrix targeted protein precursor sequence which directs the import of a nonmitochondrial protein into mitochondria (7). The loss of the hybrid protein dependent phenotype due to the F_1-lacZ hybrid is defined by deletions which retain 15 versus 16 amino terminal residues of the precursor. Since these deletions extend to the processing site for the presequence, these data suggest that the processing of the F_1-Z hybrid precursor "locks" it into the inner membrane (3) where it blocks further oxidative phosphorylation-dependent growth. Plasmid linked mutations which suppress the hybrid protein dependent phenotype in the wild type host are expected to map in this region of the ATP2 gene. It is anticipated that deletions within ATP2 near the region of codon 47 will most likely alter assembly and/or function of the correctly delivered peptide. Analysis of these constructs and second site mutations which suppress these plasmid deletions to restore growth will further define the role of these determinants in vivo.

ACKNOWLEDGMENTS

We would like to thank Laura Vallier, Marjorie Britten and Jill Siegel for their technical assistance and Janet Kendall for typing this manuscript.

REFERENCES

1. Hay R., Bohni P, Gasser S (1984). How mitochondria import proteins. Biochim Biophys Acta 779:65-87.
2. Saltzgaber-Muller J, Kunapuli S, Douglas M (1983). Nuclear genes coding the yeast mitochondrial adenosine triphosphatase complex: Isolation of ATP2 coding the F_1-ATPase β subunit. J Biol Chem 258:11465-11470.
3. Douglas M, Geller B, Emr S (1984). Intracellular targeting and import of an F_1-ATPase β-subunit β-galactosidase hybrid protein into yeast mitochondria. Proc Natl Acad Sci 81:3983-3987.
4. Emr S, Vassarotti A, Geller B, Garrett J, Takeda M, Douglas M (1985). Sequences necessary for mitochondrial targeting of the F_1-β ATPase β-subunit protein, in press.
5. Rothstein R (1983). One-step gene disruption in yeast. Methods Enzymol 101:202-211.
6. McAda P, Douglas M (1982). A neutral metallo endoprotease involved in the processing of an F_1 ATPase β subunit precursor in mitochondria. J Biol Chem 257:3177-3182.
7. Hurt E, Pesold-Hurt B, Schatz G (1984). The amino terminal region of an imported mitochondrial precursor polypeptide can direct cytoplasmic dihydrofolate reductase into the mitochondrial matrix. EMBO J 3:3149-3156.

IMPORT OF A PUT2-LACZ HYBRID PROTEIN INTO THE MITOCHONDRIAL MATRIX OF SACCHAROMYCES CEREVISIAE[1]

Marjorie C. Brandriss and Karen A. Krzywicki

Department of Microbiology
UMDNJ-New Jersey Medical School
Newark, New Jersey 07103

ABSTRACT Pyrroline-5-carboxylate dehydrogenase, the second enzyme in the S. cerevisiae proline degradative pathway and the product of the PUT2 gene, was localized in the soluble matrix space of the mitochondria using the fractionation procedures of Daum et al (1). The PUT2 gene was fused to the E. coli lacZ gene to form hybrid proteins containing, respectively, the aminoterminal 14 and 124 residues of pyrroline-5-carboxylate dehydrogenase attached to the catalytically active carboxyterminal portion of β-galactosidase. Both gene fusions caused the production of β-galactosidase that was regulated like the PUT2 gene product. The shorter gene fusion made β-galactosidase that was located in the cytoplasm of the cell. The longer gene fusion produced mitochondrially-associated β-galactosidase that was found in the matrix space and had an effect on the proper localization of the wild-type pyrroline-5-carboxylate dehydrogenase.

INTRODUCTION

The movement of proteins from their sites of synthesis to their ultimate location within or outside of cells has been the subject of intense study recently. Studies on secreted, plasma membrane, or vacuolar proteins in E. coli (reviewed in ref. 2), S. cerevisiae (reviewed in ref. 3), and in mammalian cells (4) have revealed the importance of signal sequences, carbohydrate addition and cytoplasmic factors in this process. For proteins that are to be

[1]This work was supported by Public Health Service grant GM 30405 from the National Institute of General Medical Sciences.

imported into the mitochondria, there is a signal or pre-sequence which is often, but not always, cleaved off during the import process and a requirement for energy if the ultimate location is the inner mitochondrial membrane, the intermembrane space, or the matrix space. A review on this subject has recently appeared (5).

Several laboratories have applied the gene fusion approach to the study of mitochondrial protein import in an attempt to learn more about the nature of the process. Schatz and coworkers reported the in vitro targeting and delivery of a cytosolic protein, murine dihydrofolate reductase, to the matrix compartment using amino-terminal sequences from cytochrome oxidase subunit IV (6). Hase et al. (7) fused the gene encoding the 70 kD protein of the outer mitochondrial membrane to β-galactosidase and found that it was properly delivered to the mitochondrial surface. Douglas et al (8) have reported the successful import of a large hybrid protein carrying 380 aminoterminal residues of the β subunit of the F1 ATPase attached to the catalytically active carboxyterminus of β-galactosidase. The β subunit of the F1 ATPase is part of a large complex of polypeptides that are associated with the inner mitochondrial membrane but is itself considered a soluble matrix-localized subunit. The hybrid protein was found associated with the inner mitochondrial membrane and resulted in respiratory deficiency for those cells carrying the gene fusion. From the results found in E. coli with β-galactosidase and those in yeast with the F1 ATPase hybrid protein, it had been thought that β-galactosidase could not pass through membranes, but would always get stuck due to the nature of its sequence or conformation.

We have attempted to direct β-galactosidase to the mitochondrial matrix by attaching to it a matrix pre-sequence. The pre-sequence was derived from pyrroline-5-carboxylate (P5C) dehydrogenase, the product of the PUT2 gene, that participates in the mitochondrial pathway of proline utilization in S. cerevisiae (9,10,11). P5C dehydrogenase was shown to be a mitochondrially-localized enzyme (12) and behaved as if it were soluble in the matrix space. In in vitro translation and import assays, P5C dehydrogenase was synthesized as a precursor of approximately 64 kD that was subsequently processed by removal of about 2 kD (Kaput and Brandriss, unpublished results).

We report here the localization of P5C dehydrogenase to the matrix compartment, the construction of a gene fusion between the PUT2 gene and lacZ of E. coli carrying 124 aminoterminal codons of P5C dehydrogenase and the carboxyterminus of β-galactosidase, and the localization of this hybrid protein within the mitochondrial matrix.

METHODS

Construction of PUT2-lacZ Gene Fusions

Two put2-lacZ fusions were constructed to contain the PUT2 promoter sequences and either 14, or 124, codons ligated to a truncated lacZ gene. Since the sequences of both PUT2 and lacZ have been determined (11,13), these fusions were predicted to be in frame and to cause the production of proline-inducible β-galactosidase.

Plasmid pKB1 (10) was digested with BglII and SalI to remove all PUT2 sequences except for the 5' noncoding (promoter) region and the first 14 codons of the structural gene. The 6.2 kB lacZYA fragment contained on plasmid pSKS107 (14, a gift from M. Casadaban) was removed by partial SalI and complete BamHI digestion and ligated to the BglII-SalI fragment of plasmid pKB1. The new plasmid, pKB25, contained put2-lac14, URA3 and 2μ and pBR322 DNA for selection and maintenance in either E. coli or S. cerevisiae. A yeast integrating plasmid, pKB27, was derived from plasmid pKB25 by digestion with EcoRI to remove the 2μ sequence.

Plasmid pKB11 (10) was digested with EcoRI and BglII (partially) to remove a fragment from the PUT2 gene containing 0.7 kb 5' noncoding (promoter) sequences and coding sequences up to codon 124. Plasmid pSKS107 was digested with EcoRI and BamHI and ligated with the EcoRI-BglII fragment from pKB11, so as to place the PUT2 promoter 5' to the lacZ sequences in plasmid pKB34. The put2-lacZ124 construct was removed from plasmid pKB34 by digestion with HindIII and SalI (partial) and ligated to plasmid pKB11, cut with HindIII and SalI, to form the autonomously replicating, URA3-marked plasmid pKB41, or to HindIII and SalI-cut YIp5 (15) to form the integrating plasmid, pKB42.

Strains

S. cerevisiae strain MB1433 (MATα trp1 ura3-52 PUT2) was used as the recipient for all transformation experiments. Plasmids pKB27 and pKB42 were each linearized at a unique site in the PUT2 region and used to transform strain MB1433 to Ura$^+$. Since linearized plasmid DNA is known to integrate by homologous recombination in S. cerevisiae (16), these plasmids were directed to integrate at the PUT2 region on chromosome VIII. The strains MB1452 and MB1457 resulted from such integration events and carried a single wild-type PUT2 gene in tandem with, respectively, seven copies of the put2-lacZ14 or two copies of the put2-lacZ124 gene fusion.

Each strain was then crossed to MB1057 (MATa his4-42 PUT2) and meiotic products were analyzed for the appropriate segregation of the various markers. MB643-3B (MATa his 4-42 trp1 PUT2::put2-lacZ14 URA3 and MB652-3A (MATa trp1 PUT2::put2-lacZ124 URA3) were segregants that were used in the submitochondrial fractionation experiments described below.

Subcellular and Submitochondrial Fractionations

The fractionation procedures of Daum et al. (1) were used to determine the localization of various enzymes within the cell and the mitochondria. In the subcellular localization experiments, the cytoplasmic and particulate fractions were analyzed to determine total distribution of the enzymes. Glucose-6-phosphate dehydrogenase and fumarase were used as markers of the cytoplasm and mitochondria, respectively. In the submitochondrial fractionation procedures, inner and outer membranes were not separated, but were treated as total membranes, represented by the behavior of cytochrome c oxidase. Fumarase and cytochrome b_2 (measured as lactate dehydrogenase activity) were the marker enzymes for the matrix and intermembrane space compartments, respectively. Strain MB1000 was grown either on a minimal medium containing 2% galactose and 0.1% proline or the yeast extract-lactate-ammonia medium of Daum et al. (1) to which 0.1% proline was added. Strains MB643-3B and MB652-3A were grown on the medium containing yeast extract-lactate-ammonia-proline.

Enzyme Assays

Published procedures were used for the assays of cytochrome c oxidase (17), lactate dehydrogenase (18), fumarase (19), β-galactosidase (20), glucose-6-phosphate dehydrogenase (21) and P5C dehydrogenase (9). Protein was determined by the method of Bradford (22) with crystalline bovine serum albumin as the standard.

RESULTS

Submitochondrial Localization of P5C Dehydrogenase

Mitochondria were isolated from cells of the wild-type strain, MB1000, and fractionated and analyzed according to procedures described by Daum et al (1, see Methods) into total mitochondrial membranes, intermembrane space and matrix fractions. Each fraction, as well as total unfractionated mitochondria were assayed for cytochrome c oxidase (inner membrane marker), cytochrome b_2 (intermembrane space marker), fumarase (matrix marker) and P5C dehydrogenase activities and total protein. The specific activity of each enzyme in each fraction was compared to the specific activity of the unfractionated mitochondria and that ratio ("relative specific activity") was plotted against the percent of total mitochondrial protein represented by each fraction. Figure 1 shows the result of the fractionation procedure.

Cytochrome c oxidase and cytochrome b_2 activities fractionated as expected for inner membrane and intermembrane space enzymes. Fumarase, a known matrix enzyme, reproducibly appeared predominantly in the matrix, but also in the intermembrane space. The fumarase activity in the intermembrane space probably reflects premature rupture of the inner mitochondrial membrane in an early step in the procedure in which only the outer membrane should lyse. This pattern differs somewhat from the published results of Daum et al (1), and may reflect strain differences in the osmotic sensitivity of the mitochondrial membranes. Nonetheless, the marker enzymes show distinct patterns for the various mitochondrial fractions and indicate a reasonable isolation of the various mitochondrial compartments. The results for the fractionation of P5C dehydrogenase clearly indicate that it is a matrix-localized enzyme, behaving analogously to fumarase.

Behavior of Put2-lacZ Gene Fusions

Two gene fusions between the S. cerevisiae PUT2 gene and the E. coli lacZ gene were constructed as described in Methods. The put2-lacZ14 fusion contained the PUT2 promoter region as well as the initial 14 codons of the deduced P5C dehydrogenase coding sequence ligated to a gene fragment containing the lacZ gene beginning at its eighth codon. The put2-lacZ124 fusion carried DNA corresponding to the first 124 codons of P5C dehydrogenase ligated to the same lacZ fragment. These fusions, whether located on autonomously replicating plasmids or integrated into the yeast genome, caused the production of proline-inducible β-galactosidase regulated analogously to that found for the P5C dehydrogenase (Table 1). Both fusions were integrated adjacent to the PUT2 locus on chromosome VIII (data not shown).

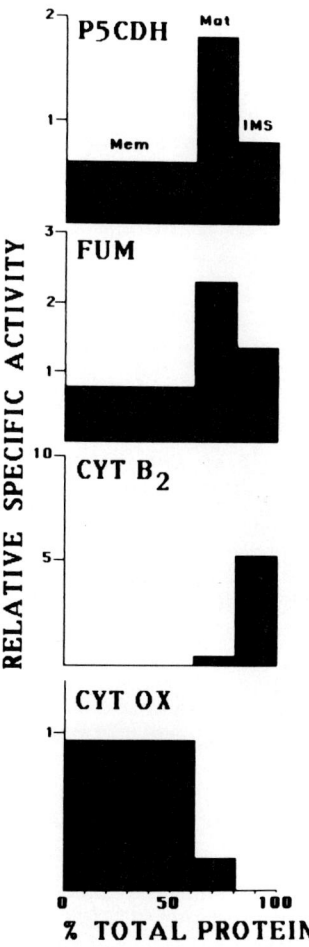

FIGURE 1. Distribution of enzymes in subfractions of yeast mitochondria in the wild-type strain. Yeast mitochondria were fractionated as described in Methods and each fraction was assayed for the indicated enzymes. The ordinate denotes the specific activity of each enzyme as a fraction of the specific activity found in intact mitochondria. The abscissa denotes the percentage of mitochondrial protein recovered in each subfraction. Mem, inner and outer membranes; Mat, matrix; IMS, intermembrane space.

TABLE 1
ENZYME LEVELS OF STRAINS CARRYING PUT2-LACZ FUSIONS[a]

Specific Activity[b]

Strain	β-galactosidase			P5CDH		
	Amm	Amm + Pro	Pro	Amm	Amm + Pro	Pro
MB643-3B	57	402	848	15	77	199
MB652-3A	46	442	885	6	36	170

[a] Growth media contained 2% glucose and 0.2% ammonium sulfate or 0.1% proline or both, supplemented with histidine (50 mg/l) and tryptophan (50 mg/l).
[b] Given as nanomoles of ONPG hydrolyzed or NADH formed per min per mg protein.

Subcellular Localization of the P5C Dehydrogenase-β-galactosidase Hybrid Proteins

Subcellular fractionation procedures were employed to determine the location of the β-galactosidase in the yeast cells carrying the gene fusions. The activities of β-galactosidase, glucose-6-phosphate dehydrogenase (G6PDH, a cytoplasmic marker) and fumarase (a mitochondrial marker) were determined in lysed spheroplasts, cytoplasm and a crude mitochondrial pellet, as described in Methods. The results are listed in Table 2. In the strain carrying the put2-lacZ14 fusion, β-galactosidase and glucose-6-phosphate dehydyrogenase behaved similarly with greater than 99% of both activities found in the cytoplasm and only 1% in the mitochondria. In the strain carrying the put2-lacZ124 fusion, the β-galactosidase and the fumarase activities behaved similarly with approximately 50-60% found in the pellet fraction. From these results, it appears that the shorter hybrid protein remains in the cytoplasm while the longer hybrid protein is delivered to the mitochondria.

Submitochondrial Localization of the Hybrid β-galactosidases

Both yeast strains carrying the hybrid β-galactosidases were

TABLE 2
SUBCELLULAR FRACTIONATION OF β-GALACTOSIDASE IN THE GENE FUSION STRAINS

Gene Fusion	Enzyme	% Total Activity in Cytoplasm	Mitochondria
put2-lacZ14	G6PDH	99	1
	Fumarase	23	77
	β-gal	99	1
put2-lacZ124	G6PDH	99	1
	Fumarase	41	59
	β-gal	48	52

fractionated as described in the first section to determine the distribution of β-galactosidase and P5C dehydrogenase within the mitochondrial compartments. Figure 2 shows the results of those experiments. In the strain with the predominantly cytoplasmic β-galactosidase, the mitochondrial enzymes fractionate as in the wild-type strain. The β-galactosidase activity that is found in the mitochondria represents only 1% of the total cellular activity and is localized predominantly in the intermembrane space. In the strain carrying the mitochondrial β-galactosidase, it is evident that the hybrid protein fractionates like fumarase, and appears to be predominantly in the matrix compartment. The specific and total activities of P5C dehydrogenase are lower but its distribution is similar to that in the wild-type or cytoplasmic fusion strains.

DISCUSSION

The addition of the 124 aminoterminal residues of P5C dehydrogenase to β-galactosidase is sufficient to deliver the enzyme to the mitochondria and to localize it properly in the matrix compartment. This is the first report of a mitochondrially-imported β-galactosidase that appears to be soluble, i.e. not membrane associated, and disputes the current notion that β-galactosidase cannot pass through membranes. The laboratories of Douglas and Emr (8) have previously reported mitochondrial targeting and delivery of hybrid β-galactosidases that contained aminoterminal residues of the β-subunit of the F1 ATPase, a protein associated

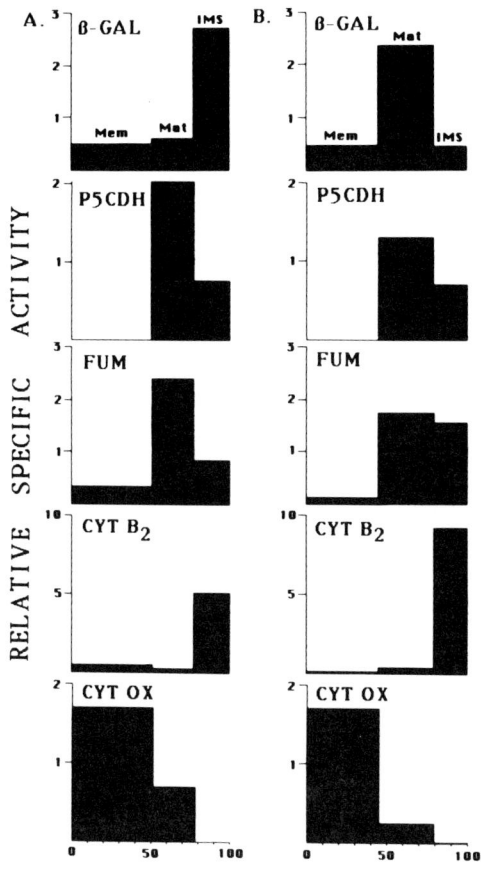

FIGURE 2. Distribution of enzymes in subfractions of yeast mitochondria in put2-lacZ gene fusion strains. A, strain MB643-3B containing the cytoplasmically localized put2-lacZ14 hybrid protein. The mitochondrial β-galactosidase activity represents less than 1% of the total activity. 99% is cytoplasmic. B, strain MB652-3A containing the mitochondrially-localized put2-lacZ124 hybrid protein. Axes and abbreviations are as in Fig. 1.

with the inner mitochondrial membrane. The β-galactosidase was shown to be tightly associated with the inner mitochondrial membrane and the presence of some of these hybrid proteins resulted in respiratory deficiency.

The β-galactosidase hybrid protein carrying only 14 residues of P5C dehydrogenase is predominantly localized in the cytoplasm. Of the activity that is mitochondrial, very little passes through the inner membrane. It is as if there is insufficient signal sequence to allow the proper sorting of this hybrid protein.

The decrease in activity of the P5C dehydrogenase in the presence of the P5C dehydrogenase-β-galactosidase hybrid protein suggests a form of competition between the two polypeptides. The nature of this competition is not understood at present but could reflect titration of import receptors recognizing this particular signal sequence, decreased processing of the precursor P5C dehydrogenase to its mature form, or perhaps a physical interaction between the dehydrogenase and the β-galactosidase.

ACKNOWLEDGMENTS

We are grateful to Dr. J. Kaput for helpful suggestions in the fractionation procedure.

REFERENCES

1. Daum G, Bohni PC, Schatz G (1982). Import of proteins into mitochondria. J Biol Chem 257:13028.
2. Silhavy TJ, Benson SA, Emr SD (1983). Mechanisms of protein localizatin. Microbiol Rev 47:313.
3. Stevens T, Esmon B, Schekman R (1982). Early stages in the yeast secretory pathway are required for transport of carboxypeptidase Y to the vacuole. Cell 30:439.
4. Gilmore R, Blobel G (1983). Transient involvement of signal recognition particle and its receptor in themicorsomal membrane prior to protein translocation. Cell 35:677.
5. Hay R, Bohni P, Gasser S (1984). How mitochondria import proteins. Biochim Biophys Acta 779:65.
6. Hurt EC, Pesold-Hurt B, Schatz G (1984). The amino-terminal region of an imported mitochondrial precursor polypeptide can direct cytoplasmic dihydrofolate reductase into the mitochondrial matrix. EMBO J 3:3149.
7. Hase T, Muller U, Riezman H, Schatz G (1984). A 70 kd protein of the yeast mitochondrial outer membrane is targeted and anchored via its extreme amino terminus.

EMBO J 3:3157.
8. Douglas MG, Geller BL, Emr SD (1984). Intracellular targeting and import of an F-ATPase β-subunit-β-galactosidase hybrid protein into yeast mitochondria. Proc Natl Acad Sci USA 81:3983.
9. Brandriss MC, Magasanik B (1979). Genetics and physiology of proline utilization in Saccharomyces cerevisiae: enzyme induction by proline. J Bacteriol 140:498.
10. Brandriss MC (1983). Proline utilization in Saccharomyces cerevisiae: analysis of the cloned PUT2 gene. Mol Cell Biol 3:1846.
11. Krzywicki KA, Brandriss MC (1984). Primary structure of the nuclear PUT2 gene involved in the mitochondrial pathway for proline utilization in Saccharomyces cerevisiae. Mol Cell Biol 4:2837.
12. Brandriss MC, Magasanik B (1981). Subcellular compartmentation in control of converging pathways for proline and arginine metabolism in Saccharomyces cerevisiae. J Bacteriol 145:1359.
13. Kalnins A, Otto K, Muller-Hill B (1983). Sequence of the lacZ gene of Escherichia coli. EMBO J 2:593.
14. Casadaban MJ, Martinez-Arias A, Shapira SK, Chou J (1983). β-galactosidase gene fusions for analyzing gene expression in Escherichia coli and yeast. Meth Enzymol 100:293.
15. Botstein D, Falco SC, Stewart SE, Brennan M, Scherer S, Stinchcomb DT, Struhl K, Davis RW (1979). Sterile host yeasts (SHY): a eukaryotic system of biological containment for recombinant DNA experiments. Gene 8:17.
16. Orr-Weaver TL, Szostak JW, Rothstein RJ (1981). Yeast recombination: a model system for the study of recombination. Proc Natl Acad Sci USA 78:6354.
17. Mason TL, Poyton RO, Wharton DC, Schatz G (1973). Cytochrome c oxidase from bakers' yeast I. J Biol Chem 248:1346.
18. Appleby CA, Morton RK (1959). Lactic dehydrogenase and cytochrome b_2 of baker's yeast. Biochem J 71:492.
19. Racker E (1950). Spectrophotomatic measurements of the enzymatic formation of fumaric and cis-aconitic acids. Biochim Biophys Acta 4:211.
20. Guarente L (1983). Yeast promoters and lacZ fusions designed to study expression of cloned genes in yeast. Meth Enzymol 101:181.
21. Clifton D, Weinstock SB, Fraenkel DG (1978). Glycolysis mutants in Saccharomyces cerevisiae. Genetics 88:1.
22. Bradford MM (1976). A rapid and sensitive method for the

quantitation of microgram quantities of protein utilizing the principle of protein dye binding. Anal Biochem 72:248.

DO MITOCHONDRIA AND NUCLEI SHARE TRANSFER RNA MODIFICATION ENZYMES?

Diana Najarian[1], Steven Ellis[1], Melitta Dihanich[2], Michael Morales[1], Anita Hopper[2], and Nancy Martin[1]

Department of Biochemistry, Southwestern Graduate School Univ. of Texas Health Science Center at Dallas[1], Texas Department of Biological Chemistry, Hershey Medical School, Penn State University, Hershey, PA[2]

Abstract: Single nuclear mutations in TRM1 and MOD5 genes abolish m_2^2G and i^6A in both cytoplasmic and mitochondrial tRNAs. To determine how such mutations can affect tRNA modification in two distinct cellular compartments we have cloned the genes and are currently examining their structure and expression.

INTRODUCTION

Many pairs of enzymes that catalyze the same reactions have been found in both mitochondrion and cytosol or nucleus. In each case, the isoenzymes are known to be coded by separate nuclear genes and the proteins differ in ways which presumably serve as the basis for their intracellular compartmentalization (1). Data in contrast to the widely held notion that mitochondrial enzymes differ genetically and structurally from their nuclear/cytoplasmic counterparts are rare. The cytosolic and mitochondrial fumarases from mouse and human cells have been suggested to be encoded by the same structural gene (1). Other apparent exceptions are tRNA processing enzymes in the yeast, Saccharomyces cerevisiae. Previous studies have shown that single nuclear mutations can abolish tRNa

---

Supported by grants from NSF to A.K.H. and N.C.M., the American Heart Association to N.C.M., the NIH to S.R.E., and the Austrian Government and the Austrian American Fullbright Commission to M.E.D.

tRNA modifications in both cytoplasmic and mitochondrial tRNAs (2,3).

Nuclear mutations in the TRM1, TRM2, and MOD5 loci abolish m_2^2G, m^5U, and i^6A respectively in cytoplasmic and mitochondrial tRNAs. The enzymes which carry out these modifications are reduced in activity in in vitro assays using whole cell (4,5) or mitochondrial extracts of the mutants (6). The most straightforward explanation is that a single nuclear gene codes for an enzyme active in both the nuclear/cytoplasmic and mitochondrial compartments. Alternatively, TRM1, TRM2, and MOD5 could code for proteins essential for the biosynthesis or regulation of tRNA modification enzymes which have different structures in the different compartments.

We have isolated the wild-type alleles of trm1 and mod5-1 to enable us to initiate genetic and biochemical experiments to determine if they are structural genes for their respective modification enzymes. If they are, our long range goal is to determine how a single nuclear gene can code for a protein that must be shared between the mitochondria and nuclear/cytoplasmic compartments.

RESULTS

Cellular Location of tRNA Modification Enzymes

The location of tRNA modification enzymes is largely unknown in any organism. In yeast, tRNA(guanosine-N^2,N^2) dimethyltransferase and tRNA(uridine-5)methyltransferase can be measured in extracts of purified mitochondria and mitochondrial tRNAs carry this modification. Since nuclear precursor tRNAs (pretRNAs) also carry both the m_2^2G and m^5U modifications in yeast, we infer that the methyltransferases are located in the nucleus. At the present time we do not have the tools necessary to determine if these transferase enzymes are excluded from the cytoplasm (ie., nuclear and mitochondrial restricted) or whether they move about the cell by some as yet undefined simple diffusion process in yeast. In contrast, the ability to inject macromolecules into the nucleus and cytoplasm of Xenopus oocytes provides the opportunity to differentiate between these two possibilities.

Substrate tRNA for the injection experiments was prepared from trm2 cells labeled with ^3H uracil. Such tRNA is m^5U deficient so it can be injected into the nuclei and cytoplasm of oocytes, recovered, and analyzed for m^5U. As can be seen in figure 1, only the tRNA that was injected

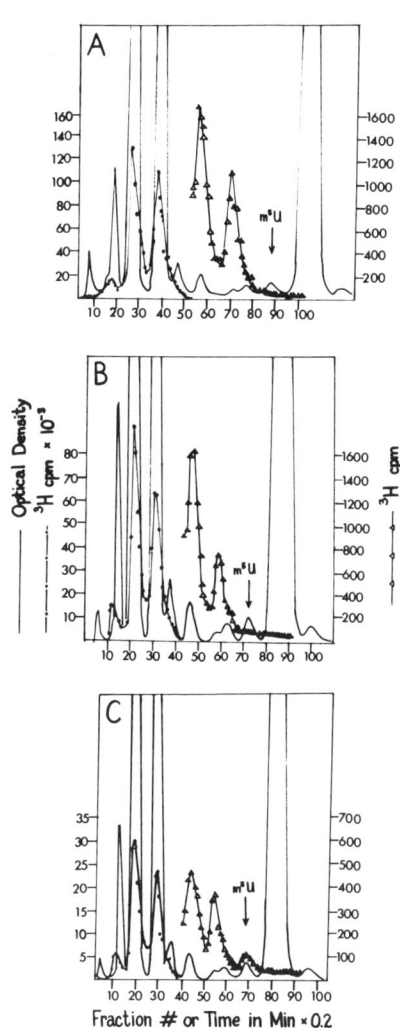

FIGURE 1: HPLC Elution Profile for Ribonucleosides Derived from m^5U Deficient ^3H Uracil-Labeled tRNA Injected into <u>Xenopus</u> Oocytes. ^3H-uracil labeled tRNA (4 x 10^5 cpm/μg) from a <u>trm1</u> <u>trm2</u> mutant yeast strain was injected into <u>Xenopus</u> oocytes. The injected tRNA was incubated for 3 hr and then isolated from the oocytes by proteinase K digestion, three phenol:chloroform extractions, and ethanol precipitation. The RNAs were purified further by DE52 chromatography and then digested to nucleosides and resolved by HPLC chromatography as described previously (2). The OD_{256} (full scale = 0.0400) was monitored and the radioactivity of collected fractions (0.2 min/fraction) determined.

A. Profile of nucleosides from 1.8 x 10^6 cpm RNA not injected into <u>Xenopus</u> oocytes.

B. Profile of nucleosides from 1.3 x 10^6 cpm RNA injected into oocyte cytoplasm. No m^5U modification was detected.

C. Profile of nucleosides from 5 x 10^5 cpm RNA injected into oocyte nuclei. Approximately 16% of the injected tRNAs contain m^5U. This value is in agreement with the extent of m^5U modification observed when DNA encoding yeast $tRNA^{Tyr}$ is transcribed, processed and modified in <u>Xenopus</u> oocytes (8).

into the nucleus of Xenopus oocytes contained m^5U. This indicates tha tRNA(uridine-5) methyltransferase is present in the nucleus, but not the cytoplasm of oocytes. The fact that modifications are found in nuclear restricted pretRNAs and that in Xenopus the methyltransferase is excluded from the cytoplasm, suggests that the tRNA methyltransferases are localized to nuclear and mitochondrial compartments. No modification was detected in similar experiments with i^6A deficient tRNA. This negative result and the data of others showing that the i^6A modification is only on mature-size tRNAs and not on pretRNAs (7,8) means that this modification could be added as a late step in the nucleus or in the cytoplasm. There is precedence for cytoplasmic addition of Q, another alteration of the anticodon loop (9).

Cloning of genes that complement trm1 and mod5-1

The only phenotype of trm1 and trm2 is loss of m_2^2G or m^5U and methyltransferase activity so transformation of genes that correct these defects could only be detected by enzyme assays. A trm1 trm2 mutant was transformed with

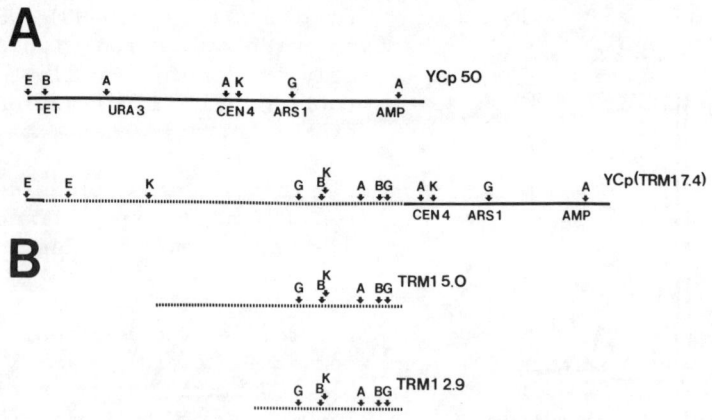

FIGURE 2: Isolation and subcloning of the TRM1 gene. A) Map of YCp50 and its derivative which restores tRNA(guanosine-N^2,N^2,)dimethyltransferase activity to trm1 trm2 cells. Solid lines are vector, stippled lines, the cloned fragment. Enzymes: A, ScaI; B, BamHl, E, EcoRI; G, BglII, and K, KpnI. B) Fragments of the 7.4 Kb TRM1 insert that were subcloned into YEp24 and restored m_2^2G dimethyltransferase activity to trm1 trm2 cells.

genomic fragments cloned into YCp50 and the transformants assayed for methyltransferase activity. Three of 5000 restored tRNA(guanosine-N^2,N^2) dimethyltransferase activity, none restored tRNA(uridine-5) methyltransferase activity. An analysis of the transformants with tRNA(guanosine-N^2,N^2) dimethyltransferase activity showed that each had at least two plasmids but all shared one plasmid type. Transformations with the common plasmid were unsuccessful due to the deletion of the URA3 gene from the YCp50 plasmid carrying the sequences that restored the dimethyltransferase activity (Figure 2). We continued to study this TRM1 clone by doing cotransformations with a "helper" YCp50. Subcloning of the TRM1 sequences (Figure 2) and integration of these sequences into the TRM1 locus have been accomplished (Ellis, unpublished).

The mod5-1 mutation decreases the efficiency of the tyrosine inserting ochre suppressor, SUP7. In a mod5-1 background, the SUP7 i^6A^- tRNA is incapable of suppressing the ade2-1 or can1-100 mutations. Transformants containing a copy of the MOD5 gene will appear wild-type with respect to ade and can markers if the i^6A is restored by complementation of mod5-1. Transformation with a yeast genomic library constructed in the vector YEp24 yielded two related plasmids that restored suppression (Figure 3). Genetic analyses of integrants demonstrated that the sequence is an allele of MOD5-1 (Dihanich, unpublished).

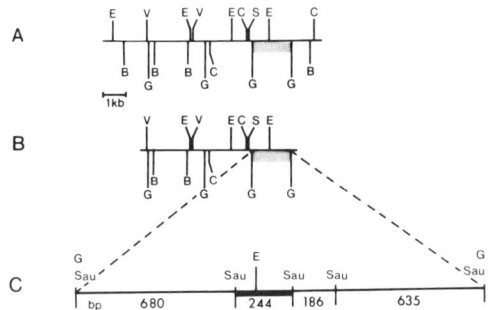

FIGURE 3: Restriction maps of recombinant DNAs which complement the MOD5-1 mutation. A) 9.9 kb insert of YEpMOD5 (9.9); B) 7.0 kb insert of YEpMOD5 (7.0); C) enlargement of 1.7 kb BglII fragment that complements the mod5-1 modification defects. Stippled area: 1.7 kb BglII fragment. Enzymes: B,BamHI; C,ClaI; E,EcoRI; G,BglII; S,SalI; Sau,Sau3A; V,PvuII.

Restoration of modification enzymes by cloned genes.

Since the original selection of the TRM1 gene required the restoration of m_2^2G dimethyltransferase activity it was clear that the YCp50 clone carried a genomic fragment that restored the enzyme activity. As a first step towards determining if the TRM1 locus codes for the tRNA(guanosine N^2,N^2) dimethyltransferase we compared the activity of a transformant carrying the TRM1 gene on YEp24 to the activity found in a wild type cell (Table 1). The fact that the activity is elevated compared to wild-type cells suggests that TRM1 is the structural gene for the transferase enzyme.

TABLE 1
tRNA(GUANOSINE $N^2,N^2,$) DIMETHYLTRANSFERASE ACTIVITY

Strain	Plasmid	pmol/100µg/90 min
MH41		20
trm1 trm2	YEp24	–
trm1 trm2	YEp(TRM12.9)	320
trm1 trm2	YEp(TRM15.0)	300

Activity - pmol methyl group donated (acid precipitable counts) from (methyl-^{3}H) S-adenosyl methionine to trm1 tRNA in 90 min at 30° using 100 µg of cell extract protein.

Cell extracts of YEpMOD5 (7.0) transformants were assayed for Δ2 IPP transferase activity. We do not detect this activity in mod5-1 cells but it is present in an integrant of YEpMOD5 (Table 2). Transformants with multi-copy plasmids show elevated activity when compared to wild type cells (Najarian, unpublished), suggesting that YEpMOD5 (7.0) does contain the structural gene for the Δ2 IPP.

Restoration of tRNA modification by the cloned genes.

A most interesting aspect of the trm1 and mod5-1 mutations is that they affect modification of both nuclear and mitochondrial tRNAs. The enzyme assays using

TABLE 2

Δ2 IPP TRANSFERASE ACTIVITY IN CELL EXTRACTS

pmoles/50 µg Yeast Protein/60 min

Strain	100,000 x g Supernatant	0-35% Ammonium Sulfate	35-55% Ammonium Sulfate	>55% Ammonium Sulfate
MH41, wild-type	12.6	9.1	33.3	0
H43, i^6A-mutant	0	-	-	-
H43 YEpMOD5 (7.0) integrant	12.5	8.9	37.3	0

Cell homogenate was assayed for ability to catalyze the transfer of the isopentenyl group of ^{14}C-IPP to i^6A-tRNA. Following a 45' incubation at 37°, the samples were extracted with phenol:$CHCl_3$(1:1) and incorporation of label into acid-precipitable material measured.

whole cell extracts show that the cytoplasmic/nuclear enzyme activities had been restored in the transformants. Since it is most unlikely that mitochondrial tRNAs exit the mitochondria to be modified, we use the presence or absence of modified bases in mitochondrial tRNA to determine if the enzyme activity has been restored in that cellular compartment.

The modified base m^2_2G was detected by HPLC in digests of whole cell and mitochondrial tRNAs isolated from TRM1 cells cotransformed with YCp(TRM1 7-4) and YCp50 (Figure 4).

i^6A is also clearly present in total cytoplasmic tRNA of the transformants because suppression has been restored. If i^6A is found in mitochondrial tRNA it will be retarded upon RPC chromatography relative to undermodified tRNA (3). The mitochondrial tryptophanyl tRNA contains i^6A and a comparison of it from wild-type, mod5-1, and transformants by RPC 5 chromatography shows that transformation of mod5-1 cells with YEpMOD5(7.0) restores i^6A (Figure 5). Cytoplasmic tryptophanyl tRNA does not contain i^6A so its mobility is unchanged.

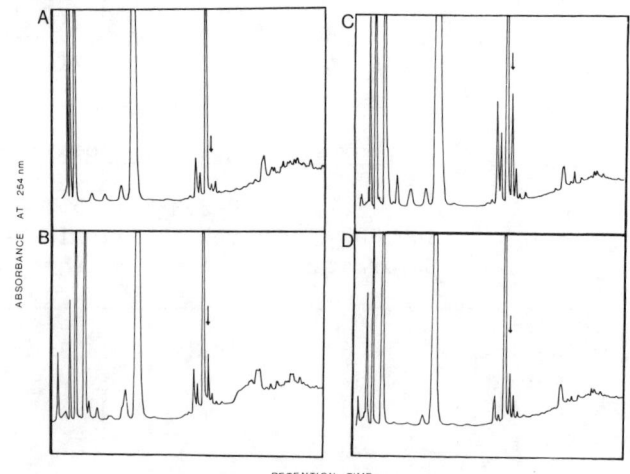

FIGURE 4: Restoration of m_2^2G modification by YCpTRM1. A: HPLC profile of nucleosides from total tRNA of a trm1 trm2 mutant, B: wild type, C: trm1 trm2 cotransformed with YCp50 and YCp TRM1, D: Mitochondrial tRNA from the cotransformant. Position of m_2^2G (↓).

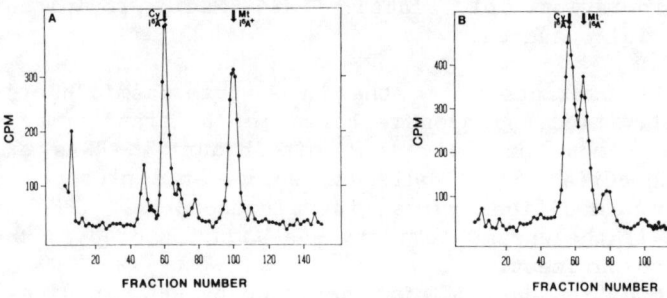

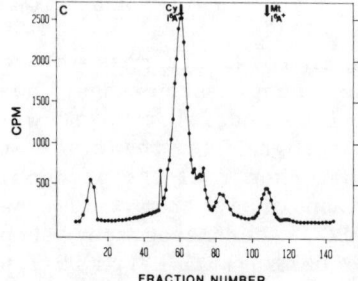

FIGURE 5: RPC-5 elution profile of cytoplasmic (Cy) and mitochondrial (Mt) $tRNA^{Trp}$. Whole cell tRNA was isolated from wild-type (A), mod5-1 (B), and mod5-1 with an integrated copy of YEpMOD5 (7,0) (C). Aminoacylation with ^{14}C-tryptophan and chromatography was performed as described previously (3).

DISCUSSION

We are interested in the TRM1 and MOD5-1 genes because mutations in them abolish tRNA modification enzyme activity in both nuclear/cytoplasmic and mitochondrial compartments. It is possible that these genes code for an element necessary to activate transcription of genes that then would code for distinct nuclear/cytoplasmic and mitochondrial tRNA modification enzymes or for proteins necessary for the posttranscriptional processing of products from different genes. Nonetheless, the most straightforward explanation of the results we have obtained is that the genes code for the tRNA modification enzymes and that the same or similar proteins are targeted to the mitochondria as well as the nucleus and/or in the case of the Δ2 isopentenyl transferase, the cytoplasm. The fact that transformants carrying multicopy plasmids contain elevated dimethyltransferase and Δ2 isopentenyl pyrophosphate transferase activity suggests that these genes are the structural genes for their respective enzyme activities.

TRM1 and MOD5 both code for poly A^+ RNA and contain open reading frames that could encode proteins of 570 and 311 amino acids respectively (unpublished). To determine if these genes are actually structural genes for the modification enzymes we are trying to purify the transferase proteins. Another approach we are using is to express TRM1 and MOD5 in expression vectors so that we can assay such engineered proteins for enzyme activity as well as use the protein products to raise antibodies.

If, as we suspect, TRM1 and MOD5 do code for tRNA modification enzymes, our future goal is to determine if and how the nuclear and mitochondrial forms of the enzymes differ from each other and how the products of these genes are localized to two different cellular compartments.

ACKNOWLEDGMENTS

We would like to thank Dr. Howard Laten for sending us the mod5-1 mutant and Dr. Doug Melton for performing the Xenopus oocyte injections, and Raquel Voss for her dedication in preparing the manuscript. Betty Zehfus and Mildred Symonette provided valuable assistance.

REFERENCES

1) Doonan S, Barra D, and Bossa F (1984) Structure and Genetic Relationships Between Cytosolic and Mitochondrial Enzymes. Int. J. Biochem. 16:1193.
2) Hopper, A, Furukawa A, Pham H, and Martin N (1982). Defects in cytoplasmic and mitochondrial tRNA modifications are caused by single nuclear mutations. Cell 28:543.
3) Martin N, and Hopper A, (1982) Isopentenylation of Cytoplasmic and Mitochondrial tRNA is Affected by a Single Nuclear Mutation. J. Biol. Chem. 257:10562.
4) Phillips J, and Kjellin-Straby K, (1967) Studies on microbial ribonucleic acid: IV Two mutants of Saccharomyces cerevisiae Lacking N^2-dimethylguanine in Soluble Ribonucleic Acid. J. Mol. Biol 26:509.
5) Laten H, Timmons R, and Suid S (1985) An Anti-suppressor Mutant of Saccharomyces cerevisiae Deficient in Isopentenylated tRNA$_2$ Has Reduced Δ^2 Isopentenylpyrophosphate: tRNA-Δ^2 Isopentenyl Transferase Activity, FEBS Letts. 179:307.
6) Smolar N, and Svensson I (1974) Transfer RNA Methylating Activity of Yeast Mitochondria. Nucl. Acids Res. 1:707.
7) Knapp G, Beckmann JS, Johnson PF, Fuhrman S.A. and Abelson J, (1978) "Transcription and Processing of Intervening Sequences in Yeast tRNA Genes: Cell 14, 221-236.
8) Nishikura K and DeRobertis EM, (1982) RNA Processing in Microinjected Xenopus Oocytes; Sequential Addition of Base Modifications in a Spliced Transfer RNA, J. Mol. Biol. 145, 405-420.
9) Carbon P, Haumont E, DeHenau S, Keith G and Grosjean H, (1982) Enzymatic Replacement In Vitro of the First Anticodon Base of Yeast tRNA^{Asp}: Application to the Study of tRNA Maturation In Vivo, After Micro-injection into Frog Oocytes, Nucleic Acids Res. 10, 3715-3732.

VIII. MODELS FOR DEVELOPMENT

SPECIALIZED CELL TYPES IN YEAST: THEIR USE IN ADDRESSING PROBLEMS IN CELL BIOLOGY

Ira Herskowitz

Department of Biochemistry and Biophysics
University of California, San Francisco
San Francisco, California 94143

ABSTRACT The specializations exhibited by the three yeast cell types (a, α and a/α) include production of peptide pheromones and response to these pheromones, a highly controlled genetic rearrangement event, and the ability to enter meiosis. Studies of these properties of yeast cells make it possible to learn about several important areas of cell biology, in particular, cell specialization, proteolytic processing, the cell cycle and growth control, protein localization, chromosome structure, meiosis, and morphogenesis.

INTRODUCTION

Each of the yeast cell types plays a distinctive role in the life of yeast. a and α cells are mating types specialized for mating with each other. The ability to mate involves an intercellular signalling system with peptide growth factors and corresponding receptors: mating factors produced by each cell type act on the other cell type to coordinate the cell cycles of the mating partners prior to cell and nuclear fusion. This is obviously a group of phenomena with rich possibilities for addressing problems in cell biology. The result of mating is the a/α cell, which is unable to mate but which has the ability to undergo meiosis and spore formation upon the appropriate nutritional signal, starvation. Yeast thus offers the opportunity to study not only regulation of the mitotic cell cycle by the mating factors, but also to study exit from the mitotic cell cycle—entry into the important shunt, meiosis. A third important facet of the yeast life cycle is the ability of the two different mating types to interconvert, to switch from one to the other, which occurs at high frequency in certain genetic backgrounds. The process of mating type interconversion is well understood at the molecular level and involves genetic rearrangement, a movement of genetic cassettes. How this process is controlled and coordinated with the cell cycle has revealed a wealth of phenomena that bear on important questions of cell and developmental biology such as periodic expression in the cell cycle and the production of specific cell lineages.

A single genetic locus, the mating type locus (MAT) is ultimately responsible for programming the three cell types. As described in some detail below, the alleles of MAT code for regulatory proteins whose targets are genes coding for the proteins that are intimately involved in processes such as pheromone production and response. These studies of cell biology thus provide information not only on a process itself but often include information on the synthesis of the proteins that are involved in carrying out the process.

The primary purpose of this paper is to describe the relevant aspects of yeast cell type specializations and how they are being used to study various important questions in cell biology. The format of this paper is to identify the problem, question, or area of cell biological interest and then describe the ways in which it is studied using yeast mating types. The intent is that this paper will serve as an entry point to the literature to provide more specific information and details. A rather extensive bibliography is provided. Inevitably, I have had to leave out direct reference to many worthy papers, for which I apologize. General reviews of yeast mating type and mating type interconversion are available (refs 1-5). The particular areas of cell biology that I shall discuss in this article are: cell specialization, proteolytic processing, the cell cycle and growth control, protein localization, chromosome structure, meiosis, and morphogenesis.

CELL SPECIALIZATION

In development of multicellular organisms, cells become differentiated and specialized to carry out different tasks. In many cases, one cell produces a characteristic set of proteins that is absent from another cell, which reflects differences in transcription of the corresponding genes. Why specialized cells express their distinctive set of genes is a major problem in cell and developmental biology. I shall first describe the specialized cell types of yeast and the proteins and genes that characterize them. Then I shall describe the regulatory locus (the mating type locus) that is responsible for differences in expression among the cell types.

Cell Types and Cell-Type-Specific Differences.

As already noted, yeast has three types of cells, **a**, α, and **a**/α. The **a** and α types are haploid and are specialized for mating; the **a**/α cell is unable to mate but is capable of sporulation upon starvation. These differences in cellular behavior are readily assayed by simple physiological tests (the confrontation and halo assays for α-factor, 6, 7; Bar assay, 8, 9) and by biochemical methods (10, 11). **a** and α cells each secrete a characteristic mating factor, **a**-factor and α-factor, respectively. These mating factors are oligopeptides, 11 amino acids for **a**-factor and 13 amino acids for α-factor. Each mating type also contains a response system (discussed in the section, 'The Cell Cycle and Growth Control') that enables it to respond to the mating factors of the opposite type. Thus, **a** cells respond to α-factor, and α cells respond to **a**-factor. The ultimate response to the mating factors is arrest in the cell division cycle at a particular point in G1. **a** and α cells also agglutinate with each other, which reflects some type of agglutinin on the cell surface, about which relatively little is known. Another difference between **a** and α cells is that **a** cells have the ability to degrade α-factor (8, 10, 12). **a**/α cells do not mate, hence it is not surprising that they do not exhibit any of specializations that are involved in mating: they do not produce mating factors, nor do they respond to the factors. Furthermore, they do not agglutinate with any cell type and do not degrade α-factor.

Regulation of Cell Type by the Mating Type Locus.

All of the properties of the three cell types just described (both macroscopic and microscopic) are under control by alleles of the mating type locus, MATa and MATα. Cells with the a allele exhibit an a phenotype, those with the α allele exhibit the α phenotype, and those with both alleles exhibit the a/α phenotype. The mating type locus is responsible for producing three regulatory activities, α1, α2, and a1-α2 (13). These three activities govern expression of four gene sets whose members are scattered throughout the genome (see Fig 1).

The α-specific gene set (αsg) is made up of genes whose transcripts are produced only in α cells (and not in a or in a/α cells). Examples of α-specific genes are STE3 (14) and MFα1 (15).

The a-specific gene set (asg) is made up of genes that are expressed only in a cells (and not in α or in a/α cells). Examples of a-specific genes are STE6 (11) and STE2 (16)

The 'haploid-specific' gene set (hsg) is made up of genes that are expressed in both a and α cells but not in a/α cells. Examples are HO (17), RME1 (18), and the Ty element (19).

The a/α-specific gene set (also termed 'sporulation-specific' gene set, ssg) is made up of genes that are expressed in a/α cells but not in a or α cells. Transcripts expressed in this manner have been identified (see 20, 21, and Kurtz et al, this volume).

MATα codes for two activities, α1 and α2: α1 is a positive regulator of the α-specific genes. It is required for production of the transcripts for these genes (14, 15) and is likely to act to stimulate transcription initiation (or possibly message stabilization). α2 is a negative regulator of the a-specific genes. It is a site-specific DNA-binding protein that recognizes a site upstream of the members of the a-specific gene set (22) whose binding presumably prevents transcription initiation. Thus in an α cell, α1 activates the α-specific genes, and α2 turns off the a-specific genes; hence, the appropriate genes are expressed in an α cell. In an a cell, the absence of α2 allows the a-specific genes to be expressed; the absence of α1 results in failure to express the α-specific genes.

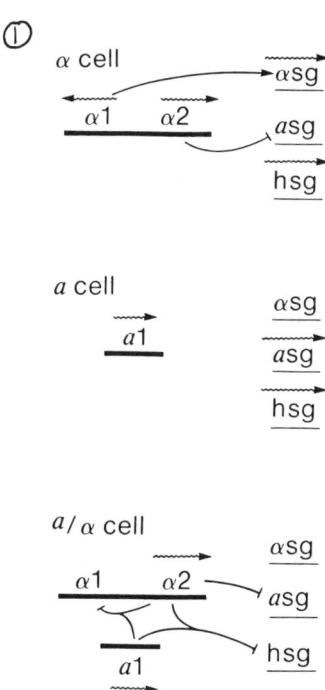

Figure 1. Control of cell-type-specific gene sets by the mating type locus. The mating type locus is drawn to the left, and the regulated gene sets (αsg, α-specific genes; asg, a-specific genes; hsg, haploid-specific genes) to the right. Sporulation-specific genes (ssg) are described in Fig 4. Wavy arrows indicate transcription; pointed arrowheads indicate stimulation of gene expression by α1; lines with blunt ends indicate inhibition of gene expression by α2 or a1-α2. Modified from ref 13.

MATa contains only one gene, MATa1, that plays a role in cell type determination (23, 24). The a1 gene has no known role in conferring the a cell phenotype, but it is crucial for the a1-α2 activity of a/α cells. a1-α2 is a negative regulator of expression of the set of haploid-specific genes; inhibiting expression of these genes requires both a1 and α2 products. a1-α2 is also formally an activator of expression of the a/α-specific genes (ssg), in that both a1 and α2 are required for expression of these genes. The actual molecular constitution of the a1-α2 activity is not known, nor is its molecular mechanism of action. By analogy with α2, it has been proposed that a1-α2 also binds to DNA (25, 16); a putative recognition site has been identified (16). a1-α might thus act as a repressor to turn off the haploid-specific genes and as an activator to turn on the sporulation-specific genes. Another possibility is that a1-α2 promotes expression of the sporulation-specific genes by repressing an inhibitor of these genes. We have recently obtained evidence (discussed below in 'Meiosis') indicating that this is the case for at least some of the sporulation-specific behaviors of an a/α cell (18).

The nucleotide sequences of MATa and MATα and the open reading frames that correspond to α1 and α2 have been identified (26). The MATa1 gene is unique in yeast in that its transcript contains two introns (27; see Guthrie et al, this volume). Whether different forms of the a1 polypeptide are produced is not known.

PROTEOLYTIC PROCESSING

Synthesis of Mating Factors.

Both mating factors are oligopeptides. In both cases, the genes coding for a putative or demonstrated precursor have been cloned. There are two genes (MFα1 and MFα2; MFα= α mating factor) coding for the α-factor precursor (28, 29), and two genes coding for the a-factor precursor (MFA1 and MFA2) (30). The MFα1 gene is the major contributor to α-factor synthesis—inactivation leads to a large decrease in total α-factor synthesis (31; K Luper and I Herskowitz, unpublished). It contains an open reading frame of 165 amino acids that has been confirmed to be a precursor in vivo (32). The structure of the gene led to a proposal of how it is processed (see Fig 2). In particular, the first half of the gene contains a hydrophobic signal as well as a segment containing three glycosylation sites. This half of the precursor is used as a 'delivery system' for secretion of the attached α-factor peptides and can also be used to secrete other polypeptides (such as somatostatin, interferon, etc) that have been joined in vitro (33, 34). The second half of the precursor contains four mature copies of α-factor separated from each other by a peptide 'spacer' which has the form, Lys-Arg-Glu-Ala-Glu-Ala-Glu-Ala. The precursor is cleaved at the spacer to liberate the mature α-factor. Two of the enzymes that carry out this processing have been identified. First, the product of the KEX2 gene cleaves after the Arg residues (35; Fuller et al, this volume). Then the product of the STE13 gene, a dipeptidyl aminopeptidase, removes the Glu-Ala residues sequentially until the first residue of mature α-factor is encountered (36). The specific enzymes responsible for removal of the Lys and Arg residues from the carboxy terminus of the α-factor have not yet been identified. Even though only α cells produce α-factor, both **a** and a/α cells contain all of the enzymatic machinery to produce α-factor; the only block to α-factor synthesis in these cell types is failure of the α-factor transcript to be produced (37).

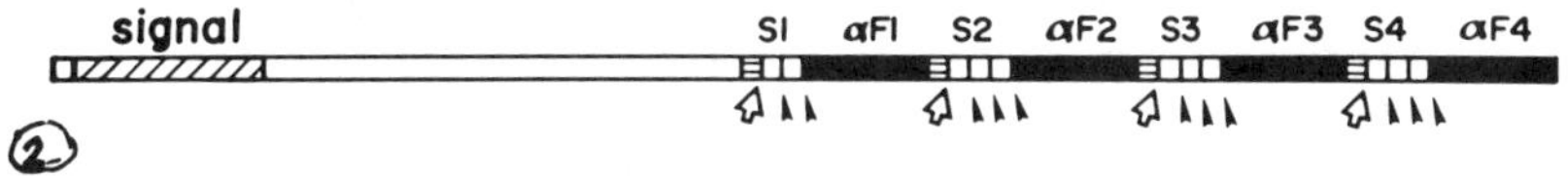

Figure 2. Structure and proteolytic processing sites in the precursor to α-factor. The precursor coded by the MFα1 gene is shown (28, 32). S1-S4 are peptide spacers and αF1-αF4 are mature α-factor segments. The open arrows indicate the positions of cleavage by the KEX2 protein (after Lys-Arg); filled arrowheads indicate the positions of cleavage by the STE13 protein (after Glu-Ala or Asn-Ala). Modified from ref 28.

Our present understanding of biosynthesis of **a**-factor is much less developed. The two **a**-factor genes contain open reading frames of 36 and 38 amino acids, each of which includes a single mature **a**-factor (30) preceded by the dibasic sequence, Lys-Lys. By analogy with α-factor, it is presumed that this is a processing signal. Three genes have been identifed that appear to be specifically defective in production of **a**-factor: STE6, STE14, and STE16 (11; K. Wilson, unpublished observations). These genes may code for processing enzymes or perhaps are involved in secretion of **a**-factor.

Specific Protein Turnover.

There are several cases—notably in mating type interconversion—in which we expect proteins to be degraded, perhaps due to specific proteolysis. The observation that cells can switch from **a** to α and from α to **a** in a single cell division (38) would seem to require that the cell-type specific products—both the apparatus for responding to the mating factors as well as the regulatory proteins—be unstable. Consider for example an α cell that switches to **a** in its next cell division. The initial α cell mates as α and therefore has the appropriate apparatus for responding to **a**-factor. This apparatus includes the STE3 protein, the likely receptor for **a**-factor, assembled into its cell surface (discussed below). When this α cell undergoes mating type interconversion, its MAT is changed to MAT**a**, and then it expresses its new mating phenotype, which must include synthesis of the new receptor, the STE2 protein. What happens to the old receptor, the STE3 protein? We presume that it is inactivated in some way, for example, degraded. One possible route for this inactivation would be for it to be taken into clathrin-coated vesicles (39, 40; Payne and Schekman, this volume) and then sent to the vacuole for degradation.

Regulatory proteins might also have to be degraded. Consider for example an α cell producing α2, the repressor of **a**-specific genes. If α2 were stable and produced in adequate amounts, then it might repress the **a**-specific genes even after the mating type locus has been removed. (This situation is analogous to the ability of free lambda repressor in the absence of a lambda prophage to confer immunity to superinfection; 41). We have noted (22) that the α2 protein appears to be subject to proteolysis <u>in</u>

vivo: this cleavage may be occurring near the carboxy terminal segment of the α2 protein that is homologous to the homeo domain. Perhaps this segment of α2 (and the homeo domain of Drosophila proteins) is a recognition site for cleavage.

There are numerous other cases where specific proteins might be targeted for destruction. The HO-encoded endonuclease need be produced only at very low level, since it is an enzyme that acts by producing only one double-stranded cut at the mating type locus each cell division cycle 42, 43). This nuclease is not present in daughter cells (38, 44). Thus it must neither be synthesized in daughter cells nor distributed to them. We suggest that HO is an unstable protein that is inactivated after it has functioned, perhaps by being degraded. In all of these cases of inactivation, it shall be interesting to determine whether ubiquitination plays a role in identifying proteins for destruction (see 45).

CELL CYCLE AND GROWTH CONTROL

The yeast cell types offer several different opportunities to study the cell cycle, how it is regulated and how events within it are programmed. I shall discuss two topics that fall into this area: regulation of the cell cycle by the mating factors and periodic expression of the HO gene.

Regulation of the Cell Cycle by Mating Factors.

Both of the mating factors, a-factor and α-factor, cause cells of opposite type to arrest in the G1 phase of the cell cycle (46, 47). This arrest point is the same as for certain cell division cycle mutants (such as those defective in the CDC28 gene) (48). Cells arrested by mating factors are blocked in duplication of the spindle pole body; they continue protein synthesis but do not undergo the initiation of DNA synthesis. Arrest leads to a morphological change in the cells so that they become initially pear-shaped, then more grossly distended. A central question is to determine how the mating factors trigger cell cycle arrest. The mating pheromones also increase expression of a variety of a- and α-specific genes, for example, the induction of STE3 expression by a-factor (49). Understanding how this specific molecular event occurs ought to shed light on the pathway leading to cell cycle arrest.

Genes and proteins involved in response have been identified by isolation of mutants that are resistant to the mating factors (50) or that are simply defective in mating. The STE2 gene appears to be a component of the receptor to α-factor because binding of α-factor to ts ste2 mutants in vitro is temperature sensitive (51). The nucleotide sequence of the STE2 gene shows that it has seven hydrophobic segments capable of spanning a membrane (52, 53). It is thus argued to be an integral membrane protein. The STE3 gene is believed to be a component of the receptor to a-factor because mutants defective in STE3 are resistant to a-factor and because it also has seven hydrophobic segments (53, 54). Binding studies with a-factor have not yet been done. Thus these gene products appear to be the external parts of the response system. The STE2 protein is needed for mating only by a cells, and STE3 only for mating by α cells. The requirements for these products are reflected by their synthesis in the different cell types: STE2 is an a-specific gene (expressed in a cells and not in the other cell types) (see ref 16), and the STE3 gene is an α-specific gene (expressed only in α cells) (14). We thus see why response to α-factor is limited to a cells: only a cells have the receptor. A similar situation exists for response of α cells to a-factor.

It is not known whether the factors are taken up into the cell or trigger their effects simply by binding to their receptor. It has been suggested that α-factor exerts its effects by reducing the activity of membrane-bound adenylate cyclase (55, 56). Further studies of adenylate cyclase have failed to confirm this initial report (57; see Bourne et al, this volume). Therefore, it is unclear at the present time whether adenylate cyclase and cyclic AMP play any role in the action of the mating pheromones. Mutants resistant to α-factor in principle ought to identify intracellular components involved in response. Recent work has shown that some genes required for response--genes STE7, STE11, and STE12--are required because they are needed for full expression of the receptor genes (15; G Sprague Jr, D Chaleff, and S Fields, personal communication). Thus these genes are unlikely to identify components of the response system. In contrast, we have recently identified a new gene, STE18, whose only defect is in response to the mating factors (Kent Matlack, unpublished observations). We anticipate that its product is involved in the response system.

a cells are able to recover from arrest by α-factor, apparently by becoming accommodated to it and not simply by inactivating it (100). The SST2 gene may play a role in this recovery process; mutants defective in SST2 are exquisitely sensitive to mating factors (101).

Periodic Expression of HO.

Cells that undergo mating type interconversion always produce pairs of cells with changed mating type (see Fig 3). Switching in pairs was argued to result from a change in MAT genotype due to genetic rearrangement early in the cell cycle, that is, before the mating type locus was repli-cated (58). Recent work indicates that the expression of the HO gene is limited to a brief window of the cell cycle, in late G1 and early S (44). Furthermore, transcription of HO requires progression through the step of the cell cycle governed by CDC28. Thus HO expression does occur early in the cell cycle, and synthesis of its protein at this time presumably leads to a change of MAT prior to its duplication (discussed also in ref 59). The upstream region of the HO region has been genetically dissected (60) and its sequence determined (60, 61), which reveals that more than 1400 basepairs prior to the HO transcript startpoint are required for proper expression. One part of this region (termed 'upstream region 2') contains a octanucleotide sequence repeated ten times that has been shown to be at least partially responsible for periodic, CDC28-dependent expression of the HO gene (62). It may be important to note that these cell cycle analyses have been carried out with cells that have been synchronized by starvation. Analysis of cell populations isolated from asynchronous vegetative cultures by elutriation has not been reported. Thus the present conclusions concerning cell cycle regulation of HO expression may be relevant only to cells leaving starvation and not necessarily valid for cells growing exponentially.

The next phase of the analysis is to identify the proteins or other factors that are responsible for triggering expression of HO and that presumably act at the octanucleotide site. We have identified five genes, SWI1-SWI5 (SWI, 'switch'), that are necessary for production of the HO transcript (63). Thus the products of these SWI genes may act at these sites to promote HO expression. These SWI genes appear to be involved in more than just inducing HO. This conclusion follows from the observation that the swi mutants are pleiotropic and that certain swi double mutants (swi4 in combination with swi1, swi2, or swi3) are inviable. Perhaps these SWI products are involved in activating expression of certain essential genes in the cell cycle. The SWI gene products may work by antagonizing an inhibitor. We have recently identified several additional genes, SIN ('SWI independent'), that may code for such an inhibitor (Paul Sternberg, Ira Clark, and Mike Stern, unpublished). Whether the SWI or SIN proteins bind to HO DNA in its upstream region is under study.

PROTEIN LOCALIZATION

There are many different places in the yeast cell where proteins must be directed: to different membrane systems (the endoplasmic reticulum and the plasma membrane, for example), to organelles (nucleus, mitochondria, vesicles, and vacuole), and from the cell itself. There are excellent examples of proteins targeted to many of these different places (the exceptions being to the vacuole or mitochondria) that are known from studies of cell specialization. In this section, I shall briefly give examples of such proteins.

Secretion from the Cell.

Both α-factor and a-factor are found in culture fluids (6, 46, 47) and are assumed to be secreted from yeast cells and not liberated by lysis of some cells in the culture. Although a mutational analysis of the α-factor gene has not yet been reported, it appears that the first half of the α-factor precursor (of the major, MFα1, gene; see Fig 2) is responsible for directing secretion from the cell. This conclusion comes from the observation that this segment is sufficient to direct secretion of other proteins from yeast (33, 34). The first 20 amino acids of the precursor polypeptide are hydrophobic and are presumed to be a signal for directing the precursor to the endoplasmic reticulum. Studies of intracellular forms of α-factor demonstrate that it follows the same pathway of secretion as other secretory proteins, using the same SEC genes and acquiring the appropriate modifications (32). Unlike many other proteins that pass through the endoplasmic reticulum, the hydrophobic segment at the amino terminus of the precursor is not removed. Whether yeast has analogues to the SRP and SRP receptor of mammalian cells is not known. An in vitro system capable of protein translocation using the α-factor precursor has been developed and should make it possible to identify the components of the yeast secretory system (64, 65; G Blobel, personal communication).

The open reading frame that presumably is the precursor to a-factor is only 36 (or 38) amino acids in length (30), which is unusually small for a precursor to a secretory protein. The putative a-factor precursor is also striking in that its amino terminal segment does not have the hydrophobic hallmarks of a signal peptide. It is not known whether fusion of any part of the a-factor precursor to a test protein will direct secretion from the cell. Because the putative precursor is so unusual, it is possible that it may exit from the yeast cell by a route different from α-factor or require some accessories for exiting by the same route (Susan Michaelis, personal communication). The yeast genes necessary for production of a-factor (STE6, STE14, and STE16) thus might be involved not only in processing but also in protein localization or secretion.

Another yeast cell-type-specific protein that is apparently secreted from the cell is the 'Bar' protein, encoded by the BAR1 gene. Bar is an activity found in a cell culture fluids that degrades α-factor (8-10, 12).

Integral Membrane Proteins.

The nucleotide sequences of STE2 and STE3 show that the corresponding polypeptides contain seven hydrophobic regions that are capable of spanning a membrane (52-54). Thus the STE2 and STE3 proteins appear to be intimately associated with the cell plasma membrane. How these proteins become directed specifically to the plasma membrane and why they do not become trapped in other membrane systems are intriguing questions.

Vesicles.

STE13 and KEX2 code for proteolytic processing enzymes that act on the α-factor precursor (35, 36). The STE13-dependent dipeptidyl aminopeptidase activity and the KEX2-dependent dibasic-cleavage activity are reported to be in fast-sedimenting particles or to be membrane-associated. Presumably these proteases are within vesicles. Whether these two proteins are located within the same cellular organelle is not presently known. A molecular characterization of yeast vesicles has not been reported.

Nuclear Localization.

There are numerous proteins involved in cell type determination or mating type interconversion that are known to act in the nucleus or that are thought to do so. These include HO and MATα2 proteins, which do act in the nucleus, and presumed regulatory proteins coded by MATα1, MATa1, STE7, STE11, STE12, RME1, the SIR genes (SIR1-SIR4), the SWI genes, and the SIN genes. Only one of these has been studied from the standpoint of nuclear localization, α2 (66; Hall, this volume).

As noted above, α2 is a site-specific DNA-binding protein, a repressor that turns off a-specific genes and that must work in the nucleus. Immunocytochemical localization of an α2-beta-galactosidase hybrid protein demonstrates that it is indeed located in the nucleus. Hybrids with as little as 13 amino acids of α2 are able to confer nuclear localization. This region thus appears to contain a signal for localization, which might be responsible for uptake of the protein into the nucleus or retention within the nucleus. It appears to be small, as in the case of the signal from SV40 T antigen (67). Because several α2-beta-galactosidase hybrids lead to cell death, it may be possible to identify mutants altered in their localization of these proteins and define the machinery of transport.

Mother-Daughter.

One of the most intriguing aspects of mating type interconversion is that mother cells are competent to switch mating types, whereas daughter cells are not (38) (see Fig 3).

Specialized Cell Types in Yeast 639

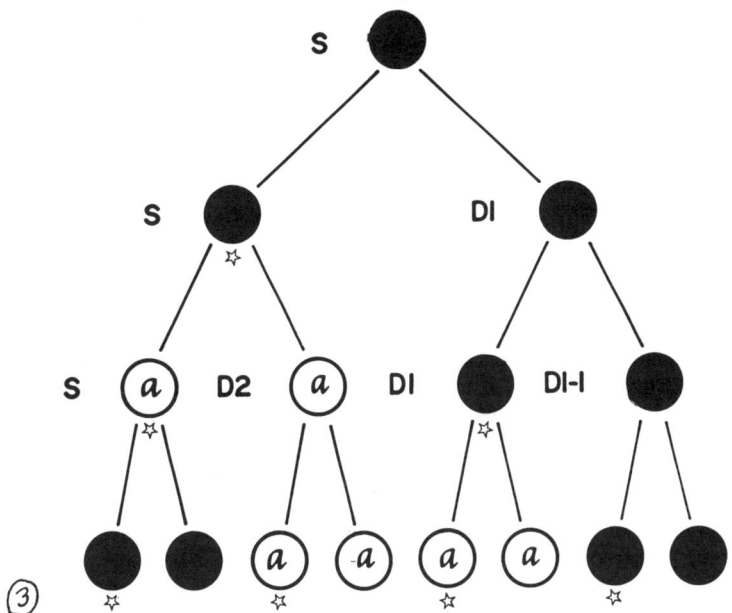

Figure 3. The pattern of switching exhibited by yeast carrying the HO gene. Black cells are α, white cells are a. The initial cell, S, is an α spore that is allowed to germinate and then undergo one cell division cycle to yield two cells, S (the spore cell) and D1 (the spore's first daughter). In the next cell division, the S cell gives rise to cells (S and D2) that have changed mating type to a. Stars indicate cells that are capable of switching mating types in their next cell division. As described in the text, only mother cells are competent to switch. The difference between mother and daughter cells (for example, between S and D1 at the two-cell stage) might reflect asymmetric distribution of material between daughter and mother cell. Modified from ref 38.

The defect in daughter cells appears to be due to failure
to express the HO gene (44); they can switch if HO is
expressed under control by the GAL promoter (59). Why do
daughter cells not produce the HO transcript? We noted
above that the expression of the HO gene requires the
products of five SWI genes. Thus mother cells might contain sufficient amounts of one of these products whereas
daughter cells do not. One specific explanation for this
type of distinction would be, for example, if the product
of the SWI5 gene was produced only early in the cell cycle
(prior to bud emergence) and remained localized with the
mother cell, for example, by binding to the nuclear membrane or spindle pole body. Another explanation for the
difference between mother and daughter cells is that the
daughter cell contains an inhibitor of HO expression,
whereas the mother cell does not. This inhibitor would be
specifically localized to the bud, as has been seen for
newly-synthesized acid phosphatase (68). This inhibitor
would have to be degraded each cell cycle, so that the
daughter cell acquires the ability to produce HO. The SIN
genes may define such inhibitors (Paul Sternberg and Ira
Clark, personal communication).

CHROMOSOME STRUCTURE

Studies of cell type determination and mating type
interconversion have turned up some phenomena whose molecular explanations might involve alterations in chromatin
structure or whose understanding might involve the overall
structure of chromosomes. The first phenomenon is negative
regulation by the products of the SIR genes, which may act
by altering chromatin structure. The second phenomenon
concerns the question of whether the arrangement and location of genetic cassettes—all on chromosome III—has functional significance. This arrangement appears to facilitate mating type interconversion and perhaps reflects a
gross conformation of this chromosome in mitotic cells.

Repression by Sir.

The silent copies that contain MAT information are
located at HML (which has a silent α cassette in standard
strains) and at HMR (which has a silent **a** cassette in

standard strains). Synthesis of the transcripts from these
loci is blocked by products of four SIR genes (also called
'MAR' genes): inactivation of any of these genes allows
expression from both HML and HMR (reviewed in 1, 4, 69).
It is not known whether Sir represses transcription initiation or whether it causes instability of these transcripts.
For the purposes of this description, I shall make the
conventional assumption—that it represses transcription
initiation. The two transcripts from MATα (for MATα1 and
MATα2) are transcribed divergently from a regulatory region
residing between the two genes (70, 71). It is thought
that Sir acts not in this region of HMLα but rather downstream of these transcripts, in sequences (the 'E' site and
'I' site) that are present at HML and HMR and not at MAT
(72, 73). This conclusion comes from analysis of gross
chromosomal rearrangements (see for example 74, 75) and
mutations produced in vitro (72, 73) that lead to constitutive expression, and from placing other genes under Sir
control by coupling them to these regions (76). Placement
of a test gene, even one that is transcribed by RNA polymerase III (a tRNA gene), into the HMR region causes this
gene to become under control by Sir (77). The fascinating
molecular question is how Sir can act more than 1 kb 3' to
the transcript initiation point. One possibility is that
Sir binds at the E and I sites and causes a local change in
chromatin conformation, a specific condensation, in the
region between the E and I sites. Another possibility is
that Sir binds across the entire E and I region and acts
like a prokaryotic repressor to prevent binding of RNA
polymerase and other proteins.

The E and I sites overlap functional ARS elements (72,
73), which are presumed origins of replication. Whether
there is any functional significance to the overlap is not
known. In studies using a sir-ts mutant shifted from high
to low temperature (78), it has been inferred that reestablishment of repression by Sir requires that cells pass into
S. The working view is that Sir affects chromatin structure and that it must be incorporated into a structure or
otherwise act to assemble such a structure during a time
when the DNA is duplicated. It may be useful to note similar temperature-shift experiments using thermosensitive

repressor (cI) mutants of phage lambda (79). In this case, the molecular explanation for failure to restore repression on shift to low temperature resulted from production at high temperature of an inhibitor of repressor synthesis (79, 80).

Sir regulates not only expression from the silent cassettes but also governs their role in the cassette transposition process. In mating-type interconversion, the cassette information from HML or HMR is transposed to MAT (see Fig 4). The distinction between donor loci (HML and HMR) and the recipient locus (MAT) occurs because the HO-encoded endonuclease produces a double-stranded break only at MAT and not at HML or HMR, even though the cutting site is present at all sites (42, 81). In sir⁻ strains, the distinction between donor and recipient cassettes is lost

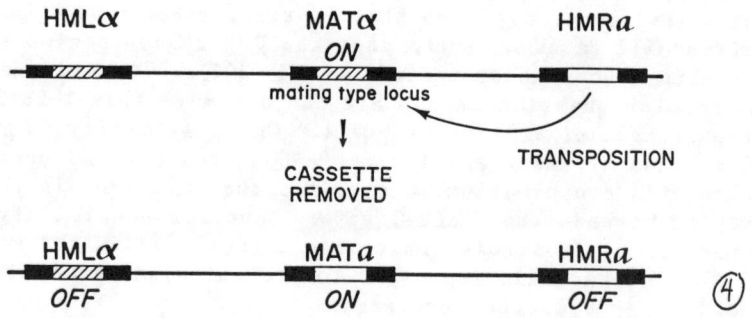

Figure 4. Mating type interconversion occurs by transposition of genetic cassettes. The top line shows the arrangement of cassettes on chromosome III in an α cell. Switching to a occurs by removal of the α cassette from MAT and its replacement by information from HMRa. The central regions of the cassettes (striped or open rectangles) represent the distinctive 'Y' regions of the cassettes. The regions flanking the Y region (black rectangles, 'X' to the left and 'Z' to the right) are involved in the transposition event. The cassettes at HMLα and HMRa are repressed by Sir. Modified from ref 1.

(82). The arguments about Sir causing a change in chromatin conformation or otherwise coating the HML and HMR regions easily accommodate the finding that they prevent cutting by the HO nuclease. Control of expression of a cassette is therefore correlated with its ability to participate in a recombinational event, in this case, as a recipient of transposition.

Gene Organization and Chromosome Behavior.

It is striking that all three of the loci that are substrates for mating type interconversion are located on chromosome III, in the order HML--MAT--HMR (83). HML and HMR lie near the ends of the chromosome, 200 kb and approximately 150 kb from MAT respectively. Thus the cassette donor loci are much farther from the recipient locus than in Schizosaccharomyces pombe (84). To determine whether there is any functional significance to having the cassettes arranged in this manner, studies of switching have been done in which HML is translocated to another chromosome (85; Jeanne Margolskee, personal communication). In the latter study, which employed a translocation separating MAT from HML, utilization of the cassette from HML was reduced nearly 10-fold. Thus physical linkage of this cassette is important for its ability to act as donor. The requirement for physical linkage might reflect a conformation of chromosome III in which the three cassette loci are near each other. (Certain gross regularities of chromosome arrangement in nuclei are known; 86, 87). The broken chromosome that is initially produced as a result of HO protein action might interact preferentially with homologous DNA sequences that are near by. Another possibility is that the broken chromosome searches for homology preferentially with DNA sequences that are physically connected to it. For example, the search might involve threading the arms of chromosome III past the gap in the chromosome at MAT.

A complete physical and functional picture of chromosome III is being developed (Newlon et al, this volume). The starting point for these studies is a large circular derivative of chromosome III, formed by recombination between HML and MAT (74). This circle could be purified because it is present as a covalently closed molecule. A physical map of this entire 200 kb region (more than half of chromosome III) has been determined. (Other segments, for example, the left telomere, have also been cloned; Caroline Astell, personal communication). The properties of the circular chromosome demonstrated that the yeast centromere is DNA and not protein, which was subsequently demonstrated by cloning centromeres (see Ng et al, this volume). Now its segments are being characterized for ARS activity and transcripts. It will therefore be possible to determine the relationship between ARS elements and replication origins and determine how the use of different replication origins is orchestrated during S phase. It has also been possible to produce linear derivatives of chromosome III that are of various sizes and study the stability of these 'native' chromosomes (88).

MEIOSIS

There are two conditions that must be met in order for yeast to undergo sporulation (meiosis and spore formation; reviewed in ref 89). First, cells must be subjected to nutritional starvation, which presumably works through depressing cyclic AMP levels (91; see Matsumoto et al and Uno et al, this volume). Secondly, cells must be a/α: a/a and α/α diploids are not able to sporulate (90). Genetic analysis (23, 92) and, more recently, biochemical analysis (18) has provided evidence that the mating type locus governs sporulation by regulating an inhibitor of meiosis, the RME1 gene product. It was first noted that a/a and α/α cells are able to sporulate in strains defective in RME1 (23) and that the rme1 mutation does not simply provide a1 and α2 activity by allowing expression of cassettes at HML and HMR (92). It was thus proposed that RME1 codes for an inhibitor of sporulation and that the RME1 product is produced in haploid cells (thereby blocking sporulation) and not in a/α cells (thereby allowing sporulation) (see Fig 5).

This hypothesis has received strong support in studies with the cloned RME1 gene (18). First of all, production of the RME1 transcript is greatly reduced in a/α cells compared to a and α cells; this reduction is due to a1-α2 activity. Secondly, synthesis of RME1 product in a/α cells (from a multi-copy plasmid) inhibits sporulation. Therefore, RME1 product is truly an inhibitor of sporulation. There are several possibilities for its mechanism of action. RME1 product might be a repressor and inhibit expression of certain key genes regulating sporulation; it might be a protease or protein kinase that covalently modifies a key protein involved in sporulation. The targets of RME1 protein action may identify important early steps in the pathway of meiosis and spore formation. One of the earliest steps in sporulation is premeiotic DNA synthesis. This may be one of the sites of action of the RME1 product.

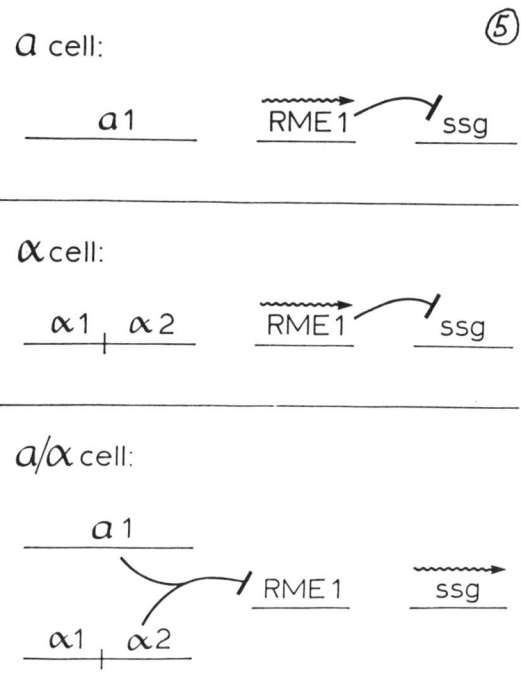

Figure 5. Regulation of meiosis and spore formation by the mating type locus occurs by regulation of the RME1 product. Conventions are the same as in Fig 1. ssg, sporulation-specific gene set. a and α cells are unable to sporulate because they produce the RME1 product. a1-α2 in a/α cells turns down synthesis of the RME1 transcript, thereby allowing sporulation. The molecular nature of the inhibition exerted by RME1 product is not known. Modified from ref 92.

MORPHOGENESIS

There are two areas in which studies of cell specialization and associated phenomena may offer opportunities to study cellular morphogenesis, both of which involve the position of budding. In one case, the position of the bud is determined by the mating type locus; in the other, the position of the bud is determined by the presence of a mating factor.

Regulation of Budding Pattern by the Mating Type Locus.

Cells that are **a** or α haploids or **a**/**a** or α/α diploids form buds from near the junction of mother and daughter cells (50, 58, 93), a pattern that is termed 'medial' budding (see Fig 6). In contrast, **a**/α cells (or haploid cells such as sir mutants, which have the phenotype of **a**/α cells) tend to bud from the end of the cell, a pattern that is termed 'polar' budding. The history of budding by a

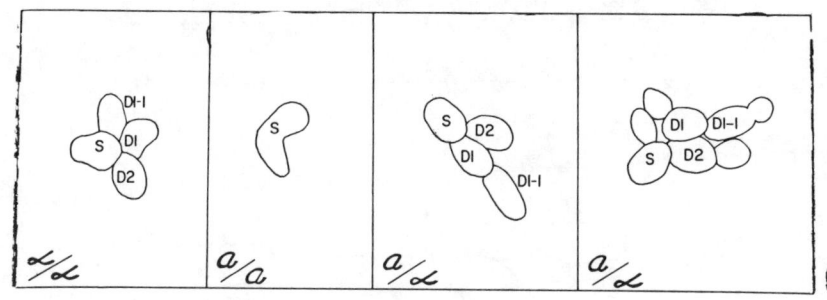

Figure 6. Regulation of budding pattern by the mating type locus. The figure shows tracings of cells after 2 or 3 cell divisions following spore germination. Cells are designated as in the legend to Fig 3. α cells (and the α/α cells shown in this figure) exhibit a medial budding pattern. **a**/α cells exhibit a polar budding pattern. The analysis shown in this figure was done in the presence of α-factor. Hence cell division by the spore of the **a** cell type (which in this experiment was an **a**/**a** diploid) is arrested. Modified from ref 58.

given cell can be seen also by examining the position of its bud scars (see for example ref 94 and Pringle et al, this volume). Thus the mating type locus determines a morphogenetic behavior: the position at which the daughter bud is initiated. Perhaps **a** and α cells produce a protein that in some manner guides the position of bud emergence; this protein is not produced by **a**/α cells. There is no clue as to what this product might be; there are no mutations known that affect the budding pattern (except for those that change the cell type).

Effect of α-factor on the Position of Budding.

When **a** cells are allowed to grow mitotically in the presence of sub-inhibitory concentrations of α-factor, they bud towards the α-factor source (Jasper Rine, personal communication). The molecular mechanism for this directional budding is not known. Apparently, the responding cells are able to identify a gradient of α-factor. Cells that are responding to α-factor undergo cell wall changes, as detected by a change in binding of concanavalin A (95), response to specific antibodies (96), and distinctive chitin deposition (97). These changes must occur initially at the position of the cell nearest the α-factor source and then result in commitment to produce a bud at that point. Studies with drugs blocking microtubule action and with mutants defective in tubulin show that microtubules are not necessary for formation of a bud (Thomas et al, this volume, Pringle et al, this volume). It is, however, possible that microtubules determine the <u>position</u> of the new bud. If we assume that this is the case, then there is an appealing hypothesis for the ability of α-factor to elicit directional budding. It has recently been shown that microtubules are in a state of dynamic instability, in which they shorten and lengthen continuously (98). Proper connections between microtubules and kinetochores are formed by a process of selection: those microtubules that have made the proper connection are stabilized (99). This selective stabilization has also been argued to be responsible for formation of intracellular structures involved in cell movement. In the case of α-factor, one could propose that α-factor itself or some change that it induces locally in the cell acts to stabilize microtubules and thereby programs bud emergence at that position.

CONCLUDING COMMENTS

Studies of yeast cell specialization have been undertaken for different reasons. For some, the initial fascination has been with the mating factors and how they act. For others, it has been with cell type determination and mating type interconversion, how regulatory proteins and genetic rearrangement govern gene expression. I hope that it has been apparent from this paper that these studies have converged and that the different aspects of mating type provide many different opportunities to learn about cell biology. These studies should contribute to the development of a complete picture of a eukaryotic cell—a picture that links structure, function, and synthesis.

ACKNOWLEDGEMENTS

It is a pleasure to acknowledge the seminal contributions by Jim Hicks to yeast work in my laboratory, in organizing the conference on Yeast Cell Biology, and in preparation of this book. I thank the people currently in my laboratory—in particular, Susan Michaelis, Paul Sternberg, Aaron Mitchell, Ken Kubo, Kent Matlack, and Ira Clark—and those that have departed for greener pastures—Jasper Rine, Kathy Wilson, Stan Fields, Karen Luper, Mike Stern, and Jeanne Margolskee—for their unpublished observations that have been cited. I also thank A. Mitchell, S. Michaelis, I. Clark, and P. Sternberg for comments on the manuscript. Our work has been supported by Research and Program Project Grants from the National Institutes of Health, by the Weingart Program in Developmental Genetics, and by postdoctoral fellowships from the California Division of the American Cancer Society, the Walter Winchell-Damon Runyon Cancer Fund, the National Institutes of Health, the Jane Coffin Childs Memorial Fund for Medical Research, and the Weingart Program.

REFERENCES

1. Herskowitz I, Oshima Y (1981). Control of cell type in Saccharomyces cerevisiae: Mating type and mating type interconversion. In Strathern JN, Jones EW, Broach JB (eds): 'The Molecular Biology of the Yeast Saccharomyces: Life Cycle and Inheritance,' Cold Spring Harbor, New York; Cold Spring Harbor Laboratory Press, pp 181-209.
2. Sprague Jr GF, Blair LC, Thorner, J (1983). Cell interactions and regulation of cell type in the yeast Saccharomyces cerevisiae. Ann Rev Microbiol 37:623-660.
3. Herskowitz I (1983). Determination of yeast cell type. Symp Soc Dev Biol 41:65-75.
4. Klar AJS, Strathern JN, Hicks JB (1984). Developmental pathways in yeast. In Losick R, Shapiro L (eds): 'Microbial Development,' Cold Spring Harbor, New York; Cold Spring Harbor Laboratory Press, pp 151-195.
5. Herskowitz I (1986). Cell type determination and mating type interconversion in yeast. In Fougereau M, Stora M (eds): 'Cellular and Molecular Aspects of Developmental Biology,' Amsterdam; Elsevier Science, pp 41-71.
6. Duntze W, MacKay VL, Manney TR (1970). Saccharomyces cerevisiae: a diffusible sex factor. Science 168:1472-1473.
7. Fink GR, Styles CA (1972). Curing of a killer factor in Saccharomyces cerevisiae. Proc Natl Acad Sci 69:2846-2849.
8. Hicks JB, Herskowitz I (1976). Evidence for a new diffusible element of mating pheromones in yeast. Nature 260:246-248.
9. Sprague Jr G, Herskowitz I (1981). Control of yeast cell type by the mating type locus. I. Identification and control of expression of the a-specific gene, BAR1. J Mol Biol 153:305-321.
10. Ciejek E, Thorner, J (1979). Recovery of S. cerevisiae a cells from G1 arrest by α-factor pheromone requires endopeptidase action. Cell 18:623-635.
11. Wilson KL, Herskowitz I (1984). Negative regulation of STE6 gene expression by the α2 product of Saccharomyces cerevisiae. Mol Cell Biol 4:2420-2427.
12. Manney TR (1983). Expression of the BAR1 gene in Saccharomyces cerevisiae: Induction by the α mating pheromone of an activity associated with a secreted protein. J Bacteriol 155:291-301.

13. Strathern J, Hicks J, Herskowitz I (1981). Control of cell type in yeast by the mating type locus: The α1-α2 hypothesis. J Mol Biol 147:357-372.
14. Sprague Jr GF, Jensen R, Herskowitz I (1983). Control of yeast cell type by the mating type locus: Positive regulation of the a-specific STE3 gene by the MATα1 product. Cell 32:409-415.
15. Fields S, Herskowitz I (1985). The yeast STE12 product is required for expression of two sets of cell-type-specific genes. Cell 42:923-930.
16. Miller AM, MacKay VL, Nasmyth KA (1985). Identification and comparison of two sequence elements that confer cell-type specific transcription in yeast. Nature 314:598-603.
17. Jensen R, Sprague Jr G, Herskowitz I (1983). Regulation of yeast mating-type interconversion: Feedback control of HO gene expression by the yeast mating type locus. Proc Natl Acad Sci 80:3035-3039.
18. Mitchell AP, Herskowitz I (1986). Activation of meiosis and sporulation by repression of the RME1 product of yeast. Nature, in press.
19. Elder RT, St John TP, Stinchcomb DT, Davis RW (1981). Studies on the transposable element Ty1 of yeast. I. RNA homologous to Ty1. Cold Spring Harbor Symp Quant Biol 45:581-584.
20. Clancy MJ, Buten-Magee B, Straight DJ, Kennedy AL, Partridge RM, Magee PT (1983). Isolation of genes expressed preferentially during sporulation in the yeast Saccharomyces cerevisiae. Proc Natl Acad Sci 80:3000-3004.
21. Percival-Smith A, Segall J (1984). Isolation of DNA sequences preferentially expressed during sporulation in Saccharomyces cerevisiae. Mol Cell Biol 4:142-150.
22. Johnson AD, Herskowitz I (1985). A repressor (MATα2 product) and its operator control expression of a set of cell type specific genes in yeast. Cell 42:237-247.
23. Kassir Y, Simchen G (1976). Regulation of mating and meiosis in yeast by the mating-type region. Genetics 82:187-206.
24. Tatchell K, Nasmyth KA, Astell C, Smith M (1981). In vitro mutation analysis of the mating-type locus in yeast. Cell 27:25-35.
25. Herskowitz I (1982). The MATα2 gene. Rec Adv Yeast Mol Biol 1:320-331.
26. Astell CR, Ahlstrom-Jonasson L, Smith M, Tatchell K, Nasmyth KA, Hall BD (1981). The sequence of the DNAs coding for the mating type loci of Saccharomyces cerevisiae. Cell 27:15-23.

27. Miller AM (1984). The yeast MATa1 gene contains two introns. The EMBO Journal 3:1061-1065.
28. Kurjan J, Herskowitz, I. (1982). Structure of a yeast pheromone gene (MFα): A putative α-factor precursor contains four tandem copies of mature α-factor. Cell 30:933-943.
29. Singh A, Chen EY, Lugovoy JM, Chang CN, Hitzeman RA, Seeburg PH (1983). Saccharomyces cerevisiae contains two discrete genes coding for the α-factor pheromone. Nucleic Acids Res 11:4049-4063.
30. Brake AJ, Brenner C, Najarian R, Laybourn P, Merryweather J (1985). Structure of genes encoding precursors of the yeast peptide mating pheromone a-factor. In Gething M-J (ed): 'Protein Transport and Secretion,' Cold Spring Harbor, New York; Cold Spring Harbor Laboratory Press, pp 103-108.
31. Kurjan J (1985). α-factor structural gene mutations in Saccharomyces cerevisiae: Effects on α-factor production and mating. Mol Cell Biol 5:787-796.
32. Julius D, Schekman R, Thorner J (1984). Glycosylation and processing of prepro-α-factor through the yeast secretory pathway. Cell 36:309-318.
33. Brake AJ, Merryweather JP, Coit DG, Heberlein UA, Masiarz FR, Mullenbach GT, Urdea MS, Valenzuela P, Barr PJ (1984). α-factor directed synthesis and secretion of mature foreign proteins in Saccharomyces cerevisiae. Proc Natl Acad Sci 81:4642-4646.
34. Bitter BA, Chen KK, Banks AR, Lai P-H (1984). Secretion of foreign proteins from Saccharomyces cerevisiae directed by α-factor gene fusions. Proc Natl Acad Sci 81:5330-5334.
35. Julius D, Brake A, Blair L, Kunisawa R, Thorner J (1984). Isolation of the putative structural gene for the lysine-arginine-cleaving endopeptidase required for processing of yeast prepro-α-factor. Cell 37:1075-1089.
36. Julius D, Blair L, Brake A, Sprague G, Thorner J (1983). Yeast α factor is processed from a larger precursor polypeptide: The essential role of a membrane-bound dipeptidyl aminopeptidase. Cell 32:839-852.
37. Ammerer G, Sprague Jr GF, Bender A (1985). Control of yeast α-specific genes: Evidence for two blocks to expression in MATa/MATα diploids. Proc Natl Acad Sci 82:5855-5859.

38. Strathern JN, Herskowitz I (1979). Asymmetry and directionality in production of new cell types during clonal growth: The switching pattern of homothallic yeast. Cell 17:371-381.
39. Mueller SC, Branton D (1984). Identification of coated vesicles in Saccharomyces cerevisiae. J Cell Biol 98:341-346.
40. Payne GS, Schekman R (1985). A test of clathrin function in protein secretion and cell growth. Science 230:1009-1014.
41. Ogawa T, Tomizawa J-I (1967). Abortive lysogenization of bacteriophage lambda b2 and residual immunity of non-lysogenic segregants. J Mol Biol 23:225-245.
42. Strathern JN, Klar AJS, Hicks JB, Abraham JA, Ivy JM, Nasmyth KA, McGill C (1982). Homothallic switching of yeast mating type cassettes is initiated by a double-stranded cut in the MAT locus. Cell 31:183-192.
43. Kostriken R, Heffron F (1984). The product of the HO gene is a nuclease: purification and characterization of the enzyme. Cold Spring Harbor Symp Quant Biol 49:89-96.
44. Nasmyth K (1983). Molecular analysis of a cell lineage. Nature 302:670-676.
45. Ozkaynak E, Finley D, Varshavsky A (1984). The yeast ubiquitin gene: head-to-tail repeats encoding a polyubiquitin precursor protein. Nature 312:663-666.
46. Bücking-Throm E, Duntze W, Hartwell LH, Manney TR (1973). Reversible arrest of haploid cells at the initiation of DNA synthesis by a diffusible sex factor. Exp Cell Res 76:99-110.
47. Wilkinson LE, Pringle JR (1974). Transient G1 arrest of Saccharomyces cerevisiae of mating type α by a factor produced by cells of mating type a. Exp Cell Res 89:175-187.
48. Hereford LM, Hartwell LH (1974). Sequential gene function in the initiation of Saccharomyces cerevisiae DNA synthesis. J Mol Biol 84:445-461.
49. Hagen DC, Sprague Jr GF (1984). Induction of the yeast α-specific STE3 gene by the peptide pheromone a-factor. J Mol Biol 178:835-852.
50. Hartwell LH (1980). Mutants of S. cerevisiae unresponsive to cell division control by polypeptide mating hormones. J Cell Biol 85:811-822.

51. Jenness DD, Burkholder AC, Hartwell LH (1983). Binding of α-factor pheromone to yeast a cells: Chemical and genetic evidence for an α-factor receptor. Cell 35:521-529.
52. Burkholder AC, Hartwell LH (1985). The yeast α-factor receptor: structural properties deduced from the sequence of the STE2 gene. Nucleic Acids Res 13:8463-8475.
53. Nakayama N, Miyajima A, Arai K (1985). Nucleotide sequences of STE2 and STE3, cell type-specific sterile genes from Saccharomyces cerevisiae. The EMBO Journal 4:2643-2648.
54. Hagen DC, McCaffrey G, Sprague Jr GF (1986). Evidence the yeast STE3 gene encodes a receptor for the peptide pheromone, a factor: Gene sequence and implications for the structure of the presumed receptor. Proc Natl Acad Sci, in press.
55. Liao H, Thorner J (1980). Yeast mating pheromone α factor inhibits adenylate cyclase. Proc Natl Acad Sci 77:1898-1902.
56. Thorner J (1982). An essential role for cyclic AMP in growth control: The case for yeast. Cell 30:5-6.
57. Casperson GF, Walker N, Brasier AR, Bourn HR (1983). A guanine nucleotide-sensitive adenylate cyclase in the yeast Saccharomyces cerevisiae. J Biol Chem 258:7911-7914.
58. Hicks JB, Strathern JN, Herskowitz I (1977). Interconversion of yeast mating types. III. Action of the homothallism (HO) gene in cells homozygous for the mating type locus. Genetics 85:373-393.
59. Jensen RE, Herskowitz I (1984). Directionality and regulation of cassette substitution in yeast. Cold Spring Harbor Symp Quant Biol 49:97-104.
60. Nasmyth K (1985). At least 1400 base pairs of 5'-flanking DNA is required for the correct expression of the HO gene in yeast. Cell 42:213-223.
61. Russell DW, Jensen R, Zoller MJ, Burke J, Smith M, Herskowitz I (1986). Structure of the yeast HO gene and analysis of its upstream region. Mol Cell Biol, in press.
62. Nasmyth K (1985). A repetitive DNA sequence that confers cell-cycle START (CDC28)-dependent transcription of the HO gene in yeast. Cell 42:225-235.
63. Stern M, Jensen R, Herskowitz I (1984). Five SWI genes are required for expression of the HO gene in yeast. J Mol Biol 178:853-868.

64. Rothblatt JA, Meyer DI (1986). Secretion in yeast: Reconstitution of the translocation and glycosylation of α-factor and invertase in a homologous cell-free system. Cell, in press.
65. Hansen W, Garcia PD, Walter P (1986). In vitro protein translocation across the endoplasmic reticulum of Saccharomyces cerevisiae: Post-translational translocation and glycosylation of the precursor to α-factor. Cell, in press.
66. Hall MN, Hereford L, Herskowitz I (1984). Targeting of E. coli beta-galactosidase to the nucleus in yeast. Cell 36:1057-1065.
67. Kalderon C, Roberts BL, Richardson WD, Smith AE (1984). A short amino acid sequence able to specify nuclear location. Cell 39:499-509.
68. Field C, Schekman R (1980). Localized secretion of acid phosphatase reflects the pattern of cell-surface growth in Saccharomyces cerevisiae. J Cell Biol 86:123-128.
69. Nasmyth KA (1982). Molecular genetics of yeast mating type. Ann Rev Genetics 16:439-500.
70. Nasmyth KA, Tatchell K, Hall BD, Astell C, Smith M (1981). A position effect in the control of transcription at yeast mating type loci. Nature 289:244-250.
71. Siliciano PG, Tatchell K (1984). Transcription and regulatory signals at the mating type locus in yeast. Cell 37:969-978.
72. Abraham J, Nasmyth KA, Strathern JN, Klar AJS, Hicks JB (1984). Regulation of mating-type information in yeast. Negative control requiring sequences both 5' and 3' to the regulated region. J Mol Biol 176:307-331.
73. Feldman JB, Hicks JB, Broach JR (1984). Identification of sites required for repression of a silent mating type locus in yeast. J Mol Biol 178:815-834.
74. Strathern JN, Newlon CS, Herskowitz I, Hicks JB (1979). Isolation of a circular derivative of yeast chromosome III: Implications for the mechanism of mating type interconversion. Cell 18:309-319.
75. Hicks J, Strathern J, Klar A, Ismail S, Broach J (1984). Structure of the SAD mutation and the location of control sites at silent mating type genes in Saccharomyces cerevisiae. Mol Cell Biol 4:1278-1285.
76. Brand AH, Breeden L, Abraham J, Sternglanz R, Nasymth K (1985). Characterization of a 'silencer' in yeast: A DNA sequence with properties opposite to those of a transcriptional enhancer. Cell 41:41-48.

77. Schnell R, Rine J (1986). A position effect on the expression of a tRNA gene mediated by the SIR genes of Saccharomyces cerevisiae. Mol Cell Biol, in press.
78. Miller AM, Nasmyth KA (1984). Role of DNA replication in the repression of silent mating type loci in yeast. Nature 312:247-251.
79. Eisen H, Brachet P, Pereira da Silva L, Jacob F (1970). Regulation of repressor expression in lambda. Proc Natl Acad Sci 66:855-862.
80. Johnson AD, Poteete AR, Lauer G, Sauer RT, Ackers GK, Ptashne M (1981). Lambda repressor and cro--components of an efficient molecular switch. Nature 294:217-223.
81. Kostriken R, Strathern JN, Klar AJS, Hicks J, Heffron F (1983). A site-specific endonuclease essential for mating-type switching in Saccharomyces cerevisiae. Cell 35:167-174.
82. Klar AJS, Strathern JN, Hicks JB (1981). A position-effect control for gene transposition: State of expression of yeast mating-type genes affects their ability to switch. Cell 25:517-524.
83. Harashima S, Oshima Y (1976). Mapping of the homothallic genes, HMα and HMa in Saccharomyces yeasts. Genetics 84:437-451.
84. Beach DH (1983). Cell type switching by DNA transposition in fission yeast. Nature 305:682-688.
85. Haber JE, Rowe L, Rogers DT (1981). Transposition of yeast mating type genes from two translocations of the left arm of chromosome III. Mol Cell Biol 1:1106-1119.
86. Byers B, Goetsch L (1975). Electron microscopic observations on the meiotic karyotype of diploid and tetraploid Saccharomyces cerevisiae. Proc Natl Acad Sci 72:5056-5060.
87. Mathog D, Hochstrasser M, Gruenbaum Y, Saumweber H, Sedat J (1984). Characteristic folding pattern of polytene chromosomes in Drosophila salivary gland nuclei. Nature 308:414-421.
88. Surosky RT, Newlon CS, Tye B-K (1986). The mitotic stability of deletion derivatives of chromosome III in yeast. Proc Natl Acad Sci 83:414-418.
89. Esposito RE, Klapholz S (1981). Meiosis and ascospore development. In Strathern JN, Jones EW, Broach JB (eds): 'The Molecular Biology of the Yeast Saccharomyces: Life Cycle and Inheritance,' Cold Spring Harbor, New York; Cold Spring Harbor Laboratory Press, pp 211-287.

90. Roman H, Philips MM, Sands SM (1955). Studies of polyploid Saccharomyces. I. Tetraploid segregation. Genetics 40:546–561.
91. Matsumoto K, Uno I, Ishikawa T (1983). Initiation of meiosis in yeast mutants defective in adenylate cyclase and cyclic AMP-dependent protein kinase. Cell 32:417–423.
92. Rine J, Sprague Jr GF, Herskowitz, I (1981). rme1 mutation of Saccharomyces cerevisiae: Map position and bypass of mating type locus control of sporulation. Mol Cell Biol 1:958–960.
93. Freifelder D (1960). Bud position in Saccharomyces cerevisiae. J Bacteriol 80:567–568.
94. Sloat BF, Pringle JR (1978). A mutant of yeast defective in cellular morphogenesis. Science 200:1171–1173.
95. Tkacz JS, MacKay VL (1979). Sexual conjugation in yeast. Cell surface changes in response to the action of mating hormones. J Cell Biol 80:326–333.
96. Lipke PN, Ballou CE (1980). Altered immunochemical reactivity of Saccharomyces cerevisiae a-cells after α-factor-induced morphogenesis. J Bacteriol 141:1170–1177.
97. Schekman R, Brawley V (1979). Localized deposition of chitin on the yeast cell surface in response to mating pheromone. Proc Natl Acad Sci 76:645–649.
98. Mitchison T, Kirschner M (1984). Dynamic instability of microtubule growth. Nature 312:237–242.
99. Mitchison TJ, Kirschner MW (1985). Properties of the kinetochore in vitro. II. Microtubule capture and ATP-dependent translocation. J Cell Biol 101:766–777.
100. Moore S (1984). Yeast cells recover from mating pheromone α factor-induced division arrest by desensitization in the absence of α factor destruction. J Biol Chem 259:1004–1010.
101. Chan RK, Otte CA (1982). Physiological characterization of Saccharomyces cerevisiae mutants supersensitive to G1 arrest by a factor and α factor pheromones. Mol Cell Biol 2:21–29.

Index

a-factor
    and protein localization, 636–640
    and yeast calcium, 483–484
    see also Mating factors
α-factor
    biosynthesis
        approaches for investigating, 471–472
        and carboxypeptidase β-like activity, 468–469
        intermediates in prepro-A-factor processing, 463–464
        and *kex2* mutants, 464–466
        potential intermediates in, 470–471
        and *stel3* mutants, 466–468
        and structure of α-factor precursors, 462–463, 469–470
    effect on budding position, 647
    precursors, gene products required for post-translational processing of, 472–473
    and protein localization, 636–640
    "-specific sterile" mutations, 464–466
    and yeast calcium, 483–484
    see also Mating factors
Actin filaments, yeast cell division and, 8–9
Actin gene of yeast
    cell-division cycle studies and, 14
    cloning studies of 17
    evidence of, 16–17
    mutations, post-translational processing of, 472–473
    nucleotide sequence of, 17
Adenylate cyclase
    control by guanine nucleotides and *ras* gene products, 82–85
    in membrane fractions of yeast *cyr1* and *ras* mutants, 118–120
    see also GTP-binding proteins and cyclic AMP in yeast cells
alg Mutants, and vacuolar protease genetics, 512–515

Amebas, endocytosis in, 444
Amino acids
    of *MFα1* gene, 462–463
    of yeast ribosomal protein L3, 385–388, 389–391
    see also Amino acid sequences
Amino acid sequences
    and protein accumulation in nucleus, 395–412
    of purified α-factor, 471
    of telomeres, 252–253
    of thymidylate kinase from mutant yeast, 174–176
    of vacuolar glycoprotein carboxypeptidase, 522–527
    of yeast and vertebrate *ras* proteins, 88–94
α-Amylase, uptake by yeast spheroplasts, 451, 454, 455–456
Antibodies
    in α-factor biosynthesis studies, 471–472
    and analysis of residual nuclei proteins, 370–372
    ant-actin, in yeast actin mutants phenotypes study, 22–25
    anti-β-galactosidase, and characterization of yeast ribosomal protein L3 genes, 382–388
    and detection of α-factor-related molecules, 463–464
    and hybrid protein detection and subcellular localization for targeting analysis during mitochondria biogenesis, 583, 584
    and identification of gene products, 70, 72–73
    and isolation of molecular clone of clathrin heavy chain gene, 434–435
    polyclonal anti-*rep1*, and identification of *rep1* protein, 329–333

of single-strand DNA binding proteins, 187
and vacuolar proteases genetics, 512–515
of yeast tubulin, 30–31
ARS elements
and mating type interconversion, 641–643
and nucleotide sequences of yeast, 202–203
plasmids, and centromere studies, 225–238
relative effcieincy of 217–219
role in chromosome stability, 221
and yeast DNA replication, 183–186, 188, 194–207
and yeast ring chromosome III, 211–221
ATP hydrolysis and cytoplasmic proteins, 572–573
ATP2 gene deletions, 593–598
Autonomously replicating sequence elements. *See* ARS elements

Benomyl-resistant yeast β-tubulin mutants, 28–29
Budding
bud growth, in yeast cell cycle, 48, 50–52
effect of α-factor on position of, 647
pattern regulation by mating type locus, 646–647

Calcium
-binding proteins in yeast
and *cdc31* protein, 490–492
control of cytosolic concentration of, 480–486
effect of mating pheromones on intracellular, 482–484
efflux of, 485–486
entry of, 480–482
function in cells, 477–480
and yeast spindle pole body duplication, 155–157
and processing of yeast pheromone precursors, 466
see also Calmodulin in yeast
Calmodulin in yeast, 486–490
and calcium function in cells, 479–480
potential targets for action, 493–496

see also Calcium, -binding proteins in yeast
Carbohydrates, yeast fermentation of. *See* Telomere fermentation genes
Carboxypeptidase β-like activity, and α-factor biosynthesis, 468–469
cdc31 protein in yeast, 490–492
Cell-division cycle. *See* Yeast cell division cycle
Cells, calcium ion function in, 477–480; *see also* Cellular morphogenesis in yeast cell cycle; Yeast, specialized cell types and cell biology
Cellular morphogenesis in yeast cell cycle, 47–73
approaches to, 48–50
effects of microtubule-disrupting drugs, 52–57
genetic approaches to, 57–60
and identification and characterization of gene products, 68–73
and microtubule and actin distribution in relation to bud growth, 50–52
and mutagenesis studies, 59–68
steps in, compared with other eukaryotic cells, 48, 49
see also Morphogenesis, and mutant yeast genes
Centromeres, 225–238
and control of 2-micron plasmid replication, 235
isolation of functional, 226–228
kinetochore microtubules and function of, 235–236, 237–238
proteins of, 235–237
role in eukaryotic chromosomes, 212
and structural studies, 228–230
structure–function studies of meiotic function, 233–235
structure–function studies of mitotic function, 230–233
Chelation, and removal of calcium from yeast cells, 480–481
Chimeric proteins, and nuclear protein localization, 417–420
Chitin ring, role in yeast cell division cycle, 22, 25, 48
Chlamydomonas

cilium of, 5
flagellar axoneme, polypeptides of, 59–60
Cholera toxin, and cyclic AMP synthesis in vertebrate membranes, 83
Chromatin footprinting experiments in yeast, 345–361
Chromosome III
  structure and organization, 211–221
    cloning and mapping methods, 214–217
    and effects of deleting *ARS* elements, 219–221
    and identification of *ARS* elements, 215–217
    and relative efficiencies of *ARS* elements, 217–219
    and mating type interconversion, 643–644
    see also Centromeres
Chromosomes
  movement during fission yeast mitosis, 279–295
    and control mechanisms in nuclear division, 294–295
    genes required for early steps of, 288–289
    and genes required for nuclear division of fission yeast, 280–282
    and isolation of *cut* and *nuc* mutants by cytological screening, 292–294
    and phenotypes of tubulin mutants, 282–285
    and temperature shift-up, 285–288
  stability, role of *ARS* elements in, 221
  see also Centromeres; DNA, topoisomerase mutants; Karyotyping of yest genome; Telomere fermentation genes
cis-Acting DNA sequences, and yeast model system for chromatin structure–function relationships, 346
Clathrin
  heavy chain gene
    function of, 438–439
    isolation of molecular clone of, 433–435
    mutating, 435–438
    role in yeast cell growth and protein transport, 429–439
    and characterization in yeast, 433–434
    and clathrin heavy chain function, 438–439
    genetic approach, 433–438
    and properties of clathrin-coated membranes, 431–432
    and transport vesicles, 430–431
Cloning studies
  of actin gene of yeast, 17
  of adenylate cyclase in yeast mutants, 84–85
  of *ARS* elements in yeast, 183–186, 188
  of *cdc31* protein in yeast, 490–492
  of centromere DNAs, 225–238
  of chomosome movement during fission yeast mitosis, 280–295
  of clathrin role in yeast cell growth and protein transport, 429–439
  and construction of yeast actin mutants in vitro, 17–19
  of expression of *cyr1* and *ras2* genes, 117–118
  of gene expression during yeast sporulation, 161–165
  of meiosis, 644–645
  of pheromone precursor maturation in yeast, 461–473
  of phospholipid synthesis, regulation in yeast, 551–563
  of small nuclear RNAs and RNA processing in yeast, 301–318
  of single-strand DNA binding protein genes, 181–183, 186–187
  of transfer RNA modification enzymes, 613–621
  of translocation, sorting and transport of vacuolar glycoproteins, 519–534
  of yeast calmodulin, 487–490
  of yeast chromosome III, 211–221
  of yeast DNA polymerase I, 177–178
  of yeast fermentation gene families, 241–248
  of yeast M1-dsRNA-encoded killer toxin, 537–549
  of yeast ribosomal protein, 380–391

Index

of yeast telomeres, 252–263
see also Genetic studies; Mutations
Complementation tests, of yeast β-tubulin gene mutations, 28–29
Conjugation, cyclic AMP and mutant yeast cell division and, 107–108
Conserved intron sequences, as binding sites for yeast snRNAs, 313–317
Cyclic AMP
and mutant yeast cell division, 101–110
and conjugation, 107–108
and initiation of meiosis, 108–109
and isolation and characterization of cyclic AMP-requiring mutants, 102–105
and mitosis, 107
and sporulation, 108, 109–110
and suppressors of cyclic AMP-requiring mutations, 105–107
production by yeast, and relationship between *cyr1* and *ras* mutations, 113–121
see also GTP-binding proteins cyclic AMP in yeast cells
cyr1 genes in yeast. *See* Cyclic AMP, and mutant yeast cell division; Yeast, *cyr1* and *ras* mutations
Cytological screening methods, and isolation of fission yeast mutants, 292–294
Cytoplasmic microtubules. *See* Microtubules
Cytoplasmic proteins, sorting for assembly in mitochondria, 569–576
and catalyzing mitochondrial compartmentation, 572–575
and coordination of mitochondrial assembly, 575
and import apparatus, 571–572
and molecular genetic analysis, 574–575
Cytoskeleton studies
biochemical analysis of, 4
functional, 4
historical background, 3–7
molecular analysis of, 5–7
relationship of in vitro activity to in vivo function, 4
see also Genetic studies, of yeast cytoskeleton; Yeast, cytoskeleton

Deletion analysis, of yeast centromere mutants, 230–231, 233–234

Density transfer experiments, 221
Detector and effector elements of *ras* proteins, 94–95
Detector-effector coupling, by GTP-binding proteins and cyclic AMP in yeast cells, 85–88
Detergent extraction, and molecular analysis of microtubules, 6–7
Dictyostelium, genetic analysis of, 5
DNA
in arrested *ndcl-l* yeast mutants, 39–40
loops, lampbrush chromosomes of, 367–368
partial sequence analysis of *rep1* protein, 331–333
polymerase I, and yeast DNA replication, 176–179
polymerase I gene
disruption, 178–179
isolation, 177
recombination. *See* Cloning studies
replication in *ndcl-l*-arrested yeast cells, 36; *see also* Yeast, DNA replication
required for yeast genome growth, 263–266
topoisomerase mutants
blockage of nuclear divisioon in, 289–290
distinct in vivo roles of types I and II, 290–291
see also Amino acid sequences; Replication
cDNA probes, and gene expression during sporulation, 161
DNase, yeast actin and, 16–17
DNase I footprinting, of *ars1* consensus sequence, 185–186
Drosophila melanogaster
centromeres, compared with yeast centromeres, 229–230
essential genes, 263–264
homeo box, 423
initiation of DNA replication in, 193–194
ras proteins of, 96
Drugs. *See* Microtubule-distrupting drugs

ecoRI endonuclease, and cleavage of nuclear DNA in vivo,

402–404; *see also* Yeast, nucleus, entry of procaryotic endonuclease into
Effector component, interaction with *ras* molecules, 96
Electron microscopy
  and identification of *ARS* elements on ring chromosome III, 215–216
  of *rep1* protein, 333–339
  and replication of spindle pole bodies, 152
  of yeast bud growth during cell cycle, 51
  of yeast cell cycle, 37
  of yeast life cycle phases, 27
  of yeast mutants defective in endocytic content, 448
  of yeast nuclear protein matrix, 327–329
"Emmenthaler body," 448–449
Endocytosis in yeast, 443–450
  and clathrin role in membrane vesiculation, 439
  functions of, 444
  and internalization of soluble and particulate markers into cells and spheroplasts, 451–458
  and FITC-dextran staining of vacuoles, 456–457, 458
  and subcellular fractionation, 455–456
  and isolation of mutants defective in accumulation of endocytic content, 447–448
  mechanism of, 444–445
  and secretory mutants, 446–447
Endonuclease. *See* ecoRi endonuclease; Yeast, nucleus, entry of procaryotic endonuclease into *entries*
Enveloped viruses, as markers to study endocytosis in yeast spheroplasts, 451–458
Enzymes
  and calmodulin action in yeast, 494–496
  required for processing yeast pheromone precursors, 465–466
  sporulation-specific, 160
  see also α-Amylase; Phospholipid synthesis regulation in yeast; Transfer RNA modification enzymes
Escherichia coli cells
  in cloning studies of yeast DNA polymerase I, 177–178

expression of yeast *ras2* genes in, 126–129
β-galactosidase, in study of yeast ribosomal protein, 380–391
genetic analysis of, 5
and isolation of molecular clone of clathrin heavy chain gene, 435
lacZ gene from, and hybrid protein detection and subcellular localization for targeting analysis during mitochondria biogenesis, 581–590
yeast *cyr1* and *ras* genes expressed in, 113–121
Eukaryotic cells
  calcium ion transport, 485
  chromosome structure, 212–213
  compartmentalization, 452
  endocytosis, 444
  ribosomes, 379–380
Eukaryotic microorganisms, calmodulin-containing, 487
Eukaryotic telomeres, properties of 252
Exocytosis, in mammalian cells, 452
Exoproteases, and processing of yeast pheromone precursors, 467

Fatty acids, and modification of *ras* proteins, 143
Fermentation genes. *See* Telomere fermentation genes
Fission yeast. *See* Chromosomes, movement during fission yeast mitosis; DNA topoisomerase mutants
FITC-dextran, and uptake of vesicular stomatitis virus by yeast vacuole, 456–457, 458
Fluorescence microscopy
  of endocytosis in yeast, 445–446
  of fission yeast mutants, 292–293
  and hybrid protein detection and subcellular localization for targeting analysis during mitochondria biogenesis, 583, 584
  and phenotypes of actin mutants, 22
Fluorescent dextran. *See* FITC-dextran
Fluorescent dye, and studies of endocytosis in yeast, 445–446
Fluoride ion, and cyclic AMP synthesis in vertebrate membranes, 83

Index

Forskolin, and cyclic AMP synthesis in vertebrate membranes, 83
Fungi, filamentous, conserved intron sequences in, 316

Gal1 promoter, and production of *ecoRI* endonuclease in yeast, 397–399
gal4 gene, and nuclear protein localization, 416–420
β-Galactosidase, and nuclear protein localization, 416–420; *see also Escherichia coli* β-galactosidase; *PUT2-lacZ* hybrid protein import into yeast mitochondrial matrix
GDP, and action of *ras* proteins, 136, 138, 139; *see also* Guanine nucleotides
Gene disruption experiments
  and nuclear protein localization, 416–420
  and yeast DNA polymerase I gene, 178–179
  and yeast snRNA gene, 309
  and yest β-tubulin gene mutations, 29
Gene fusion
  and clathrin role in yeast cell growth and protein transport, 433
  and hydrid protein detection and subcellular localization for targeting analysis during mitochondrial biogenesis, 581–590
  see also PUT2-lacZ hybrid protein import into yeast mitochondrial matrix
Gene(s)
  and chromosome movement during fission yeast mitosis, 279–295
  encoding α-factor precursors of yeast pheromones, 462–463
  products
    identification and characterization, cellular morphogenesis in yeast cell cycle and, 68–73
    in sporulating yeast cells, 159–168
  -replacement techniques, and cellular morphogenesis in yeast cell cycle, 59–68
  tcm1, and yeast ribosomal protein, 380–391
  telomere length and, 259–260
  see also Cloning studies *entries*; Genetic studies *entries*; Karyotyping of yeast genome; Telomere fermentation genes; Yeast, nucleus, entry of procaryotic endonuclease into; *and entries under* specific genes
Genetic studies
  of cellular morphogenesis in yeast cell cycle, 57–60
  of clathrin role in yeast cell growth and protein transport, 432–439
  of cytoplasmic protein assembly in mitochondria, 574–575
  of genotype variation in fermentation genes, 241–248
  linkage tests of yeast β-tubulin gene mutations, 28–29
  mapping of yeast fermentation genes, 241–248
  of nuclear protein localization signals in yeast, 421–423
  organisms suitable for, 4–5
  and *PUT2-lacZ* hybrid protein import into yeast mitochondrial matrix, 601–610
  of small nuclear RNAs. *See* Small nuclear RNAs and RNA processing in yeast
  of transfer RNA modification enzymes, 613–621
  of yeast cytoskeleton, 13–40
    and actin and β-tubulin genes, 14, 16–32
    approaches to, 9
    and cell-division cycle, 13–17
    classical genetic method and recombinant DNA method, 15–16, 32–40
  of yeast specialized cell types and cell biology, 625–648
  of yeast spindle-pole body regulation, 151–157
    and calcium-ion binding pockets, 155–157
    and negative regulation by mutant yeast, 154–155
  of yeast sporulation, 159–168
    and characterization of sporulation-specific messages, 165–168
    and sporulation-specific enzyme activities, 160

and studies of gene expression,
160–165
see also Actin gene of yeast; Mating type interconversion; Yeast, actin gene mutations; Yeast, genome structure and genetics; Yeast, vacuole, protease genetics
1,3-β-Glucanase, and sporulation, 160
β-1,4-Glucosidase, and sporulation, 160
GTP
and action of *ras* proteins, 136, 138, 139
-binding proteins and cyclic AMP in yeast cells, 81–96
and control of adenylate cyclase by guanine nucleotides and *ras* gene products, 82–85
detector–effector coupling by, 85–88
and detector and effector elements associated with *ras* proteins, 94–95
and evolution of GTP-binding signal transducers, 97
and functional and structural domains of *ras* molecules, 95–96
-dependent adenylate cyclase, reconstitution from *cyr1* or *ras* mutations, 118–120
and structural comparison of retinal transducin subunits and *ras* proteins, 88–94
GTPase activity, of *ras* proteins, 126, 129–133
Guanine nucleotides, binding by *ras* proteins, 95, 126, 128, 129
Guanylate cyclase, and *Tetrahymena* calmodulin, 494

Haploidy, and yeast model system for chromatin structure–function relationships, 347
Heat shock gene expression, 161–165
HO gene
and cell cycle, 635–636
and endonuclease, 633
and mating type interconversion, 638–640
Homeo box, and nuclear protein localization signals in yeast, 423
Hormone response, and endocytosis, 444

hsp82 gene, chromatin footprinting, 350–360
hxk fermentation genes, 247
Hybrid proteins
detection and subcellular localization for targeting analysis during mitochondria biogenesis, 581–590
and nuclear protein localization, 406–408, 416–423

Immunoblot analysis of *rep1* protein, 329–333
Indirect immunofluorescence studies
of nuclear protein localization signals in yeast, 421–423
of β-tubulin mutant cells, 285–288
of yeast ribosomal protein, 383–388
Inositol-1-phospholipid synthase. *See* Phospholipid synthesis regulation in yeast
Integrative transformation, disruption of actin gene by, 18–21
Intermediate filaments, function of, 4
Intron sequences. *See* Conserved intron sequences

Karyotyping of yeast genome, 271–278
application of OFAGE to tetrad analysis, 274–278
and typical OFAGE pattern of *S. cerevisiae*, 272–273
kex2 mutants, and α-factor biosynthesis, 464–466
Kinetochore microtubules, and centromere function, 235–236, 237–238
Kluyveromyces lactis α pheromones gene, 463

Lampbrush chromosomes DNA loops, 367–368, 375
Lucifer yellow CH, and endocytosis studies, 445–446

mal gene family, 245–246
Mammalian cells
endocytosis, 451, 452
microtubule organization in, 151–152
Mammalian clathrin, compared with yeast clathrin, 433–434
matα cell. *See* α-factor, biosynthesis

Mating factors, regulation of cell cycle by, 633–635
Mating pheromones of yeast. *See* Pheromones, production in yeast
Mating type interconversion
 and chromosome structure, 640–644
 and gene organization and chromosome behavior, 643–644
 and *HO* gene, 638–640
 and *sir* genes, 640–643
Meiosis in yeast
 centromeres and, 233–235
 cyclic AMP as negative effector on initiation of, 108–109
 and specialized cell types and cell biology, 644–645
mel gene family, 246–247
Metazoan snRNAs, compared with yeast snRNAs, 303–304
^{35}S-Methionine labeled vesicular stomatitis virus. *See* Vesicular stomatitis virus
Methylbenzimidazole-2-y1-carbamate (MBC). *See* Microtubules, -disrupting drugs
mfa Genes, encoding α-factor precursors of yeast pheromones, 462–463
mg1 Gene family, 246–247
Microfilaments, function of, 4
Microtubles
 in arrested *ndcl-l* yeast mutants, 39–40
 and bud growth during yeast cell cycles, 50–52
 -disrupting drugs, effect on cellular morphogenesis in yeast cell cycle, 52–57
 function of, 4
 molecular analysis of, 6–7
 role in yeast cell division cycle, 29–31
 of yeast, 27
 of yeast cytoskeleton, functions of, 8
 see also Genetic studies, of yeast spindle pole body regulation; β-Tubulin fission yeast mutants
Minichromosome maintenance defective yeast mutants
 phenotypes of, 199–205
 in yeast DNA replication initiation, 195–207

Mitchondria
 biogenesis, 569–570
 import of proteins, transfer RNA modification enzymes and, 613–621
 protein encoding and synthesis, 593–598
 targeting and assembly of F1 ATPase β-subunit deletions, 593–598
  see also Cytoplasmic proteins, sorting for assembly in mitochondria; Hybrid proteins, detection and subcellular localization for targeting analysis during mitochondria biogenesis
Mitosis
 Centromeres and, 230–233
 and cyclic AMP and mutant yeast cell division, 107
 function of centromere during, 226
 see also Chromosomes, movement during fission yeast mitosis
Morphogenesis
 and mutant yeast genes, 60–68
 and yeast specialized cell types and cell biology, 646–647
Mutation studies in yeast
 of adenylate cyclase, 84–85
 of ATPase deletions, 593–598
 of calcium requirements for growth, 482
 of calmodulin, 486–490
 of cell-division cycle, 13–17, 33–40
  cellular morphogenesis in, 59–68
 of centromere proteins, 236–237
 of chromatin structure–function relationships, 346–347
 of chromosome movement during fission yeast mitosis, 279–295
 of DNA replication, 173–189
 of DNA replication initiation, 193–207
 of fermentation ability, 242
 of endocytosis, 443–450
 of genetics of spindle pole body regulation, 151
 of nuclear protein localization, 415–420
 of pheromone precursor maturation, 461–473
 of production of *ecoRI* endonuclease, 397–399
 of role of centromeres in meiosis, 233–235

of role of centromeres in mitosis,
    230–233
of small nuclear RNAs and RNA processing, 301–318
of telomere, 254–263
of vacuolar glycoproteins, 519–534
of vacuolar protease genetics, 505–515
see also Genetic studies *entries*; Nuclear protien localization signals in yeast; Yeast, actin gene mutations

ndcl-l Gene mutation in yeast, cell division cell and, 33–40
Neurospora crassa enzyme, and efflux of calcium from yeast, 485
Nocadazole. *See* Microtubule-disrupting drugs
Northern blot analysis
    M1-dsRNA encoded yeast killer preprotoxin, 540
    and phospholipid synthesis regulation in yeast, 557–558
Nuclear division arrest mutants, 280–295
Nuclear migration, microtubules and, 55–57
Nuclear protein localization signals in yeast, 417–423
    and homeo box, 423
    see also Yeast nucleus, entry of procaryotic endonuclease into
Nucleus. *See* Yeast, nucleus; Nuclear *entries*

OFAGE
    application to tetrad analysis of yeast, 274–278
    of *S. cerevisiae*, 272–273
Oligonucleotide probes
    and structure of α-factor precursors, 470
    and yeast calmodulin, 487–490
Oligopeptides, and mating factors of specialized yeast cells, 630–633

3H-Palmitic acid labeling of *ras2* proteins, 143
pep gene mutations, 505–515
Peptide(s)
    amphipathic, and inhibition of calmodulin action, 496
    and yeast pheromone precursors, 461–473

    see also Pheromones
Phalloidin
    and localization of actin in yeast cells, 53
    rohdamine-tagged, in yeast actin mutant phenotypes study, 22, 24
Phallotoxins, and localization of yeast actin, 52
Phenothiazines, and yeast calmodulin, 488–489
Phenotypes
    of *ecoRI*-producing cells, 404–405
    of minichromosomes maintenance defective yeast mutants, 199–205
    of *prc1* propeptide deletions, 526–527
    of tubulin fission yeast mutants, 282–285
    of yeast actin gene mutations, 21–25
    of yeast β-tubulin gene mutations, 29–31
Pheromones
    effect on intracellular calcium, 482–484
    production in yeast, 461–473
        and greater α-factor biosynthesis, 462–473
Phospholipid synthesis regulation in yeast, 551–563
Plasmids
    containing telomeres of *Oxytricha* or *Tetrahymena*, transformation of yeast with, 252–253
    cross-complementation, and suppressors of morphogenetic mutant yeast genes, 67–68
    expressing *ecoRI* endonuclease in yeast, 395–412
    and hybrid protein detection and subcellular localization for targeting analysis during mitochondria biogenesis, 581–590
    2-micron, centromere control of replication of, 235
    and vacuolar protease genetics, 505–515
Polyclonal anti-*rep1* antibodies, and identification of *rep1* protein, 329–333
Polypeptides
    of *Chlamydomonas* flagellar axoneme, 59–60
    required for processing yeast pheromones precursors, 465–466
    see also Peptide(s)

Index

prc1 gene encoding vacuolar glycoprotein carboxypeptidase Y, 519–534
Prepro-α-factor precursor in yeast, posttranslational processing, 462–469
Procaryotic endonuclease. *See* Yeast nucleus, entry of procaryotic endonuclease into
Protease β activity. *See* Yeast vacuole, protease genetics
Proteases
   required for processing yeast pheromone precursors, 465–466
   and sporulation, 160
Protein(s)
   of centromeres, 235–237
   import into mitochondria, 593–598
   integral membrane, 637
   localization in specialized yeast cells, 636–640
   localization with vesicles, 638
   microtubule-associated in yeast, 7, 9
   nuclear localization, 638–639
   PUT2-lacZ hybrid, 601–610
   of residual nuclei, 370–372
   secreted from cell, 636–637
   sequence-specific in yeast replication, 183–186
   specific turnover in specialized yeast cells, 632–633
   see also Calcium, -binding proteins in yest; Cytoplasmic proteins; Gene(s), products; GTP, -binding proteins and cyclic AMP in yeast cells; hybrid proteins; *ras* proteins; *rep1* protein of yeast plasmid 2 micron circle; *PUT2-lacz* hybrid protein import into yeast mitochondrial matrix; Single-strand DNA binding proteins; Vacuolar glycoproteins; Yeast, nucleus, entry of procaryotic endonuclease into; Yeast, ribosomal protein; Yeast, vacuole, glycoproteins
Protein kinase, cyclic AMP dependent, of yeast mutants, 103–104
Poteinase K digestion, and spehroplast internalization of α-amylase, 454
Pseudorevertants
   and analysis of morphogenetic mutant yeast genes, 62–67
   and identification of proteins interacting with tubulin, 31
PUT2-lacZ hybrid protein import into yeast mitochondrial matrix, 601–610

Rabbit muscle actin, compared with yeast actin, 17
ras genes in yeast
   function of, 126
   product of, control of adenylate cyclase in yeast and vertebrate cells, 82–85
   see also Yeast, *cyr1* and *ras* mutations; *ras* proteins of yeast and vertebrates
ras proteins of yeast and vertebrates compared with retinal transducin, 88–94
detector–effector coupling by GTP-binding proteins and, 85–88
detector and effector elements associated with, 94–95
functional and structural domains of, 95–96
ras2 protein
   conversion from GTP- to GDP-bound form, 133
   mutant, 133, 136
   see also GTP-binding proteins and cyclic AMP in yeast cells; Yeast, *ras1* and *ras2* protein biochemistry
Recombinant DNA studies of yeast, 15–16, 264–266; *see also* Cloning studies
Reconstitution studies of yeast DNA replication, 176–189
rep1 protein of yeast plasmid 2-micron circle
   association with yeast nuclear protein matrix, 232–341
   identification of, 239–333
   copurification with yeast nuclear matrix, 333–339
   DNA partial sequence analysis of, 331–333
Replication
   2-micron plasmid, centromeres and, 235
   of telomeres, 260–263
   see also DNA, replication; Yeast, DNA replication; Yeast DNA replication initiation

Residual nucleus, 368–370
Restriction enzymes, 380
  and karyotyping of yeast genome, 276–277
  and mapping of yeast ring chromosome III, 214–215
Retinal transducin, compared with *ras* proteins, 88–94
Ribonuclease, and sporulation, 160
Ribosomal protein. *See* Yeast ribosomal protein
RNAs, yeast total cellular, in study of gene expression during sporulation, 161–168; *see also* mRNAs; Small nuclear RNAs; Transfer RNA modification enzymes
mRNAs
  of *Drosophilia*, 263–264
  splicing in yeast, and snRNAs, 312–313
  translation, and detection of α-factor-related moleucles, 463–464
  yeast total cellular, in study of gene expression during sporulation, 161–168
sac-gene products, and suppression of yeast actin mutants, 26–27
Saccharomyces cerevisiae. *See* Yeast *entires*
Sarcomere. *See* Yeast, Sarcomere
Schizosaccharomyces pombe, conserved intron sequences in, 316; *see also* Chromosome movement during fission yeast mitosis; Genetic studies, of yeast cytoskeleton
SDS-Polyacrylamide gel electorphoresis in gene expression during sporulation study, 163
  purification of *ras2* fusion protein, 129, 130
  purification of single-strand DNA binding proteins from yeast, 181
  and *ras* proteins, 14, 142
Secretory mutants
  and endocytosis in yeast, 446–447
  and translation, sorting and transport of vacuolar glycoproteins, 519–520
Sedimentation analysis, of β-galactosidase from cells synthesizing *L3-β-gal* proteins, 386–388

Single-strand DNA binding protein, and yeast DNA replication. 179–183, 186–187
sir genes, and mating type analysis, 640–643
Slime mold endocytosis, 444
Small nuclear RNAs and RNA processing in yeast, 301–318
  and conserved intron sequences as binding sites for yeast snRNAs, 313–317
  and extragenic suppressors of substrate mutations, 317–318
  and growth, 307–312
  and identification of yeast snRNA, 303–307
  and mRNA splicing, 312–313
SnRNAs. *See* Small nuclear RNAs
Southern blot analysis
  and assay of tract length of yeast telomere, 254–259
  in yeast DNA growth study, 265–266
Spheroplasts
  and hybrid protein detection and subcellular localization for targeting analysis during mitochondria biogenesis, 581–590
  internalization of α-amylase, 451, 454, 454–456
  internalization of vesicular stomatitis virus, 451, 452–453, 455–456
Spindle pole bodies. *See* Yeast, spindle-pole bodies
Sporulation. *See* Single-strand DNA binding protein
sta gene family, 246–247
stel3 Mutation, 466–468
Subcellular fractionation, and internalization of α-amylase and vesicular stomatitis virus by yeast spheroplasts, 455–456
suc gene family, molecular basis for variation in genotype, 243–245
Suppressors
  of cyclic AMP-requiring mutations. 105–107
  of morphogenetic mutant yeast genes
  and plasmic cross-complementation, 67–68
    pseudorevertants and analysis of, 62–67
  of yeast actin gene mutations, 25–27

of yeast β-tubulin gene mutations, 31

Telomere fermentation genes, 241–248
   advantages of, 247–248
   chromosomal locations of, 247–248
   mal gene family, 245–246
   mel, *mgl*, and *sta* gene families, 246–247
   mutants, 254–263
   suc gene family, 243–245
Telomeres
   expected properties of, 252
   replication of, 260–263
   sequencing analysis of, 252–253
   structural and genetic analysis of, 252–263
   see also Telomere fermentation gene
Temperature
   and *ARS* element specificity of yeast minichromosome maintenance deficient yeast mutants, 197–199
   and morphogenetic mutant yeast genes, 60–68
   and *ndcl-l* mutations in yeast, 33–35
   and β-tubulin yeast mutants, 28–29
Tetrahymena calmodulin, and guanylate cyclase, 494
Thymidylate kinase from mutant yeast
   amino acid sequences of, 174–176
   and role of *cdc8* protein in yeast DNA replication, 174–176
TPCK inhibition, effects on yeast killer preprotoxin, 545–546, 548
Trans-acting factors, and yeast model system for chromatin structure–function relationships, 346
Transducin
   and cell proliferation, 82
   and detector–effector coupling by GTP-binding proteins in yeast, 85–88
   interaction with *ras* molecules, 95–96
   see also Retinal transducin
Transfer RNA modification enzymes, 613–621
   cellular location of, 614–616
   cloning studies, 616–621
Trypsin, and amino acid sequences of yeast and vertebrate *ras* protein, 92–93
β-Tubulin fission yeast mutants
and chromosome movement during fission, 279–295
   spindles of, 285–288
β-Tubulin gene of yeast, 27–32
   cell-division cycle studies and, 14
   see also Yeast, β-tubulin gene mutations

U2-containing snRNP, and metazoan pre-mRNA splicing, 312
^{3}H-Uracil, and cellular localization of transfer RNA modification enzymes, 614–616

Vacuole. *See* Yeast, vacuole *entries*
Vesicular stomatitis virus (VSV), uptake by yeast spheroplasts, 451, 452–453, 455–456
vpl mutants, and vacuolar glycoproteins, 519–534

Western blot analysis
   of hybrid *ecoRI3*, 408–410
   and hyrid protein detection and subcellular localization for targeting analysis during mitochondria biogenesis, 583, 584
Wild-type fission yeast, mitotic chromosome movements in, 285–287
Wild-type yeast
   andenylate cyclase activity in, 85
   killer preprotoxin, 539–549
   localization of actin in, 52, 53
   and pseudorevertants, 63–68
   telomeres, 257

Xenopus oocyte
   and cellular localization of transfer RNA modification enzymes, 614–616
   entry of protein into nucleus, 415
   nuclear protein localization signals in, 421

Yeast
   actin
      compared with rabbit muscle, 17
      distribution during cell cycle and bud growth, 52–54
      functions of, 17, 25

Index

actin gene, mapping of, 31-32
actin gene mutations
　conditional, construction of, 19-21
　construction in vitro, 17-19
　osmotic sensitivity, 25
　phenotypes of, 21-25
　suppressors of, 25-27
cell-division cycle
　and adenylate cyclase, 81-96
　cyclic AMP and. See Cyclic AMP and mutant yeast cell division
　and cytoskeleton, phenotypes of mutations and, 13-17, 21-25
　and ndcl-l gene mutations, 33-40
　steps in, 37
　tubulin and, 27-32
　see also Cellular morphogenesis in yeast cell cycle; Centromeres; Genetic studies, of yeast spindle-pole body regulaton
cell proliferation, cyclic AMP and, 81-82
chromosome III. See Chromosome III
chromosome centromeres. See Centromeres
chromosome, electorphoretic karyotyping of, 271-278; see also Telomere fermentation genes
cyrl and ras mutations, 113-121
　and cloning and expression of cyrl and ras2 genes, 117-118
　and reconstitution of GTP-dependent adenylate cyclase in vitro, 118-120
　relationship between, 114-117
cytoskeleton
　compared with animal cell cytoskeleton, 8
　genetic studies of. See Genetics of yeast cytoskeleton
　molecular analysis of, 5-7
　ndcl-l mutation and function of, 33-40
　reasons for studying, 3, 5-9
　see also Cellular morphogenesis in yest cell cycle; Cytoskeleton studies; Genetic studies, of yeast cytoskeleton
DNA replication, 173-189
　and ARS elements, 212-213

　and DNA polymerase I, 176-179
　and role of cdc8 protein, 174-176
　role of specific sequences in, 183-186
　and single-strand DNA binding protein, 179-183, 186-187
DNA replication initiation, 193-207
　and ARS elements, 194-207
　minochromosome maintenance defective yeast mutants in, 195-207
　models for, 194, 205-207
endocytosis. See Endocytosis in yeast
genome
　DNA fraction required for growth of, 263-266
　electrophoretic karyotyping of, 271-278
　structure and genetics, 251-266
　see also Telomeres
growth, and small nuclear RNAs and RNA processing, 307-312
killer preprotoxin, M1-dsRNA encoded, 537-549
　and cleavage of preprotoxin leader peptide, 543-544
　effects of TPCK inhibition on, 545-546, 548
　intracellular protoxin and secreted toxin in fusion gene transformants, 540-543
　pho5 regulation of cDNA copy of, 540
　secretion pathway, 544-548
model system for chromatin structure-function relationships, 345-361
　and basics of yeast chromatin, 347-348
　and chromatin structure of hsp82 locus, 350-360
　and isolation of yeast nuclei, 348-350
mutations. See Cyclic AMP and mutant yeast cell division; Mutation studies in yeast Yeast, cyrl and ras mutations
nuclear protein matrix
　copurification with repl protein, 333-339
　electron microscopy of, 327-329
　preparation of, 326-327

Index

and *rep1* protein of yeast plasmid 2-micron circle, 323–341
nucleus, DNA sequences, 367–375
   advantages of yeast for study of, 370–372
   and "residual nucleus," 368–370
   and studies in other organisms, 372–373
nucleus, entry of procaryotic endonuclease into, 395–412
   and *ecoRI* cleavage of nuclear DNA, 402–404
   and *ecoRI* endonuclease activity in yeast lysates, 400–402
   and hybrid proteins, 406–408
   and phenotypes of cells producing *ecoRI*, 404–405
   and production of *ecoRI* endonuclease in yeast, 397–399
   and selection of mutations in nuclear localization, 408–411
   transfer RNA modification enzymes in, 613–621
   see also Yeast, ribosomal protein
pheromones. *See* Pheromones
plasmid 2-micron circle. *See* Plasmids; *rep1* protein of yeast plasmid 2-micron circle
ras1 and *ras2* protein biochemistry, 125–146
   and characterization of bacterially produced *ras2* protein, 129
   and conversion from GTP- to GDP-bound form, 133, 136, 138, 139
   and exchange of bound GDP with free GTP, 133, 135
   and expression of yeast *ras2* gene in *E. coli*, 126–129
   and GTPase activity of *ras2* protein, 129–133
   and model for action of *ras* proteins, 136, 138, 139
   and modification by fatty acid and association to membranes, 143
   and mutant *ras2* protein, 133, 136
ribosomal protein
   and characterization of L3-β-galactosidase genes, 381–383

nuclear localization signal, 379–391
   and subcellular localization of hybrid proteins, 383–388
sarcomere molecular analysis, 5–7
specialized cell types and cell biology, 625–648
   and cell-type-specific differences, 627
   and chromosome structure, 640–644
   and integral membrane proteins, 637
   and meiosis, 644–645
   and morphogenesis, 646–647
   and mother-daughter cell proteins, 638–640
   and nuclear proteins, 638
   and protein localization, 636–640
   and proteins secreted from cell, 636–637
   and proteins within vesicles, 638
   and proteolytic processing, 630–633
   and regulation of cell cycle by mating factors, 633–635
   and regulation by mating type locus, 628–630
   and specific protein turnover, 632–633
   and synthesis of mating factors, 630–632
spindle-pole bodies, 27
   and bud growth during yeast cell cycle, 50–52
   duplication of, 37–38
   see also Genetic studies, of yeast spindle-pole body regulation
spindles, of β-tubulin mutant fission yeast, 285–288
sporulation
   conditions for, 644
   and cyclic AMP and mutant yeast cell division, 108, 109–110
   see also Genetic studies, of yeast sporulation
telomeres. *See* Telomere fermentation genes; Telomeres
β-tubulin gene mutations
   drug-resistant, 28–29
   phenotypes of, 29–31
   and role of microtubules in yeast cell division cycle, 29–31

suppressors of, 31
β-tubulin genes, mapping of, 31–32
vacuole
    glycoproteins, translocation, sorting, and transport of, 519–534
    protease genetics, 505–515
    protein localization mutants. *See vpl* mutants

vesicular stomatitis virus and α-amylase uptake, 456–457, 458
vesicles. *See* Clathrin, role in yeast cell growth and protein transport; Endocytosis in yeast; Wild-type yeast